Advances in Ephemeroptera Biology

Advances in Ephemeroptera Biology

Edited by

John F. Flannagan

and

K. Eric Marshall

Freshwater Institute
Winnipeg, Manitoba, Canada

Plenum Press · New York and London

Library of Congress Cataloging in Publication Data

International Conference on Ephemeroptera, 3d, Winnipeg, Man., 1979.
Advances in ephemeroptera biology.

Includes indexes.
1. Mayflies – Congresses. I. Flannagan, John F. II. Marshall, K. E. III. Title.
QL505.I56 1979 595.7'34 79-27713
ISBN 0-306-40357-9

Third International Conference On Ephemeroptera
July 4th-10th
1979
Winnipeg
Canada

Proceedings of the Third International Conference on Ephemeroptera, held in Winnipeg, Manitoba, Canada, July 4–10, 1979.

Printed in the United States of America

Preface

Following the first Conference in Florida, U.S.A. (1970) and the second in Krakow, Poland (1975), the Third International Conference on Ephemeroptera was held in Winnipeg, Manitoba, Canada from the fourth to the tenth of July, 1979, at the invitation of the International Permanent Committee on Ephemeroptera. (W.L. Peters, U.S.A. (Chairman); I. Müller-Liebenau, West Germany; H.J. Schoonbee, South Africa; G.F. Edmunds, Jr., U.S.A.; E.F. Riek, Australia; R. Sowa, Poland). The Conference was hosted by the Freshwater Institute (Department of Fisheries and Oceans, Canada) and the University of Manitoba and sponsored by the International Union of Biological Science. Sixty delegates from 11 countries attended.

Registration for the Conference took place on July 4th, and in the evening the delegates got acquainted, or renewed old acquaintances, at an informal gathering in the Hospitality Room at Mary Speechly Hall, the University residence which accomodated most of the delegates. Next morning, in the Fletcher-Argue theatre, the Conference was opened by the Honorary Chairman, Dr. F.P. Ide, who welcomed the delegates, then introduced the guest speakers: G.H. Lawler, Director-General of the Western Region of the Canada Department of Fisheries and Oceans; P. Jarvis, Deputy Minister, Manitoba Department of Mines, Natural Resources and Environment, and D.R. Campbell, President, University of Manitoba, who on behalf of their various organizations, welcomed the delegates to Canada and Manitoba, and extended good wishes for the success of the Conference. Following this, and commencing the scientific program, D. Dudley Williams gave the Invitational Lecture - Applied Aspects of Mayfly Biology - which is reproduced as the first paper in the Proceedings.

All but two of the papers presented at the Conference are reproduced here. The papers have been grouped together under broad subject headings. The editors would like to point out that these groupings had to be carried out in a rather arbitrary fashion, since many papers cross the subject boundaries.

We have not appended a discussion or question section to each paper since each paper was reviewed by a delegate, or member of the editorial committee during the Conference and each author had an opportunity during the Conference to adjust his or her paper in light of the questions asked after review and presentation. These adjustments have, in some cases, resulted in minor changes to the titles (and authorship) of some of the papers compared with those given in the Program. In addition, a number of small changes have been made during the editorial process. If these changes have affected the meaning or intent of the paper the editors accept responsibility, since, in some cases, the authors have not seen the final copy.

The submitted papers presented at the Conference have been, as stated above, grouped into seven sections. Within each section we have attempted to arrange the papers in either an historical or a general to specific sequence. The seven sections: Phylogeny and Systematics, Faunistics, Biology and Ecology, Behaviour, Methods, Environmental Impact and Toxicology, and Reviews and Historical Aspects of Mayfly Biology, are indicative of the wide range of topics covered at the Conference, almost every facet of Mayfly Biology being expanded on and/or reviewed. This volume therefore represents the state-of-the-art of Ephemeroptera biology at this point in time.

At the end of the volume are three indexes, prepared by K.E.M. The first is an index to the authors contributing papers to this volume. The Taxonomic Index provides a key to the major references in the text to families, subfamilies, genera and species. The Subject Index lists the main topics covered by the papers. This index is by no means exhaustive but hopefully provides a useful aid for the reader in locating pertinent information which cannot be located using the list of contents.

Dr. Bill Peters, on behalf of the Permanent Committee, and Dr. Fred Ide, the Honorary Chairman, closed the Conference by thanking the organizers and reviewing the achievements, both scientific and social, of this Third Conference.

The Convenor would like to thank the following individuals and organizations for their assistance in organizing, running and funding the Conference:

Contributing individuals:

W. Brown, W. Burton, S. Bubbs, P. Chang, D. Cobb, I. Davies, G. Decterow, M. Friesen, F. Frittaion, F.W.I. Women's Committee (chaired by D. Marshall), T. Galloway, J. Gee, D. Iredale, H. Klause, A. Kooyman, P. Laufersweiler, S. Lawrence, M. Layton, S. Leonhard, D. Malley, B. de March, E. Marshall, J. Mathias, G. McRae, A. Olivier, K. Patalas, D. Rosenberg, A. Salki, E. Scherer, L. Taite, B. Townsend, F. Ward, A. Wiens, H. Welch, R. Wickstrom, R. York, S. Zettler and M. Ziprick.

Sponsoring and contributing Organizations:

Freshwater Institute, Canada Dept. of Fisheries and Oceans
Dept. of Mines, Natural Resources and Environment, Manitoba
City of Winnipeg
University of Manitoba
International Union of Biological Sciences
Mobil Oil, Canada Ltd.
Fisher Scientific Co. Ltd.
International Centre (folk dancers)
Great Northern Fish Co. Ltd. (folk singers)

John F. Flannagan
K. Eric Marshall

Contents

INVITATIONAL LECTURE

Applied Aspects of Mayfly Biology 1
D.D. Williams

PHYLOGENY AND SYSTEMATICS

Permian Mayfly Nymphs: New Taxa and Systematic Characters 19
M.D. Hubbard and J. Kukalova-Peck

Phylogeny of the Leptophlebiidae (Ephemeroptera): An Introduction 33
W.L. Peters

Phylogenetic Relationships and Zoogeography of Cool-Adapted Leptophlebiidae (Ephemeroptera) 43
M.L. Pescador and W.L. Peters

Phylogenetic Relationships of the Leptophlebiidae of New Zealand (Ephemeroptera) 57
D.R. Towns and W.L. Peters

Geographic Distribution and Reclassification of the Sub-Family Ephemerellinae (Ephemeroptera, Ephemerellidae) 71
R.K. Allen

A Review of the Nearctic *Heptagenia* (Heptageniidae, Ephemeroptera) 93
R.W. Flowers

Jubabaetis gen.n. and *Platybaetis* gen.n., Two New Genera of the Family Baetidae from the Oriental Region 103
I. Müller-Liebenau

Un Changement dans la Taxonomie de la Famille Baetidae 115
M. Keffermüller

FAUNISTICS

The Habitats of British Ephemeroptera 123
P.S. Maitland

La Zoogeographie, l'Ecologie et la Protection des Ephéméroptères en Pologne, et leur Utilisation en tant qu'Indicateurs de la Pureté des Eaux Courantes 141
R. Sowa

The Distribution of Ephemeroptera in Northern Canada 155
D.G. Cobb and J.F. Flannagan

The Occurrence and Biology of *Siphlonisca aerodromia* Needham (Ephemeroptera: Siphlonuridae) in Maine, U.S.A. (Abstract) 167
K.E. Gibbs

Abundance and Altitudinal Distribution of Ephemeroptera in a Rocky Mountain Stream 169
J.V. Ward and L. Berner

BIOLOGY AND ECOLOGY

Mayfly Strategies in a Norwegian Lake 179
J.E. Brittain

Some Aspects of the Biology of *Ephemera danica* Müll. (Ephemeridae, Ephemeroptera) in Irish waters 187
K.F. Whelan

Regime Alimentaire des Larves de Deux Genres d'Ephéméroptères: *Raptobaetopus* Müller-Liebenau, 1978 et *Prosopistoma* Lattreille, 1833 201
J. Fontaine

Life History, Developmental Processes and Energetics of the Burrowing Mayfly *Dolania americana* 211
R.S. Harvey, R.L. Vannote and B.W. Sweeney

The Nutritional Dynamics of the Burrowing Mayfly, *Hexagenia limbata* 231
M.C. Zimmerman and T.E. Wissing

Overwintering Strategy of *Cloeon dipterum* L. Larvae 259
B.H. Nagell

Stages and Physiological Periods in the Development of *Cloeon dipterum* L. (Baetidae) 265
L.S. Cianciara

BEHAVIOUR

Predation, Climate and Mating of Mayflies 277
G.F. Edmunds, Jr. and C.H. Edmunds

Diel Emergence Patterns of Some Mayflies (Ephemeroptera) of the Roseau River (Manitoba, Canada) 287
M.K. Friesen, J.F. Flannagan and P.M. Laufersweiler

Diurnal Variation in the Activity of *Mirawara purpurea* Riek (Ephemeroptera, Siphlonuridae) in the Aberfeldy River, Victoria, Australia 297
I.C. Campbell

The Mechanisms that Maintain Population Stability of Selected Species of Ephemeroptera in a Temperate Stream 309
E.C. Gyselman

Importance of Behaviour to the Re-establishment of Drifting Ephemeroptera 321
J.J.H. Ciborowski and L.D. Corkum

The Importance of Species Associations and Substrate Types to Behavioural Drift 331
L.D. Corkum and H.F. Clifford

METHODS

Observations on the Use of the Radio-Isotope ^{32}P in the Study of Food Uptake by Some Mayflies and Other Benthic Macroinvertebrates in a Laboratory Stream Ecosystem 343
H.J. Schoonbee and J.H. Swanepoel

Ultrastructural Studies on Oligoneuriidae - Taxonomic Applications 353
J.D. Agnew

A Comparison of Mayfly (Ephemeroptera) Instar Determination Methods 367
T.J. Fink

The Effect of Mesh Size on the Interpretation of the Life History of Two Mayflies from South Australia 381
P.J. Suter and J.E. Bishop

ENVIRONMENTAL IMPACT AND TOXICOLOGY

Taxocenes of Ephemeroptera in Unpolluted and Polluted Streams of the Tatra Mountains 405
A. Kownacki

A Preliminary Report from a Study on the Environmental Impact of Pyrite Mining and Dressing in a Mountain Stream in Norway 419
K.J. Aanes

Effects of Crude Oil Contamination on Ephemeroptera in the Trail River, Northwest Territories, Canada 443
D.M. Rosenberg, A.P. Wiens and J.F. Flannagan

Evaluation of the Acute Toxicity of the Heavy Metal Cadmium to Nymphs of the Burrowing Mayfly, *Hexagenia rigida* 457
S.L. Leonhard, S.G. Lawrence, M.K. Friesen and J.F. Flannagan

REVIEWS AND HISTORICAL ASPECTS OF MAYFLY BIOLOGY

Online Retrieval of Information on Ephemeroptera: A Comparison of Different Sources 467
K.E. Marshall

Mermithids (Nematoda) and Mayflies (Ephemeroptera) 491
W.M. Hominick and H.E. Welch

Numerical Abundance Values of Mayfly Nymphs from the Holarctic Region 503
H.F. Clifford

The Mayfly Family Baetiscidae (Ephemeroptera) Part 1 511
L. Berner and M.L. Pescador

Première Description d'un Ephéméroptère en Angleterre: L'Ephemeron de Harris (1730-1788) 525
L. Arvy

Pictet et Eaton: Les Premiers Spécialistes des Ephéméroptères. Pictet and Eaton: The First Mayfly Specialists 531
J. Peters, L. Arvy and W.L. Peters

Author Index 541

Taxonomic Index 543

Subject Index 551

APPLIED ASPECTS OF MAYFLY BIOLOGY

D. Dudley Williams

Division of Life Sciences
Scarborough College
University of Toronto
West Hill, Ontario, Canada

ABSTRACT

Mayflies have been of interest to man for centuries. This paper will trace the history of this interest from its earliest beginnings in crude piscatorial entomology, through more recent application of the group as water quality indicators and research tools to present and future uses in modern aquatic technology.

Frost and Brown (1967) state that the use of artificial baits to catch trout by means of rod and line dates back to twelfth century Europe, although Leonard and Leonard (1962) report that it was practised by the ancient Greeks. Documentation of this occurred first in 1496 in the form of "The Treatyse of Fysshynge wyth an Angle" by Dame Juliana Berners. Doubtless, tempting fish with freshly caught aquatic insects would have preceded this, but the delicate nature of insect bodies would probably soon have led to the making of more robust imitations from a variety of readily available materials such as fur, feathers and wool. Presumably, the more realistic the artificial insect, the more successful its user would be. This fact alone probably led to a more careful scrutiny of lake or riverside insects and their habits, bringing the first aquatic entomologists into being.

A considerable literature on this topic, in both scientific journals and the popular press, now exists. I do not propose to review the state of the art of angling here, but to summarize some of the basic good descriptive entomology that has led to the refinement of fly fishing as illustrated by the Ephemeroptera. Figure 1 shows the stages in the life cycle of a typical mayfly (left) and the artificial flies that are meant to imitate these stages (right). The aquatic nymphs is fished as a wet fly beneath the water surface. Dominant features of the nymph, such as the cerci, segmented abdomen, legs and darkened, pre-emergent wingpads are all duplicated in the fly. The emerging subimago is also fished as a wet fly, though just below the surface; most mayflies emerge at or near the water surface (Needham *et al.* 1935). The characteristic rumpled wings are simulated on the artificial fly by a small portion of feather. The fully emerged subimago, now a terrestrial stage, is represented by a dry fly fished on the water surface. The subimago is similar in appearance to the adult, but has duller colours, the legs and cerci are shorter, and heavy pigmentation along the veins may produce a dark pattern on the wings (Leonard and Leonard 1962). The latter rarely persists in the imago but is faithfully copied in the fly (called a dun) by use of a mottled feather. The features of the imago are seen in its counterpart - the spinner - which is again fished on the water surface. The colours in this lure are brighter than in the dun, and the wings are made from a non-mottled feather. Segmentation in the abdomen is duplicated by a silk thread binding. Many hundreds of patterns are known for this fly depending on the species of mayflies occurring in local waters. The final stage of mimicry is that of the prostrate, spent female floating on the water surface. This change in posture is again reflected in the dry fly counterpart.

Why should the art of deceiving fish have developed to such a high degree? Tebo and Hasler (1963) state that availability and not abundance is the most important factor determining what foods are eaten by trout. The Ephemeroptera are most easily accessible as prey at emergence (Frost 1939). This accounts for the findings of Frost and Went (1940) that young atlantic salmon stomachs (*Salmo salar* L.) contained larger numbers of *Baetis* spp. than *Ephemerella* spp. even though their densities in the benthos were almost identical; *Baetis* has a longer emergence period than *Ephemerella*. A certain amount of selectivity while feeding therefore appears evident. Bryan and Larkin (1972) reached a similar conclusion for brook trout (*Salvelinus fontinalis* (Mitchill)), cutthroat trout (*Salmo clarki* Richardson) and rainbow trout (*Salmo gairdneri* Richardson). This specialization is more evident in fish with full stomachs than in those with only a few items in their guts where feeding is more random (Allen 1941). Food constancy is of advantage to a fish in that it enables it to temporarily set its feeding behaviour for a transient but abundant supply of identical organisms (Frost and Brown 1967). This promotes efficiency in foraging, with a considerable safety benefit and has parallels in flower selection constancy documented for worker honeybees (Michener 1974). A fish may reset its feeding for successively available insect species (Williams and Coad 1979 have shown this for cyprinids).

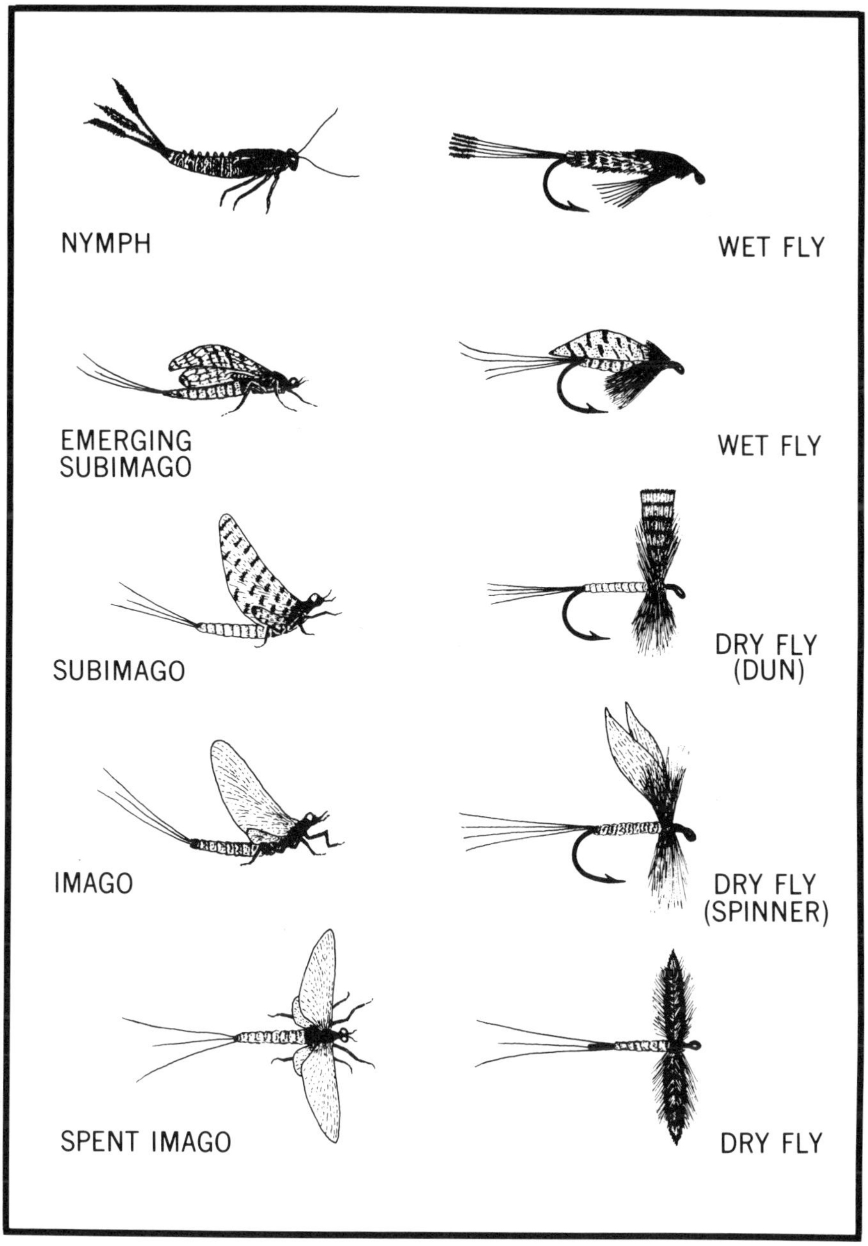

Figure 1. Comparison of the stages in the life cycle of a mayfly and the artificial flies that imitate them.

The angler takes advantage of this fact by presenting a suitable imitation of a particular mayfly during the time when the fish's feeding behaviour is set for that species; the so-called technique of "matching the hatch".

More recently, mayflies, along with other benthic macroinvertebrates, have been looked to as possible indicators of aquatic pollution (e.g. Britt 1975). Studies have shown that many pollutants have a marked affect on mayfly abundance and this can readily be detected. Table 1 cites some examples of mayfly abundance above and below various sources of pollution. It shows different degrees of of response by nymphs of different families. Even the same genus, *Baetis*, responded differently to organic pollution in the Rivers Dee and Tees. This could well have been due to a difference in species or in the exact nature of the pollutant, however. In one case, that of *Tricorythodes* in Deer Creek, a build up of quarry stonedust on a riffle resulted in an increase in density. This is perhaps not surprising as the genus typically prefers a silty substrate (Burks 1953). Other studies have found species apparently preferentially selecting polluted conditions. Snow and Rosenberg (1975), for example, in an experimental study of colonization of artificial substrates coated with crude oil, found Baetidae only in oily substrates. These examples clearly indicate that mayflies do respond to environmental change, but because of insufficient identification the conclusions are somewhat limited, in application, to the individual studies. Although mayflies are generally considered to be very sensitive to pollutants, particularly those of organic origin, Roback (1974) provided data to show that this, as a generality, is not so. He cites examples of common genera, like *Stenonema*, that may occasionally be found in extreme conditions, and concludes that although there may be differences at the species level, nymphal taxonomy is not sufficiently advanced, at present, to allow reliable prognostications to be made. Resh and Unzicker (1975) similarly point out the importance of species identification in the meaningful biological assessment of water quality.

Roback's argument, would, therefore, seem to preclude widespread application of mayflies as biological indicators in North America at present. However, where the nymphs of certain nearctic genera are separable, e.g., *Stenonema*, *Cloeon*, *Heterocloeon*, *Ephemerella*, *Neoephemera*, and *Hexagenia* (Edmunds *et al.* 1976), or where keys to local species are available (e.g., Flowers and Hilsenhoff 1975) then the group becomes valuable as a biotic index. Hilsenhoff (1977), for example, used a modified version of Chutter's (1972) empirical biotic index to evaluate the water quality of Wisconsin's streams. Familiarity with the local fauna allowed him to assign values to species, many of them mayflies. These values ranged between 0 and 5, with 0 assigned to species collected only in streams of very high water quality, and 5 assigned to species collected in badly polluted streams. Table 2 gives some examples

Table 1. Examples of mayfly abundance above and below various sources of stream pollution.

Taxon	Number/sample upstream pollution(x)	Number/sample immediately downstream of pollution (y)	Type of Pollutant	Location	Data source
Ephemeridae	237	188	Sewage	Red Cedar River	Leonard (1962)
Baetidae	16	2.5	"	Michigan, U.S.A	"
Heptageniidae	2.6	0	"	"	"
Baetis	5	4	"	River Tees	Butcher *et al.* (1937)
Ephemerella	3	4	"	England	"
Ecdyonurus	18	7	"	"	"
Rhithrogena	33	4	"	"	"
Baetis	205	36	Mild organic pollution	River Dee Wales	Hynes(1960)
Ecdyonurus venosus	5	4			"
Rhithrogena semicolorata	13	0	"	"	"
Heptagenia sulphurea	8	2	"	"	"
Stenonema frontale	195	0	Acid mine waste	Margaret Creek /Sandy Run, Ohio, U.S.A.	Napier and Hummon (1976)
Baetisca callosa	35	0			
Caenis	y/x ratio = 0.1		Deposited quarry stonedust	Deer Creek Indiana, U.S.A.	Gammon (1970)
Baetis	0.7				"
Tricorythodes	1.7		"	"	"

of the values he assigned to the mayflies. The biotic index was calculated according to the formula:

$$B.I. = \frac{\Sigma n_i a_i}{N}$$

where n_i is the number of individuals in each species, a_i is its assigned value, and N is the total number of individuals in the sample. The biotic index for a given stream was then compared with a standard scale of values calculated for Wisconsin streams. Values of less than 1.75 indicated excellent water quality, characterizing clean, undisturbed streams; values of between 2.25 and 3.00 indicated fair quality, associated with moderate enrichment or disturbance; and greater than 3.75 indicated very poor water quality with gross enrichment or disturbance. Not only can mayflies be used to detect and categorize pollution, but, as Edmunds *et al.* (1976) have suggested, they may help to remove some of the organic material by incorporating it into their body tissues which may later decompose on land.

An increasing nymber of laboratory studies are being reported in the literature in which mayfly nymphs are used as experimental animals. For example, Fremling and Schoening (1973) advocated the use of the burrowing nymphs of *Hexagenia* for the study of behaviour and for bioassay work, and designed a special artificial substrate for these purposes. Mayflies may have considerable potential in this kind of research, particularly as many species are easy to collect and maintain in the laboratory. These facts may also make them useful educational material in the classroom (Needham *et al.* 1935).

Mayflies provide an important food source for a great many predators, in both aquatic and terrestrial environments. Some, such as fish, are immediately obvious, others are perhaps not so obvious. For example, Leonard and Leonard (1962) cite adult dragonflies, hornets and spiders as well as a variety of birds, bats and other mammals, many of which are attracted to areas of mass emergence. In the aquatic phase, mayfly nymphs are eaten by waterfowl and may be an important component of the diet of their young (Krull and Boyer 1976). By far the most studied aspect of the food potential of the Ephemeroptera is that for fish, and salmonids in particular. New ideas are presently opening up in this area and attempts are being made to apply principles of stream ecology to increasing salmonid stocks.

As part of the Canadian Federal Government's Pacific Salmon Enhancement Programme, Mundie (1974) proposed the feasibility of raising salmon smolts (primarily coho salmon, *Oncorhynchus kisutch* (Walbaum)) in high density in seminatural streams. The approach was to combine some of the desirable features of rivers with the

Table 2. Examples of water quality values assigned to mayflies by Hilsenhoff in a study of Wisconsin streams (0 = excellent water quality, 5 = extremely poor water quality).

0	1	2	3	4	5
		BAETIDAE			
	Cloeon alamance *Heterocloeon curiosum*	*Baetis frondalis* *B. vagans* *Pseudocloeon dubium*	*B. levitans* *B. pygmaeus*		
		HEPTAGENIIDAE			
Heptagenia pulla *S. rubrum*	*Stenonema fuscum* *H. lucidipennis*	*S. terminatum* *S. mediopunctatum*	*S. exiguum* *H. diabasia*		
		EPHEMERIDAE			
	Ephemera simulans	*Hexagenia limbata*			
		EPHEMERELLIDAE			
Ephemerella attenuata *E. bicolour* *E. excrucians* *E. subvaria* *E. funeralis*	*E. simplex* *E. needhami*			*E. temporalis*	

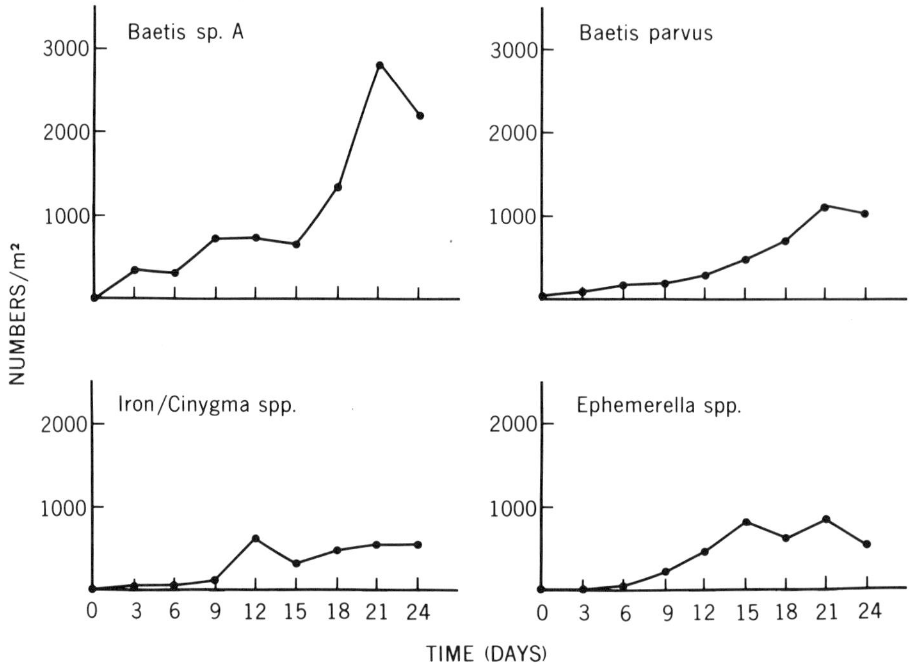

Figure 2. Colonization curves for mayfly nymphs.

productive capacity of hatcheries, but at a reduced cost. He considered the basic environmental unit for young coho to be a riffle/pool sequence at a meander. The unit provides an area of fast-moving, shallow water that supports benthic invertebrates, an area of deeper, slow-moving water for the fish, and an overhanging bank which provides their shelter. Central to the theory is the use of benthos on the riffles as part of the fry's diet, the invertebrates conveniently being transported from the riffles to the waiting fish in the pools by drift, and the recycling of excess manufactured food and fish faeces into the benthic food chain.

A prototype channel was constructed in 1975 alongside the Big Qualicum River, Vancouver Island, British Columbia. It is 400 m long by 4.5 m wide and consists of 25 riffles and 25 pools arranged alternately. The pools are 9 m long by 1 m deep and the riffles are 5 m long by 15 cm deep. An optimum discharge of 0.4 m^3/s gives a surface velocity of 60 cm/s on the riffles and 9 cm/s in the pools. Water is drawn from the river at the top end of the channel and is returned at the bottom end. Screens prevent the loss of fry and the entry of predatory fish at either end. Other screens divide the channel into five sections so that the effects of downstream accumulation of wastes can be assessed.

In coastal streams of B.C., several factors limit natural smolt production. Foremost amongst these are extremes of discharge, limited benthic production, and predation. Discharge is controlled in the channel and predation from birds and mink has been countered with a fence and net canopy. The fry are fed on prepared pellet food, in order to increase the capacity of the system, but receive a natural supplement of benthic invertebrates and aerial insects (Mundie and Mounce 1978).

The contribution potential of benthos to the diet of the fry and the possibility of recycling fish wastes and excess pellet food into edible benthos were evaluated experimentally. Much of this work has already been published (Williams *et al.* 1977) but I propose to present here data specifically involving the role of mayflies in these processes.

Drifting mayflies often form a substantial part of the natural diet of salmonids (e.g. Elliott 1973). Not only may a steady intake of these insects provide vitamins and essential trace elements, but a necessary natural feeding experience that may be lacking in hatchery-reared fish (Fenderson *et al.* 1968). As drift occurs mostly at night, fish cannot fully exploit this food source. In a semi-natural channel, however, insects can be dislodged artificially during the day. In order to establish a regime for doing this without completely depleting areas of benthos, the cropping potential of insects on a unit area of riffle must be calculated. Williams *et al.* (1977) set out troughs of sterile gravel and collected all the invertebrates that had colonized at three day intervals, up to 24 days. Colonization curves were then drawn for dominant groups. Figure 2 shows the curves for the four main groups of mayflies. The response was not the same in each group, but within 2-3 weeks the standing crop of mayflies had levelled out. The results suggested that under winter conditions, small areas of the channel's riffles could be disturbed every 15 days to yield about 80,000 invertebrates/m^2, weighing about 0.62 g dry weight; this contains a mayfly component of around 2,400/m^2. This yield should be sustainable provided that undisturbed areas are left to serve as production sites.

The possibility of recycling wastes in the channel was investigated as follows: troughs of gravel were loaded with different amounts of food pellets (Oregon Moist Pellets, Formula II) and pure fish faeces and left for 4 weeks. The response of the mayflies to O.M.P. is shown in Figure 3. *Baetis* sp. A responded favourably initially with a 36% increase in its numbers between the clean

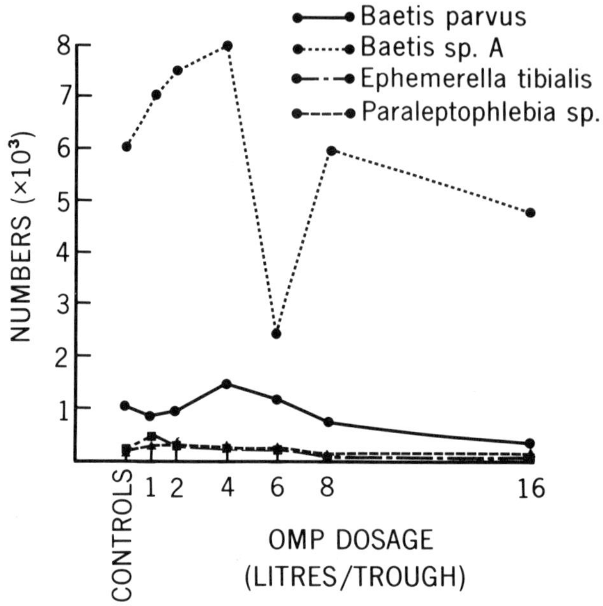

Figure 3. Response of mayflies to fish pellets (O.M.P.).

gravel control trough and the trough loaded with 4 L of pellets. A gradual decline followed. The other mayflies varied in their initial responses but gradually declined with increasing dosage. *Rhithrogena*, *Cinygma* and *Iron* were eliminated entirely. Figure 4 shows the response of *Baetis parvus* Dodds and *Baetis* sp. A to fish faeces. A steady decline is evident.

The response of the total benthos to these wastes was more one of increased numbers (Figure 5), although this was mostly due to chironomids. It appears, therefore, that the mayflies would only play an important part in recycling wastes in the channel if the wastes consisted primarily of a limited excess of food pellets and provided that the level of fish faeces could be kept to a minimum. The standing crop of mayflies on the channel riffles could, however, be increased by other means. For example, Mundie *et al.* (1973) obtained an increase in the numbers of several species after enriching gravel with wheat grain, and Williams and Mundie (1978) showed that selection of a suitable size for the riffle gravel could promote desirable elements in the fauna.

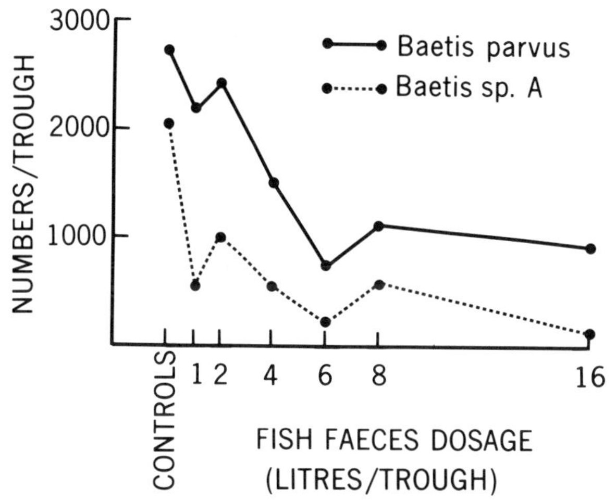

Figure 4. Response of mayflies to fish faeces.

Warren *et al.* (1964) produced an increased mayfly biomass in a small Oregon creek by adding sucrose, primarily by increasing the growth of *Sphaerotilus* as food for the herbivorous insects. Along the same lines, additions of soluble plant nutrients, such as nitrates and phosphates, to channel water may, where light and other nutrients are non-limiting, increase periphyton which in turn would promote the mayfly fauna. It would, of course, be desirable to maintain high water quality in the system as this would probably promote mayflies naturally.

The results of the Big Qualicum channel's first few years of operation are encouraging. It was stocked with 400,000 swim-up fry in May, 1976 and these were released to the sea as 15 g smolts in May, 1977. During this time, they obtained sufficient natural food (from unaided drift) to necessitate a reduction to 25% hatchery rations of their pelleted food to prevent waste. The percentage of mayflies in their diet is given in Table 3. As might be expected, there were seasonal differences in the numbers of mayflies eaten. 1800 h seemed to be the time when most were eaten but this peak was not well marked. Only total numbers were assessed with the result that the size difference between those drifting in the day (very small ones) and those at night (large

Table 3. Percentage of mayfly nymphs in the diet of coho fry (based on sampling 25 fish four times in 24 hours in each section of the channel).

Date	Section 2 (upstream)	Section 3	Section 4	Section 5 (downstream)
May 11, 1976	13	20	8	13
June 24	15	2	8	6
July 15	2	4	3	3
August 23	9	9	13	12
September 2	8	13	7	12

ones) was not adequately recorded (see Allan 1978). Fish in all sections of the channel were eating mayflies. Even though the percentages in the guts were never high (20% maximum) each fish had the chance to develop a normal feeding behaviour that would help prepare it for a free-ranging existence once released from the channel.

The 1977 released fish returned to the river in the fall of 1978 and results indicate a return percentage similar to that of a local hatchery. However, this was the first complete cycle and increased yields can reasonably be expected in subsequent returns. The first batch of smolts was, however, raised at substantially less cost than hatchery fish and some of this is attributable to natural food (Mundie, personal communication).

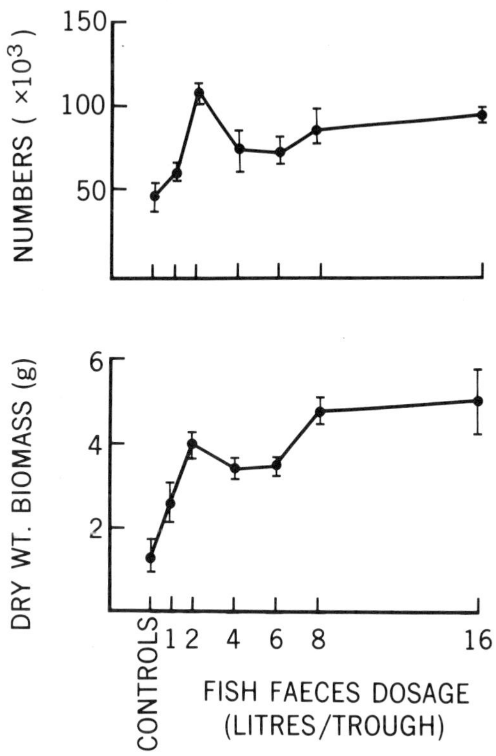

Figure 5. Response of total benthos to fish faeces (vertical lines represent range of three subsamples).

To conclude, a possible future use of mayflies is given. Frey (1964) reviewed the potential of using animal remains in Quaternary lake and bog sediments as an aid to reconstructing past environments and their ecological interrelationships. He listed a great many groups of insects and other invertebrates but indicated that information on many of them is limited. One of the most commonly used groups is the Coleoptera and this is probably because they may constitute over 85% of insect remains in Pleistocene sediments (Shotton 1959). So far as mayflies are concerned, Frey listed three records of either nymphal or adult remains (Brehm *et al.* 1948; Swain 1961 and Tasch and Zimmerman 1961). He concluded by saying that "this general dearth of records suggests that fragments of nymphs either are not generally preserved or else have not yet been generally recognized. I suspect that the latter is correct, not only for this order but also for a number of other orders in which the immature forms are aquatic." The Ephemeroptera, with its ancient lineage and specialization to different aquatic habitats could probably make significant contributions to paleoecology. This has certainly proved to be the case for the Trichoptera as they are now being used with considerable success (e.g., Williams and Morgan 1977, Moseley 1978).

In retrospect, mayfly biology has the potential to be applied in many areas. Perhaps the inability to identify many of the nymphs is the chief factor currently impeding this.

ACKNOWLEDGMENTS

I thank N.E. Williams for reading the manuscript and J.H. Mundie for allowing me to use his unpublished data.

RESUME

Cela fait des siècles que l'homme s'intéresse aux éphéméroptères. L'auteur retrace l'histoire de cet intérêt dupuis les premières tentatives entomologiques de l'homme se rapportant à la pêche jusqu'aux usages présents et futurs de ce groupe d'insectes en hydrologie, en passant par ses applications plus récentes comme indicateur de la qualité des eaux et comme outil de recherches.

ZUSSAMENFASSUNG

Eintagsfliegen sind seit mehreren Jahrhunderten für den Menschen von Interesse. Die vorliegende Arbeit verfolgt die Geschichte dieses Interesses von den ersten Anfängen in einfacher, früher Fischerei-Entomologie über die neuzeitlichere Verwendung der Fliegen als Indikatoren für die Qualität des Wassers und ferner als Forschungswerkzeuge für gegenwärtige und zukünftige Nutzung in der modernen aquatischen Technologie.

REFERENCES

Allan, J.D. 1978. Trout predation and the size composition of stream drift. *Limnol. Oceanogr.* 23: 1231-1238.

Allen, K.R. 1941. Studies on the biology of the early stages of the salmon *(Salmo salar)*. 2. Feeding habits. *J. Anim. Ecol.* 10: 47-76.

Berners, Dame Juliana. 1496. The Treatyse of Fysshynge wyth an Angle. Reprinted from the book of St. Albans. London. Pickering, 1827.

Brehm, V., G. Krasske and W. Krieger. 1948. Subfossile tierische Reste und Algen im Schwarzsee bei Kitzbühel. *Oesterr. bot. Z.* 95: 74-83.

Britt, N.W. 1975. Mayflies (Ephemeroptera) as biological indicators of water quality. *Ohio Biol. Surv., Inf. Circ.* 8: 52-54.

Bryan, J.E. and P.A. Larkin. 1972. Food specialization by individual trout. *J. Fish. Res. Board Can.* 29: 1615-1624.

Burks, B.D. 1953. The mayflies, or Ephemeroptera, of Illinois. *Ill. Nat. Hist. Surv. Bull.* 26: 1-216.

Butcher, R.W., J. Longwell and F.T.K. Pentelow. 1937. Survey of the river Tees, III. The non-tidal reaches. Chemical and Biological. *Tech. Pap. Wat. Pollut. Res. Lond.* 6: 1-187.

Chutter, F.M. 1972. An empirical biotic index of the quality of water in South African streams and rivers. *Water Res.* 6: 19-30.

Edmunds, G.F., Jr., S.L. Jensen and L. Berner. 1976. The Mayflies of North and Central America. Univ. Minnesota Press, Minneapolis.

Elliott, J.M. 1973. The food of brown and rainbow trout *(Salmo trutta* and *S. gairdneri)* in relation to the abundance of drifting invertebrates in a mountain stream. *Oecologia* 12: 329-347.

Fenderson, O.C., W.H. Everhart and K.M. Muth. 1968. Comparative agonistic and feeding behaviour of hatchery reared and wild salmon in aquaria. *J. Fish. Res. Board Can.* 25: 1-14.

Flowers, R.W. and W.L. Hilsenhoff. 1975. Heptageniidae (Ephemeroptera) of Wisconsin. *Great Lakes Entomol.* 8: 201-218.

Fremling, C.R. and G.L. Schoening. 1973. Artificial substrates for *Heptagenia* mayfly nymphs. p. 209-211. *In:* Proc. 1st Int. Conf. Ephemeroptera. W.L. Peters and J.G. Peters, eds. E.J. Brill, Leiden.

Frey, D.G. 1964. Remains of animals in Quaternary lake and bog sediments and their interpretation. *Ergeb. Limnol.* 2: 1-114.

Frost, W.E. 1939. River Liffey Survey II. The food consumed by the brown trout *(Salmo trutta* Linn.) in acid and alkaline

waters. *Proc. R. Irish Acad.* 46B. 139-206.
Frost, W.E. and M.E. Brown. 1967. The Trout. Collins, London.
Frost, W.E. and A.E.J. Went. 1940. River Liffey Survey III. The growth and food of young salmon. *Proc. R. Irish Acad.* 46B: 53-80.
Gammon, J.R. 1970. The effect of inorganic sediment on stream biota. *Rep. Wat. Qual. U.S.E.P.A.* No. 5501-0074, 141 p.
Hilsenhoff, W.L. 1977. Use of arthropods to evaluate water quality of streams. *Wis. Dep. Nat. Resour. Tech. Bull.* 100: 1-15.
Hynes, H.B.N. 1960. The Biology of Polluted Waters. Liverpool Univ. Press.
Krull, J.N. and R.L. Boyer. 1976. Abundance and diversity of benthos during the spring waterfowl migration. *Am. Midl. Nat.* 95: 459-462.
Leonard, J.W. 1962. Environmental requirements of Ephemeroptera. *In:* Biological Problems in Water Pollution. Third seminar. *U.S. Public Health Serv. Publ.* 999-WP-25: 110-117.
Leonard, J.W. and F.A. Leonard. 1962. Mayflies of Michigan trout streams. *Cranbrook Inst. Sci. Bull.* 43: 1-139.
Michener, C.D. 1974. The Social Behaviour of the Bees: a Comparative Study. Belknap Press, Harvard Univ., Massachusetts, 404 p.
Moseley, K.A. 1978. A preliminary report on Quaternary fossil caddis larvae (Trichoptera) *Quat. Newsletter* 26: 2-12.
Mundie, J.H. 1974. Optimization of the salmonid nursery stream. *J. Fish. Res. Board. Can.* 31: 1827-1837.
Mundie, J.H. and D.E. Mounce. 1978. Application of stream ecology to raising salmon smolts in high density. *Int. Verein. theor. angew. Limnol. Verh.* 20: 2013-2018.
Mundie, J.H., D.E. Mounce and L.E. Smith. 1973. Observations of the response of zoobenthos to additions of hay, willow leaves and cereal grain to stream substrates. *Fish. Res. Board Can. Tech. Rep.* 387: 1-123.
Napier, S., Jr. and W.D. Hummon. 1976. Survival of mayfly larvae under mine acid conditions. *Int. Rev. gesamten Hydrcbiol.* 61: 677-682.
Needham, J.G., J.R. Traver and Y.C. Hsu. 1935. The Biology of Mayflies with a Systematic Account of North American Species. Comstock, Ithaca.
Resh, V.H. and J.D. Unzicker. 1975. Water quality monitoring and aquatic organisms: the importance of species identification. *J. Water Pollut. Contr. Fed.* 47: 9-19.
Roback, S.S. 1974. Insects (Arthropoda: Insecta). p. 313-376. *In:* C.W. Hart, Jr. and S.L.H. Fuller, eds. Pollution Ecology of Freshwater Invertebrates. Academic Press, N.Y.
Shotton, F.W. 1959. Insects in Pleistocene deposits. *J. Inst. Biol.* 6: 63-65.
Snow, N.B. and D.M. Rosenberg. 1975. The effects of crude oil on the colonization of artificial substrates by zoobenthos

organisms. *Can. Fish. Mar. Serv. Res. Dev. Tech. Rep.* 553: 1-35.

Swain, F.M. 1961. Limnology and amino-acid content of some lake deposits in Minnesota, Montana, Nevada, and Louisiana. *Bull. geol. Soc. Am.* 72: 519-546.

Tasch, P. and J.R. Zimmerman. 1961. Fossil and living conchostracan distribution in Kansas-Oklahoma across a 200-million year time gap. *Science (Wash. D.C.)* 133: 584-586.

Tebo, L.B. and W.W. Hassler. 1963. Food of brook, brown and rainbow trout in streams of W.N. Carolina. *J. Elisha Mitchell Sci. Soc.* 79: 44-53.

Warren, C.E., J.H. Wales, G.E. Davis and P. Doudoroff. 1964. Trout production in an experimental stream enriched with sucrose. *J. Wildlf. Manage.* 28: 617-660.

Williams, D.D. and B.W. Coad. 1979. The ecology of temporary streams III. Temporary stream fishes in southern Ontario, Canada. *Int. Rev. gesamten Hydrobiol.* (In press).

Williams, D.D. and J.H. Mundie. 1978. Substrate size selection by stream invertebrates and the influence of sand. *Limnol. Oceanogr.* 23: 1030-1033.

Williams, D.D., J.H. Mundie and D.E. Mounce. 1977. Some aspects of benthic production in a salmonid rearing channel. *J. Fish. Res. Board Can.* 34: 2133-2141.

Williams, N.E. and A.V. Morgan. 1977. Fossil caddisflies (Insecta: Trichoptera) from the Don Formation, Toronto, Ontario, and their use in paleoecology. *Can J. Zool.* 55: 519-527.

PERMIAN MAYFLY NYMPHS: NEW TAXA AND SYSTEMATIC CHARACTERS

Michael D. Hubbard and Jarmila Kukalová-Peck*

Laboratory of Aquatic Entomology
Florida A&M University
Tallahassee, Florida 32307 USA

ABSTRACT

Fossil insect nymphs with well-developed wings described as Ephemeroptera from the Lower Permian Period were later referred to the Archodonata. In view of the great amount of phylogenetic information that these nymphs can yield, their taxonomic placement is important. We discuss the systematic characters available in these nymphs and their relationships with the Ephemeroptera and the Archodonata. We consider these nymphs to be true Ephemeroptera and transfer them all to the Protereismatoidea. One new family and one new genus are established and three new species are described.

INTRODUCTION

Six fossil insect nymphs described by Kukalová (1968) from Lower Permian beds in Moravia and Oklahoma were referred to the Ephemeroptera, five to the Protereismatoidea and one to a distinct but unnamed related family. The wing venation of most of these nymphs is remarkably well preserved and even the corrugations, or fluting, of the veins is discernable. Tracheal gills are present

* Department of Geology, Carleton University, Ottawa, Ontario K1S 5B6 Canada

on abdominal segments I to IX* and the nearly homonomous thoracic wing pads are attached to the thorax only along the articular area; they are independent of each other, bent, and project posteriorly at an oblique angle to the body. The mouthparts consist of large broad mandibles, with well-developed teeth. These nymphs could be of great importance in the study of mayfly evolution, and therefore, an understanding of their systematic position is crucial in evaluating the data which can be obtained from their study.

In a critical review of these nymphs, Demoulin (1970) removed them from the Ephemeroptera and considered them to belong in the order Archodonata. His argument that they are not Ephemeroptera depended primarily on the presence of segmented tarsi and double tarsal claws, and details of the wing venation, in particular a simple MA, separate R_4 and R_5, and the existence of a connecting vein between MP and CuA. He regarded this connecting vein between MP and CuA as a precursor of the capture of MP by CuA as described in the Permian archodonate family Permothemistidae. Demoulin also takes into consideration the factor of the numerous small marginal forks of the longitudinal veins similar to those observed in adult archodonate wings.

However, to transfer these Permian nymphs from the order Ephemeroptera to <u>any</u> other pterygote order would necessitate the refutation of two basic characters which are generally accepted as being uniquely ephemeropteran: the presence of 1) three caudal filaments, and 2) a well developed "costal brace". While the former character is plesiomorphic (primitive) and is shared with the most "primitive" insect order, the apterygote Archaeognatha, the costal brace may well be considered as one of the basic derivative (apomorphic) characters of the order Ephemeroptera. The history of the costal brace is that of an ancient structure which became emphasized because of the involvement in a new special function. Morphologically, the brace is the convex anterior branch of the subcostal vein system that became engaged in the flight mechanics peculiar to the Ephemeroptera. It is presumed that the manner of flight is one of the basic agents, if not <u>the</u> basic agent,

---

* Appendages on the genital segments VIII and IX appear to be serial with the other tracheal gills. Recently, however, <u>two</u> pairs of gonopods complete with terminal double claws have been found in adult *Protereisma* by one of us (JKP). There is therefore, a certain possibility that the last pair, or last two pairs, of "tracheal gills" in the Paleozoic mayfly nymphs may belong to the gonopods protruding laterally from beneath the genital segments.

that triggered the early radiation of taxa in the Paleozoic. While the more distal venation of the wing is often subject to convergence, the basal venation and the details of wing articulation are the most dependable and informative landmarks for discerning relationships at the higher taxonomic levels. Quite clearly, the presence of three caudal filaments and the costal brace within the Pterygota point to only one order: the Ephemeroptera.

Demoulin pointed out that the segmented tarsi and double tarsal claws of these Permian nymphs are not known to occur in any Recent mayfly nymphs. As is generally known, the embryonic, the juvenile, and the adult stages of the Insecta undertake different developmental pathways in the process of adapting to their special needs and conditions. Consequently, the more specialized, and the "younger" (in a phylogenetic sense) the stages are, the more liable they are to be mutually different. In the fossil record, the embryonic stage cannot be documented. However, there are certainly two distinctive, progressively diverging developmental lines, one for the juveniles and another for the adults, whenever both stages are found. In the Pterygota, the differences in morphology between juveniles and adults eventually become so deep that they have to be bridged by a metamorphic stage. In the Ephemeroptera, the origin of this metamorphic stage came only after the Paleozoic (Kukalová-Peck 1978). In the light of this fossil evidence, the difference in the leg morphology, viz., the presence of five tarsal segments and of double tarsal claws in the Permian nymphs, as well as the reduction to unsegmented tarsi and one claw in the modern mayfly nymphs is fully consistent with the other knowledge of the developmental process. It contributes evidence that the simple tarsi and single claw of the Recent mayfly nymphs are the derived and not the primitive condition. The reduction in tarsal segmentations and the presence of a single unguis in modern mayfly nymphs are undoubtedly apomorphic characters derived from the primitive condition of double claws and segmented tarsi still manifested in the related Odonata, and are of more recent origin that the Permian Period.

RESIDUAL PRIMITIVE FEATURES IN LIVING MAYFLIES

It is often the case, in the insects, that the ancestral condition documented by the fossils are somehow expressed, in a subtle and inconspicuous way, in at least some living members of the same phylogenetic lineage. The reconstruction of the Permian nymph no. 1 from Oklahoma (*Kukalova americana* Demoulin; Fig. 1) shows three pairs of thoracic wings. The rudimentary prothoracic wings are still in existence in a few modern mayflies: they were first recognized by Ide (1936) in *Ecdyonurus venosus* (Fabricius) and are apparently also found in the nymph of *Dolania americana* Edmunds and Traver.

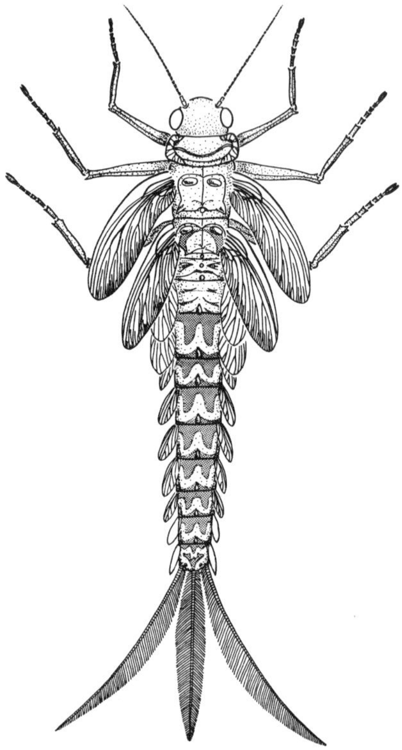

Figure 1. Nymph of *Kukalova americana*.

The movable meso- and metathoracic wing pads of these Permian mayfly nymphs testify that the wing appendages of the Ephemeroptera in the Paleozoic were narrowly attached and independently articulated (Kulaková-Peck, in prep.). In modern nymphs, however, the attachment is usually very broad along the whole tergum and the articulation is completely obliterated. Thus, the wing pads are fused with the tergum into one inseparable unit. This is not, however, always the case. For example, the nymph of *Analetris* Edmunds has the wing pads narrowly attached, similar to those of the Paleozoic mayflies, and the only difference is that the articulation in the Recent mayflies is obliterated.

In regard to residual primitive features in the legs, while no known Recent mayfly nymphs possess 5-segmented tarsi and double tarsal claws, this is a common condition in the adults. Rudimentary tarsal segmentation is found in some nymphs of the modern Siphlonuridae (Edmunds 1972). The large mandibles of the Permian Oklahoma nymph are reminiscent of the mouthparts of some modern generalist feeders.

THE VENATION OF PERMIAN MAYFLY NYMPHS

Demoulin argued that the permian nymphs could not belong to the Ephemeroptera because of the differences in venation, namely the simple MA, separate R_4 and R_5, and the presence of numerous small marginal forks (Recent mayflies have a forked MA, simple R_{4+5}, and few marginal forks). This argument cannot be valid on the ordinal level, because of the reasons discussed above. It could, however, be of importance on sub-ordinal levels.

The vein systems are derived from the blood lacunae and were primitively multiple-branched to supply blood to the alar appendage, long before it became the wing. They also provided the "tunnels" for the tracheae that grew into the richly tracheated alar appendage. Since the original function of the venation was not to support the aerofoil but to evenly distribute the blood, the primitive condition of the venation in all pterygote orders is always the (symmetrically) branched venation. This includes even such notoriously simple veins as the costa, the sub-costa and the radius I.

Some of the primitively dichotomously symmetrical branching was lost under the rigid requirements of flight mechanics (Kukalová-Peck 1978) soon after the Paleoptera and the Neoptera started using their alar appendages for forward movement. However, remnants of the primitively branched condition of <u>all</u> veins are scattered throughout the ancient Paleozoic pterygotes. This residual branching gives evidence as to the original state and is well known to paleoentomologists (see Carpenter 1966, Sharov 1966, Kukalová-Peck 1978, and Wootton 1976 for summaries). Therefore, since the appearance of Redtenbacher's paper (1886) most paleoentomologists have insisted that the richer branching of <u>any</u> particular vein system is principally the more primitive condition. Also, the loss of one branch is usually considered to be a one-way evolutionary step. Under these premises it is unlikely that the simple MA in any paleopterous lineage would precede the primarily forked MA in their descendants. The Carboniferous *Triplosoba* Handlirsch with its simple MA is most likely the representative of a blind evolutionary line (Carpenter 1963). Clearly, the Permian nymphs described by Kukalová (1968) either had forked MA, or they probably do not belong to the Protereismatoidea.*

* It should be noted that several of the Permian nymphs are older nymphs, the venation of which is expected to be reasonably close to that of the adults.

The small branches along the posterior margin, on the other hand, do not seem to be of much phylogenetic importance in the Paleozoic insects, certainly not on higher than generic levels. The use of wing venation in taxonomy would be much easier if the wing membrane between the basic hemocoel - based veinal systems was not capable of forming secondary connective veins, cross veins, braces, forks, and secondary branches. These last are sometimes formed along the crests of aerodynamic folds and flexion lines, like the well-known "spurious vein" of the Diptera, or, they may simply connect together two neighboring veins and simulate the capture of veins (Fig. 2, cr). Occasionally, these secondary vein "additions" are so inconspicuous that they blend perfectly with the primary venation. If this is the case, argument may arise and the phylogeny of a particular group might linger unsolved for decades. The fossil record is of invaluable help, because it brings direct evidence of the ancestral stage, and may solve the problem.

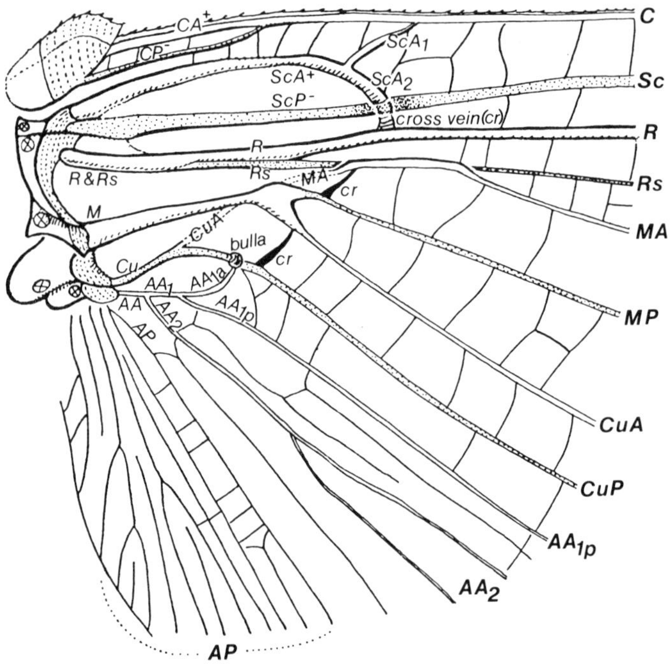

Figure 2. Adult *Protereisma latum* (hind wing). Concave veins dotted. Veinal bases striped. (After J. Kukalová-Peck, unpublished figure from study on the Paleozoic wing bases, in prep.).

The puzzle of the inconsistent venation pattern of the Permian nymphs clearly has two different phylogenetic levels: the supposedly simple MA and separate (i.e., branched) R_4 and R_5 concern the primary branching of the primitive vein systems and reflect on the higher systematic categories such as superfamilies of the Recent Ephemeroptera; the numerous small marginal branches, forks, and secondary veinlets are less important and apparently operate on the specific or, at most, generic levels. This dilemma raises one obvious question: was the venation of the Permian nymphs correctly identified by Kukalova in 1968? Regrettably, the answer must be that it was not.

The Oklahoma Permian nymph (*Kukalova americana* Demoulin) is superbly three-dimensionally preserved with the natural pigment intact. Compared to it, the Permian nymphs from Obora are only flat and colorless imprints of the body structures and venation in the soft mud, while the chitin is completely dissolved. The venation of the freely articulated wing pads of *K. americana* Kukalova 1968, Pl. 29), is darkly pigmented, distinct, and even fluted. However, the wing pads are damaged by the irregular splitting of the rock. The venation shown by Kukalova (1968, Fig. 2) is the composite of the obverse and the (unfigured) reverse. In spite of the fact that the venation on the photograph looks like that figured, we believe that the split of the rock and perhaps an indetectable slip within the wing membrane may have masked the true, typically protereismatoid venation.

The protereismatoid wing base and articulation of the nymphs and adults, in comparison with those of the modern "primitive" mayflies, have been studied lately by one of us (JKP) and a paper is currently in preparation. The wing base pattern of adult *Protereisma latum* Sellards (Fig. 2) and 25 related forms provides new aspects for a better understanding of the Permian nymphal venation and, perhaps, offers a solution to this problem.

As shown in Fig. 2, the radius in adult *Protereisma* starts as a common stem with R and Rs. Soon, Rs is visible adjacent to R, but running distinctly at a lower level. After Rs diverges from R, it is met by MA which lies on top of it and completely covers Rs for a short distance. After than, Rs and MA separate and diverge as regular convex (MA) and concave (Rs) veins. The media also starts as a common and well-defined stem. MA separates inconspiciously; the portion between the stem of M and the point of its superimposition on Rs is weak and is discernable only in some specimens. There is a strong auxiliary cross vein which parallels distally the weak portion of MA and apparently takes over the bracing. The cubitus also starts as a common stem which is well-defined. CuA separates early and inconspicuously, and fuses laterally with the stem of M, leaving no trace of fusion. Like the initial portion of MA, CuA is also very weak and paralleled by an auxiliary cross vein which acts as the mechanical brace. CuP carries a distinct bulla at the point where it is met by the anal brace. The anterior anal vein starts immediately from the basianale, following the general trend of pterygote anal veins to form a fan-line pattern. It divided into two branches, AA_1 and AA_2, AA_1 and its upwardly continuing branch AA_{1a} form the mechanically important structure called the anal brace which prevents the anal area of the wing from buckling. AA_{1b} and AA_2 are branched and run parallel with CuP. The posterior anal vein (AP) has lost its basal stem and starts immediately as a group of branches. The number of AP branches seems to be variable on the specific level. There may be a special jugal vein independent of the anals and starting from an inconspicuous basijugale.

A detailed comparison of the wing bases of the nymphal *Kukalova americana* (Fig. 3) and the adult *Protereisma latum* (Fig. 2) shows a close similarity, probably at the family level. A major peculiarity of apparently all of the nymphal wings (not suspected when the wings were first described) is that MA seems to stay on top of Rs for a surprisingly long distance. This deviation from the adult pattern of all known mayflies, extinct or extant, is apparently due to the fact that the wings were simultaneously moving and bent backwards at a seemingly awkward angle. If the wing venation is reconstructed with this new interpretation of MA in mind, as is done here in Fig. 3, it becomes quite similar to the pattern of the adult Protereismatoidea.

SYSTEMATIC POSITION OF THE ARCHODONATA

The order Archodonata was established by Martynov (1932) to harbor dipterous Paleoptera with a primitive venation pattern, nine apparent abdominal segments, no prothoracic lobes, and a richly branched anal area. Nevertheless, as additional data on the group are added, it becomes more and more evident that the Archodonata are synonymous with the Palaeodictyoptera. The Archodonata have been found to have a prominent haustellate beak (Sharov 1971). This sucking beak, composed of mandibular stylets, maxillary stylets, and the hypopharynx, all resting on the trough of the labial ligula (JKP, unpublished), is one of the basic apomorphic characters of the palaeodictyopteroid orders: the Palaeodictyoptera, the Megasecoptera, and the Diaphanopterodea. It is of interest that these mouthparts are derived from the "mayfly pattern" as preserved in Recent mayfly nymphs and Paleozoic nymphs and adults. However, the elongation of the parts into stylets is very distinctive and no palaeodictyopteroid is known so far to carry the "regular" jaws. Apparently, the adaptation of the basic paleopterous mouthparts (which resembled those of living mayfly nymphs) for sucking, happened very early. The sucking beak became very narrowly bound with the palaeodictyopterous flight mechanics and with typically palaeodictyopterous wing articulation. As shown by Kukalová-Peck (1974) the articulation is

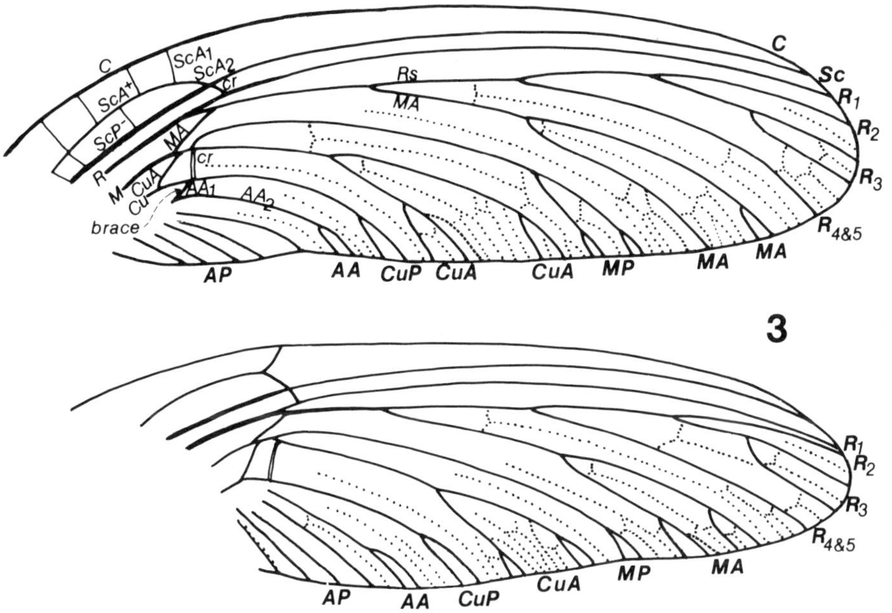

Figure 3. Fore and hind wings of nymph of *Kukalova americana*.

quite distinctively different from the mayfly base. Therefore, either the beak, or the peculiar articulation, are decisive enough for the classification of a fossil. On the other hand, the wing venation by itself can be very misleading because the ephemeropteran "pattern" occurred repeatedly in different groups. Thus, *Lithoneura* Carpenter, which has often been considered a mayfly of a monotypic superfamily, is a Palaeodictyoptera. The holotype of *L. lameerei* Carpenter has been studied by one of us (JKP) and found to be strongly convergent in venation and in the body shape to mayflies. However, the head is palaeodictyopteroid with long heavy antennae, bulging eyes, outlines of the swollen clypeus where the cibarial muscles were attached, and with small remnants of a beak; the prothorax carried a pair of large prothoracic winglets of the palaeodictyopteroid type; and the sclerites at the base of the wings were simple small specialized veinal bases characteristic of the Palaeodictyoptera. The Triassic *Litophlebia* Hubbard and Riek (=*Xenophlebia* Riek) also has wings with strikingly mayfly-like venation. They have been placed as a separate family. Since no body is attached and the articulation is missing, there are few clues for a decision. However, one important apomorphic ephemeropteroid feature is missing in both *Lithoneura* and *Litophlebia* - the costal brace. Consequently, we think that the association of the Litophlebiidae with the Ephemeroptera is doubtful and that it has to be referred to the Megasecoptera. One very important circumstance should be mentioned here: the recent discovery of a beautifully preserved wing of a member of the primitive Erasipteridae, an early odonatoid, from the Namurian (early Upper Carboniferous) of Argentina, the documentation of which one of us (JKP) was given the opportunity to study (courtesy of E.F. Riek). It becomes quite clear that the basic ephemeroid pattern is also at the base of the odonatoid venation. Indeed, there is a striking similarity and parallel between the Paleozoic mayfly wing (Fig. 2) and the Argentine Erasipterid. A paper discussing consequent changes in the odonatoid venational interpretation is in preparation by Riek.

Also of interest is the repeated tendency towards diptery which occurs both in the Ephemeroptera and in the Palaeodictyoptera (Kukalová-Peck, 1969a, b, 1970). Unlike the neopterous Diptera, the switch to two-winged flight did not bring major changes in the flight-related morphological structures and in the flight pattern in the Paleoptera. It is therefore doubtful whether the diptery of the Archodonata justifies the separation of this group into a separate order apart from the Palaeodictyoptera.

TAXONOMY (see Table 1)

Nymph no. 1 of Kukalova (1968) was placed as the type-species, *Kukalova americana*, of a new genus, *Kukalova*, by Demoulin (1970). Nymph no. 2 of Kukalova was described as *Kukalova moravica*. Demoulin also referred nymphs nos. 3, 4, and 5 of Kukalova to this genus (as *Kukalova* spp.). The family Kukalovidae was established to contain this genus.

Nymph no. 6 of Kukalova was made the type-species, *Jarmila elongata*, of a new genus, *Jarmila*, by Demoulin, for which the family Jarmilidae was established. Demoulin suggested that all of these nymphs belonged to the Archodonata.

In light of the previous discussion, we refer all of these nymphs to the Ephemeroptera and place them in the superfamily Protereismatoidea. *Kukalova* is placed in the Protereismatidae as a monotypic genus containing only the type-species, *Kukalova americana*, and the family Kukalovidae is synonymized with the Protereismatidae. The genus *Kukalova* may prove to be synonymous with *Protereisma* when further specimens are available for study.

Kukalova moravica is placed as the type-species of a new genus,

Table 1. Taxonomic changes in the Protereismatoidea.

Protereismatidae Lameere, 1917.
= Kukalovidae Demoulin, 1970. NEW SYNONYMY.

Kukalova Demoulin, 1970.
Type-species: *Kukalova americana* Demoulin. (original designation)

Kukalova americana Demoulin.
Protereisma sp. (nymph no. 1) Kukalová, 1968:313, figs. 1, 2, pl. 29.
Kukalova americana Demoulin, 1970:6.

Jarmilidae Demoulin, 1970. NEW PLACEMENT.

Jarmila Demoulin, 1970.
Type-species: *Jarmila elongata* Demoulin. (original designation)

Jarmila elongata Demoulin.
Ephemeroptera *inc. fam.* (nymph no. 6) Kukalová, 1968:320, fig. 7.
Jarmila elongata Demoulin, 1970:7.

Oboriphlebiidae Hubbard & Kukalová-Peck. NEW FAMILY.

Oboriphlebia Hubbard & Kukalová-Peck. NEW GENUS.
Type-species: *Kukalova moravica* Demoulin.

Oboriphlebia moravica (Demoulin).
Protereisma sp. (nymph no. 2) Kukalová, 1968:316, fig. 3.
Kukalova moravica Demoulin, 1970:6.

Oborphlebia tertia Hubbard & Kukalová-Peck. NEW SPECIES.
Protereisma sp. (nymph no. 3) Kukalová, 1968:318, fig. 4.
Kukalova sp., larve no. 3 Demoulin, 1970:7.

Oboriphlebia quarta Hubbard & Kukalová-Peck. NEW SPECIES.
Protereisma sp. (nymph no. 4) Kukalová, 1968:318, fig. 5.
Kukalova sp., larve no. 4 Demoulin, 1970:7.

Oboriphlebia quinta Hubbard & Kukalová-Peck. NEW SPECIES
Protereisma sp. (nymph no. 5) Kukalová, 1968:318, fig. 6.
Kukalova sp., larve no. 5 Demoulin, 1970:7.

Oboriphlebia (named for Obora in Moravia), for which the family Oboriphlebiidae is established. Nymphs nos. 3, 4, and 5 of Kukalova (1968) are named as new species in *Oboriphlebia* (*O. tertia*, *O. quarta*, and *O. quinta*, respectively). The family Oboriphlebiidae and its included genus, *Oboriphlebia*, can be distinguished from the remainder of the Protereismatoidea by the combination of the nymphal mesothorax slightly longer than the metathorax and the divergent wing pads. The holotypes of the new species are the specimens figured by Kukalova (1968).

The monobasic family Jarmilidae is also transferred to the Protereismatoidea, bringing the total number of families in this superfamily to five.

ACKNOWLEDGMENTS

This research was partially supported by a research program (FLAX 79009) of SEA/CR, U.S.D.A., at Florida A&M University, William L. Peters, Research Leader.

RESUME

Des fossiles de nymphes d'insectes à ailes bien dévelopées, d'abord classés parmi les éphéméroptères du bas Permien ont subséquemment été rattachés aux archodonates. Étant donné la profusion de renseignements phylogénétiques que ces larves fournissent, leur classement taxonomique est important. Nous parlerons donc de leurs caractéristiques systématiques disponibles et des rapports qui existent entre les éphéméroptères et les archodonates. Nous considérons ces nymphes comme de véritables éphéméroptères et les rattachons toutes aux protéreismatoïdés. Une nouvelle famille et un genre nouveau sont reconnus dont trois nouvelles espèces seront décrites.

ZUSSAMENFASSUNG

Fossil Insektennymphen mit gut entwickelten Flügeln, beschrieben als Ephemeropteren der "Unteren Perm Periode", wurden später zu den Archodonaten gezählt. Im Hinblick auf die große Menge der phylogenetischen Information, die diese Nymphen liefern können, ist ihre taxonomische Einordnung von großer Bedeutung. In der vorliegenden Arbeit erörtern wir die systematischen Merkmale, die bei diesen Nymphen auftreten, sowie ihr Verhältnis zu den Ephemeropteren und Archodonaten. Wir halten die beschriebenen Nymphen für echte Ephemeropteren und rechnen sie alle zu den Protereismatoiden. Eine neue Familie und eine neue Gattung werden eingeführt, und drei neue Arten dargestellt.

REFERENCES

Carpenter, F.M. 1963. Studies on Carboniferous insects from Commentry, France: Part IV. The genus *Triplosoba*. *Psyche* 70: 120-128.

Carpenter, F.M. 1966. The Lower Permian insects of Kansas. Part II. The orders Protorthoptera and Orthoptera. *Psyche* 73: 46-88.

Demoulin, G. 1970. Remarques critiques sur des larves "Ephemeromorphes" du Permien. *Bull. Inst. R. Sci. Nat. Belg.* 46(3): 1-10.

Edmunds, G.F., Jr. 1972. Biogeography and evolution of Ephemeroptera. *Ann. Rev. Entomol.* 17: 21-42.

Ide, F.P. 1936. The significance of the outgrowths on the prothorax of *Ecdyonurus venosus* Fabr. (Ephemeroptera). *Can. Entomol.* 68: 234-238.

Kukalová, J. 1968. Permian mayfly nymphs. *Psyche* 75: 310-327.

Kukalová, J. 1969a. Revisional study of the order Palaeodictyoptera in the Upper Carboniferous shales of Commentry, France. Part I. *Psyche* 76: 163-215.

Kukalová, J. 1969b. Revisional study of the order Palaeodictyoptera in the Upper Carboniferous shales of Commentry, France. Part II. *Psyche* 76: 439-486.

Kukalová, J. 1970. Revisional study of the order Palaeodictyoptera in the Upper Carboniferous shales of Commentry, France. Part III. *Psyche* 77: 1-44.

Kukalová-Peck, J. 1974. Pteralia of the Paleozoic insect orders Palaeodictyoptera, Megasecoptera, and Diaphanopterodea (Paleoptera). *Psyche* 81: 416-430.

Kukalová-Peck, J. 1978. Origin and evolution of insect wings and their relation to metamorphosis, as documented by the fossil record. *J. Morphol.* 156: 53-126.

Martynov, A.V. 1932. New Permian Palaeoptera with the discussion of some problems of their evolution. *Trudy Paleozool. Inst. Acad. Sci. USSR* 1: 1-44.

Redtenbacher, J. 1886. Vergleichende Studien über das Flügelgeäder der Insekten. *Ann. K.K. Natur. Hofmus* 1: 153-232.

Sharov, A.G. 1966. Basic arthropodan stock with special reference to insects. Pergamon Press, Oxford.

Sharov, A.G. 1971. Morfologicheskie osobennosti i obraz zhizin paleodiktiopter. *Dok. 24 Chteniya Pamyati N.A. Khalodkovskogo Akad. Nauk SSSR:* 49-63.

Wootton, R.J. 1976. The fossil record and insect flight. *Symp. R. Entomol. Soc. Lond.* 7: 235-254.

PHYLOGENY OF THE LEPTOPHLEBIIDAE (EPHEMEROPTERA): AN INTRODUCTION

William L. Peters

Laboratory of Aquatic Entomology
Florida A&M University
Tallahassee, Florida 32307 USA

ABSTRACT

Work has been done to revise the genera of the Leptophlebiidae and to determine the higher classification and phylogeny of monophyletic groups within the family. Detailed criteria must be imposed if such studies are to be consistent. Our phylogenies are based on the possession of several, presumed derived, character states for each monophyletic daughter line. The most primitive furcation in the evolution of the extant Leptophlebiidae is discussed. Based on this phylogeny, the need for and usefulness of subfamilies and tribes within the family are also discussed. A new subfamily: the Atalophlebiinae is proposed.

INTRODUCTION AND METHODS

In 1960, while a graduate student at the University of Utah, I began to revise the generic classification of the Leptophlebiidae on a worldwide basis. I also hoped to study the phylogeny and zoogeography of the family.

At that time only 44 genera and 13 subgenera were established in the literature. Except for 6 Holarctic genera, all were known from tropical Asia and the Southern Hemisphere. Of these 38 latter genera, exactly one-half were established on adults alone and 17 were monotypic. Further, none of the generic descriptions were consistent and any two descriptions had little comparative value.

While I had at my disposal unstudied collections of Leptophlebiidae from tropical Asia, South America, Australia and a few other areas, most material was not reared and contained only nymphs or adults. Since 1960, I and others have reared Leptophlebiidae all over the world. Based on these collections and others either loaned or given to me, I estimate there may be some 300 genera and 4,000 species of Leptophlebiidae. Except for possibly the Baetidae, the Leptophlebiidae are the most diversified family, occurring in all areas of the world except for a few distant oceanic islands.

An example of this diversity occurs in the Leptophlebiidae of New Caledonia. This Pacific island some 400 km long x 40 km wide contains a mayfly fauna consisting entirely of Leptophlebiidae (except for some introduced Baetidae) representing about 20 new genera and over 65 species. The nymphs of these mayflies have adapted to many river and stream habitats and some superficially resemble nymphs of other mayfly families, such as Heptageniidae, Baetidae, Ephemerellidae and Ephemeridae. These Leptophlebiidae are being described in a series of papers (Peters *et al.* 1978, Peters and Peters, in press).

Many of the earlier publications concerned the leptophlebiid genera of the Eastern Hemisphere excluding Australia (Peters *et al.* 1964, Peters and Edmunds 1964, 1970). In these papers and others, consistency is stressed so that all generic descriptions are consistent and comparable; however, for certain phyletic lineages additional characters have been added when needed. Illustrations are also consistent and comparable. When interpretation of morphological characters has been better understood with further study, additional detail figures have been added. For example, many small structures of nymphal mouthparts are now used in our phylogenetic analyses.

The first phylogenetic analysis of any group of Leptophlebiidae was published in Peters and Edmunds (1970). Our phylogenetic diagram represented the probable phylogeny of the Eastern Hemisphere genera, excluding Australia. While this phylogeny was developed by evaluating weighted phenetic similarities, it is surprising how close this diagram is to those that have recently been developed in our studies using presumed shared, derived character states.

Although the phylogeny of any group is incomplete until all Recent species are described and studied, this task is often larger than the time one can apply to it. Sometimes in a large monophyletic group such as Leptophlebiidae with a high diversity of genera and species, time is not available in any one person's lifetime to describe and study all species. However, based on the studies of some species and the delineation of genera and subgenera, the phylogeny of the inter-group ancestors of the Leptophlebiidae can be determined. For example, Peters and Edmunds (1972) studied the generic limits of certain cool-adapted Leptophlebiidae in southern South

America based on a review of the literature and available species, and later Pescador (1976) completely revised the species of these genera and others. The delineation of genera and the phylogenetic conclusions on the inter-group ancestors are similar in both studies.

Since a number of persons are studying various groups of Leptophlebiidae, or Leptophlebiidae of specific geographical areas, detailed criteria must be imposed if phylogenetic and classificatory studies are to be consistent. Such criteria have been developed and examples of our criteria and methods will be presented at this conference and discussed in a series of papers in press or in preparation.

The three basic systems of phylogenetic reconstruction are: (1) methods of phenetics, (2) methods of cladistics, and (3) evolutionary methods. The basic system of phenetics was developed by Sokal and Sneath (1963) and is not used in our work. The cladistic approach was best developed by Hennig (1966), while the evolutionary or morphological system was outlined in detail by Mayr (1969). Recently Ross (1974) developed a system of phylogenetic reconstruction which incorporated much of the thinking of both the cladistic and evolutionary systems. While other series of methods have been proposed, most are combinations of the three basic systems.

Hennig (1966), Mayr (1969), and Ross (1974) all claimed to develop phylogenies based on shared, derived characters. While Hennig (1966) developed the rule of shared, derived characters, Ross (1937) used the same rule and similar methods some 29 years earlier. Hennig (1966) developed a set of rigid principles and methods for phylogenetic reconstruction and classification; however, many of these methods remain in a theoretical stage of development. Mayr (1969) contended that precise and rigid methods for phylogenetic reconstruction were nearly impossible to write and would not allow taxonomists to express clearly their ideas concerning evolution of groups. Therefore, much of Mayr's system involves evaluating weighted phenetic similarities.

The methods of Ross (1974) gave a precise but flexible system for phylogenetic reconstruction and classification. Basically, these methods, with some modifications, are used for our phylogenetic studies of the Leptophlebiidae.

It is Ross's tenet that only characters with known ancestral and derived states should be used in phylogenetic analysis. In all cases, derived character states arise from ancestral ones, not vice versa. Taxa that share particular derived states are grouped together on the assumption that they arose from a common ancestor possessing that state. As groups are joined together by this method, there should result a gradual progression of groups having more and more derived states arising from the group ancestor in which all states are ancestral (Fig. 1). Hennig (1966) restricts this method of shared, derived character states so that each daughter line of any furcation, even at the level of the recent taxa, possesses at least one derived state (Fig. 2).

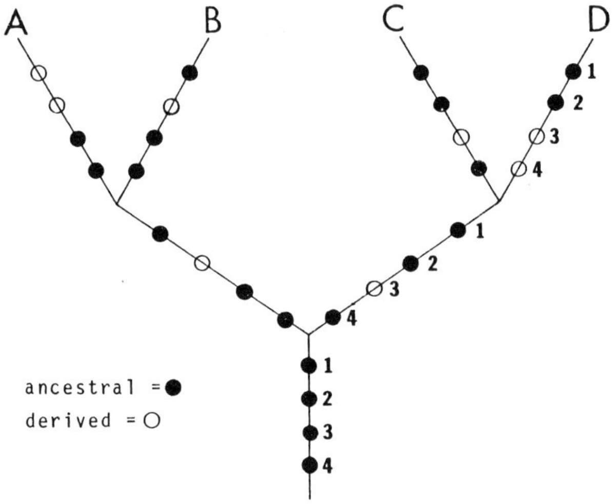

Figure 1. Phylogenetic analysis as developed by Ross (1974).

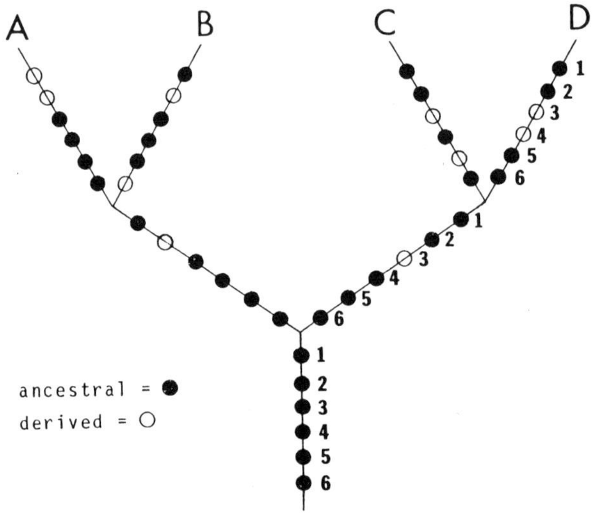

Figure 2. Phylogenetic analysis as developed by Hennig (1966).

In our phylogenetic studies of the Leptophlebiidae, we analyze 300 to 500 morphological characters from both the nymphs and adults for any assumed furcation. Based on such a large number of characters, our phylogenetic diagrams approach more the rigid Hennig diagrams than those of Ross. Sometimes for a given furcation, a derived character state is not known for one daughter line. The lack of a particular derived character state should not delay publication of a phylogenetic analysis, but the search for the derived character state should continue. Further, in some more primitive and basal furcations, not all taxa of a large monophyletic group might possess a given derived character state which we use, although other derived character states can demonstrate the monophyletic relationship. Often a character state which is not possessed by all members of a group needs only redefinition to be useful in phylogenetic analysis (or the exceptions clearly noted).

Once a formal phylogenetic analysis has been developed by using shared, derived character states, then the evolution of character complexes within the groupings should be discussed in detail. If the rate of evolution is hypothesized based on these data and shown in the phyletic diagram, then such data should be discussed for each furcation. In actuality, the phenetic similarities and differences among taxa can be given in a matrix, table, or diagram. In both the formal phylogenetic analysis and the statement of phenetic similarities and differences, it is important that all data be given in the publication.

To date, the higher classificatory limits within the Leptophlebiidae have been confined to genera and subgenera. A higher classification should be expressed as simply as possible and correspond to the phylogenetic analysis in order to express various sets of phylogenetic and evolutionary data.

Generic and subgeneric limits for the Leptophlebiidae are defined by criteria given in Peters and Edmunds (1970). These criteria use the degree of morphological difference in both nymphs and adults to differentiate genera and subgenera.

ANCESTRAL EVOLUTION

McCafferty and Edmunds (1979) recently published a phylogeny of the extant superfamilies and suborders of Ephemeroptera and the resulting higher classification. They stated that the most recent common ancestor of the Pannota, Leptophlebioidea, and Ephemeroidea was a pre-leptophlebiid; however, this ancestor could be a leptophlebiid. Further they stated that the highly derived superfamily Ephemeroidea had its origin within the Leptophlebiidae, thus making the Leptophlebiidae paraphyletic. While I agree that the most recent common ancestor of Leptophlebioidea and Ephemeroidea was at least leptophlebiid-like, the paraphyletic nature of the Leptophlebiidae remains problematical and will be considered in detail, in a future paper. Herein, I will concern myself only with the most primitive leptophlebiid ancestors within the Leptophlebiidae.

Peters and Edmunds (1970) discussed the uniqueness of the *Paraleptophlebia*-type genera with respect to all other genera of the Eastern Hemisphere Leptophlebiidae based on morphology of both the nymphs and adults. Based on my additional studies, the *Paraleptophlebia*-type genera represent the basal most primitive furcation within the Leptophlebiidae. Zoogeographical studies indicate this furcation occurred before the breakup of Laurasia and Gondwanaland during the middle Cretaceous. The group presently occurs only in the Northern Hemisphere and includes *Paraleptophlebia*, *Leptophlebia*, *Habroleptoides*, *Habrophlebia*, *Calliarcys*, *Habrophlebiodes*, *Dipterophlebiodes*, and *Gilliesia*.

Herein I designate the above eight genera to represent the subfamily Leptophlebiinae Banks. The remaining known extant genera of Leptophlebiidae are herein placed in a NEW SUBFAMILY Atalophlebiinae Peters (Fig. 3). Table 1 presents character states delineating each daughter line leading to these subfamilies.

Table 1. Character states of Figure 1.

Character	Character state in "A"	Character state in "B"
Styliger plate of ♂	● deeply cleft or deeply cleft ventrally.	O fused.
Composition of brush on anterior margin of maxillae.	● hairs or spines scattered or unevenly arranged.	O hairs or spines evenly arranged in rows.
Lingua of hypopharynx:		
lateral margins	● without lateral projections.	O with lateral projections (except *Castanophlebia*, *Terpides*, *Fittkaulus*).
venter	O with submedian, subapical patches of hair (except *Paraleptophlebia*).	● with to without scattered hairs. (Hair patterns, if present, occur on apical margin or dorsum.)
Anteromedian emargination of labrum:		
anterolateral margins	O with heavy, thickened, pointed to blade-like setae or spines.	● without modification of marginal setae.
shape of emargination	● smooth, without denticles; shallowly to moderately curved.	O with denticles or with acute median cleft; straight to acutely cleft.*

* Character is best seen in freshly molted nymphs other than those in last instar. Denticles are frequently worn (*Thraulus*, *Choroterpes*) and secondarily lost in some genera with straight labral margins (*Choroterpides*, *Kirrara*).

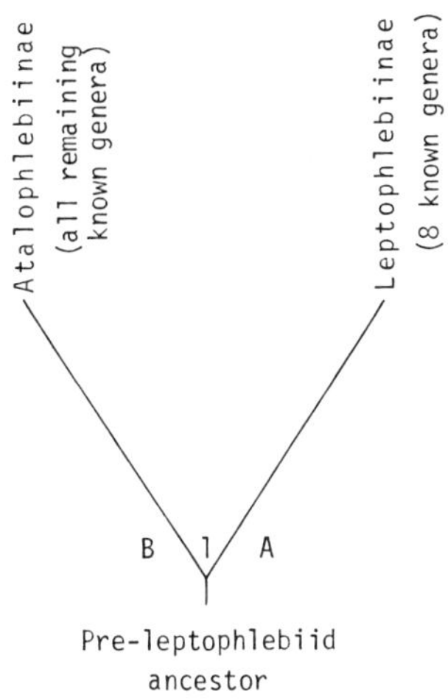

Figure 3. Phylogenetic diagram of most basal primitive furcation within the recent Leptophlebiidae (see Table 1 for character states).

The Atalophlebiinae are composed of a good number of primitive to highly specialized phyletic lineages. All of these lineages represent distinct assemblages of genera and each lineage occurs over more than one continent. While a higher classification should express various phylogenetic entities, recognition of each of these lineages as subfamilies would lead to extreme fragmentation at the subfamilial level. Therefore I propose that since each of these lineages are distinctly delineated in the literature, they be called tribes. While tribes are not now commonly used in the classification of the Ephemeroptera, I feel the tribal level will become more important as we better understand the evolution of various groups of Ephemeroptera.

ACKNOWLEDGMENTS

This research was supported by a research program (FLAX 79009) of SEA/CR, U.S.D.A., to Florida A&M University, William L. Peters, Research Leader.

RESUME

Les chercheurs qui révisent les genres des leptophlébiidés et cherchent à déterminer une classification et une phylogénie mieux articulées des groupes monophylétiques au sein de cette famille, doivent pouvoir se baser sur des critères détaillés pour que les résultats de leurs études s'accordent avec les faits. Nos théories actuelles sur la phylogénèse reposent sur un certain nombre de caractéristiques découlant d'observations portant sur chacune des lignées monophylétiques. Il est question de la plus ancienne ramification dans l'évolution des leptophlébiidés qui existent encore. En se basant sur cette phylogénèse, l'auteur traite de la nécessité et de l'utilité de sousordres et de tribus rattachés aux ordres.

ZUSSAMENFASSUNG

Um die Leptophlebiidengattungen revidieren und die höhere Klassifizierung und Phylogenie monophyletischer Gruppen innerhalb der Familie bestimmen zu können, werden detaillierte Kriterien benötigt, damit solche Studien Konsistenz erlangen. Unsere Phylogenien basieren auf dem Besitz einiger, vermutlich abgeleiteter Merkmale für jede monophyletische Tochterlinie. Die aller primitivste Gabelung bei der Evolution der noch vorhandenen Leptophlebiident wird in der vorliegenden Studie abgehandelt. Gegründet auf diese Phylogenie, wird ebenfalls der Bedarf an und die Nützlichkeit von Unterfamilien und Klassen innerhalb der Familie erörtert.

REFERENCES

Hennig, W. 1966. Phylogenetic Systematics. Univ. Illinois Press, Urbana.

Mayr, E. 1969. Principles of Systematic Zoology. McGraw-Hill, New York.

McCafferty, W.P. and G.F. Edmunds, Jr. 1979. The higher classification of the Ephemeroptera and its evolutionary basis. *Ann. Entomol. Soc. Am.* 72: 5-12.

Pescador, M.L. 1976. Systematics and zoogeography of cool-adapted Leptophlebiidae (Ephemeroptera) in Southern South America. Ph.D. diss., Florida State Univ.

Peters, W.L. and G.F. Edmunds, Jr. 1964. A revision of the generic classification of the Ethiopian Leptophlebiidae (Ephemeroptera). *Trans. R. Entomol. Soc. Lond.* 116: 225-253.

Peters, W.L. and G.F. Edmunds, Jr. 1970. Revision of the generic classification of the Eastern Hemisphere Leptophlebiidae (Ephemeroptera). *Pac. Insects* 12: 157-240.

Peters, W.L. and G.F. Edmunds, Jr. 1972. A revision of the generic classification of certain Leptophlebiidae from Southern South America (Ephemeroptera). *Ann. Entomol. Soc. Am.* 65: 1398-1414.

Peters, W.L., M.T. Gillies and G.F. Edmunds, Jr. 1964. Two new genera of mayflies from the Ethiopian and Oriental Regions (Ephemeroptera: Leptophlebiidae). *Proc. R. Entomol. Soc. Lond.*, B, 33: 117-124.

Peters, W.L., J.G. Peters and G.F. Edmunds, Jr. 1978. The Leptophlebiidae of New Caledonia (Ephemeroptera). Part I. — Introduction and systematics. *Cah. O.R.S.T.O.M., sér. Hydrobiol.* 12: 97-117.

Peters, W.L. and J.G. Peters. In press. The Leptophlebiidae of New Caledonia (Ephemeroptera). Part II. — Systematics. *Cah. O.R.S.T.O.M. sér. Hydrobiol.*

Ross, H.H. 1937. A generic classification of the Nearctic sawflies (Hymenoptera: Symphyta). *Ill. Biol. Monogr.* 15: 1-173.

Ross, H.H. 1974. Biological Systematics. Addison-Wesley, Reading.

Sokal, R.R. and P.H.A. Sneath. 1963. Principles of Numerical Taxonomy. W.H. Freeman, San Francisco.

PHYLOGENETIC RELATIONSHIPS AND ZOOGEOGRAPHY OF COOL-ADAPTED LEPTOPHLEBIIDAE (EPHEMEROPTERA) IN SOUTHERN SOUTH AMERICA

Manuel L. Pescador and William L. Peters

Florida A&M University
Tallahassee, Fl
U.S.A. 32307

ABSTRACT

Twelve cool-adapted leptophlebiid genera are known to occur in southern South America. Analysis of the external morphology of both adults and nymphs indicated that cool-adapted South American leptophlebiids represent five distinct phyletic lineages, namely: the *Hapsiphlebia*, *Penaphlebia*, *Atalonella*, New Genus A and *Meridialaris* lineages. All of these, except the *Penaphlebia* lineage, presently include some genera from other areas of the Southern Hemisphere. Phylogenetic evidence indicates that cool-adapted South American Leptophlebiidae are related to confamilials occurring in the mountain waters of Southeastern Australia (including Tasmania), New Zealand, New Caledonia, Madagascar and South Africa. This study revealed that most genera from South America are more closely related to genera from Australia than to those occurring in the other continents.

INTRODUCTION

The Leptophlebiidae comprise one of the largest and most distinctive families of mayflies. The family is cosmopolitan in distribution and reaches its maximum diversity in the Southern Hemisphere.

The taxonomy of Leptophlebiidae is inadequately known in many parts of the world, particularly in southern South America. This area has been geographically defined by Hubbard and Peters (1977) as south of the Tropic of Capricorn, a section of Continental South America comprising two major biogeographic divisions; viz., a cool-adapted group and warm-adapted group, although these divisions are not confined to southern South America. The cool-adapted group, which Illies (1969) discussed and referred to as the oligostenothermal mountain fauna, inhabits the Andes, Plains of Patagonia, Tierra del Fuego, and to a lesser extent the southern coastal mountains of Brazil. Cool mountain waters and regular periods of freezing weather prevail throughout the year. The warm-adapted group, referred to as the polystenothermal lowland fauna by Illies (1969), occupies all tropical and other areas where a stable high water temperature is maintained.

Prior to this investigation, Peters and Edmunds (1972) reported six cool-adapted leptophlebiid genera occurring in southern South America, three of which were newly established. We discovered six additional new genera making a total of twelve known to occur in this area. Pending publication of the descriptions, these six new genera are herein designated as Genera A-F.

PHYLOGENETIC RELATIONSHIPS

Detailed analysis of the external morphology of both adults and nymphs indicated that the cool-adapted South America leptophlebiids are related to several genera in the mountain streams of southeastern Australia (including Tasmania), New Caledonia, New Zealand, Madagascar and Africa. Representative genera from these areas related with the cool-adapted leptophlebiid fauna of South America, except for New Caledonia, are included in the present discussion (Table 2). One of us (WLP) is presently working on the systematics and zoogeography of the New Caledonia leptophlebiids and preliminary data show some genera that share common characters with the cool-adapted Leptophlebiidae of South America.

Figure 1 shows the proposed phylogeny of the cool-adapted leptophlebiid genera of southern South America. The sequence of branching is determined by shared possession of derived character states. The common derivation of character states used to interpret phylogeny is based on the concept that ancestral character states are generally widespread throughout the family and often throughout the order. Although the holomorphology of both nymphs and adults, when the latter were available, was analyzed, the nymphs were found to have more morphological characters applicable for phyletic and taxonomic analysis. Hence the phylogenetic reconstruction was based primarily on nymphal characters with a few complementing adult characters. Table 1 together with Figure 1 show the different phyletic lineages and corresponding character states used to interpret the phylogeny of the cool-adapted South American leptophlebiids

The cool-adapted Leptophlebiidae of South America represent five phyletic lineages, namely: the *Hapsiphlebia*, *Penaphlebia*, *Atalonella*, New Genus A, and *Meridialaris* lineages (Fig. 1). All

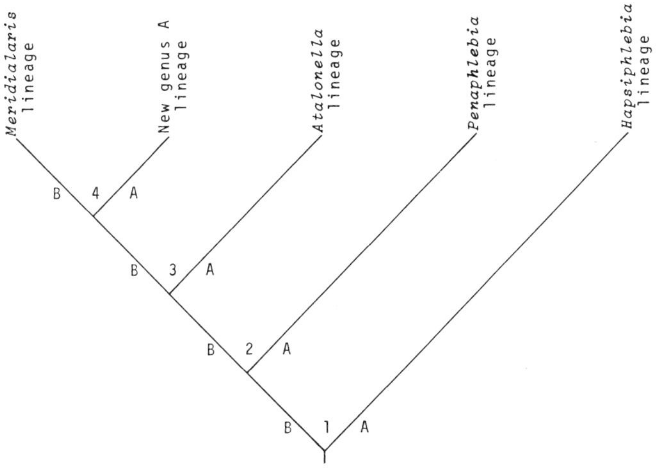

Figure 1. Phylogenetic diagram of cool-adapted Leptophlebiidae of Southern South America.

these lineages except the *Penaphlebia* lineage presently include some genera from other areas of the Southern Hemisphere (Table 2). The different lineages are arbitrarily named as such, both for convenience and since the emphasis of the present study is South American Leptophlebiidae. Therefore, it is most appropriate to simply use generic names of some of the better known leptophlebiids of southern South America.

The *Hapsiphlebia* lineage includes *Hapsiphlebia*, *Atalophlebia*, *Atalomicria*, *Jappa*, *Zephlebia cruentata* and *Aprionyx* (Table 2). Towns and Peters (personal communication) are presently revising the genus *Zephlebia* and noted that *Zephlebia cruentata* and the subgenus *Neozephlebia* not only represent two different genera but two different lineages as well. Their data and ours suggest that *Z. cruentata* and the subgenus *Neozephlebia* belong to the *Hapsiphlebia* and *Atalonella* lineages respectively. The nymphs of the *Hapsiphlevia* lineage have highly derived denticulation of incisors of the right mandible (Fig. 8) and prominent lateral hairs on the abdominal terga. Other characters differentiating this lineage from the other lineages retained the primitive states (Table 1).

The *Penaphlebia* lineage presently includes *Massartella* and *Penaphlebia* although, a few undescribed genera from Australia available to us belong to this lineage. The nymphs have derived pectinate setae on the maxillary palpi and tassel-like setae on the abdominal terga (Table 1). A phenocline ranging from almost straight outer lateral margins of the mandibles in the *Hapsiphlebia* lineage

Table 1. Character states of Figure 1. D, derived; P, primitive.

Furcation 1:	A	B
Clypeus	(P) wider than labrum (Fig. 15).	(D) narrower than labrum (Fig. 16-22).
Labrum: lateral margins	(P) subparallel (Fig. 15).	(D) rounded to angular (Fig. 16-22).
Mandible: outer margin	(P) almost straight (Fig. 7).	(D) rounded to angular (Fig. 9-14).
right outer incisor	(D) with broad subapical denticles (Fig. 8).	(P) smooth or spinose (Fig. 10, 12).
Abdominal terga: lateral margins	(D) with prominent setae.	(P) bare or with minute setae.
Furcation 2:	**A**	**B**
Clypeus	(P) subequal to slightly narrower than labrum (Fig. 16).	(D) narrower than labrum (Fig. 17-22).
Maxillary palpi: inner margin of 2	(D) with pectinate setae.	(P) with non-pectinate setae.
Abdominal terga: posterior margins	(D) with prominent spines and long tassel-like setae.	(P) with spines and scattered fine hairs.
Fore wings: vein MP_2	(P) strongly recurved (Fig. 2).	(D) moderately recurved (Fig. 3).
costal cross veins	(P) more than 25 (Fig. 2).	(D) less than 25 (Fig. 3).
Furcation 3:	**A**	**B**
Clypeus: lateral margins	(P) parallel (Fig. 17).	(D) divergent (Fig. 19,20,22).
width/labrum width	(P) less than 4/5 (Fig. 17).	(D) more than 4/5 (Fig. 19,20,22).
Labrum: lateral margins	(P) smoothly curved (Fig. 17).	(D) abruptly to angularly curved (Fig. 19,20,22).
anteromedian emargination	(D) broad, with prominent denticles (Fig. 17-18).	(P) narrow (but may be cleft or hooded), denticles reduced (Fig. 19-23).
Mandible: outer margin	(P) smoothly curved (Fig. 11).	(D) angularly curved (Fig. 13-14).
Ninth sternum of ♀	(P) deeply cleft.	(D) entire or shallowly concave.
Furcation 4:	**A**	**B**
Labrum: length/width	(P) 2/3-3/4 (Fig. 19).	(D) 1/3-1/2 (Fig. 20,22).
anteromedian emargination	(D) cleft, V-shaped dorsally and ventrally (Fig. 19).	(P) shallow U-shaped or deep V-shaped ventrally (Fig. 21-23).
Mandible: outer margin	(P) slightly angular (Fig. 13).	(D) distinctly angular (Fig. 14).
Maxilla: galea-lacinia	(P) narrow at apex (Fig. 24).	(D) broad at apex (Fig. 25).
subapical pectinate setae	(P) equal to or less than 15 (Fig. 24).	(D) more than 15 (Fig. 25).
Labium: submentum (postmentum)	(P) with thick lateral setae (Fig. 28).	(D) lateral margins bare (Fig. 29).
Body of nymph	(D) robust.	(P) flattened.

Table 2. Phyletic lineages representing the cool-adapted Leptophlebiidae of Southern South America and selected related genera from other areas of the Southern Hemisphere.

Lineage	Genera Included	Continent
1. *Hapsiphlebia*	*Hapsiphlebia*	South America
	Atalophlebia, Atalomicria Jappa, Ulmerophlebia	Australia "
	"*Zephlebia*" *cruentata*	New Zealand
	Aprionyx	Africa
2. *Penaphlebia*	*Penaphlebia, Massartella*	South America
3. *Atalonella*	*Atalonella*	South America, Australia
	New genus C, New genus D	South America
	Zephlebia (Neozephlebia)	New Zealand
4. New genus A	New genus A, New genus B	South America
	"*Atalophlebioides*" *sepia*	New Zealand
5. *Meridialaris*	*Meridialaris, Massartellopsis,* New genus E	South America " "
	Atalophlebioides s.s.	New Zealand
	"*Atalophlebioides*"	Australia
	Petersophlebia	Madagascar

(Fig. 7), through the intermediate in the *Penaphlebia* (Fig. 9), and *Atalonella* lineages (Fig. 11), to the angularly curved lateral margins in the Genus A (Fig. 13) and *Meridialaris* lineages (Fig. 14) suggest that the straight outer margin represents the primitive character state. A similar phenoclinal pattern is observed in the clypeus as apical enlargement gradually occurs from the *Atalonella* lineage (Fig. 17) reaching its maximum in the *Meridialaris* lineage (Figs. 20 and 22) with the Genus A lineage (Fig. 19) as an intermediate stage.

The *Atalonella* lineage includes *Atalonella*, two new genera (Genus C and Genus D) and *Zephlebia* (*Neozephlebia*) (Table 2). Although nymphs of both Genus C and Genus D are still unknown, the deeply cleft ninth sternum of female imagos (Fig. 4) suggest their inclusion in this lineage. Except for the derived broad anteromedian emargination with prominent denticles (Fig. 17-18), nymphal and adult characters differentiating the *Atalonella* lineage from its

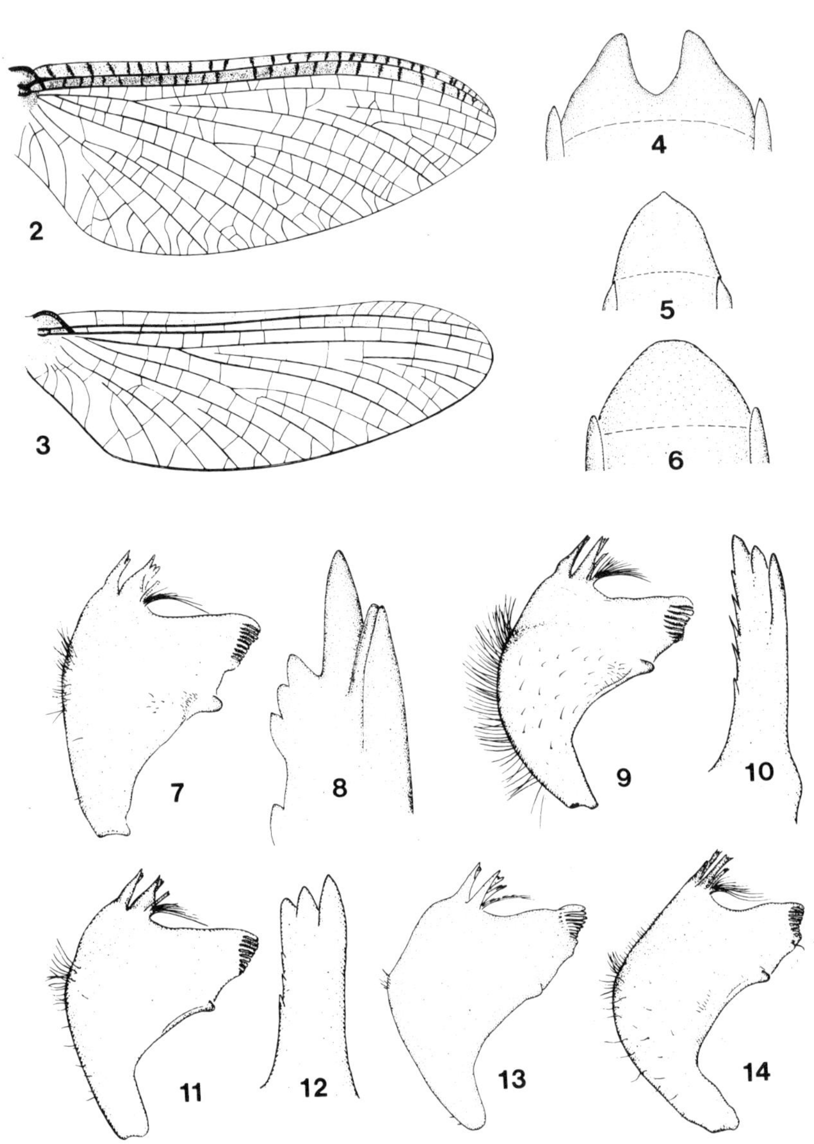

Figures 2-3, fore wing. Figures 4-6, ninth sternum of ♀ imago. Figures 7-14, left mandible and detail of outer incisor of right mandible. 2, 9, 10) *Penaphlebia*; 3, 4, 11, 12) *Atalonella*; 5, 14) *Meridialaris*; 6) *Massartellopsis*; 7, 8) *Hapsiphlebia*; 13) New genus A.

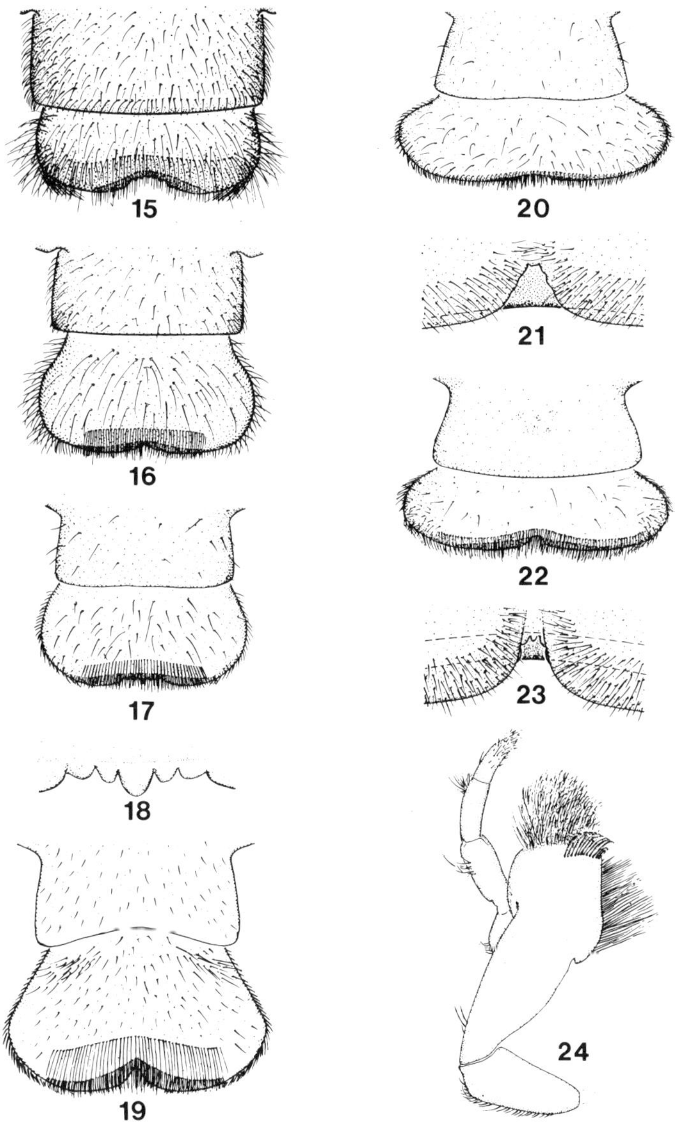

Figures 15-23, clypeus and labrum with detail of anteromedian emargination (21, 23, ventral views): 15) *Hapsiphlebia*; 16) *Penaphlebia*; 17-18) *Atalonella*; 19) New genus A; 20-21) *Meridialaris*; 22-23) *Massartellopsis*. Figure 24, maxilla of New genus A.

sister group retained the primitive states (Table 1). Progressive marginal curvature of the labrum in the five lineages forms a phenocline ranging from subparallel in the *Hapsiphlebia* lineage (Fig. 15) through rounded in the *Penaphlebia* (Fig. 16) and *Atalonella* (Fig. 17) lineages, to angular in the Genus A (Fig. 19) and *Meridialaris* (Figs. 20 and 22) lineages, suggesting that the character state in the *Meridialaris* is highly derived. A similar phenocline can be seen in the glossae: which are curved over ventrally in *Hapsiphlebia* and *Penaphlebia* (Figs. 26 and 27), knobbed in Genus A (Fig. 28) and straight in *Meridialaris* (Fig. 29).

The Genus A lineage includes two new genera (Genus A and Genus B) and the "*Atalophlebioides sepia* group" from New Zealand. Towns and Peters (in press) found two new genera that represent what Pescador (1976) referred to as "*Atalophlebioides sepia* group". Such findings complement Pescador's argument that the *Atalophlebioides* is polyphyletic. In this lineage, the nymphs have a short and chunky body, a large labrum (Fig. 19) with heavily sclerotized anterior margin and deep, V-shaped anteromedian emargination, and an apically narrow galea-lacinia of the maxillae (Fig. 24). Other characters that may have split this lineage and its sister group from a common ancestor are shown in Table 1. Genus A and Genus B are known only from nymphs.

The *Meridialaris* lineage includes *Meridialaris*, *Massartellopsis*, new Genus E, *Atalophlebioides* and the Australian "*Atalophlebioides*". Compared to the other lineages, the nymphs of the *Meridialaris* lineage possess the more derived, apically divergent clypeus (Figs. 20 and 22), short labrum (Figs. 20 and 22), distinctly angular mandibles (Fig. 14), apically broad galea-lacinia of maxillae (Fig. 25), and bare submentum (Fig. 29). Similarly, female adults have a derived ninth sternum (Figs. 5 and 6). Based on the direction of the phenocline formed by the clypeus, labrum and mandibles, the *Meridialaris* lineage appears to be the most derived among the southern South American leptophlebiid lineages.

There is one new genus (Genus F) that we consider a phyletic enigma. Pescador (1976) earlier placed it in the *Atalophlebioides* lineage which had almost the same generic composition as the *Meridialaris* lineage in this paper. However, there are also a few derived nymphal characters to include this new genus in the *Zephlebia* grouping of Towns and Peters (in press). We intend to study the morphology of the eggs of all genera representing the above phyletic lineages to determine if the egg data will offer a more concrete phyletic placement of this enigmatic genus.

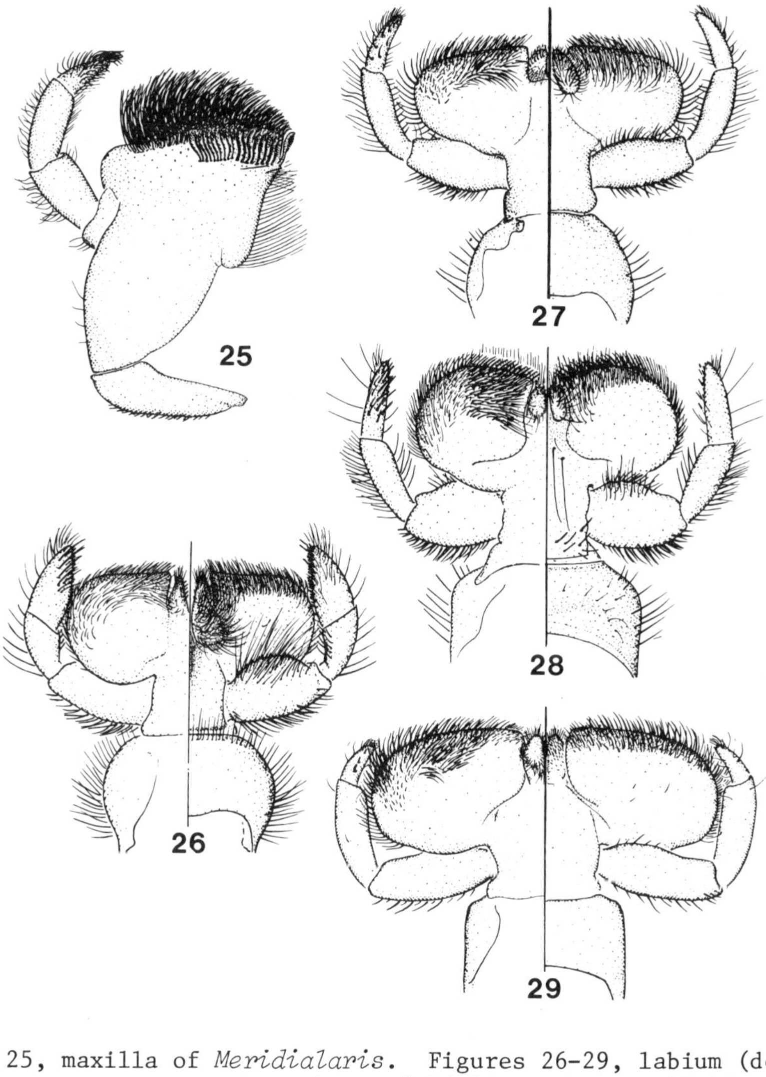

Figure 25, maxilla of *Meridialaris*. Figures 26-29, labium (dorsum on left, venter on right); 26) *Hapsiphlebia*; 27) *Penaphlebia*; 28) New genus A; 29) *Meridialaris*.

ZOOGEOGRAPHY

Phylogenetic evidence indicates that the cool-adapted South American Leptophlebiidae are closely related to confamilials occurring in the mountain waters of southeastern Australia (including Tasmania), New Zealand, New Caledonia, Madagascar and South Africa. South America and Australia still share congeneric leptophlebiid fauna, for example the genus *Atalonella*. Since mayflies have a conservative history of dispersal (Edmunds 1972), a more plausible explanation for such propinquity of a good number of extant Southern Hemisphere leptophlebiid genera leads to the argument for ancient connections or close proximity of these continents that made faunal dispersal possible. The reality of ancient Gondwanaland or a southern landmass can hardly be questioned anymore in light of overwhelming geological (Axelrod 1972, Crawford 1971, Griffiths and Varne 1972, Markl 1974, Raven and Axelrod 1972, Smith and Hallam 1970) and biological (Cracraft 1972, 1973, 1974, 1975; Edmunds 1972, 1975; Jardine and McKenzie 1972, Keast 1972, 1973; Raven and Axelrod 1972, 1974, 1975; Schlinger 1974, and Tsui and Peters 1975) data.

The five phyletic lineages represented by the cool-adapted Leptophlebiidae of southern South America includes genera from other Southern Hemisphere continents except for the *Penaphlebia* lineage (Table 2). The *Hapsiphlebia* lineage presently includes one genus from Chile, four from Australia, and one each from New Zealand and South Africa (Table 2). Genera from South America and Australia share the greatest number of shared, derived characters. For example, *Hapsiphlebia*, *Jappa* and *Ulmerophlebia* share not only morphological but to a great extent behavioral and ecological similarities. Nymphs of *Hapsiphlebia* are "semi-burrowers", partially burrowing in silt and sand. Such behavior is probably a precursor to the evolution of the burrowers as in *Jappa* and *Ulmerophlebia*. *Jappa* has derived long frontal tusks that undoubtedly improved the nymphs' burrowing efficiency.

The *Penaphlebia* lineage is presently confined to South America (Table 2), although preliminary study of some undescribed genera from Australia available to us suggests their close affinity to *Massartell* and *Penaphlebia*.

The *Atalonella* lineage presently includes genera from South America, Australia and New Zealand (Table 2). *Atalonella* occurs in South America and Australia while its sister group *Zephlebia* (*Neozephlebia*) is found in New Zealand. Two new genera (Genus C and Genus D) from South America are only known from adult specimens but available data suggest their inclusion in this lineage.

The Genus A lineage which presently includes two new genera from South America, and New Zealand, is the only cool-adapted souther South American leptophlebiid lineage not represented in Australia.

More collections and complete examination of specimens from this area available to us may give a more accurate assessment of this lineage's non-Australian representation. Genus A and Genus B from South America are only known from nymphs but they share a good number of derived characters with the "*Atalophlebioides sepia* group" from New Zealand.

The *Meridialaris* lineage presently includes three genera from South America (one new genus), and one each from Australia, New Zealand and Madagascar (Table 2). The genus *Atalophlebioides* is presently distributed in Australia and New Zealand although the Australian species were found to have more characters in common (viz. imaginal claws, ninth sternum of female imagos, nymphal clypeus and labrum) with *Meridialaris* and *Massartellopsis* than with the genus from New Zealand. Pescador (1976) indicated the Australian "*Atalophlebioides*" as a sister group of *Massartellopsis* and *Meridialaris*. Towns and Peters (1978) recently reported *Atalophlebioides*, a monotypic genus endemic to New Zealand to be the true *Atalophlebioides* and the Australian "*Atalophlebioides*" to be a different and undescribed genus. Available but limited data suggest the inclusion of *Petersophlebia* from Madagascar in the *Meridialaris* lineage, although further study is needed to ascertain its phyletic placement.

In summary, the cool-adapted leptophlebiid fauna of southern South America is related to a number of genera occurring in other areas of the Southern Hemisphere. We found most genera from South America are most closely related to genera from Australia, followed by genera from New Zealand, and Africa and Madagascar in that order. Phyletic analysis of New Caledonia Leptophlebiidae however, is yet to be completed. As mentioned earlier, such close affinities of the fauna of these areas is best explained by the presumably ancient connections or close proximity of these continents that may have facilitated faunal dispersal. Based on the degree of phylogenetic relationships of the leptophlebiid fauna of these continents, the suggested sequence of the break-up of Gondwanaland is as folows: Africa along with Madagascar, and possibly India first, followed by New Zealand with New Caledonia, and finally South America and Australia broke away from Antarctica. Inadequate fossil records make it difficult to establish the time sequence of the break-up. Geological data however, indicate the same sequential pattern of continental separation including the same approximate time of separation (see references given above).

ACKNOWLEDGMENTS

This research was supported by a research program (FLAX 79009) of SEA/CR and USDA, to Florida A&M University, W.L. Peters, Research Leader.

Our sincere thanks to Mrs. Janice G. Peters who kindly prepared the tables, phylogenetic diagram and most of the illustrations. For the illustrations of the structures of New Genus A, gratitude is expressed to Mr. David Harlos. We also thank Mrs. Rosalie P. Myers for typing the drafts of the manuscript.

RESUME

Il y a douze genres de leptophlébiidés s'adaptant au froid en Amérique du Sud méridionale. L'analyse de la morphologie externe des adultes et des nymphes indique que ces leptophlébiidés représentent cinq arbres phylétiques distincts, à savoir: les familles *Hapsiphlebia*, *Penaphlebia*, *Atalonella*, le nouveau genre A et *Meridialaris*. Tous à l'exception de la famille *Penaphlebia* englobent présentement des genres provenant de régions situées en dehors de l'hémisphère sud. Les données phylogénetiques indiquent que les leptophlébiidés d'Amérique du Sud qui s'adaptent au temps froid ont des liens de parenté avec des espèces qui se rencontrent dans les eaux des montagnes d'Australie du sud-est (y compris la Tasmanie), de Nouvelle Zélande, de Nouvelle Calédonie, de Madagascar et d'Afrique du Sud. L'étude a révélé que la plupart des genres provenant d'Amérique du Sud sont plus étroitement apparentés à ceux d'Australie qu'à ceux d'autres continents.

ZUSSAMENFASSUNG

Im südlichen Sudamerika kennt man zwölf an Kälte gewöhnte Leptophlebiiden Gattungen. Eine Analyse der externen Morpholgie, sowohl von Nymphen als auch voll entwickelten Tieren ergab, daß kältegewohnte, südamerikanische Leptophlebiiden fünf verschiedene phyletische Stämme repräsentieren, nämlich: *Hapsiphlebia*, *Penaphlebia*, *Atalonella*, Neue Gattung A und *Meridialaris* Stämme. Alle diese, außer dem *Penaphlebia* Stamm, umfassen zur Zeit einige Gattungen von anderen Gebieten der südlichen Hemisphäre. Phylogenetisches Beweismaterial zeigt an, daß kältegewohnte, südamerikanische Leptophlebiiden verwandt sind mit Familienmitgliedern, die in den Gebirgsgewässern Sud-Ostaustraliens (Tasmanien eingeschlossen), Neuseelands, Neukaledoniens, Madagaskars und Südafrikas vorkommen. Die vorliegende Arbeit ergab, daß die meisten Gattungen Südamerikas näher verwandt sind mit Gattungen von Australien, als mit solchen, die in andern Kontinenten auftreten.

REFERENCES

Axelrod, D.I. 1972. Ocean-floor spread in relation to ecosystematic problems. *Occas. Pap. Univ. Arkansas Mus.* 4: 15-68.

Crawford, A.R. 1971. Gondwanaland and the growth of India. *J. Geol. Soc. India* 12: 205-221.

Cracraft, J. 1972. Continental drift and Australian avian biogeography. *Emu* 72: 171-174.

Cracraft, J. 1973. Continental drift, paleoclimatology, and the evolution and biogeography of birds. *J. Zool. (Lond.)* 169: 455-545.

Cracraft, J. 1974. Continental drift and vertebrate distribution. *Annu. Rev. Ecol. Syst.* 5: 215-261.

Cracraft, J. 1975. Historical biogeography and earth history: Perspectives for a future synthesis. *Ann. Mo. Bot. Gard.* 62: 227-250.

Edmunds, G.F., Jr. 1972. Biogeography and evolution of Ephemeroptera. *Annu. Rev. Entomol.* 17: 21-24.

Edmunds, G.F., Jr. 1975. Phylogenetic biogeography of mayflies. *Ann. Mo. Bot. Gard.* 62: 251-263.

Griffiths, J.R. and R. Varne. 1972. Evolution of the Tasman Sea, Macquarie Ridge and Alpine Fault. *Nature Phys. Sci. (Lond.)* 235: 83-86.

Hubbard, M.D. and W.L. Peters. 1977. Ephemeroptera. p. 165-169 *in* S.H. Hulbert ed. Biota Acuatica de Sudamerica Austral. San Diego State University, San Diego, California.

Illies, J. 1969. Biogeography and ecology of Neotropical freshwater insects; especially those from running water. p. 685-708. *in* E.J. Fittkau et al. ed. Biogeography and Ecology in South America. Vol. 2. Dr. W. Junk, The Hague.

Jardine, N. and D. McKenzie. 1972. Continental drift and the dispersal and evolution of organisms. *Nature (Lond.)* 235: 20-24.

Keast, A. 1972. Continental drift and evolution of the biota on southern continents. p. 23-87 *in* A. Keast, F. Erk and B. Glass ed. Evolution, Mammals and Southern Continents. State Univ. New York Press, Albany, N.Y.

Keast, A. 1973. Contemporary biotas and the separation sequence of the southern continents. p. 309-343 *in* D.H. Tarling and S.K. Runcorn ed. Implication of Continental Drift to the Earth Sciences. Academic Press, London & New York.

Markl, R.G. 1974. Evidence for the break-up of eastern Gondwanaland by the early Cretaceous. *Nature (Lond.)* 251: 196-200.

Pescador, M.L. 1976. Systematics and zoogeography of cool-adapted Leptophlebiidae (Ephemeroptera) in southern South America. Ph.D. Dissertation. Florida State Univ.

Peters, W.L. and G.F. Edmunds, Jr. 1972. A revision of the generic classification of certain Leptophlebiidae from southern South America (Ephemeroptera). *Ann. Entomol. Soc. Am.* 65: 1398-1414.

Raven, P.H. and D.I. Axelrod. 1972. Plate tectonics and Australasian paleobiogeography. *Science (Wash. D.C.)* 176: 1379-1386.

Raven, P.H. and D.I. Axelrod. 1974. Angiosperm biogeography and past continental movements. *Ann. Mo. Bot. Gard.* 61: 539-673.

Raven, P.H. and D.I. Axelrod. 1975. History of the flora and fauna of Latin America. *Am. Sci.* 63: 420-429.

Schlinger, E.I. 1974. Continental drift, *Nothofagus*, and some ecologically associated insects. *Annu. Rev. Entomol.* 19: 323-343.

Smith, A.G. and A. Hallam. 1970. The fit of the southern continents. *Nature (Lond.)* 225: 139-144.

Towns, D.R. and W.L. Peters. 1978. A revision of genus *Atalophlebioides* (Ephemeroptera: Leptophlebiidae). *N.Z. J. Zool.* 5: 607-614.

Towns, D.R. and W.L. Peters. 1979. Three new genera of Leptophlebiidae (Ephemeroptera) from New Zealand. *N.Z. J. Zool.* (In press).

Tsui, P.T.P. and W.L. Peters. 1975. The comparative morphology and phylogeny of certain Gondwanian Leptophlebiidae based on the thorax, tentorium and abdominal terga (Ephemeroptera). *Trans. Am. Entomol. Soc.* 101: 505-595.

PHYLOGENETIC RELATIONSHIPS OF THE LEPTOPHLEBIIDAE OF NEW ZEALAND (EPHEMEROPTERA)

D.R. Towns* and William L. Peters

Laboratory of Aquatic Entomology
Florida A&M University
Tallahassee, Florida 32307 USA

ABSTRACT

The leptophlebiid mayfly fauna of New Zealand at present consists of three named genera, of which two are clearly polyphyletic. Analysis of these elements and addition of undescribed genera indicate the presence of at least 14 genera. The genera can be divided into five lineages which exhibit varying degrees of radiation within New Zealand and which show evolutionary progression in a number of character states. Four genera appear to be most closely related to genera in New Caledonia, two to generic groups in Chile, and the remainder to other genera in New Zealand, except for one genus for which the nearest known relatives are in New Guinea.

INTRODUCTION

The Leptophlebiidae are the largest mayfly family in New Zealand and commonly dominate the fauna of small, undisturbed stony streams and rivers. Despite this, the family is poorly known taxonomically and the phylogenetic relationships of its elements remain unclear. The discussion of the phylogeny of the Gondwanian Leptophlebiidae by Tsui and Peters (1975) included New Zealand representatives, but the poorly defined limits of these genera

---

* Department of Zoology, University of Adelaide, G.P.O. Box 498, Adelaide, South Australia 5001, Australia

hampered an accurate determination of their phylogenetic relationships. However, Tsui and Peters (1975) did recognize the polyphyletic nature of some New Zealand genera.

The following account presents the results of a preliminary investigation of the phylogeny of the leptophlebiids of New Zealand and discusses the relationships and possible origins of elements of the fauna. A list of some generic groups with broad austral distributions is given in Pescador and Peters (1979). Only those genera with close relationships to the New Zealand fauna are mentioned here. A series of papers revising the group are being published or are in press (Towns and Peters 1978, 1979a, b). Since none of the names of new genera we have established have appeared in print, specie groups will be discussed here either in the genera to which they have been traditionally assigned, or, in the case of undescribed genera numbered as Genus I, II, etc.

THE FAUNA

To date, three New Zealand leptophlebiid genera have been recognized: *Deleatidium* Eaton, *Atalophlebioides* Phillips and *Zephlebia* Penniket. All three genera are endemic. Species previously placed in *Atalophlebioides* from Chile and Australia do not fit the present definition of the genus (Towns and Peters 1978).

Deleatidium seems to be the largest New Zealand genus and is the only one in which the species groups appear to be related. *Atalophlebioides* and *Zephlebia* are both polyphyletic and provide the largest amount of information on the relationships of the New Zealand fauna. *Atalophlebioides* consists of three species including *A. cromwelli* (Phillips), *A. sepia* (Phillips) and *A. aucklandensis* Peters. The latter two species clearly differ from *A. cromwelli* at the nymphal and adult stages and are being placed in separate genera (Towns and Peters 1979a). *Zephlebia* consists of at least three genera. When Penniket (1961) established this genus he divided it into two subgenera: *Zephlebia* s.s. and *Zephlebia (Neozephlebia)*. *Neozephlebia* is not closely related to *Zephlebia* and should be raised to generic rank. "*Zephlebia*" *cruentata* (Hudson), referred by Penniket to *Zephlebia* s.s., has few characters in common with either subgenus of *Zephlebia*, and should have a genus of its own. A generic revision of *Zephlebia*, including the latter two species groups, is being prepared for publication (Towns and Peters in prep.). The material available to us also includes at least 6 undescribed genera. With these and analysis of the species groups of *Atalophlebioides* and *Zephlebia*, 14 genera of Leptophlebiidae can be identified from New Zealand.

A feature of the fauna is the small number of species per genus. Of the 14 genera, 11 are monotypic, and as far as we know, only two,

Deleatidium and *Zephlebia*, contain five or more species.

PHYLOGENY

The leptophlebiid fauna of New Zealand consists of five lineages shown as endpoints in Fig. 1. Derived and primitive character states are given for each furcation in Table 1. At each furcation the stated primitive condition can be assumed to occur in all other lineages unless otherwise stated. Several derived character states are given at each furcation. These include characters which will separate New Zealand genera from overseas representatives of the same lineage, as well as characters which separate the five lineages listed. However, as more overseas genera are added to the lineages shown here, modification of the terminology may, in some cases, become necessary.

The analysis is based on a combination of nymphal and adult morphological characters. Because some adult characters, which appear to be useful in a regional sense, apparently exhibit extensive parallel evolution when more widely applied, some furcations lack adult characters.

An example of parellelism of adult characters occurs in coloration of the imaginal fore wings and the structure of the imaginal claws. Imagos of *Zephlebia*, *Z. (Neozephlebia)*, and *"Zephlebia" cruentata* have claws of a pair similar, hooked with an opposing hook, and pigmentation in the fore wings. *Atalophlebioides* s.s. and *"Atalophlebioides" sepia* have the clearly derived condition of dissimilar claws and hyaline fore wings. These two sets of character states are useful in subdividing the New Zealand fauna (see Fig. 1). However, genera from outside New Zealand added to their relatives, shown as endpoints in Fig. 1, indicate that loss of pigmentation and development of dissimilar claws has developed independently on several occasions. These characters therefore become mosaics when the hypothesis is tested by adding more genera.

The suggested phylogeny presented here shows evolutionary progression of a number of characters. In *"Zephlebia" cruentata* the clypeus is broader than the labrum (Fig. 3) and the clypeal margins are subparallel. *Zephlebia (Neozephlebia)* retains the parallel margins of the clypeus, but the labrum is slightly broader than the clypeus (Fig. 4). The margins of the clypeus become more divergent and the labrum becomes broader with each lineage (Fig. 5-6) with the ultimate development in the *Atalophlebioides* lineage of a shallow broad labrum and a clypeus with strongly divergent margins (Fig. 7).

A similar phenocline occurs in the size and shape of the prosthecal tuft and incisors of the mandibles (Fig. 8-12). In

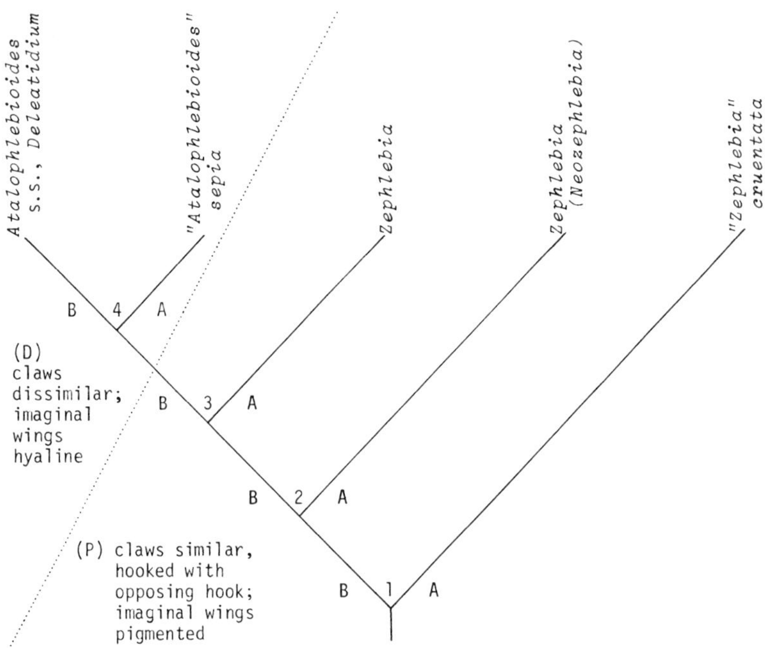

Figure 1. Phylogenetic diagram of the main generic lineages of Leptophlebiidae of New Zealand.

"*Zephlebia*" *cruentata* the prosthecal tuft has the most primitive condition, which is with the base flap-like, and hairs on the apical and lateral margins (Fig. 8); whereas, in *Z. (Neozephlebia)* it is spine-line with hairs on the anterolateral margin (Fig. 9). This spine-line shape is retained in all other lineages, but the entire prosthecal structure becomes progressively smaller (Fig. 10-11). Maximum reduction occurs in some species of *Deleatidium*, where the prosthecal tuft is reduced to a few hairs (Fig. 12). The incisors also change shape, from short and stout in "*Zephlebia*" *cruentata* to long and slender in *Atalophlebioides* (Fig. 13-17). These phenoclines generally hold when genera from outside New Zealand are included in the phylogeny (e.g. Pescador and Peters 1979). Apparent exceptions usually are clearly the result of parallel evolution. For example, a broad shallow labrum occurs outside the *Atalophlebioides* lineage in genera from New Caledonia. However, the labrum and clypeus in these genera have a number of morphological differences which separate them from the structures in the *Atalophlebioides* group.

Table 1. Character states used in proposed phylogeny of New Zealand Leptophlebiidae (Fig. 1). D, derived; P, primitives; *, occurs in furcations 1 and 2.

A	B
Furcation 1	
(D) Outer incisor of right mandible with denticles (Fig. 13).	(P) Outer incisor of right mandible smooth or spinose (Fig. 14-17).
(D) Segment 2 of maxillary palpi broad at apex.	(P) Segment 2 of maxillary palpi slender.
(P) Labrum narrower than clypeus (Fig. 3).	(D) Labrum broader than clypeus (Fig. 4-7).
(P) Base of prosthecal tuft flap-like (Fig. 8).	(D) Base of prosthecal tuft spine-like (Fig. 9-12).
(P) Segment 3 of labial palpi triangular and with stout marginal spines (Fig. 18). *	(D) Segment 3 of labial palpi slender, without stout marginal spines (Fig. 20-22).
Furcation 2	
(D) Denticles of labrum set into rectangular cavity (Fig. 4).	(P) Denticles of labrum not set into rectangular cavity (Fig. 5-7).
(P) Labrum and clypeus subequal in width (Fig. 4).	(D) Labrum broader than clypeus (Fig. 5-7).
(D) Glossae of labium with enlarged peg-like spines.	(P) Glossae of labium with pointed spines or fine hairs.
(D) Mandibles with few scattered hairs on outer margin.	(P) Mandibles with numerous hairs on outer margin.
(D) Margins of clypeus subparallel (Fig. 4).	(P) Margins of clypeus divergent (Fig. 5-7).
Furcation 3	
(D) Penes with hairs at margins of penis openings.	(P) Penes without hairs at margins of penis openings.
(P) Genital extension present to very well developed (also in furcation 1A).	(D) Genital extension absent.
(P) Labrum slightly broader than clypeus (Fig. 5).	(D) Labrum significantly broader than clypeus (Fig. 6-7).
Furcation 4	
(D) Labrum with extremely broad-based denticles on anterior margin and with deeply cleft anteromedian emargination (Fig. 6).	(P) Labrum with narrow-based denticles on anterior margin (Fig. 7).
(P) Margins of clypeus slightly divergent apically (Fig. 6).	(D) Margins of clypeus strongly divergent apically (Fig. 7).
(P) Galea-lacinia of maxillae narrow.	(D) Galea-lacinia of maxillae broad.
(P) Apex of incisors smooth (Fig. 16).	(D) Apex of incisors serrated (Fig. 17).
(P) Prosthecal tuft well developed (Fig. 11).	(D) Prosthecal tuft greatly reduced (Fig. 12).

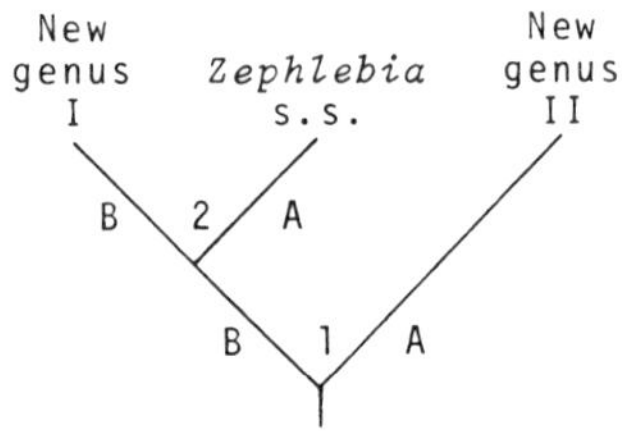

Figure 2. Phylogenetic diagram of the genera of the *Zephlebia* lineage in New Zealand.

Table 2. Character states of Fig. 2. Conventions as in Table 1.

A	B
Furcation 1	
(D) Length of penes greater than length of forceps segment 1.	(P) Length of penes less than length of forceps segment 1.
(D) Gill lamellae with fringed margins.	(P) Gill lamellae with margins entire.
(D) Female imago with long ovipositor or egg guide.	(P) Female imago with small genital extension.
(P) Abdominal gill 7 not reduced.	(D) Abdominal gill 7 reduced (single, plate-like; or paired or single thread-like filaments).
(P) Forceps of male imago without accessory lobe on segment 1.	(D) Forceps of male imago with accessory lobe on segment 1.
Furcation 2	
(D?) Projections on posterior margin of abdominal terga large.	(P) Projections on posterior margin of abdominal terga small.
(P) Femora of nymphs broad.	(D) Femora of nymphs extremely long and thin.
(P) Dorsal and ventral portions of abdominal gills similar.	(D) Dorsal and ventral portions of abdominal gills dissimilar.
(P) Labrum with well developed anteromedian denticles.	(D) Labrum with shallow, broad-based, anterosubmedian denticles.
(P) Hind wing of imago not greatly reduced (> 1/5 length of fore wing).	(D) Hind wing of imago reduced (< 1/5 length of fore wing).
(P) Mandibles with margin smoothly curved.	(D) Mandibles with apical half of outer margin straight.

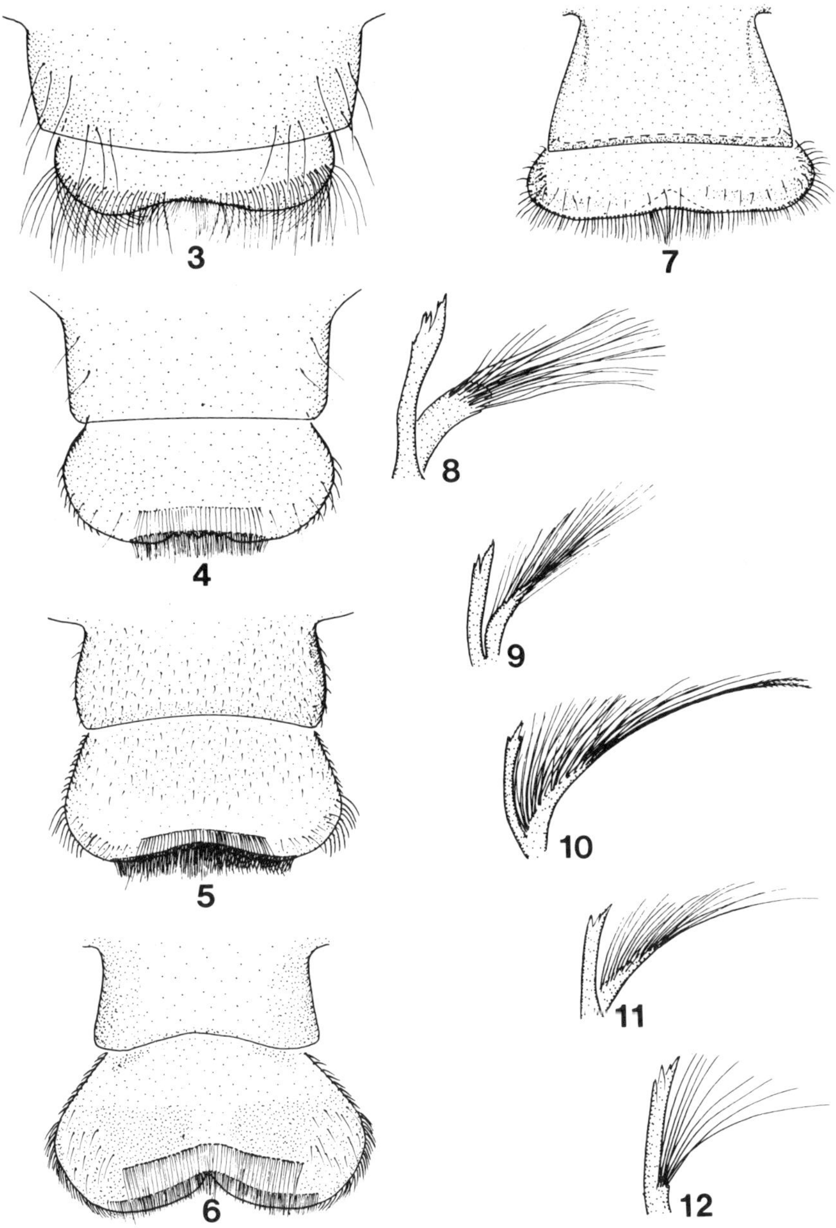

Fig. 3-7, clypeus and labrum. Fig. 8-12, prosthecal tuft of left mandible. 3,8) "*Zephlebia*" *cruentata;* 4,9) *Z.* (*neozephlebia*) group; 5,10) *Zephlebia* s.s.; 6,11) "*Atalophlebioides*" *sepia*-group; 7,12) *Atalophlebioides* s.s. group.

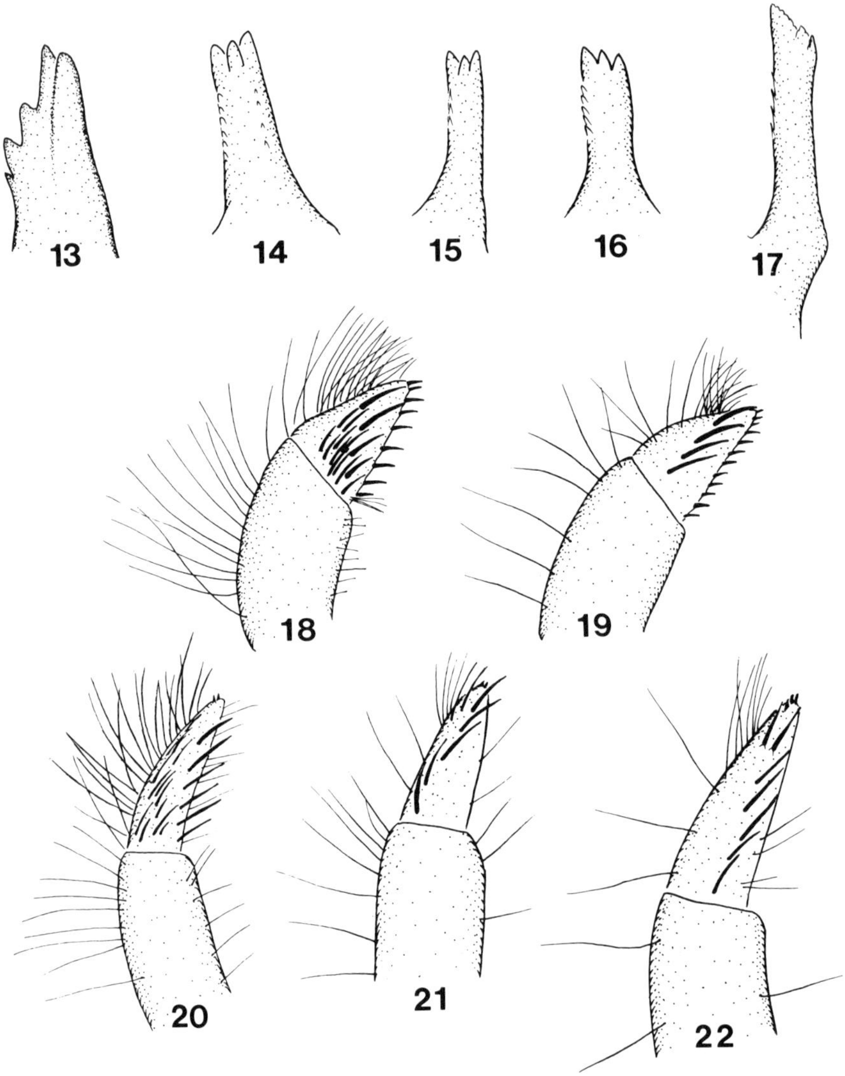

Fig. 13-17, outer incisor of right mandible. Fig. 18-22, dorsal surface of third labial palpi. 13,18) *"Zephlebia" cruentata;* 14,19) *Z. (Neozephlebia)* group; 15,20) *Zephlebia* s.s.; 16, 21) *"Atalophlebioides" sepia*-group; 17-22) *Atalophlebioides* s.s. group.

Two primitive character states link *"Zephlebia" cruentata* and *Zephlebia (Neozephlebia)*: both genera retain the triangular shaped third labial palpal segment with heavy spines along the inner margin (Fig. 18-19) and both also have hairs or spines along the outer margin of the stipes of the maxilla. All other New Zealand genera have a labium with the third palpal segment slender without stout marginal spines (Fig. 20-22), and rarely more than a single spine near the base of the stipes of the maxilla.

Only two of the lineages listed here show any marked degree of radiation within New Zealand. The *Atalophlebioides* s.s. lineage has clear relationships outside New Zealand, but the five New Zealand genera so far identified are most closely related to each other, although at least two genera have relationships which remain unclear.

Conversely, the *Zephlebia* lineage has undergone considerable radiation in the New Zealand-New Caledonia region. This lineage, in New Zealand, is represented by three groups. The *Zephlebia* group contains *Zephlebia* s.s. and a closely related undescribed genus. The remaining two groups each contain a single new genus (Fig. 2, Table 2). The most curious of the new genera (Genus II) has many characters in common with the Palearctic, Oriental and Ethiopian genus *Thraulus*, but at the same time shares derived characters with *Zephlebia*. Nymphs are difficult to distinguish from *Thraulus* in gill shape and structure of the mouthparts. Characters in common with *Zephlebia* occur in the adults and include the presence of a genital extension in the females (greatly enlarged in the new genus), hairs around the penis openings and outline shape and venation of the wings. A second new genus from New Zealand (Genus I), has nymphs with distinctive thin femora, nymphal mouthparts similar to genera in New Caledonia and male genitalia similar to *Zephlebia* of New Zealand. Both Genus I and II are being described by Towns and Peters (1979b).

ZOOGEOGRAPHIC RELATIONSHIPS

The zoogeographic relationships of New Zealand members of a variety of the older aquatic families have been investigated recently. These include the mayfly family Siphlonuridae (Edmunds 1975), Plecoptera (McLellan 1975), the chironomid subfamilies Podonominae and Diamesinae (Brudin 1966) and Trichoptera (Cowley 1978). Most of these show sister group relationships between New Zealand and temperate South America, and between temperate South America and Australia. In a few cases the picture appears to be confused by more recent wind drift between New Zealand and Australia (Plecoptera and Trichoptera) and possibly some dispersal from South-East Asia (Trichoptera). So far, in none of these groups has it been possible to investigate potential relationships between New Caledonia and New Zealand in detail. This is because stoneflies, siphlonurids and cold-adapted chironomids are unknown from New Caledonia (although the amphinotic relationships between the latter two groups of insects are well established), while Trichoptera, which have some links between New Zealand and New Caledonia (McFarlane 1976), are apparently too poorly known for their southern relationships to be fully understood (Cowley 1978). By contrast, leptophlebiid mayflies are well represented in New Caledonia and a close relationship between this and the New Zealand fauna has been suggested by Edmunds (1972).

Our data so far suggest that the relationship between the New Zealand and New Caledonia Leptophlebiidae is complex. Both old and recent groups (based on broad austral vs restricted distribution) are represented in New Zealand and New Caledonia, but some groups well represented in New Zealand, Chile and Australia are unknown in New Caledonia. Based on an extensive analysis of morphological characters, *"Zephlebia" cruentata* is most closely related to an undescribed genus in New Caledonia and both genera are part of a lineage which includes *Hapsiphlebia* (Chile), and *Atalophlebia* s.s., *Atalomicria* and other genera in Australia. A similar relationship occurs in *Zephlebia (Neozephlebia)*, which has affinities with genera in New Caledonia and possibly also with genera in Chile and Australia. Relationships of the *Zephlebia* group remain unclear, although there appear to be a number of representatives in New Caledonia and possibly one in Chile (Pescador and Peters 1979). The *"Atalophlebioides" sepia* group has clear affinities with genera in Chile, but dubious links with genera in New Caledonia. Similarly, *Atalophlebioides* s.s. is related to *Deleatidium* of New Zealand, *Massartellopsis* and *Meridialaris* of Chile and *"Atalophlebioides"* of Australia but appears to lack representatives in New Caledonia. One of the new genera mentioned previously (Genus I) has close affinities with genera in New Caledonia, but no clear relationships outside this area, while the *Thraulus*-like genus (Genus II), although showing some affinities with New Caledonian genera, appears to be most closely related to a species group of *Thraulus* in New Guinea. The latter genus is particularly interesting because it is the first member of the group identified from the temperate Southern Hemisphere. It appears to share common ancestry with *Zephlebia* and related genera thus supporting the suggestion of Peters and Edmunds (1970, p. 235) that *Thraulus* is part of a group of genera "which represents ancient dispersal in the Southern Hemisphere."

The relationships between New Zealand and New Caledonia have not yet been completely analyzed. However, preliminary data suggest the New Zealand fauna <u>does</u> consistently show closest relationships with either New Caledonia or Chile, but not with Australia. Where distribution of sister groups is inconsistent with our proposed phylogeny (*Atalophlebioides* s.s., *Zephlebia* and *"Atalophlebioides" sepia*), the difficulties may be resolved once the leptophlebiid fauna outside New Zealand has been analysed in more detail, or the problems may be partly the result of extinction in New Caledonia of cold-adapted forms (cf Edmunds 1972).

The small number of species per genus at present shown by New Zealand Leptophlebiidae requires some comment. This situation is not unique to leptophlebiid mayflies in New Zealand, but appears also in siphlonurids and apparently in other aquatic orders as the limits of genera become more clearly defined (e.g. McFarlane 1976 - Trichoptera, McLellan 1977 - Plecoptera). Both leptophlebiid mayflies and Plecoptera in Chile have a similar low degree of speciation (M.L. Pescador pers. comm., Illies 1969). To some extent the number of species per genus is related to the level of taxonomic knowledge of the group. This applies to the mayflies of New Zealand, since the country has not been collected extensively. Even allowing for this, the level of speciation seems to be low and is about the same as the level reported by Illies (1969) for cool-adapted Plecoptera of South America. Illies (1969) compared the plecopteran fauna of South America with that of Europe and suggested that the considerable differences in levels of speciation between the two continents are the differential effects of glaciation on highland faunas. This hypothesis seems to have limited relevance to the mayflies of New Zealand. It seems more likely that the low level of speciation shown by the New Zealand mayflies can be attributed to a long isolation of an ancient specialized fauna which has since undergone a low level of radiation and/or a high level of extinction. The latter two influences could be associated with the considerable geological and climatic fluctuations to which the New Zealand landmass has been subjected (cf Fleming 1975).

ACKNOWLEDGEMENTS

This research was supported by a research program (FLAX 79009) of the Science and Education Administration/Cooperative Research, United States Department of Agriculture, to Florida A&M University, William L. Peters, Research Leader.

RESUME

Le groupe des éphéméroptères leptophlébiidés de Nouvelle Zélande comprend présentement trois genres répertoriés, dont deux sont de toute évidence polyphylétiques. L'analyse de ces éléments et l'addition de genres non encore étudiés indique la présence d'au moins quatorze genres. Les genres peuvent être divisés en cinq familles qui rayonnent à des degrés divers en Nouvelle Zélande et qui montrent des signes d'évolution progressive dans un certain nombre d'états caractéristiques. Quatre genres semblent être étroitement apparentés à des genres qui se trouvent en Nouvelle Calédonie, deux à des groupes génériques du Chili, et les autres à d'autres genres qui se trouvent en Nouvelle Zélande, à l'exception d'un genre dont les spécimens les plus rapprochés que l'on connaisse sont en Nouvelle Guinés.

ZUSSAMENFASSUNG

Die leptophlebiide Eintagsfliegen Fauna Neuseelands besteht zur Zeit aus drei benannten Gattungen, von denen zwei eindeutig polyphyletisch sind. Die Analyse dieser Elemente und das Hinzufügen unbeschriebener Gattungen lassen die Existenz von mindestens vierzehn Gattungen erkennen. Diese Gattungen können wiederum unterteilt werder in fünf Stämme, die unterschiedlichen Streu-Radius innerhalb Neuseelands und evolutionäre Progression bei einer Anzahl von Merkmalen aufweisen. Vier genera scheinen ganz nah verwandt zu sein mit solcher in Neukaledonien und zwei mit generischen Gruppen in Chile. Die restlichen Gattungen sind mit anderen in Neuseeland verwandt, bis auf eine Gattung, deren nächste, bekannte Artverwandte sich in Neuguinea befinden.

REFERENCES

Brundin, L. 1966. Transantarctic relationships and their significance as evidence by chironomid midges. *K. Sven. Vetenskapsakad Handl.* 11: 1-472.

Cowley, D.R. 1978. Studies on the larvae of New Zealand Trichoptera *N.Z.J. Zool.* 5: 639-750.

Edmunds, G.F., Jr. 1972. Biogeography and evolution of Ephemeropter *Annu. Rev. Entomol.* 17: 21-42.

Edmunds, G.F., Jr. 1975. Phylogenetic biogeography of mayflies. *Ann. Mo. Bot. Gard.* 62: 251-263.

Fleming, C.A. 1975. The geological history of New Zealand and its biota, p. 1-86, *in* Kuschel, G. (ed.), Biogeography and Ecology in New Zealand. Dr. W. Junk, The Hague.

Illies, J. 1969. Biogeography and ecology of Neotropical freshwater insects, especially those from running waters, p. 685-708, *in* Fittkau, E.J., *et al.* (eds.), Biogeography and Ecology in South America. Vol. 2. Dr. W. Junk, The Hague.

McFarlane, A.G. 1976. A generic revision of New Zealand Hydropsychinae (Trichoptera). *J.R. Soc. N.Z.* 6: 23-35.

McLellan, I.D. 1975. The freshwater insects, p. 537-559, *in* Kuschel G. (ed.), Biogeography and Ecology in New Zealand. Dr. W. Junk The Hague.

McLellan, I.D. 1977. New alpine and southern Plecoptera from New Zealand, and a new classification of the Gripopterygidae. *N.Z.J. Zool.* 4: 119-147.

Penniket, J.C. 1961. Notes on New Zealand Ephemeroptera I. The affinities with Chile and Australia, and remarks on *Atalophlebia* Eaton (Leptophlebiidae). *N.Z. Entomol.* 2: 1-11.

Pescador, M.L. and W.L. Peters. 1979. Phylogenetic relationships and zoogeography of cool-adapted Leptophlebiidae (Ephemeroptera) in southern South America. This volume.

Peters, W.L. and G.F. Edmunds, Jr. 1970. Revision of the generic classification of the Eastern Hemisphere Leptophlebiidae (Ephemeroptera). *Pac. Insects* 12: 157-240.

Towns, D.R. and W.L. Peters. 1978. A revision of genus *Atalophlebioides* (Ephemeroptera: Leptophlebiidae). *N.Z.J. Zool.* 5: 607-614.

Towns, D.R. and W.L. Peters. 1979a. Three new genera of Leptophlebiidae (Ephemeroptera) from New Zealand. *N.Z.J. Zool.* 6: (in press).

Towns, D.R. and W.L. Peters. 1979b. New genera and species of Leptophlebiidae (Ephemeroptera) from New Zealand. *N.Z.J. Zool.* 6: (in press).

Tsui, P.T.P. and W.L. Peters. 1975. The comparative morphology and phylogeny of certain Gondwanian Leptophlebiidae based on the thorax, tentorium, and abdominal terga (Ephemeroptera). *Trans. Am. Entomol. Soc.* 101: 505-595.

GEOGRAPHIC DISTRIBUTION AND RECLASSIFICATION OF THE SUBFAMILY EPHEMERELLINAE (EPHEMEROPTERA: EPHEMERELLIDAE)

Richard K. Allen

22021 Jonesport Lane
Huntington Beach, California
U.S.A. 92646

ABSTRACT

The geographic distributions of the genera are included on maps and 9 latitudinal distributional zones are established, revising the system proposed by Allen and Brusca. The genera are included in 2 tribes, Ephemerellini and Hyrtanellini n. tribe, and *Acerella, Attenella, Caudatella, Cincticostella, Crinitella, Dannella, Drunella, Ephemerella* s.s., *Eurylophella, Hyrtanella, Serratella, Teloganopsis, Timpanoga* and *Torleya* are treated as genera. *Drunella* is composed of five subgenera: s.s., *Eatonella, Myllonella* n. subgen., *Tribrochella* n. subgen., and *Unirhachella* n. subgen.; *Cincticostella* of 3: s.s., *Rhionella* n. subgen., and *Vietnamella* NEW COMBINATION; and *Dannella* of 2: s.s. and *Dentatella* n. subgen.

GEOGRAPHIC DISTRIBUTION

Allen & Brusca (1973), in an attempt to understand the geographic distribution and distributional limits of the mayfly genera of Mexico, plotted collection records of all species on maps. It was observed that genera of both austral and boreal origins reached their most northern and southern limits in the same narrow latitudinal zones. For example, the boreal *Ephemerella* s. l. and *Centroptilum*, and the austral *Baetodes, Campsurus, Leptohyphes*, and *Thraulodes* all reach their distributional limits in North America between 30°-40° north latitude; and the boreal genera *Iron**, *Hepta-*

* The heptageniid taxa *Iron* and *Ironopsis* are herein considered to have generic rank.

genia, *Isonychia*, *Rhithrogena* and *Stenonema*, and the austral genus *Euthyplocia* all reach their distributional limits in southern Mexico and Central America between 15°-25° north latitude. Also, it was observed that five genera of austral origin, *Dactylobaetis*, *Lachlania*, *Tortopus*, *Traverella* and *Tricorythodes*, all reach their most northern limits in southern Canada between 45°-55° north latitude. A nomenclature was established which was based on the presumed origin of the genus and the latitudinal zone on earth where they reach their most northern or southern limits. Seven distributional zones were established to accommodate the distribution of the 21 Mexican genera.

Additional study of the geographic distribution of more than 100 genera from all parts of the world reveals that nine distributional zones must be established (Fig. 1). The proposed zones and their latitudinal limits are as follows:

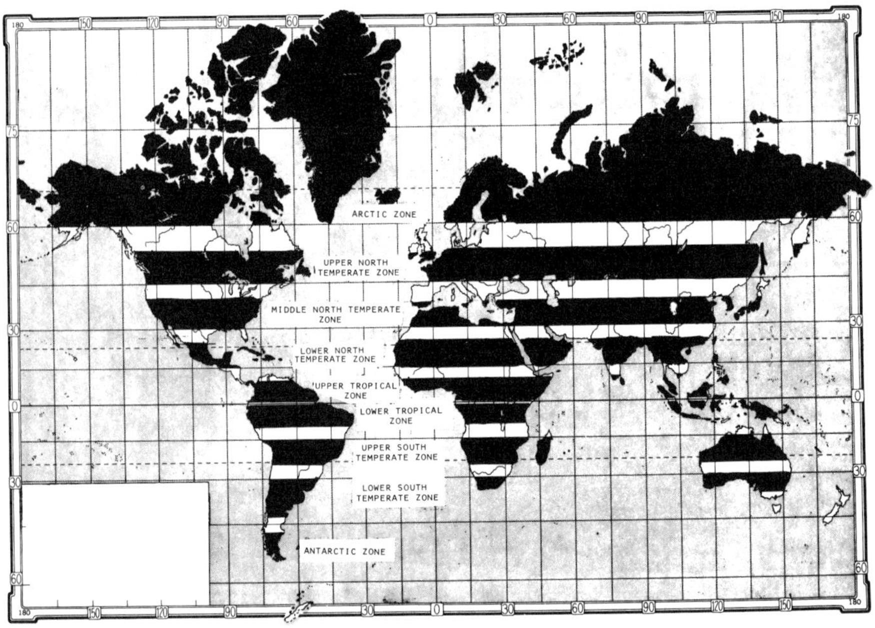

Figure 1. Mayfly distribution zones of the world.

I. ARCTIC ZONE - above 60° north latitude.
II. UPPER NORTH TEMPERATE ZONE - between 45°-55° north latitude.
III. MIDDLE NORTH TEMPERATE ZONE - between 30°-40° north latitude.
IV. LOWER NORTH TEMPERATE ZONE - between 15°-25° north latitude.
V. UPPER TROPICAL ZONE - between 0°-10° north latitude.
VI. LOWER TROPICAL ZONE - between 0°-10° south latitude.
VII. UPPER SOUTH TEMPERATE ZONE - between 15°-25° south latitude.
VIII. LOWER SOUTH TEMPERATE ZONE - between 30°-40° south latitude.
IX. ANTARCTIC ZONE - below 50° south latitude.

The paper by Allen & Brusca (1973) established the first system by which the geographic distribution and distributional limits of mayflies could be expressed in simple terms. A more exact system would designate the presumed origin, and both the northern and southern limits of the taxon. For example, the genera *Ephemerella* and *Serratella* are boreal in origin and both reach their most southern limits in the MIDDLE NORTH TEMPERATE ZONE in North America. But the northern limits of *Ephemerella* is in the UPPER NORTH TEMPERATE ZONE.

A close examination of the distributional limits of the Ephemerellinae reveals that not all species fit exactly within the designated zones dictated by the distributional limits of all the remaining species. Distributional records are good for North America, but good distributional data are lacking for most of Europe and Asia. There are 160 valid species of Ephemerellinae and only ten (*Serratella micheneri**, *S. thailandensis*, *Torleya major**, *T. nepalica*, *T. padunica*, *Drunella flavilinea**, *D. gilliesi*, *Acerella uenoi*, *A. percuita*, and *Eurylophella lithuanica*) do not "fit" the proposed distributional zones. Seven species, however are known from Europe or Asia and only from the type locality. The three species with good distributional data (marked * above) do not occur more than 2-3° from the limits established by other species.

TAXONOMY

A reevaluation of nymphal and adult characters and a detailed study of geographic distribution of the Ephemerellinae by the author, and the recent publication of the descriptions of new taxa by Allen (1971, 1975, 1977), Allen & Edmunds (1976), Gose (1969), and Tshernova (1972), strongly indicate a need for reappraisal of the status of the taxa. The subfamily is divided into 2 tribes, 14 genera, and 10 subgenera.

Subfamily Ephemerellinae

The subfamily is herein divided into the tribes Ephemerellini and Hyrtanellini n. tribe.

Tribe Ephemerellini Lameere 1917

Adult Forewing with 2 intercalaries between IMP and MP_2, and between MP_2 and CuA; numerous short intercalaries along entire outer wing margin; hind wing comparatively large with abundant venation; hind wing with small, blunt, subbasal costal projection; male genital forceps with heavy, short, basal segment; long, slightly concave second segment; and long slender to heavy, short, terminal segment.

Nymph Paired dorsal abdominal gills segments 3-7 or 4-7.

Genus *Ephemerella* Walsh 1862

Type Species *excrucians* Walsh 1862

Remarks North American workers consider *Chitonophora* to be synonymous with *Ephemerella* while most Eurasians treat *E. aurivillii*, *E. krieghoffi*, and *E. mucronata* by the former generic name.

The penes of the male imago usually possess spines, but they are lacking in some species. Also, the form of the penes is variable. The lobes are rounded with a shallow notch in most, but they are elongate with a deep notch in *E. aurivillii*, *E. krieghoffi*, *E. mucronata*, *E. needhami*, and *E. septentrionalis*. The penes of *E. needhami*, *E. krieghoffi*, and *E. mucronata* are without spines, while *E. maculata* has only a shallow notch and is without spines. The nymphs of all species, except *E. septentrionalis* and *E. berneri*, possess femoral spines on the anterior surface of the fore femora, and all species have some vestige of dorsal abdominal tubercles. The nymphs of some species appear to be without tubercles, but all have at least discernible protuberances. The nymph of *E. septentrionalis* is unique in that the abdomen possesses small median tubercles.

Ikonomov (1961) described *Chitonophora* sp. *nympha unicolorata*, and in accordance with Article 5 of ICZN the name is emended to *Ephemerella unicolorata*.

Distribution *Ephemerella* is the most widely distributed genus in the subfamily as it occurs in North America, the British Isles, northern Africa, Europe, Asia and Japan (Fig. 2). In western North America, Europe, and Asia the genus has a wide latitudinal distributional range from the ARCTIC ZONE to the MIDDLE NORTH TEMPERATE ZONE. In western North America species occur from

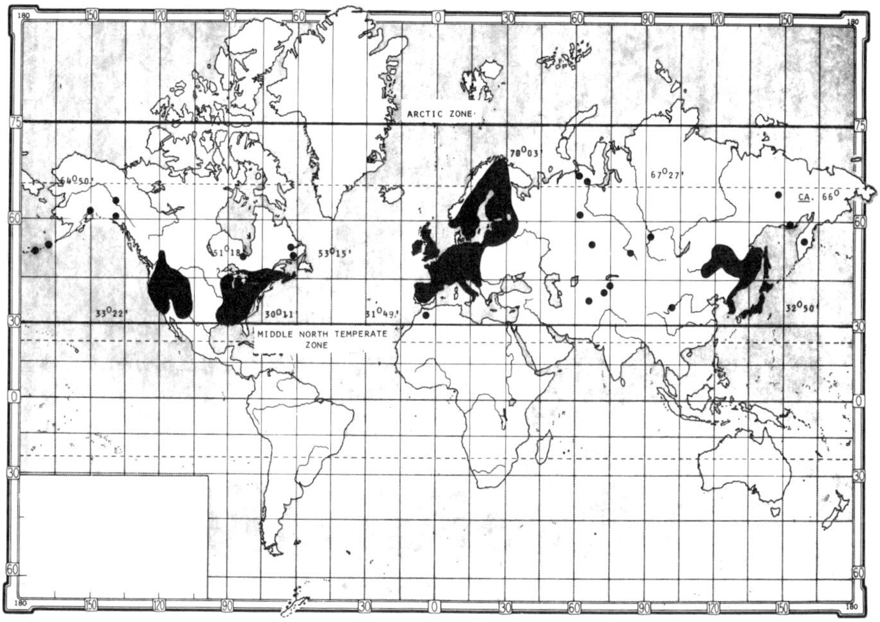

Figure 2. Distribution of the genus *Ephemerella*.

64° 50' to 33°22'; in Europe from 70° 03' to 38° 21'; and in Asia from 67° 27' to 32° 50'. In Africa the genus occurs at about 31° 50'N. latitude. The latitudinal range is narrower in eastern North America as species occur from the UPPER NORTH TEMPERATE ZONE, 53°15', to the MIDDLE NORTH TEMPERATE ZONE, 30°11', and the distribution is disjunct in North America, Europe and Asia.

Genus *Serratella* Edmunds 1959

Type Species *serrata* Morgan 1911.

Remarks The penes of the male imago resemble those of *Torleya* as they possess subapical lateral projections, except for *S. tibialis* and *S. teresa*. The nymphal head and thorax are without tubercles, except for *S. carolina*. The nymphs of *S. spiculosa* and *S. serrata* have spicules on the occiput. All nymphs, except *S. levis*, possess paired dorsal abdominal tubercles. Abdominal tubercles are barely discernible in *S. deficiens*.

Gose (1969) described and figured a nymph from Thailand which he labelled *Ephemerella* sp. TEB. This species belongs in *Serratella* and is the second known species from Asia. It is morphologically and distributionally distinct from *S. subsolana*. I formally name it *S. thailandensis*, and designate the nymph from Chantaburi, Thailand, 20-VI-61, as the type.

Distribution *Serratella* is known only from North America and Asia (Fig. 3). The North American species are widely distributed between the UPPER NORTH TEMPERATE ZONE, 53°34', and the MIDDLE NORTH TEMPERATE ZONE, 30°26', except for one western North American species. *Serratella micheneri* nymphs have become adapted to warm water and occur in central Baja California, 27°35'N. latitude. The two Asian species are known only from their respective type localities.

Genus *Torleya* Lestage 1917

Type Species *major* Klapalek 1905

Distribution The five species of *Torleya* are known from a narrow latitudinal belt in Europe and Asia (Fig. 4). In Europe the genus occurs from the UPPER NORTH TEMPERATE ZONE, 53°03', to just

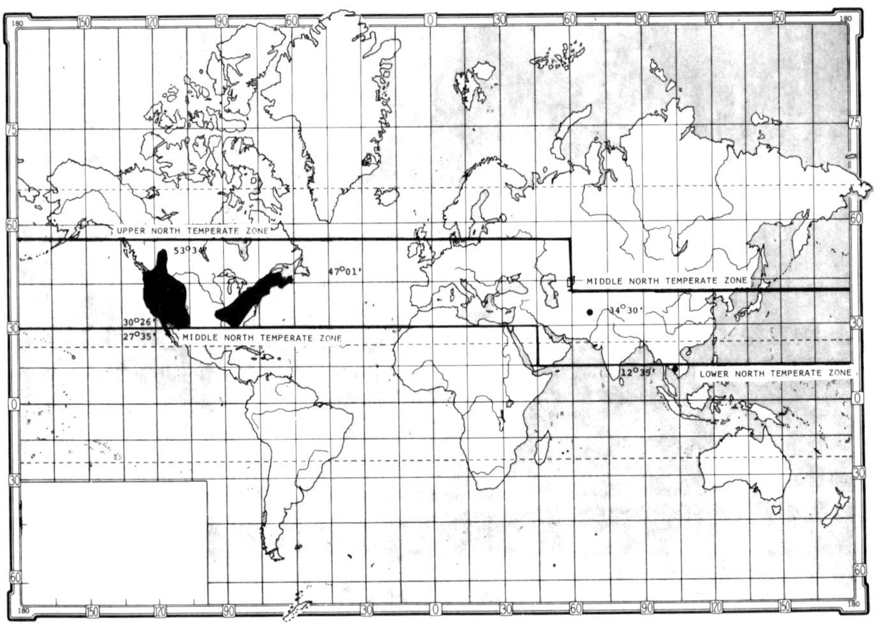

Figure 3. Distribution of the genus *Serratella*.

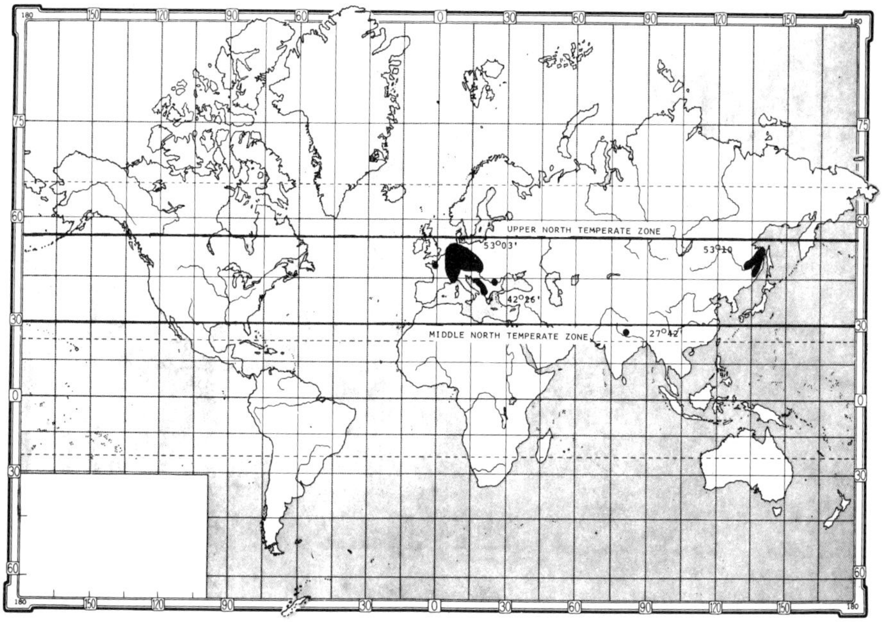

Figure 4. Distribution of the genus *Torleya*.

above the northern limits of the MIDDLE NORTH TEMPERATE ZONE, 42°26', and in Asia from the UPPER NORTH TEMPERATE ZONE, 53°10', to just below the MIDDLE NORTH TEMPERATE ZONE, 27°42' N. latitude.

Genus *Teloganopsis* Ulmer 1939

Type Species *media* Ulmer 1939

Distribution The two described species of *Teloganopsis* are allopatric (Fig. 5). The genus is restricted to eastern Asia. One species is known from the UPPER NORTH TEMPERATE ZONE, 50°12'N. latitude, and the other from the LOWER SOUTH TROPICAL ZONE, 6°57' S. latitude.

Genus *Caudatella* Edmunds 1959

Type Species *heterocaudata* McDunnough 1929

Distribution *Caudatella* is restricted to western North America, and occurs from the UPPER NORTH TEMPERATE ZONE, 50°12', to the MIDDLE NORTH TEMPERATE ZONE, 34°07' (Fig. 6).

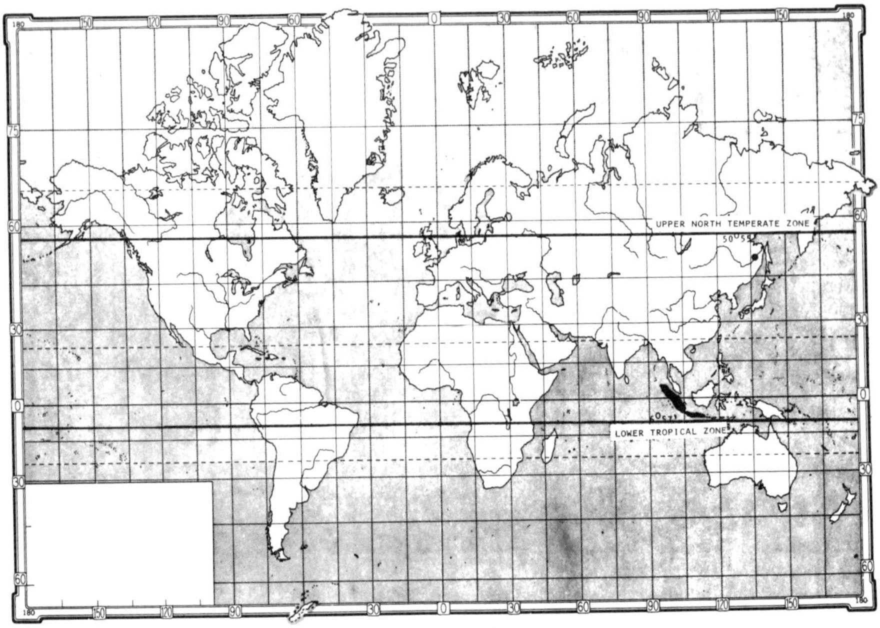

Figure 5. Distribution of the genus *Teloganopsis*.

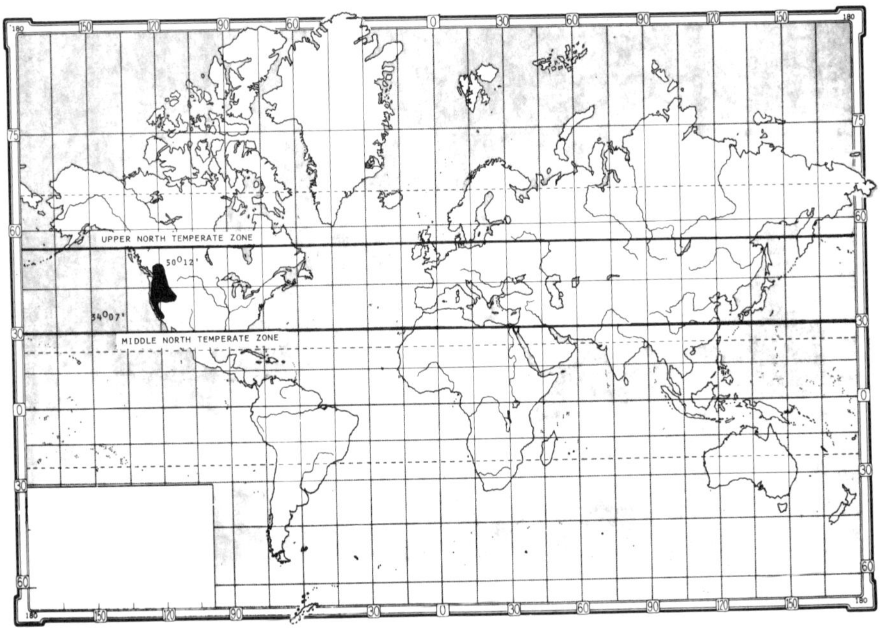

Figure 6. Distribution of genus *Caudatella*.

Genus *Drunella* Needham 1909

<u>Type Species</u> *grandis* Eaton 1884

<u>Remarks</u> The nymphs of *Drunella* are diverse in structure as species have adapted to a variety of stream habitats. Five subgenera are recognized in the genus.

<u>Distribution</u> The genus occurs in only North America, Asia and Japan (Fig. 7), and it has a wide latitudinal range on both continents. In North America and Asia the latitudinal range is from the ARCTIC ZONE to the MIDDLE NORTH TEMPERATE ZONE. In western North America it is known from Alaska, 61°10', to Baja California, 31°03', and northern Sonora, 30°59'. In Asia the range is from Kamchatka, 59°20', to Nepal, 27°42', and to 34°41' in Japan. The latitudinal range is narrower in eastern North America, from only the UPPER NORTH TEMPERATE ZONE, 51°18', to the MIDDLE NORTH TEMPERATE ZONE, 35°03'.

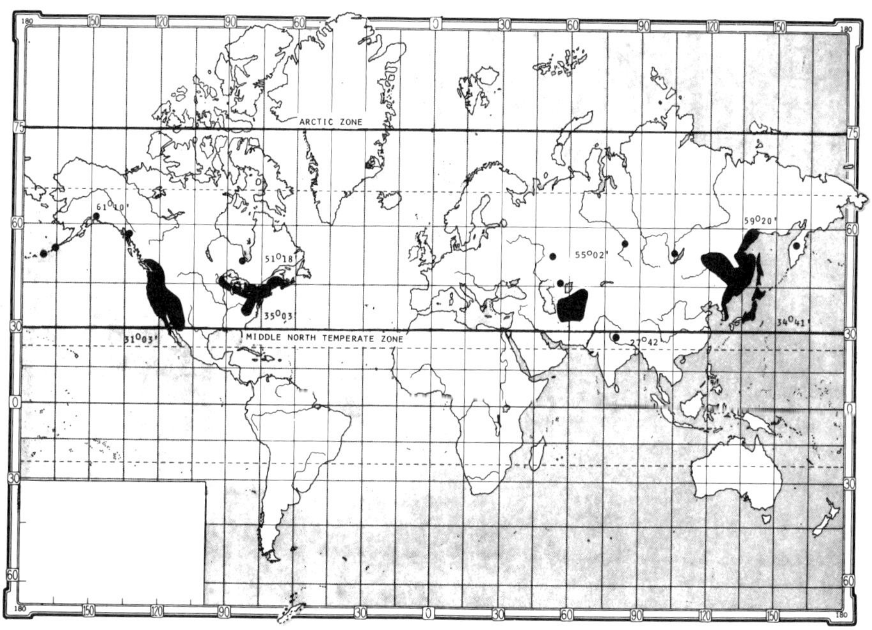

Figure 7. Distribution of the genus *Drunella*.

Subgenus *Drunella* s.s.

Remarks *Drunella pelosa* is the most aberrant species in the genus. The nymph lacks body tubercles, but possesses paired submedian tufts of setae on the occiput, thorax, and abdomen. The femora are without tubercles, but the ventral margins are lined with thick setae, and the sternum has a partial adhesive disc. The inclusion of *D. pelosa* in this subgenus is appropriate as the nymphs have features that are most similar to the other *Drunella* s.s. nymphs.

Subgenus *Tribrochella* Allen n. subgen.

trispina group Imanishi 1940: 193.

Type Species *trispina* Imanishi 1940.

Nymph Head with anteromedian and lateral frontoclypeal projections; abdominal tubercles absent, with only paired submedian ridges.

Remarks *Drunella (Tribrochella) kohnoi* is emended to *D. (T.) kohnoae* as R.W. Baumann (Brigham Young University) has informed the author that Dr. Mitsuko Kohno is female. This subgenus is the largest in the genus with 12 species, and it is the only one in which the great majority of species are in Asia.

Subgenus *Unirhachella* Allen n. subgen.

Type Species *tuberculata* Morgan 1911

Nymph Head with well developed occipital tubercles, large median metanotal tubercle, and small paired dorsal abdominal tubercles.

Subgenus *Myllonella* Allen n. subgen.

Type Species *coloradensis* Dodds, 1923

Nymph Head with small occipital tubercles, without thoracic tubercles, and with small abdominal tubercles.

Subgenus *Eatonella* Needham 1927

Type Species *doddsi* Needham 1927

Remarks McDunnough (1931) acknowledged the close relationship of *D. doddsi* and *D. walkeri* (as *fuscata*) nymphs, but he rejected *Eatonella* as a taxon. He stated that *D. walkeri* and *D. tuberculata* were closely related in the adult stage, and he was unable to understand the great diversity of characters in the nymphs of the latter and the other two species. Of course, *D. walkeri* and *D. tuberculata* are not closely related. Allen & Edmunds (1962b) recognized the common characters of the nymphs of *D. conestee*, *D. doddsi*, *D. wayah*, and *D. walkeri* and placed them together in a species group. I herein resurrect *Eatonella* as a subgenus for *D. doddsi* and the other three species.

Genus *Crinitella* Allen & Edmunds 1963

<u>Type Species</u> *coheri* Allen & Edmunds 1963

<u>Remarks</u> It seems advisable to elevate this taxon to full generic rank along with the others that were formerly placed in *Ephemerella* s. l., but this placement is tentative until the male imago is described.

<u>Distribution</u> *Crinitella* is known from India, 31°34' N. latitude, in the lower limits of the MIDDLE NORTH TEMPERATE ZONE to Malaysia, 3°08' N. latitude, in the UPPER TROPICAL ZONE (Fig. 8).

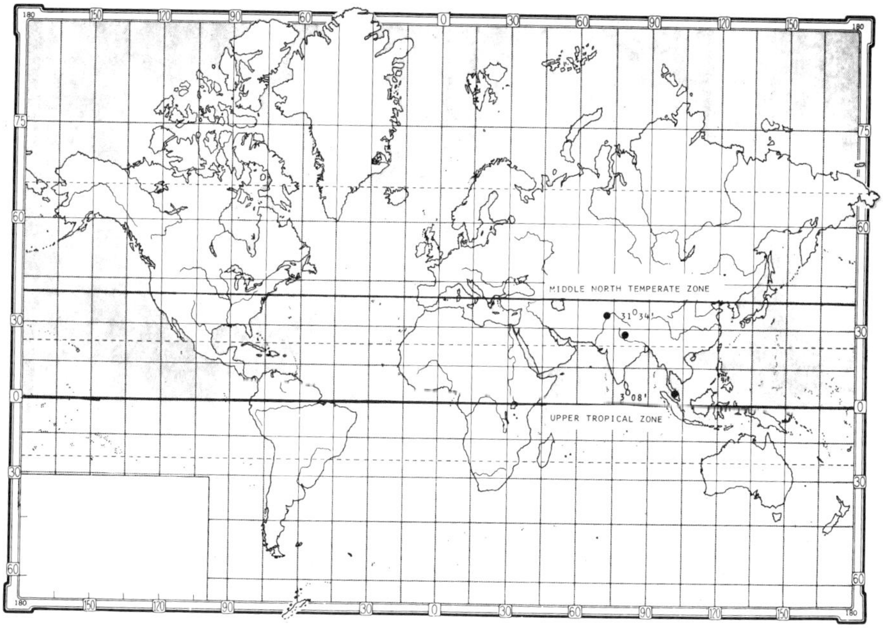

Figure 8. Distribution of the genus *Crinitella*.

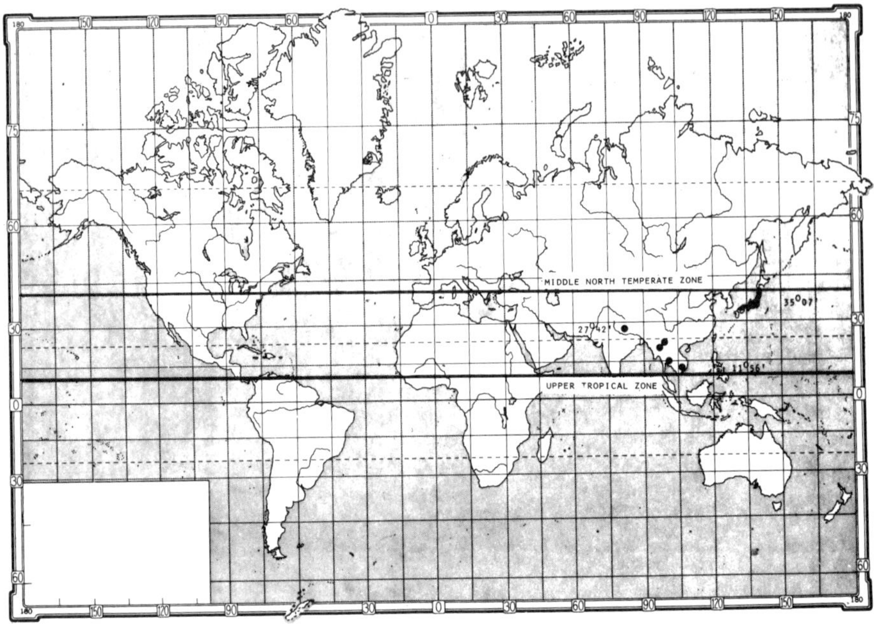

Figure 9. Distribution of the genus *Acerella*.

Genus *Acerella* Allen 1971

<u>Type Species</u> *longicaudata* Ueno 1928

<u>Distribution</u> This genus is known only from southeast Asia and Japan from the MIDDLE NORTH TEMPERATE ZONE, 35°07', to near the upper limits of the UPPER TROPICAL ZONE, 11°56' (Fig. 9).

Genus *Cincticostella* Allen 1971

<u>Type Species</u> *nigra* Ueno 1928

<u>Remarks</u> *Cincticostella* nymphs are diverse in structure and 3 subgenera are recognized, including *Vietnamella*, NEW COMBINATION.

<u>Distribution</u> This genus is known from western and southwestern Asia, and Japan, from the UPPER NORTH TEMPERATE ZONE, 53°44', to the LOWER NORTH TEMPERATE ZONE, 18°48' (Fig. 10).

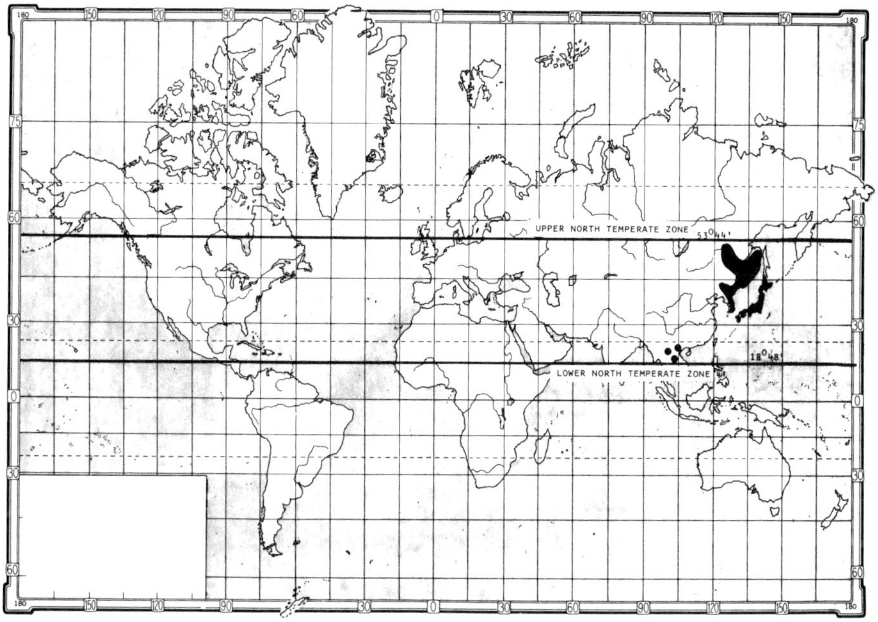

Figure 10. Distribution of the genus *Cincticostella*.

Subgenus *Cincticostella* s.s.

nigra group Allen 1975: 17.

<u>Nymph</u> Head without tubercles; femora narrow and entire.

Subgenus *Rhionella* Allen n. subgen.

insolta group Allen 1975: 17.

<u>Type Species</u> *insolta* Allen 1971.

<u>Nymph</u> Head with occipital or suboccipital tubercles; femora with serrations and protuberances; middle and hind femora expanded.

Subgenus *Vietnamella* Tshernova 1972

Vietnamella Tshernova 1972: 609. NEW COMBINATION

<u>Type Species</u> *thani* Tshernova 1972.

<u>Nymph</u> Head with long frontoclypeal projections and paired frontal tubercles; fore femora with tubercles on ventral margin and fore femora expanded; middle and hind femora without tubercles and not expanded.

Genus *Attenella* Edmunds 1971

Attenella Edmunds 1971: 152 (=*Attenuatella* Edmunds 1959: 465, *nec* Stehli 1954: 343)

<u>Type Species</u> *attenuata* McDunnough 1925

<u>Distribution</u> *Attenella* is known only from North America (Fig. 11). Species occur from the UPPER NORTH TEMPERATE ZONE to the MIDDLE NORTH TEMPERATE ZONE, between 46°11' and 30°27' N. latitude in the east and between 53°54' and 35°34' N. latitude in the west (Fig. 11).

Genus *Eurylophella* Tiensuu 1935

<u>Type Species</u> *karelica* Tiensuu 1935

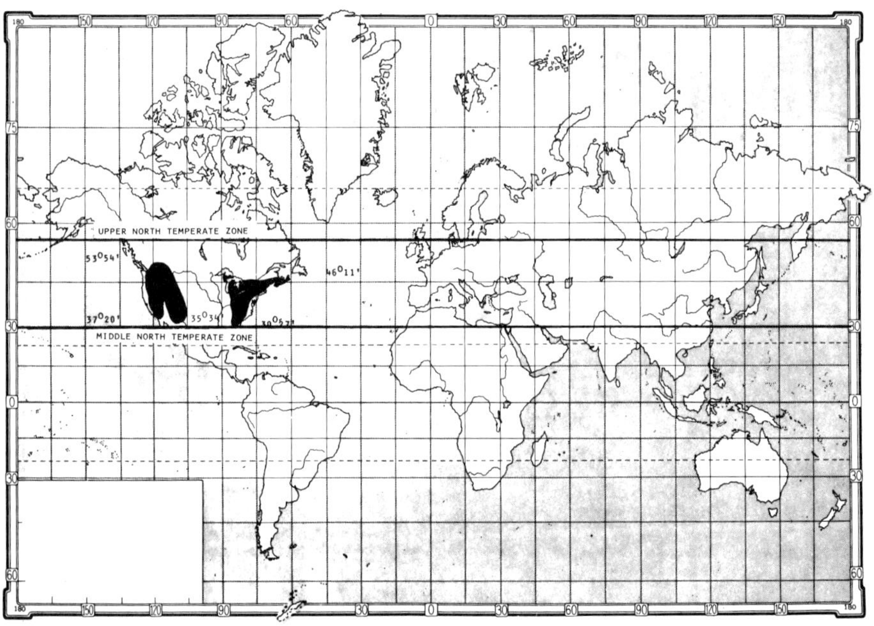

Figure 11. Distribution of the genus *Attenella*.

Remarks Edmunds, Jensen & Berner (1976) incorrectly list *E. brunnescens* Tiensuu 1935 as the type of this taxon. The type is *E. karelica* which was established by original designation.

Distribution Eleven of the 15 species of *Eurylophella* occur in eastern North America, one is known from western North America, and three from Europe. In North America the genus occurs from the ARCTIC ZONE, 59°47', to the MIDDLE NORTH TEMPERATE ZONE, 29°37'. In Europe, it is known from the ARCTIC ZONE, 64°32', to near the lower limits of the UPPER NORTH TEMPERATE ZONE, 42°31' (Fig. 12).

Genus *Dannella* Edmunds 1959

Type Species *simplex* McDunnough 1925

Remarks Allen & Edmunds (1962a) characterized the adult and nymphal stages of *Dannella* based on the characters of known species. Allen (1977) recharacterized the taxon to include *D. bartoni*, a newly described nymph, which possesses paired dorsal abdominal tubercles and dentition on the claws. McCafferty (1977) failed to include *D. bartoni* in *Dannella*, as, in his opinion, the species

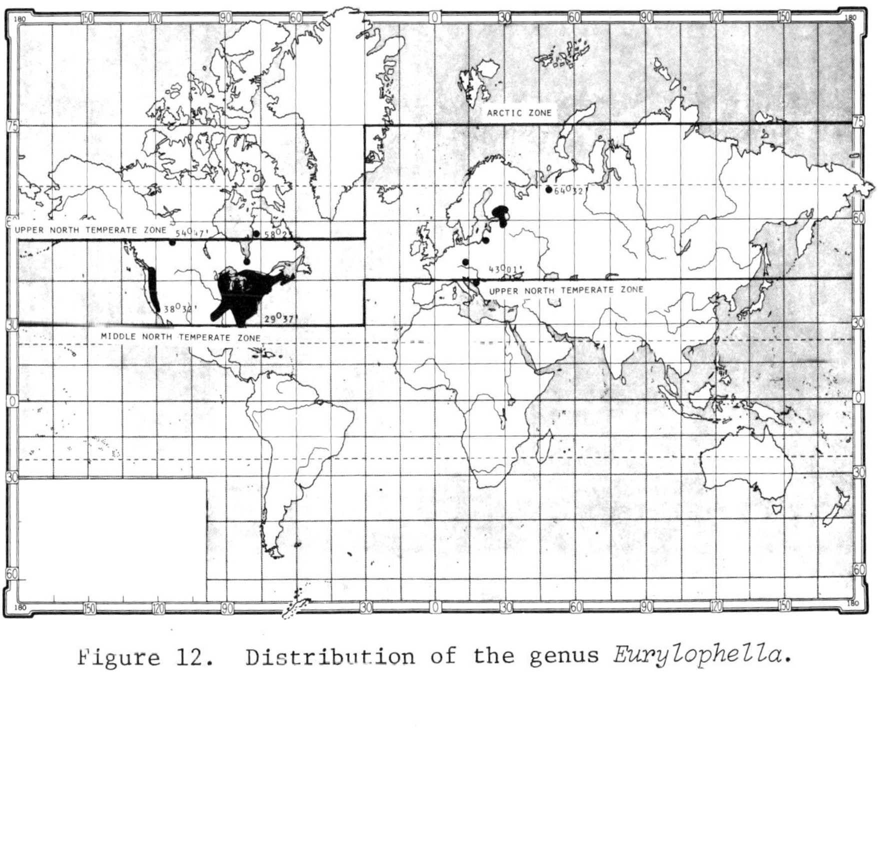

Figure 12. Distribution of the genus *Eurylophella*.

changed the concept of the taxon. Recharacterization of taxa to accommodate additional species has been done several times recently in the Ephemeroptera (see Allen & Edmunds 1962b; Allen 1967; and Allen & Brusca 1973). *Drunella* s.l. was characterized as having only head and body tubercles until *D. pelosa* was described with tufts of setae in place of tubercles. *Leptohyphes* was characterized as being without head and body tubercles, or protuberances, until *L. undulatus*, *L. dicinctus*, and *L. melanobranchus* were discovered; *Tricorythodes* was characterized as being without body tubercles until *T. bullus* and *T. cristatus* were described; and *Leptohyphes* and *Tricorythodes* were characterized as possessing only small compound eyes in both sexes until *L. mirus* and *T. dimorphus* were found.

McCafferty (MS) places *D. bartoni* in *Eurylophella* as he states, "....*E. bartoni* possesses three derived larval character states that are found in the *Eurylophella* lineage." "This evolution included the loss of maxillary palpi, the apical rounding of the gills, and the reduction of the midlength of abdominal segments 5, 6, and 7." The maxillary palpi are well developed in *D. provonshai*, reduced in *D. lita* and *D. simplex*, and absent in *D. bartoni*. Obviously there is a tendency for the reduction of this structure in *Dannella*, as is the case in the nymphs of *Acerella*, *Cincticostella*, *Serratella*, and *Torleya*. The taxonomic significance of the rounding of the gills has not been investigated by the author. The reduction of the length of abdominal segments 5-7 does not reflect a relationship to *Eurylophella* as the nymphs of all ephemerellid genera with operculate gills possess this character, including *Hyrtanella*.

Dannella nymphs are not easily recognized since the characters which distinguish them are not as well defined as they are in *Eurylophella*. The body is extremely depressed, like *Timpanoga*, the abdomen is broad and the posterolateral projections are well developed; the body is heavily setaceous; and the abdominal segments 8-9 are subequal (Fig. 13). *Eurylophella* nymphs (Fig. 14) on the other hand, are the most easily recognized in the subfamily because of two consistent characters. First, abdominal segment 9 is considerably longer than 8, and second, the paired dorsal abdominal tubercles on segments 1-3 are blunt and those on 4-9 are sharp.

The morphological characters of the nymph of *D. bartoni* suggest that it be placed in a separate subgenus and *Dentatella* is erected to accommodate this species.

Distribution *Dannella* occurs only in North America from the ARCTIC ZONE, 58°45', to the MIDDLE NORTH TEMPERATE ZONE, 35°22'. Four species occur in eastern North America, and one undescribed species is known from Colorado (Fig. 15).

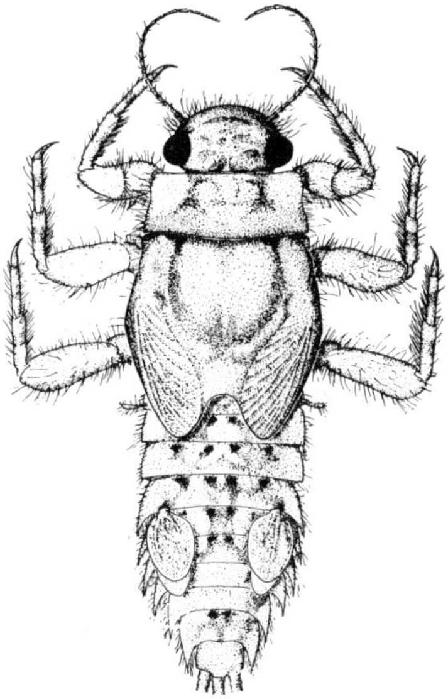

Figure 13. Nymph of *Dannella*.

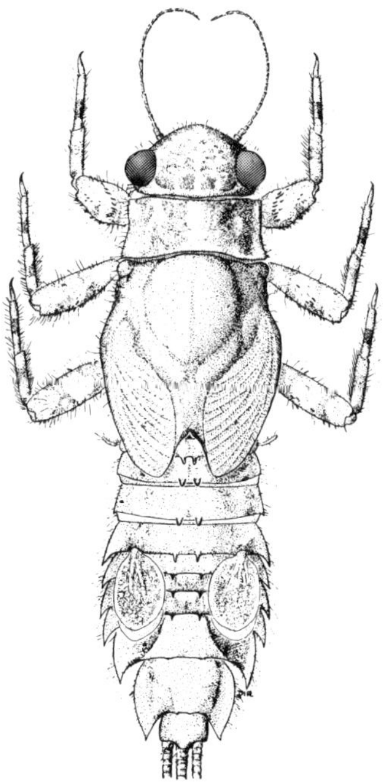

Figure 14. Nymph of *Eurylophella*.

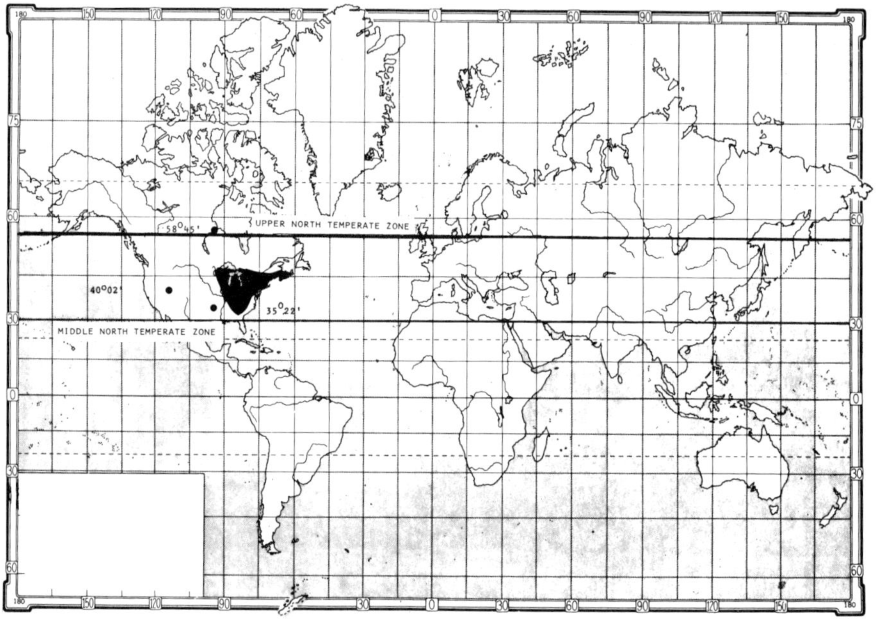

Figure 15. Distribution of the genus *Dannella*.

Subgenus *Dannella* s.s.

<u>Nymph</u> Abdominal terga without tubercles; claws without denticles.

Subgenus *Dentatella* Allen n. subgen

<u>Type Species</u> *bartoni* Allen 1977

<u>Nymph</u> Abdominal terga with tubercles; claws with denticles.

Genus *Timpanoga* Needham 1927

<u>Type Species</u> *hecuba* Eaton 1884

<u>Distribution</u> *Timpanoga* is known only from a single species in western North America from the UPPER NORTH TEMPERATE ZONE, 50°03', to the MIDDLE NORTH TEMPERATE ZONE, 35°34' (Fig. 16).

<u>Tribe Hyrtanellini Allen n. tribe</u>

<u>Adult</u> Forewing with 1 intercalary between IMP and MP_2, and

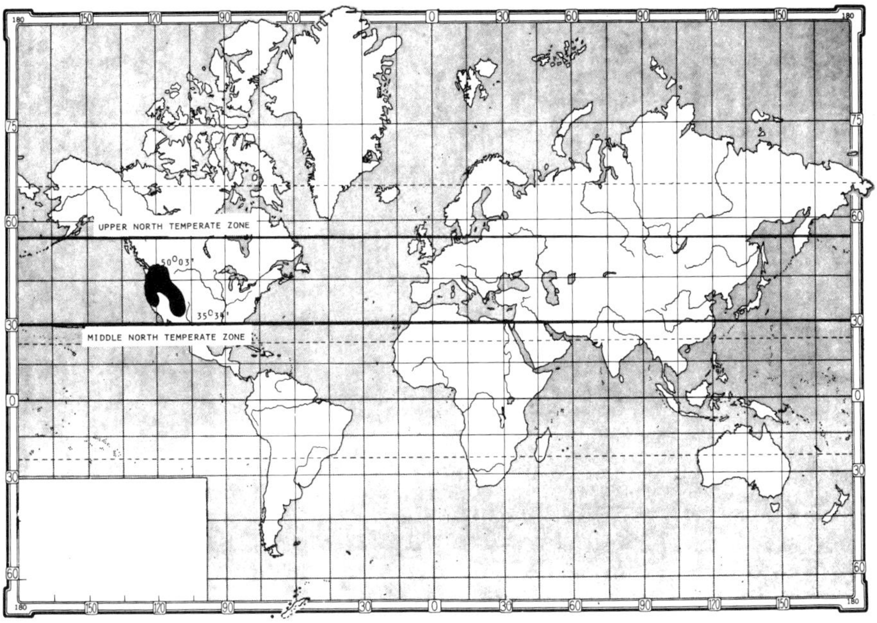

Figure 16. Distribution of the genus *Timpanoga*

without intercalaries between MP_2 and CuA; with few intercalaries along outer margin; hindwing small with sparse venation; male genitalia unknown.

Nymph Abdominal gills segments 3-6; gill operculate segment 3, imbricated 4-6; abdominal terga 4-8 depressed laterally; maxillary palpi vestigial; claws with palisade subapical denticles; caudal filaments with setae.

Genus *Hyrtanella* Allen & Edmunds 1976

Type Species *christineae* Allen & Edmunds 1976

Distribution *Hyrtanella* is known only from Malaysia, *ca.* 5° N. latitude in the UPPER TROPICAL ZONE.

RESUME

Le rapport détermine la répartition géographique des genres du sous famille des éphémérellinés ainsi que l'établissement de neuf zones de répartition latitudinales et révise le système proposé par Allen et Brusca. Les genres font partie de deux tribus, les Ephemerellini et les Hyrtanellini, n. tribus, et *Acerella*, *Attenella*, *Caudatella*, *Cincticostella*, *Crinitella*, *Dannella*, *Drunella*, *Ephemerella* au sens strict, *Eurylophella*, *Hyrtanella*, *Serratella*, *Teloganopsis*, *Timpanoga* et *Torleya* sont considérés comme des genres. *Drunella* comprend cinq sous-genres: sens strict, *Eatonella*, *Myllonella*, n. subgen. *Tribrochella* n. subgen. et *Unirhachella* n. subgen.; *Cincticostella* comprend trois sous-genres: sens strict, *Rhionella* n. subgen., et *Vietnamella*, n. comb.; *Dannella* comprend deux sous-genres; sens strict et *Dentatella*, n. subgen.

ZUSSAMENGASSUNG

Diese Arbeit befaßt sich mit der geographischen Verteilung von Gattungen der Unterfamilie *Ephemerella* und stellt neun latitudinale Verteilungszonen auf, wobei das von Allen & Brusca vorgeschlagene System revidiert wird. Die n. Sippe, sowie *Acerella*, *Attenella*, *Caudatella*, *Cincticostella*, *Crinitella*, *Dannella*, *Drunella*, *Ephemerella*, *Eurylophella*, *Hyrtanella*, *Serratella*, *Teloganopsis*, *Timpanoga* und *Torleya* werden als Gattungen behandelt. *Drunella* besteht aus fünf subgenera: s.s., *Eatonella*, *Myllonella* n. subgen., *Tribrochella* n. subgen. und *Unirhachella* n. subgen., *Cincticostella* von 3: s.s., *Rhionella* n. subgen., und *Vietnamella* Neue Kombination. *Dannella* von 2: s.s. und *Dentatella* n. subgen.

LITERATURE CITED

Allen, R.K. 1967. New species of New World Leptohyphinae (Ephemeroptera: Tricorythidae). *Can. Entomol.* 99: 350-375.

Allen, R.K. 1971. New Asian *Ephemerella* with notes (Ephemeroptera: Ephemerellidae). *Can. Entomol.* 103: 512-528.

Allen, R.K. 1975. *Ephemerella (Cincticostella)*: a revision of the nymphal stages (Ephemeroptera: Ephemerellidae). *Pan-Pac. Entomol.* 51: 16-22.

Allen, R.K. 1977. A review of *Ephemerella (Dannella)* and the description of a new species. *Pan-Pac. Entomol.* 53: 215-217.

Allen, R.K. and R.C. Brusca. 1973. New species of Leptohyphinae from Mexico and Central America (Ephemeroptera: Tricorythidae). *Can. Entomol.* 105: 83-95.

Allen, R.K. and G.F. Edmunds, Jr. 1962a. A revision of the genus *Ephemerella*. IV. The subgenus *Dannella* (Ephemeroptera: Ephemerellidae). *J. Kans. Entomol. Soc.* 35: 333-338.

Allen, R.K. and G.F. Edmunds, Jr. 1962b. A revision of the genus *Ephemerella*. V. The subgenus *Drunella* in North America (Ephemeroptera: Ephemerellidae). *Misc. Publ. Entomol. Soc. Am.* 3: 145-179.

Allen, R.K. and G.F. Edmunds, Jr. 1976. *Hyrtanella*: A new genus of Ephemerellidae from Malaysia (Ephemeroptera). *Pan-Pac. Entomol.* 52: 133-137.

Edmunds, G.F., Jr., S.L. Jensen and L. Berner. 1976. The mayflies of North and Central America. Univ. Minnesota Press, Minneapolis.

Gose, K. 1969. Mayflies (Ephemeroptera) from Thailand. *Nat. Life Southeast Asia* 6: 125-138.

Ikonomov, P. 1961. Eintagsfliegen (Ephemeroptera) Macedoniens fam. Ephemerellidae. *Frag. Bal. Mus. Macedonici Sci. Nat.* 8: 53-74.

Imanishi, K. 1940. Ephemeroptera of Manchouchuo, Inner Mongolia and Chozen. *Rep. Limnobiol. Surv. Kwantung and Manchuochuo* 1940: 169-263.

McCafferty, W.P. 1977. Biosystematics of *Dannella* and related subgenera of *Ephemerella* (Ephemeroptera: Ephemerellidae). *Ann. Entomol. Soc. Am.* 70: 881-889.

McCafferty, W.P. MS. A natural subgeneric classification of *Ephemerella bartoni* and related species (Ephemeroptera: Ephemerellidae).

McDunnough, J. 1931. The eastern North American species of the genus *Ephemerella* and their nymphs (Ephemeroptera). *Can. Entomol.* 63: 187-197, 201-216.

Tshernova, O.A. 1972. Some new species of mayflies from Asia (Ephemeroptera: Heptageniidae, Ephemerellidae). *Entomol. Rev. (Engl. Transl. Entomol. Obozr).* 51: 604-617.

A REVIEW OF THE NEARCTIC *HEPTAGENIA* (HEPTAGENIIDAE, EPHEMEROPTERA)

R.W. Flowers

Laboratory of Aquatic Entomology
Florida A&M University
Tallahassee, FL 32307

ABSTRACT

Nearctic *Heptagenia* is a polygeneric group consisting of true *Heptagenia* and a complex allied to Old World ecdyonurids. Nymphal mouthparts, adult eyes, genitalia and tarsi separate these ecdyonurids from true *Heptagenia*. The ecdyonurid complex is shown to consist of two new genera. One new genus is identical to Traver's *maculipennis* group while the other corresponds to the combined *lucidipennis* and *simpliciodes* groups. Since the *lucidipennis* and *simplicioides* groups are distinct only in the adult stage, each should receive subgeneric rank.

REVIEW

The world-wide genus *Heptagenia* Walsh contains 37 Nearctic species (Edmunds *et al.* 1976). Although widespread, often abundant, and familiar to most aquatic entomologists, they present many unresolved problems for the taxonomist, and the non-systematic biologist still cannot reliably identify many species.

The existence of species groups within Nearctic *Heptagenia* has long been known (Traver 1933). Based on the shape of the male genitalia, Traver recognized six groups: *flavescens*, *pulla*, *elegantula*, *lucidipennis*, *simplicioides* and *maculipennis*. She divided the nymphs into two groups; *flavescens-pulla-elegantula* lacking pectinate claws and with filaments on the 7th pair of gills, and *maculipennis-lucidipennis* with pectinate claws but lacking filaments on gill 7. Traver also noted variability in other nymphal characters, notably the mouthparts and abdominal spines. Although these groups have been used in keying out species, their taxonomic significance has not been fully appreciated.

Based on detailed morphological studies, Bogoescu and Tabacaru (1962) concluded that a number of European mayflies then assigned to *Heptagenia* should be placed in *Ecdyonurus*. They did this on the basis of nymphal mouthpart structure and of adult genitalia and tarsal ratios of the middle and hind legs. *Ecdyonurus* nymphs differ from *Heptagenia* nymphs by having the paraglossae of the labium and the superlinguae of the hypopharynx distinctly expanded. Adults of these two genera are distinguished by structure of the male genitalia and by tarsal ratios. Jensen (1972) used mouthpart characters, those listed above and others, to define an Old World phyletic line of five genera: *Afronurus*, *Ecdyonurus*, *Cinygmina*, *Thalerosphyrus*, and *Compsoneuriella*. Adult characters were not used in his phylogenetic analysis.

It is thus apparent that prior to the work of Bogoescu and Tabacaru, European *Heptagenia* was polygeneric. The structural diversity of Nearctic *Heptagenia* also indicates a polygeneric group and the present research is an attempt to clarify the composition of this assemblage. A detailed external morphological study was undertaken on representatives of all Traver's species groups in all life stages. The results show an interesting parallel between New and Old World *Heptagenia:* the same characters that led to the transfer of some European *Heptagenia* species to *Ecdyonurus* are found in many Nearctic *Heptagenia*. This clearly indicates that two distinct phyletic groupings, unnoticed for many years, existed within the taxon *Heptagenia*.

One phyletic grouping is represented by the *flavescens-pulla-elegantula* group which, as it includes the genotype of *Heptagenia* (*H. flavescens* (Walsh)) is *Heptagenia* s.s. As characterized below, *Heptagenia* s.s. agrees well with *Heptagenia* as defined by Bogoescu and Tabacaru (1962). The second phyletic grouping, represented by the *maculipennis*, *lucidipennis* and *simplicioides* groups, differs substantially from *Heptagenia* s.s. in both adult and nymphal characters. This grouping is part of the Old World ecdyonurid complex (*Afronurus*, *Ecdyonurus*, etc.), based on the condition of nymphal mouthparts and claws and the middle and hind tarsi of the adult. There are also striking similarities in nymphal color

---

Fig. 1-10. Nymphal mouthparts: 1-5, *Heptagenia flavescens;* 6-10, "A" *hebe*. 1,6, tip of left mandible, enlarged; 2,7, tip of right mandible, enlarged; 3,8, hypopharynx (a-ventral, b-dorsal); 5,9, left maxilla; 4,10, labium, palpi detached (a-ventral, b-dorsal).

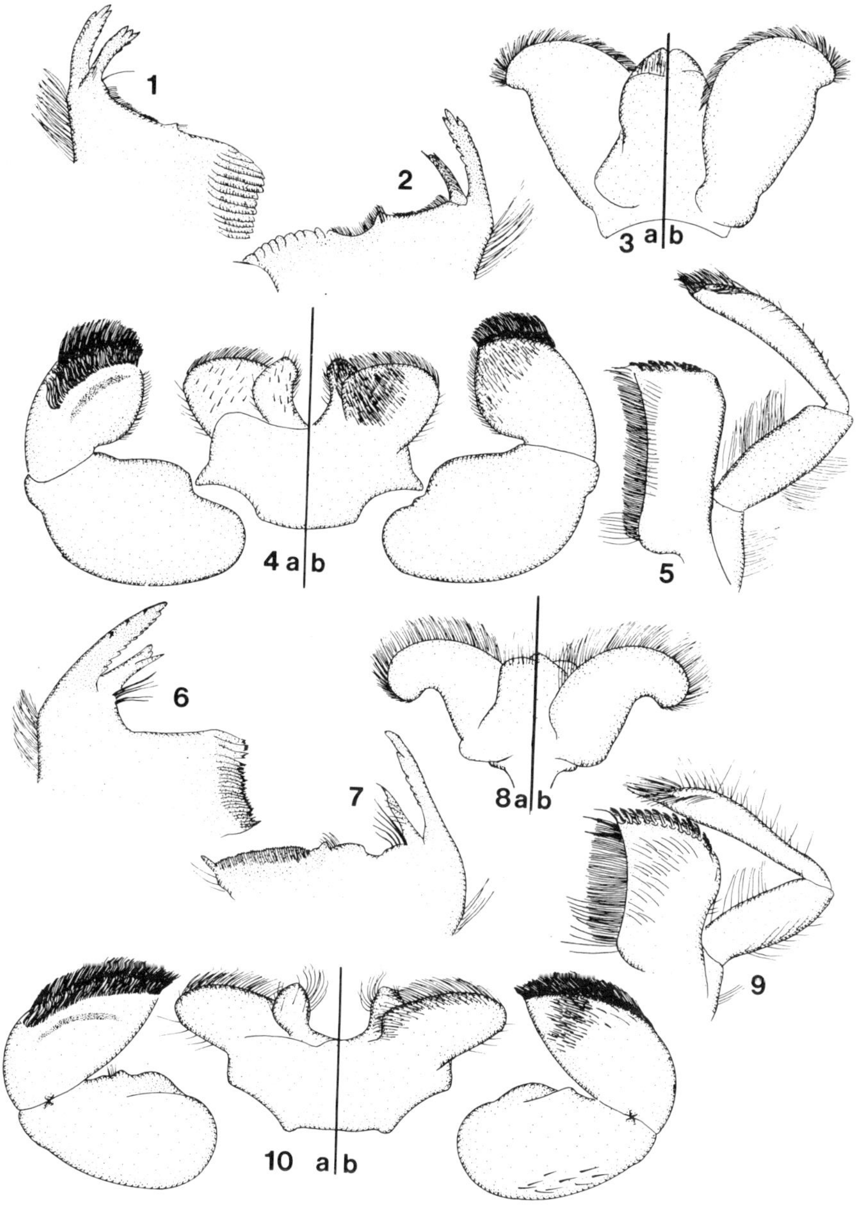

Figures 1 - 10

patterns between many New and Old World members of this complex.

The New World ecdyonurid complex differs from *Heptagenia* s.s. by the following combination of characters: in the adult, (1) first segment of hind tarsus subequal to second, (2) eyes of male meeting on vertex or separated by 3-5 times the width of median ocellus (Fig. 12,13), (3) male genitalia with large or small dorso-lateral spines, if large, discal spines present (Fig. 17-21); in the nymph, (1) mandible with 4-5 prosthecal setae and lacking small setae between molar and incisor areas (Fig. 6,7) (2) maxilla with ventral setae of galea-lacinia scattered (Fig. 9), (3) superlinguae of hypopharynx laterally elongated and curved (Fig. 8), (4) glossae oval, stalked, not bent in middle, paraglossae strongly expanded laterally, apical segment of labial palp elongated and acutely pointed (Fig. 10), (5) 7th abdominal gill lacking filamentous tuft, (6) tarsal claws pectinate, (7) postero-lateral spines on abdominal segments 3-8. In New World *Heptagenia* s.s., these characters are as follows: in the adult, (1) first segment of hind tarsus usually shorter than second, if subequal, then first segment of middle tarsus shorter than second, (2) eyes of male separated on vertex by 1-1.5 times the width of median ocellus (Fig. 11), (3) male genitalia with large dorso-lateral spines but no discal spines (Fig. 14-16); in the nymph, (1) mandible has one or no prosthecal setae and a row of small setae between incisor and molar areas (Fig. 1,2), (2) maxilla with ventral setae of galea-lacinia in a sub-median row (Fig. 5), (3) superlinguae of hypopharynx short, notched on lateral margin (Fig. 3), (4) glossae curved at middle, not stalked, paraglossae moderately expanded laterally, apical

Table 1. Composition of *Heptagenia* s.s. and the ecdyonurid complex in North America

Heptagenia s.s. - (*flavescens-pulla-elegantula* group)

Ecdyonurid complex

- New Genus A. - (*maculipennis* group)
- New Genus B. - (*lucidipennis* - *simplicioides* group)
  - New Subgenus 1. - (*lucidipennis* group)
  - New Subgenus 2. - (*simplicioides* group)

segment of labial palp not elongated (Fig. 4), (5) 7th abdominal gill with filamentous tuft, (6) tarsal claws lacking pectines, (7) postero-lateral abdominal spines absent or on segment 8 only.

The New World ecdyonurid complex divides into two new genera; one (New Genus A) represented by Traver's *maculipennis* group and the other (New Genus B) by her *lucidipennis* and *simplicioides* groups (Table 1). Formal descriptions of these genera are in preparation for publication elsewhere.

New Genus A adults can be recognized by the wide separation of the male eyes (Fig. 12) and the presence of discal and large dorso-lateral spines on the penes (Fig. 17,18). In all species except "A" *umbratica* the costal and sub-costal cross veins are heavily margined with brown. Eggs of this genus are similar to those of *Heptagenia* s.s. (Fig. 22) in having knob-terminated coiled threads (KTC's of Koss and Edmunds 1974) either concentrated at the poles or evenly distributed; the chorion has fine granular sculpture (Fig. 23, 24). Nymphs of New Genus A differ from those of the other New World ecdyonurids by the absence of setae of the caudal filaments. The head capsule often has dark granulations on the frontal margin and is slightly wider than the pronotum.

New Genus B can be recognized in the adult stage by the male eyes which meet on the vertex (Fig. 13). The penes have small dorso-lateral spines (Fig. 20,21) and segments 3 and 4 of the forceps are longer than in either New Genus A or *Heptagenia* s.s. (Fig. 19). The eggs of New Genus B (Fig. 25-29) have the KTC's evenly distributed, sometimes densely packed, with the chorion sculptured with ridges or small knobs. Nymphs of this genus have fine setae on the caudal filaments and the head capsule does not exceed the width of the pronotum or have dark granulations on the frontal margin.

Except for color patterns, no reliable characters distinguish the nymphs of the *lucidipennis* and *simplicioides* groups. However, these two groups are differentiated by characters of the eggs and genitalia. The penes of the *lucidipennis* group have non-divergent lobes and thick, spiculate median titilators (Fig. 19,20). The penes of the *simplicioides* group are divergent and the titilators are slender and smooth (Fig. 21). All *lucidipennis* group eggs studied so far are readily distinguished by the large-mesh reticulation of the chorion (Fig. 25-27). This feature has been previously known only from the *Hexagenia* complex in North America (Koss 1968). Unlike *Hexagenia*, however, *lucidipennis* eggs have a KTC in each mesh. The *simplicioides* group eggs lack a reticulated mesh. One species ("B" *simpliciodes)* has an irregular system of ridges (Fig. 28), while another species ("B" *criddlei)* has only small knobs between densely packed KTC's (Fig. 29). Based on the criteria of Edmunds (1962) and Peters and Edmunds (1970), the *lucidipennis* and *simplicioides* groups should be considered separate subgenera of New Genus B (Table 1).

Except for the California "B" *kennedyi* (provisionally included here on the basis of adult male characters), Subgenus 1 occurs east of the Great Plains. Subgenus 2 is found from the Rocky Mountain region westward with one species ("B" *salvini*) in northern Mexico.

New Genus A is distributed throughout North America.

As far as can be determined from available specimens and the literature the genera proposed for the New World ecdyonurid complex are endemic to North America. European *Ecdyonurus* have some derived characters in common with New Genus B, and *Afronurus* nymphs (particularly some small Asian species) are strikingly similar to nymphs of New Genus A, a similarity that even extends to color patterns. However, in both these cases the New World genera differ in other important characters and there is at present no reason for classifying New Genus A or B with any currently known Old World genera. A geographic region worthy of particular study is the eastern Siberian and Japanese area. Several *Ecdyonurus* illustrated by Imanishi (1936) could possibly be allied to the New World fauna, and Bajkova (1972), describing an adult Siberian *Rhithrogena*, gave a figure of the male genitalia which suggests that her species may be allied with the *simplicioides* group.

Until more is known about Old World *Ecdyonurus* and allied genera, it is premature to speculate further on the generic affinities and phylogeny of the New World ecdyonurids.

ACKNOWLEDGMENTS

I sincerely thank Drs. David R. Towns, Manuel L. Pescador and William L. Peters and Mrs. William L. Peters for their help and useful suggestions. I also thank Mr. William Miller, Electron

---

Fig. 11-13. Eyes of male, dorsal view: 11, *Heptagenia flavescens;* "A" *hebe;* "B" *lucidipennis* group.
Fig. 14-21. Male genitalia and lobes of penes. Fig. 14,17,19, ventral view of genitalia: 14, *Heptagenia flavescens;* 17, "A" *thetis;* 19, "B" *lucidipennis* group.
Fig. 15-16,18,20-21, lobes of penes, enlarged: 15-16, *Heptagenia pulla* (lateral and ventral views); 18, "A" *hebe;* 20, "B" *horrida;* 21, "B" *simplicioides* all ventral views).

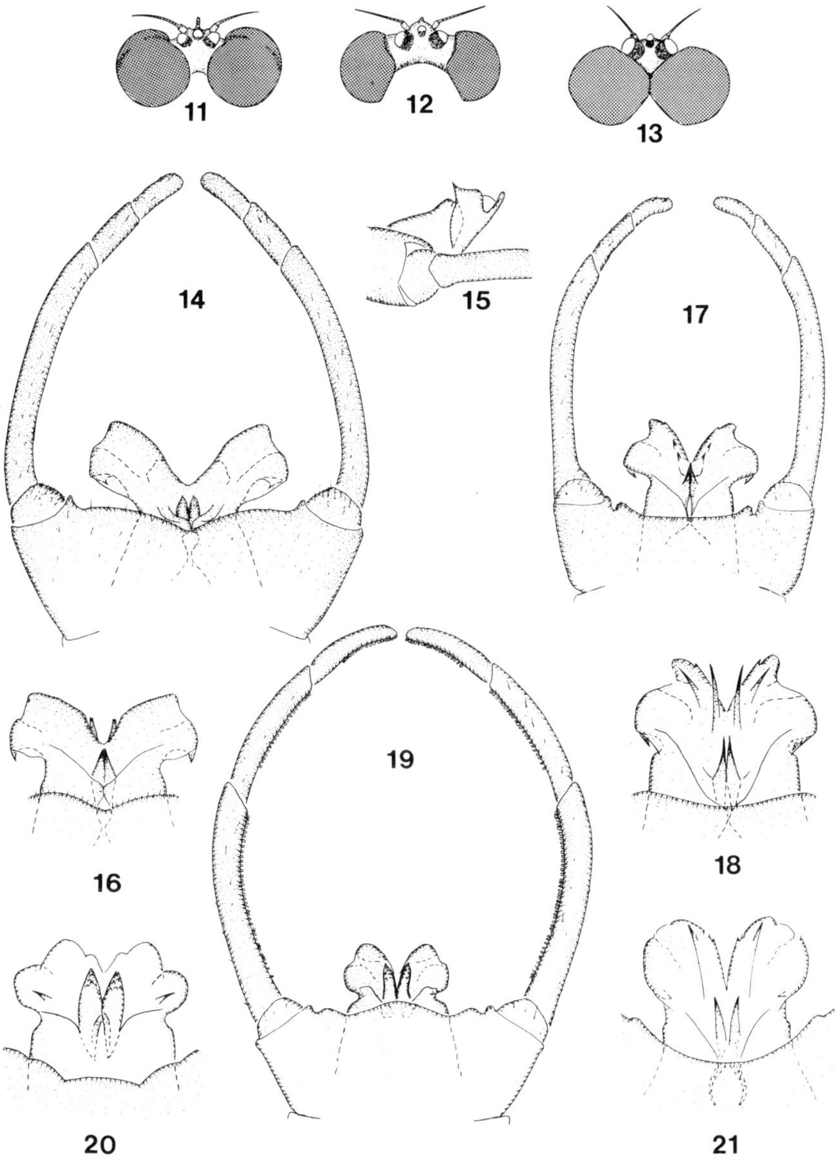

Figures 11 - 21

Microscope Laboratory, Florida State University, for his assistance in the SEM work. This research was supported by a research project (FLAX 79009) of SEA/CR, USDA, to Florida A&M University, William L. Peters, Research Leader.

RESUME

L'*Heptagénia* néarctique est un ordre polygénérique comprenant l'*Heptagénia* véritable et d'autres genres associés aux ecdyonurides de l'ancien monde. Les pièces buccales des larves, les yeu des adultes, les organes génitaux et les tarses distinguent ces ecdyonurides de l'*Heptagénia* véritable. Le complexe ecdyonuride compte deux genres. L'un, récent, est identique au groupe *maculipennis* de Traver, l'autre correspond au tandem de *lucidipennis* et *simplicioides*. Comme ces deux derniers ne se différencient qu'à l'état adulte, ils devraient tous deux être classés dans un sous-genre.

ZUSSAMENFASSUNG

Die nearktische *Heptagenia* ist eine polygenerische Gruppe, die aus echten *Heptagenia* und einem weiteren Komplex besteht, der verwandt ist mit Ekdyonuriden der alten Welt. Mundteile, der Nymphen, Augen der reifen Tiere, Genitalien und Fußwurzeln unterscheiden diese Ekdyonuriden von den echten *Heptagenia*. An Hand dieser Arbeit wird gezeigt, daß der Ekdyonuriden Komplex aux zwei neuen Gattungen besteht. Eine dieser neuen Gattungen ist identisch mit Travers *maculipennis* Gruppe, während die andere mit den kombinierten *lucidipennis* und *simplicioides* Gruppen übereinstimmt. Da die *lucidipennis* und *simplicioides* Gruppen nur im ausgewachsenen Stadium verschieden sind, sollte jede ven ihnen den Rang einer Untergattung erhalten.

Fig. 22-29. Eggs: 22, *Heptagenia flavescens* (425X); 23, *aphrodite* (550X); 24, "A" *juno* (525X); 25, "B" *inconspicua* (475X); 26-27 "B" *lucidipennis* group (450 and 800X); 28, "B" *simplicioides;* 29; "B" *criddlei.*

Figures 22 - 29

REFERENCES

Bajkova, O. Ya. 1972. K poznaniyu podenok (Ephemeroptera) basseina Amura: II. Imagines (*Rhithrogena, Heptagenia*). *Izv. Tikhookean. Nauchno-Issled. Inst. Rybn. Khoz. Okeanogr.* 77: 207-232.

Bogoescu, C. and I. Tabacaru. 1962. Beitrage zue Kenntnis der Untersuchungsmerkmale zwischen den Gattungen *Ecdyonurus* und *Heptagenia* (Ephemeroptera). *Beitr. Entomol.* 12: 273-291.

Edmunds, G.F., Jr. 1962. The principles applied in determining the hierarchic level of the higher categories of Ephemeroptera. *Syst. Zool.* 11: 22-31.

Edmunds, G.F., Jr., S.L. Jensen and L. Berner. 1976. The mayflies of North and Central America. Univ. Minnesota Press, Minneapolis.

Imanishi, K. 1936. Mayflies from Japanese torrents VI. Notes on the genera *Ecdyonurus* and *Rhithrogena*. *Annot. Zool. Jpn.* 15: 538-549, pl. 39.

Jensen, S.L. 1972. A generic revision of the Heptageniidae of the world (Ephemeroptera). Ph.D. thesis. Univ. of Utah.

Koss, R.W. 1968. Morphology and taxonomic use of Ephemeroptera eggs. *Ann. Entomol. Soc. Am.* 61: 696-721.

Koss, R.W. and G.F. Edmunds, Jr. 1974. Ephemeroptera eggs and their contribution to phylogenetic studies of the order. *Zool. J. Linn. Soc.* 55: 267-349.

Peters, W.L. and G.F. Edmunds, Jr. 1970. Revision of the generic classification of the Eastern Hemisphere Leptophlebiidae (Ephemeroptera). *Pac. Insects.* 12: 157-240.

Traver, J.R. 1933. Heptagenine mayflies of North America. *J. N.Y. Entomol. Soc.* 41: 105-125.

JUBABAETIS GEN. N. AND *PLATYBAETIS* GEN. N., TWO NEW GENERA OF THE FAMILY BAETIDAE FROM THE ORIENTAL REGION

Ingrid Müller-Liebenau

Max-Planck-Institut für Limnologie
Abteilung Allgemeine Limnologie
August-Thienemann-Str. 2, D-232o Plön

ABSTRACT

Nymphs of two new genera are described in detail. *Jubabaetis* gen.n. collected by M.L. Pescador from the Philippines represents a more or less isolated genus within the family. *Platybaetis* gen.n. seems to be closely related to *Pseudocloeon* and is represented by three species. One species is found in Dr. Pescador's Philippine material. The second species was described in the nymphal stage by Ueno in 1955 from Nepal as "*Baetis* sp. 2". The third species is contained in a collection of baetid nymphs from Malaysia (Coll. Bishop, not yet published). Thus, this second new genus seems to be widespread but restricted to the oriental region.

The two new genera of Baetidae, *Jubabaetis* gen.n. and *Platybaetis* gen.n. discussed herein are contained in a small collection made by Dr. M.L. Pescador on the island of North Luzon (Philippines).

Jubabaetis Müller-Liebenau gen.n.

Type species: *Jubabaetis pescadori* Müller-Liebenau sp.n.

Mature nymph: Mouthparts and antennae covered by prolonged front margin of head. Front margin of prolongation with a dense row of posteriorly curled, completely unpigmented, fine bristles (hair-line and much finer than shown in Fig. 1b and c). Abdominal terga I - IX each with conspicuous, caudally directed, median hook near hind margin. Terminal filament reduced to 1 segment.

Imago ♂ and ♀ not known.

Etymology: The name is derived from the Latin word juba = mane of hair.

Jubabaetis pescadori Müller-Liebenau sp.n.

Material: 2 nymphs. Philippines: Mountain Prov., Sumigar stream, Sumigar Banaue, 3.X.1967, leg. M.L. Pescador.

Description: For morphological details see Fig. 1 - 4.

Body length: 11 mm, Cerci 14 mm, with very fine bristles at inner margin, terminal filament reduced to 1 segment. - Head: with shield-like prolongation on front margin entirely covering mouth parts including labrum. - Antennae inserted beneath "shield". - Abdomen: Terga I - IX each with a caudally directed hook. Longitudinal row of tube-like bristles similar to those on metatergum extending from anteromedial margin of each segment to median hook. Upper surface and hind margin of terga as in Fig. 3. - Legs: with dense row of long, fine bristles on outer margin of all three segments; all bristles very finely feathered (Fig. 4). Inner margin of tarsus apically with two conspicuous bristles of different lengths. - Gills: on segments 1 - 7; rounded posteriorly, with weak, smooth margin.

Holotype: 1 dissected nymph on three slide preparations. - Philippines: Mountain Prov., Sumigar stream, Sumigar Banaue, 3.X.1967, leg. M.L. Pescador; det. I. Müller-Liebenau 1980, Zool. Staatsinstitut und Zool. Mus., Abt. Entomol., Hamburg.

Paratype: 1 partly dissected nymph in alcohol, same locality as Holotype. Coll. M.L. Pescador.

I take pleasure in dedicating the new taxon to the collector.

Platybaetis Müller-Liebenau gen.n.

Type species: *Platybaetis edmundsi* Müller-Liebenau sp.n.

Figure 1. Nymph of *Jubabaetis pescadori* sp.n. a) nymph; b) anterolateral view of head; c) ventral view of head; d) metathorax with small hind wing pads; e) caudally directed median hooks of three abdominal segments; f) part of ventral surface of tibia; g) part of dorsal surface of tibia; h) gill; i) leg; j) tarsus and claw; k) paraproct; m) base of cerci.

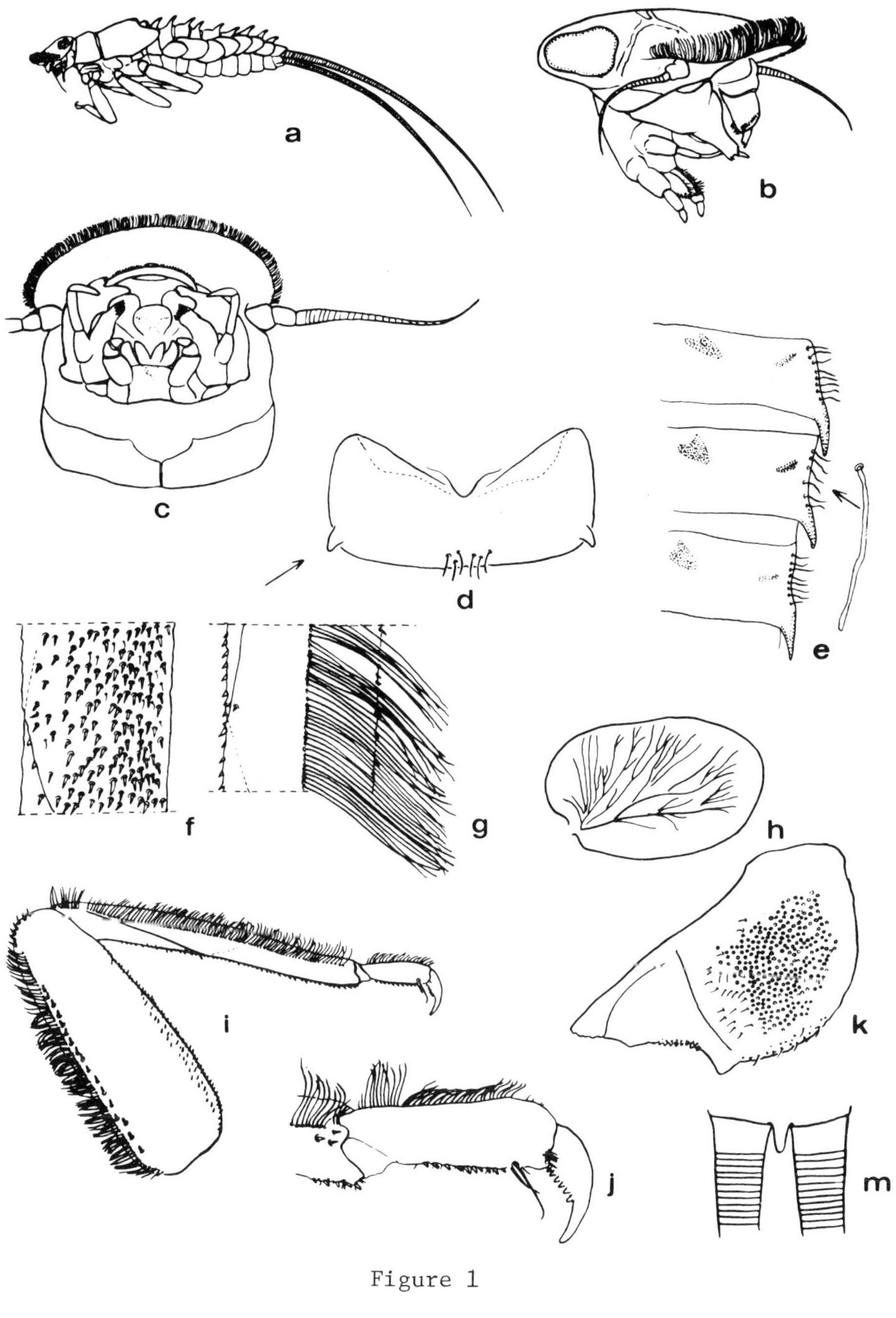

Figure 1

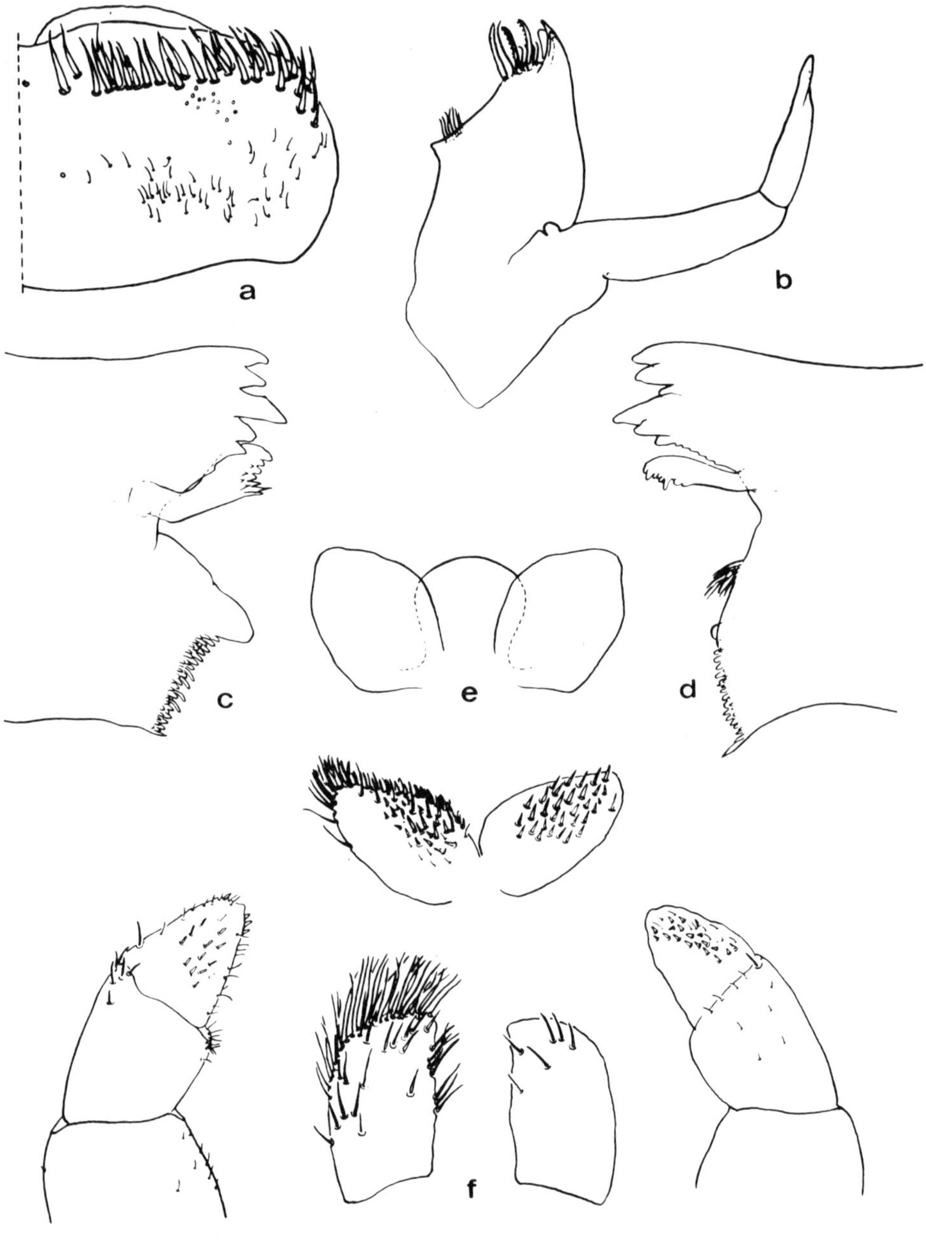

Figure 2. Nymph of *Jubabaetis pescadori* sp.n. a) right half of labrum; b) maxilla; c) canini and molar area of left mandible; d) canini and molar of right mandible; e) hypopharynx; f) dissected parts of labium, left dorsal, right ventral.

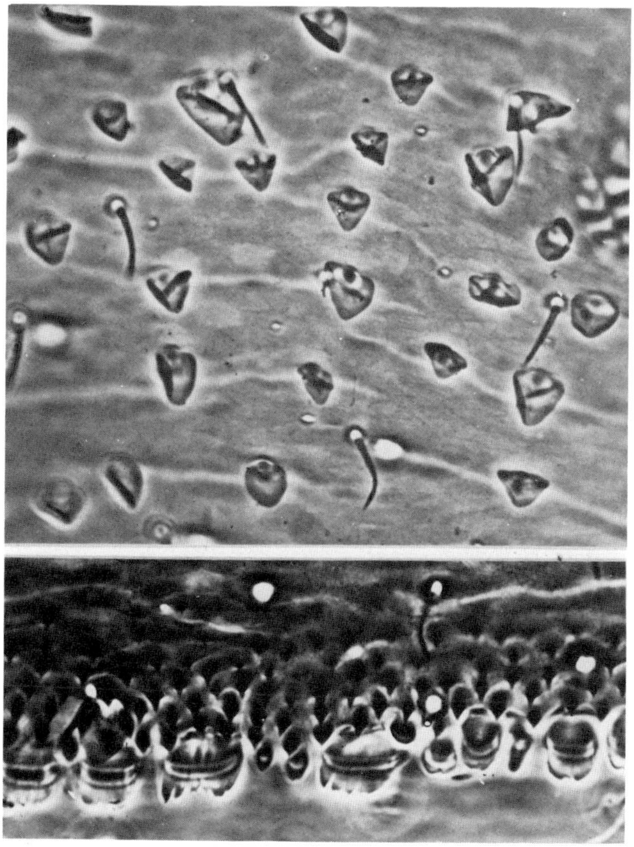

Figure 3. Surface and hind margin of nymphal tergum of *Jubabaetis pescadori* sp.n.

Mature nymph: Head barely wider than long, nearly quadrangularly shaped, nearly as wide as pronotum. - Antennae comparatively thick and short, a little longer than head capsule. Scape a little broader than long, pedicel cylindrical, considerably longer than broad, first segment of flagellum also cylindrical, about twice as long as broad. - Labrum very small compared to head capsule, about twice as wide as long, ventrally directed. - Maxillary palpus with three segments, not distinct in all specimens. - Glossae and Paraglossae stout, about twice as long as broad, glossa only a little narrower than paraglossa. Glossae near apex with two large clavate spines. - Pronotum about twice as wide as long, nearly as broad as head capsule, rounded laterally and on outer hind margin. - Head wing pads very small or lacking. - Caudal filaments fringed only on inner side. Terminal filament reduced to a few segments

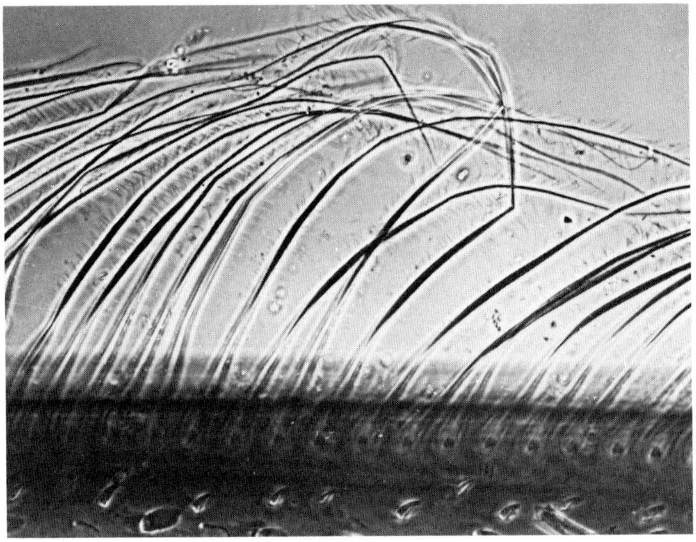

Figure 4. Nymph of *Jubabaetis pescadori* sp.n. section of outer margin of femur with feathered bristles.

only (from 1 segment up to 6 or 7; 1 segment in the species mentioned from Malaysia). - Paraprocts smooth at inner margin, without teeth. - Legs: relatively short, front leg shorter than mid- and hind-leg, tarsal segment and claw outwardly directed due to a considerable twisting along longitudinal seam on tibia. Dense row of long bristles along outer margin of femur, similar row on tibia and tarsus. Tarsus with a single long bristle near inner margin near apex. Claws with about 7 teeth. Body somewhat flattened, with gills somewhat projected laterally. Gills smooth, without teeth, only a few fine bristles, mostly on hind margin; with distinctly branched trachea.

Figure 5. Nymph of *Platybaetis edmundsi* sp.n. a) dorsal view of head; b) labrum; c) canini and molar area of left and right mandible, note mandibles of subsequent instar within worn down older pair just before moulting; d) base of antenna; e) dissected parts of labium left dorsal, right ventral; f) hypopharynx.

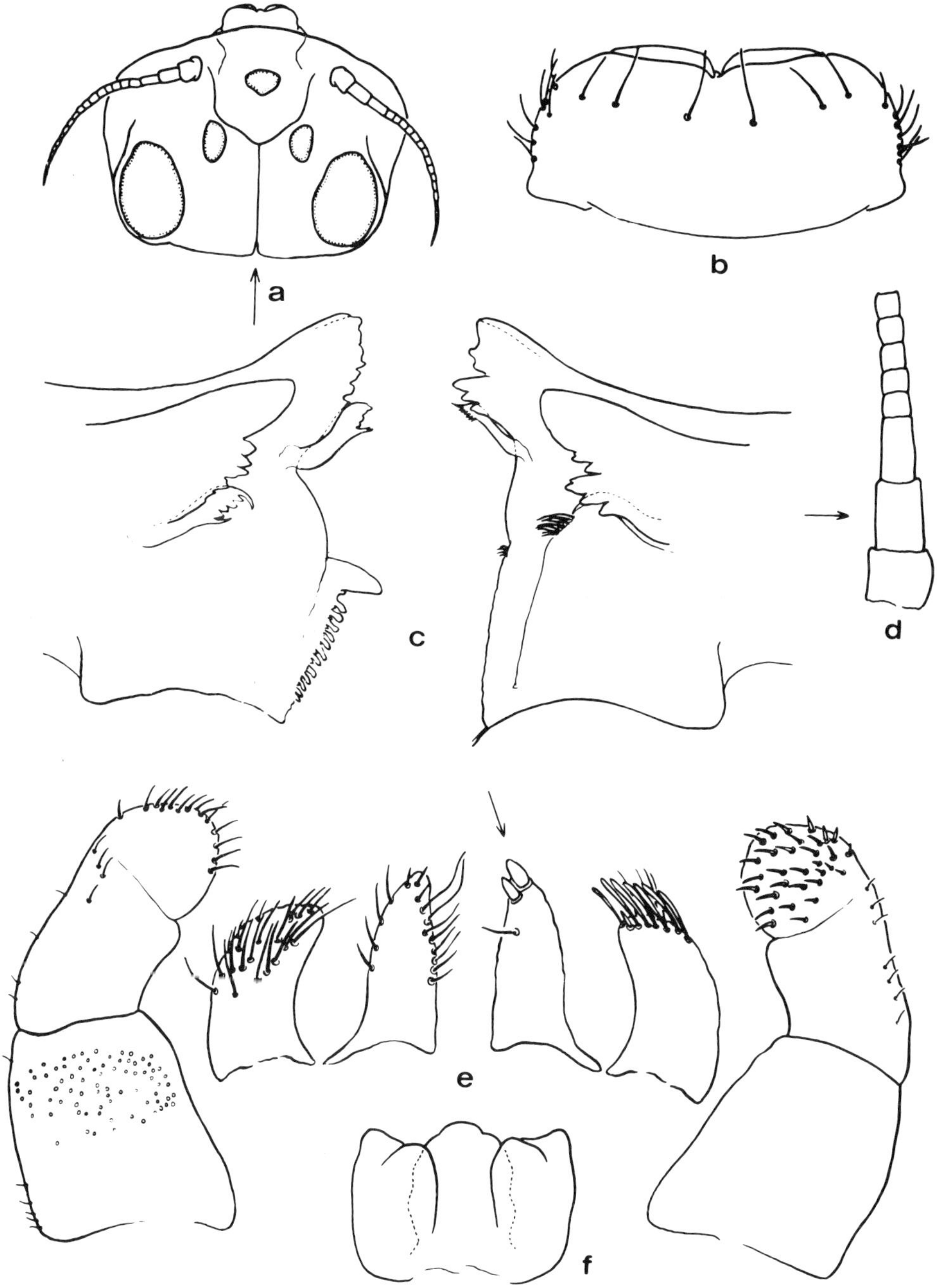

Figure 5

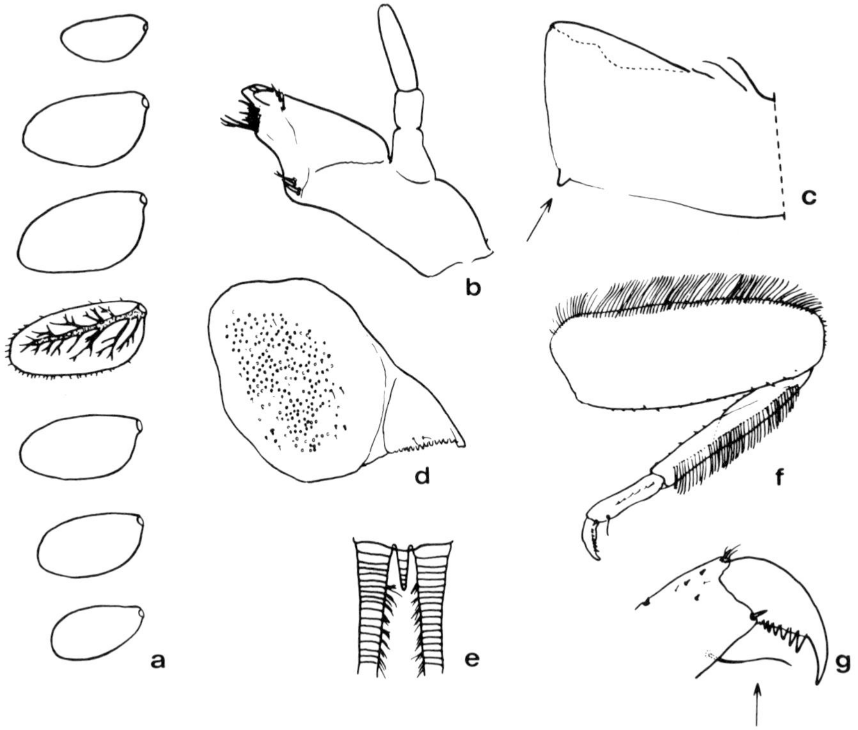

Figure 6. Nymph of *Platybaetis edmundsi* sp.n. a) gills I to VII, all gills tracheated as shown in gill IV; b) maxilla; c) left half of metathorax with small hind wing pad; d) paraproct; e) base of cerci; f) leg; g) apex of tarsus with claw.

Imago ♂ and ♀ not known.

Etymology: The name is derived from the Greek word platy = flat, wide.

Platybaetis edmundsi Müller-Liebenau sp.n.

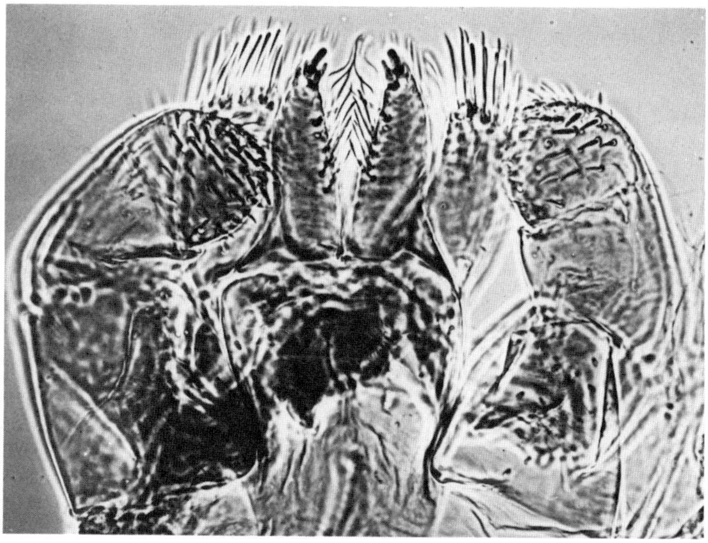

Figure 7. Nymph of *Platybaetis edmundsi* sp.n. labium (note large clavate spines near apex of glossae).

Material: 14 nymphs. Philippines: Mountain Prov., Sumigar stream, Sumigar Banaue, 3.X.1967, leg. M.L. Pescador.

Description: For morphological details see Fig. 5-9.

Body length: mature nymph 11.5 mm, Cerci 9.8 mm, terminal filament reduced to 5 segments. - Head: nearly quadrangularly shaped with a clear incision in the middle of hind margin; labrum small.

I would like to name the new species of the genus *Platybaetis* after the well respected mayfly worker Dr. G.F. Edmunds Jr. and will call it *Platybaetis edmundsi*.

Holotype: 1 nearly mature nymph in alcohol. - Philippines: Mountain Prov., Sumigar stream, Sumigar Banaue, 3.X.1967, leg. M.L. Pescador, det. I. Müller-Liebenau 1980. - Zool. Staatsinstitut und Zool. Mus., Abt. Entomol., Hamburg. - Paratypes: 11 nymphs of different sizes in the collection of Dr. M.L. Pescador; 2 slide preparations in my collection. All of same locality as Holotype.

Dr. Pescador gave me the following information on the collecting site: small, swift flowing, rock-bottom mountain stream at an approximate altitude between 1,600 m - 2,200 m above sea level.

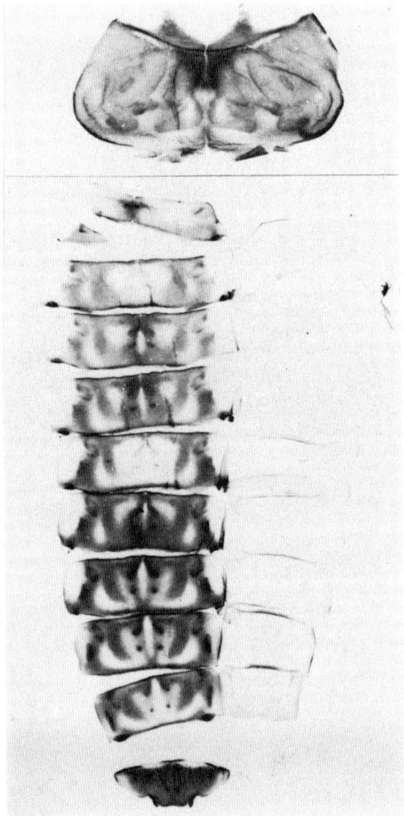

Figure 8. Nymph of *Platybaetis edmundsi* sp.n., pronotum and abdomen (pronotum at higher magnification).

The nymphs were found in cracks of submerged logs on debris trapped between rocks. No records of water temperature during time of collection are available. The whole Mountain Province is the coldest region of the Philippines with an annual average of ca. 18°C air temperature.

Two more species of this new genus are known. One was described from Nepal in a paper by Ueno (1955: 305/306) dealing with the results of a Japanese scientific expedition to the Himalayas in 1952-1953. Ueno gave a detailed description of a nymph, which he left unnamed and referred to as "*Baetis* sp. 2". - In honour of the aforementioned author the species is named *Platybaetis uenoi* sp.n.

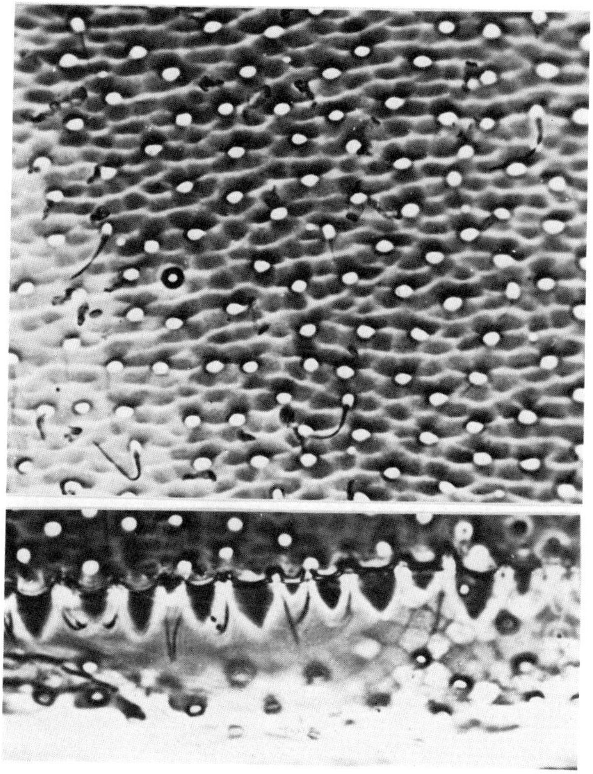

Figure 9. Surface and hind margin of nymphal tergum of *Platybaetis edmundsi* sp.n.

The third species which was collected by Dr. J.E. Bishop in the Gombak river, Malaysia, will be treated in a later paper.

DISCUSSION

It is my opinion, that the three known species of this new genus do not fit into the genus *Pseudocloeon* or into any other related group (e.g. within the genus *Baetis*). As is well known, baetid species whose nymphs more or less resemble the genus *Pseudocloeon* established by Klapalek in 1905 are for many reasons very problematic with regard to generic integration. Therefore I think it is convenient to establish a new genus for these three species according to a few remarkable characters common to them.

ACKNOWLEDGMENTS

I am very thankful to Dr. M.L. Pescador, Florida A&M University who generously gave me his material from the Philippines for study. I also want to express my thanks to Dr. G.F. Edmunds Jr., University of Utah, for helpful comments.

RESUME

Les larves de deux nouveaux genres sont traitées en détail. La *Jubabaetis* gen.n. recueilli par M.L. Pescador aux Philippines, représente un genre plus ou moins isolé appartement à la famille. Le *Platybaetis* gen.n. semble être apparenté de près au *Pseudocloeon* et est représenté par trois espèces. La première espèce se trouve dans le matériel glané aux Philippines par M. Pescador. La seconde a été décrite en 1955 par Ueno (du Népal) à son stade larvaire comme "*Baetis* sp. 2". La troisième espèce fait partie d'une collection de larves de baetidés de la Malésie (collection Bishop, inédite). Ainsi, ce genre nouveau semble être répandu mais limité à la région orientale.

ZUSSAMENFASSUNG

Nymphen zweier neuer Gattungen werden in dieser Arbeit im Detail beschrieben. *Jubabaetis* gen.n., von M.L. Pescador auf den Philippinen gesammelt, stellt eine mehr oder weniger isolierte Gattung innerhalb der Familie dar. *Platybaetis* gen.n. scheint mit *Pseudocloeon* eng verwandt zu sein und wird durch drei Arten vertreten. Eine Art kommt in Dr. Pescadors philippinischem Material vor. Die zweite Art wurde im Nymphenstadium 1955 von Ueno aus Nepal als "*Baetis* sp. 2" dargestellt. Die dritte Art ist in einer Sammlung von Baetidennymphen von Malaysia enthalten. (Sammlung Bishop, noch nicht veröffentlicht). So hat es den Anschein, daß diese zweite, neue Gattung weitverbreitet aber auf den Orient beschrankt ist.

REFERENCES

Klapalek, F. 1905. Plecopteren und Ephemeriden aus Java, gesammelt von Prof. K. Kraepelin 1904. *Mitt. Naturhist. Mus. Hamburg* 22: 103-107.

Ueno, M. 1955. Mayfly nymphs. *In:* Fauna and Flora of Nepal Himalaya. - *Scientific Results of the Japanese Expedition to Nepal Himalaya 1952-1953.* 1: 301-316.

UN CHANGEMENT DANS LA TAXONOMIE DE LA FAMILLE BAETIDAE

Maria Keffermüller

Institute of Systematic Zoology
Adam Mickiewicz University
Fredry 10, 61-701 Poznan, Poland

RESUME

Il faut supprimer le genre *Pseudocloeon* Klapalek et inclure son espèce dans le genre *Baetis* Leach. En effet, toutes les caractéristiques reconnues comme typiques de la *Baetis* tant chez les nymphes que chez les formes ailées se retrouvent dans le *Pseudocloeon*, sauf qu'elles ont en plus des ailes postérieures. Ces ailes postérieures se retrouvent aussi dans toutes les formes intermédiaires depuis celles de la *Baetis* qui sont parfaitement développées jusqu'à leur absence complète chez le *Pseudocloeon*. Il faut donc exclure les ailes postérieures des caracteristiques diagnostiques du genre *Baetis*.

Chez les éphéméroptères ♂♂ il existe, en principe, un organe extérieur de copulation dont la construction constitue un trait taxonomique important. Cependant chez les genres peu nombreux, comme *Baetis* p. ex., on ne trouve pas cet organe, et des organes génitaux on ne prend en considération que la forme des forceps. Néanmoins, ici aussi, il existe des sclérites intérieurs génitaux, dit les "bras péniaux" (Grandi 1960), constituant la base de cet organe. D'un bout ils sont attachés dans le coin postérieur du IXe segment où ils se touchent au coin extérieur de la base des forceps, leur autre bout, chez le genre *Baetis* est placé librement, paraît-il, dans la cavité du corps (Fig. 1). Chez les deux espèces *Baetis* du groupe *atrebatinus* j'ai constaté (Keffermüller 1972) qu'avec l'écartement des "bras" en arrière du corps, leurs bouts libres s'avancent à l'extérieur étendant sur eux-mêmes un sac membraneux, au sommet duquel s'évadent les *ducti ejaculatorii*. C'est pourquoi j'ai considéré cet organe comme le pénis.

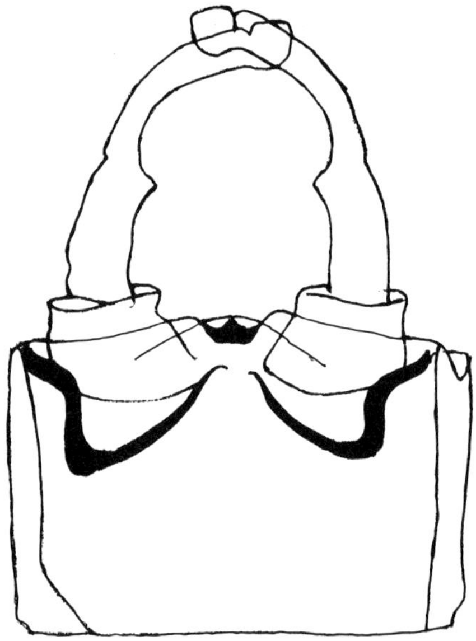

Figure 1. *Baetis calcaratus* Keffermüller
appareil genital de ♂ imago, face ventrale
genitalia of ♂ imago, ventral view

Dans le travail déjà mentionné de Grandi (1960), il y a une revue illustrée de la construction des organes génitaux intérieurs de 23 espèces européennes des éphéméroptères, faisant partie de 8 familles. On y voit que la forme des bras péniaux est différente chez les différentes familles, mais elle est semblable à l'intérieur des genres. Autant que je sache, les différences ci-dessus n'étaient pas prises en considération jusqu'ici dans la taxonomie des éphéméroptères.

Parmi les représentants régionaux de la famille Baetidae, l'organe extérieur de copulation n'est pas formé chez les espèces du genre *Baetis* Leach*, chez le *Pseudocloen* Klapalek avec une seule espèce européenne *inexpectatum* Černova (Tshernova) ainsi que, peut-être, dans les genres *Baetopus* Keffermüller et *Raptobaetopus* Müller-Liebenau. A ce problème qui touche les deux dernières espèces je pense consacrer un travail à part, d'ailleurs ces deux genres apparentés entre eux prochement, se distinguent clairement de tous les autres, tant au stade ailé qu'au stade de larve. Par contre,

* à l'exception de *B. muticus* (L.) = *B. Pumilus* (Durmeister) (Grandi 1960).

Pseudocloeon inexpectatum dans les deux stades montre tous les traits du genre *Baetis*, duquel il se distingue seulement par manque des ailes postérieures. Cependent dans de certains groupes des espèces du genre *Baetis*, des ailes postérieures sont déjà bien fortement arriérées, surtout chez les ♀♀, et à leur tour chez les larves ♂ *Pseudocloeon inexpectatum* Černova - Sowa (1962) a trouvé là, où sont placées normalement des ailes postérieures, de petites écailles constituantes des noeuds réduits de ces ailes, qui ne se développent plus chez l'individu ailé. Par contre Edmunds *et al.* (1976) mentionnent l'existence en Amérique des espèces Baetidae dont les ♂♂ possèdent des ailes postérieures, tandis que des ♀♀ non. Je crois que la présence ou la manque de l'organe étant à l'étape de la réduction, ne peut pas décider de la distinction de genre. Dans le genre *Baetis* le paracerque démontre pourtant toutes les étapes de la réduction, jusqu'au dépérissement y compris.

Déjà Bogoescu et Tabacaru (1957) ont fait remarquer une grande ressemblance des larves *Acentrella sinaica* Bogoescu et *Pseudocloeon hyalopterum* Bogoescu (Fig. 2-4), et avancent même une hypothèse que le genre *Acentrella* Bengtsson tire son origine du genre *Baetis* Leach. Le genre *Acentrella* a été éliminé par Müller-Liebenau (1965) et ses deux espèces ont été englobées au genre *Baetis*, où ils forment

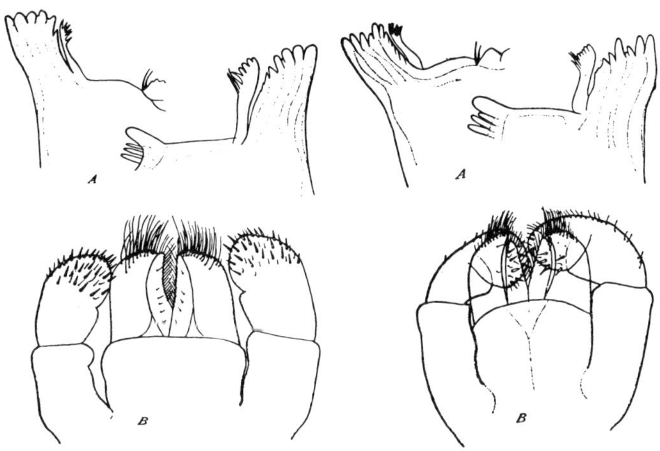

Figure 2a. *Baetis sinaicus* (Bogoescu)
2b. *Pseudocloeon inexpectatum* Černova
- nymphe (selon Bogoescu et Tabacaru):
A = caninens des mandibules; B = labium.
- nymph (by Bogoescu and Tabacaru):
A = mandibular canine teeth; B = labium.

le groupe *lapponicus* (Bengtsson), (Müller-Liebenau 1969). Par contre *Pseudocloeon hyalopterum* a été jugé par Sowa (1975) comme plus jeune synonyme de *P. inexpectatum*. La connaissance de l'imago ♂ *P. inexpectatum* (Kazlauskas 1964; Keffermüller 1964) a montré, que ses forceps ressemblent beaucoup aux forceps chez les espèces *Baetis* du groupe *lapponicus*, surtout chez *B. sinaicus* (Fig. 5). En outre, les bras péniaux sont construits ici d'une façon typique pour le genre *Baetis*. Il me semble que justement ce trait-là doit être considérer comme une preuve importante de l'appartenance de ces trois espèces à un genre.

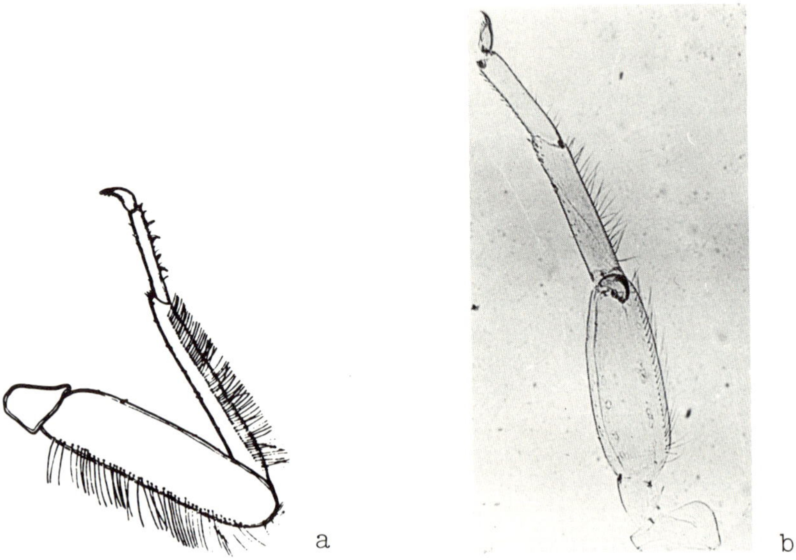

Figure 3a. *Baetis sinaicus* (Bogoescu), (selon Müller-Liebenau), (by Müller-Liebenau)
3b. *Pseudocloeon inexpectatum* Černova
- nymphe, patte
- nymph, leg

La construction des organes de copulation chez les insectes est généralement considérée comme un trait taxonomique très important, ce qui prouve Mayr (1974), p. 241: "...Les structures composèes ont la valeur plus grande que les structures simples, même s'il y a davantage de ces dernières. C'est l'une des raisons pour lesquelles les organes de copulation des Arthropodes ont une si grande importance taxonomique....La probabilité de la ressemblance convergente de ces organes chez les espèces peu apparentées est très petite."

Sur les image des génitaux ♂ des espèces de l'Amérique du Nord du genre *Pseudocloeon* Klapalek (Needham *et al.* 1935), on voit aussi de "bras péniaux" construits de la même facon que dans le genre *Baetis*, et la variabilité de la forme des forceps de ces espèces ne dépasse pas non plus des limites de variabilité qui se produisent dans ce genre-là.

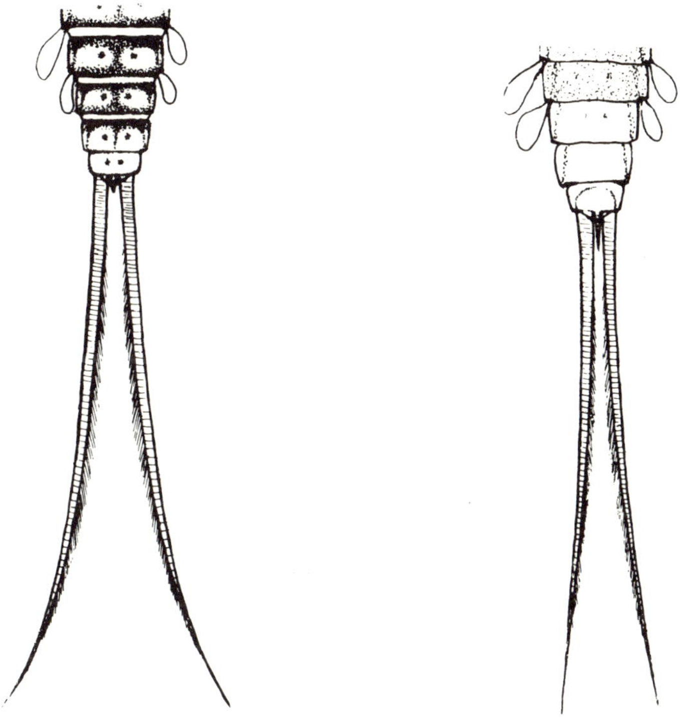

Figure 4a. *Baetis sinaicus* (Bogoescu)
4b. *Pseudocloeon inexpectatum* Černova
- nymphe, bout distal de l'abdomen
- nymph, end of abdomen

Donc, je suis d'avis qu'il faut éliminer le genre *Pseudocloeon* et faire entrer ses espèces dans le genre *Baetis*, ou bien le dégrader au moins au range de sous-genre: *Baetis* (*Pseudocloeon*) qui comprendrait aussi des espèces de l'ex-genre *Acentrella*. En tout cas, l'espèce européenne *inexpectatum* faut-il faire enter dans le group *lapponicus*, parce que toutes les deux espèces de ce group: *B. lapponicus* et *B. sinaicus* démontrent plus de traits communs avec lui (*inexpectatum*), qu'avec d'autres espèces du genre *Baetis*.*

* "Céla a remarqué déjà Kazlanskas (1972)."

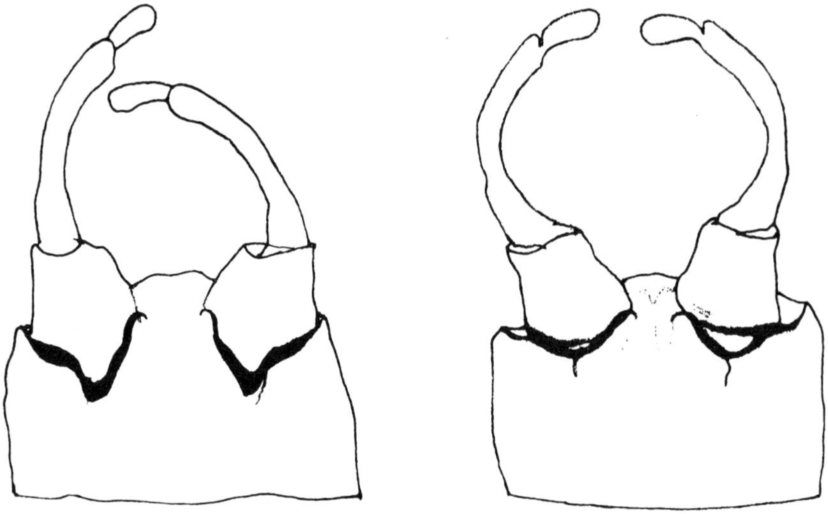

Figure 5a. *Baetis sinaicus* (Bogoescu), (selon Müller-Liebenau) (by Müller-Liebenau)
5b. *Pseudocloeon inexpectatum* Černova
- appareil genital de ♂ imago, face ventrale
- genitalia of ♂ imago, ventral view

Par conséquent, la présence des ailes postérieures doit être éliminée de l'ensemble des traits caractéristiques de genre *Baetis*.

ABSTRACT

It has become necessary to eliminate the genus *Pseudocloeon* Klapalek and include its species in the genus *Baetis* Leach. Since: all the characters known to be typical for *Baetis* in both nymphs and winged forms are also found in *Pseudocloeon*, with the exception of the existence of hind wings. All intermediate forms exist from the fully developed hind wings in *Baetis* to their complete reduction in *Pseudocloeon*. Thus hind wings must be excluded from the diagnostic characters for the genus *Baetis*.

ZUSSAMENFASSUNG

Die vorliegende Arbeit weist nach, daß der Gattungsname *Pseudocloeon* Klapalek eliminiert und die darin eingeordneten Arten in die Gattung *Baetis* Leach eingegliedert werden müssen. Alle Merkmale, die als typisch gelten für *Baetis*, sowohl fur Nymphen als auch geflügelte Formen, kommen auch bei *Pseudocloeon* vor, mit

Ausnahme der Hinterflügel. Diese existieren in allen Zwischenformen, von den vollentwickelten Hinterflügeln bei *Baetis*, bis zu ihrer völligen Reduzierung bei *Pseudocloeon*. Deshalb müssen die Hinterflügel ausgeschlossen werden von den diagnostischen Merkmalen der Gattung *Baetis*.

OUVRAGES CITES

Bogoescu, C. et I. Tabacaru. 1957. Etude comparée des nymphes d'*Acentrella* et de *Pseudocloeon*. *Beitr. Entomol.* 7: 483-492.

Edmunds, G.F., Jr., S. Jensen et L. Berner. 1976. The Mayflies of North and Central America. Univ. Minnesota Press, Minneapolis.

Grandi, M. 1960. Ephemeroidea (Fauna d'Italia, V. III), Bologna.

Kazlauskas, R. 1964. Materialy k poznaniju reki Oki. *Tr. Zool. Inst. Akad. Nauk SSSR.* 32: 163-176.

Kazlauskas, R. 1972. Neues über das system der Eintagsfliegen der familie *Baetidae* (Ephemeroptera). Proc. XIII Int. Entomol. Congr., Moscow, 1968, Leningrad, 111, p.337-338.

Keffermüller, M. 1964. Uzupelnienie bedań nad fauna jetek (Ephemeroptera) Wielkopolski. *Bad. fizj. nad Polska Zach., Poznań* 14: 69-86.

Keffermüller, M. 1972. Badania nad fauna jetek (Ephemeroptera) Wielkopolski. IV. *Pr. Kom. biol. Pozn. T.P.N., Poznań* 35(4): 1-30.

Mayr, E. 1974. Podstawy systematyki zwierzat. Warszawa.

Müller-Liebenau, I. 1965. Revision der von Simon Bengtsson aufgestellten *Baetis*-Arten (Ephemeroptera). *Opusc. Entomol.* 30: 79-123.

Müller-Liebenau, I. 1969. Revision der europäischen Arten der Gattung *Baetis* Leach, 1815 (Insecta, Ephemeroptera), *Gewässer Abwässer* 48/49: 1-214.

Needham, J., J. Traver et Y. Hsu. 1935. The Biology of Mayflies, with a systematic account of North American species. Ithaca, New York.

Sowa, R. 1962. Materialy do poznania Ephemeroptera i Plecoptera w Polsce. *Acta Hydrobiol.* 4: 205-224.

Sowa, R. 1975. Ecology and biogeography of mayflies (Ephemeroptera) of running waters in the Polish part of the Carpathians. 1. Distribution and quantitative analysis. *Acta Hydrobiol.* 17: 223-297.

THE HABITATS OF BRITISH EPHEMEROPTERA

Peter S. Maitland

Institute of Terrestrial Ecology
78 Craighall Road
Edinburgh, Scotland

ABSTRACT

Forty-seven species of Ephemeroptera occur in Great Britain. This is many fewer than in Europe as a whole. During synoptic studies of freshwater ecosystems in Britain over 1000 different waters have been surveyed. Collections of zoobenthos were made at all sites and many included Ephemeroptera. A selection of the data have been analysed to describe quantitatively the habitats of British Ephemeroptera and the nature of the communities and ecosystems in which they do and do not occur. A small number of species are so rare that it is not possible to describe their habitats. Of the commoner species, some occur in a variety of waters, others in one particular type. The defining factors appear to be related to water flow, temperature regime, chemistry, exposure and other complex variables, some of them biotic. Association and other multiple analyses show that, although there are transitions, certain communities of Ephemeroptera can be defined. The value of the study in both the applied and the theoretical fields is discussed.

INTRODUCTION

This paper is an attempt to describe how the habitats of the British species of Ephemeroptera may be defined in quantitative terms, using data available from a large number of sites all over Great Britain. There is now, too, a considerable literature available on the ecology of the British Ephemeroptera covering their taxonomy (e.g. Macan 1949, 1950, 1951, etc.), distribution (e.g. Maitland 1965, Wise 1976), life cycles (e.g. Elliott 1967, Brittain 1972), food (e.g. Brown 1961) and general ecology (e.g. Minshall and Kuehne

1969). General descriptions of the habitat are included in many of these papers, but there is rarely any attempt to quantify these for individual species. The recent key to the nymphs of British Ephemeroptera by Macan (1979) summarises much of the available information and describes the habitat of each species in a general way.

METHODS

This study forms a small part of an extensive synoptic survey of the fresh waters of Great Britain, which is being carried out at present (Maitland 1979). The primary aim of this research is to provide a scientific background for the interpretation and classification of aquatic ecosystems in Great Britain, but there are a number of additional valuable facets to the work. These include the development of autecological and synecological studies of various important groups of organisms such as Ephemeroptera.

The initial phases of this research, which have mostly been completed, are desk studies aimed at describing the number and variety of fresh waters (Smith and Lyle 1979), the range of species of plants and animals which inhabit them (e.g. Maitland 1977), and the published literature relating to freshwater sites and organisms in Great Britain. Thus a comprehensive background is now available for detailed field studies of the type described below. For example, it is now known that the total number of standing waters in Great Britain is approximately 60,000, and of running waters approximately 8,000. About 3,800 species of freshwater animals (excluding Protozoa) occur in Great Britain, 47 of which are Ephemeroptera.

The field data for this paper origniate from three separate synoptic surveys of British fresh waters. The first of these followed a decision by the Nature Conservancy to make a thorough scientific appraisal of ecosystems of all types over the whole of Great Britain, in order to identify as certainly as possible those sites of outstanding value (Ratcliffe 1977). The initial selection of freshwater sites was based on a desk study of existing information on sites of all kinds. These were then classified according to size, depth and trophic status and most were visited between 1968 and 1970 to collect new data.

In 1974, a synoptic survey of the fresh waters of Shetland was initiated. Following a comprehensive study of topographical (1:63, 360) and geological maps, which located 1,596 standing waters and 1,970 running waters, a principal component analysis was carried out using the correlation matrix derived from values of a number of measured variables (e.g. altitude, size, catchment size, etc.). This analysis gave 16 clusters (plus 14 aberrant sites) and 14 clusters (plus 9 aberrant sites) for standing and running waters respectively. For field study a random sample of three sites from each cluster together with the aberrant sites gave a total of 113 sites which were visited for sampling during the spring and autumn of 1974.

A third synoptic study was carried out during 1977 and 1978, on this occasion to survey the fresh waters of the Tayside Region of central Scotland. Initially an analysis and count of all the waters shown on 1:50,000 maps of the area concerned was carried out. This gave a total of about 950 standing waters and 10,300 stream segments. These were then classified using the following variables: geology (base poor, intermediate, base rich), size (five categories) and altitude (four categories). For field study, random selections of two sites from each class were made and this gave a total of 125 sites which were visited for sampling in the spring and autumn of 1977. A further 19 sites were studied during 1978.

In addition to comparable map information assembled for all the sites mentioned above, standard techniques were used in the field to collect data or samples covering a wide number of variables. These included physical features (e.g. substrate type, current speed, etc.), water chemistry (samples were deep frozen and taken back to the laboratory for the analyses of main ions), phytoplankton, macrophytes, zooplankton and zoobenthos. Samples of the latter were made in shallow water with a standard hand-net for a known period of time using a consitent collecting technique (Maitland 1966). In deeper water, an Ekman grab or a small dredge was used. These samples were preserved in the field using 4% formaldehyde and then sorted, counted and identified in the laboratory using, for Ephemeroptera, mainly the keys provided by Macan (1949, etc; 1961).

RESULTS

Even for a small group like Ephemeroptera it is only possible within the confines of this short paper to give a brief outline of some of the results of these studies and how they are being interpreted. As explained above, the main objectives of the overall programme are a comprehensive interpretation and classification of aquatic ecosystems in Great Britain using physical, chemical, botanical and zoological features. A computer study is at present being organised with these objectives. Selected data, such as those for Ephemeroptera, are also being subjected to a variety of individual analyses.

General Ephemeroptera Habitats

Although entirely aquatic for most of their life history, Ephemeroptera as a group are very versatile and occupy a wide range of freshwater habitats, both running and standing. However,

Table 1. The percentage occurrence of Ephemeroptera at freshwater sites in Great Britain as a whole, and in two specific areas of Great Britain - Tayside and Shetland.

Area	Number of Ephemeroptera Species Recorded															Number of Sites
	0	1	2	3	4	5	6	7	8	9	10	11	12	13	14	
Great Britain (38 species)	19.8	18.5	15.4	16.4	9.7	8.1	2.7	2.7	2.7	1.7	1.0	0.3	0.3	0.7	0	300
Tayside (28 species)	11.2	19.2	14.4	12.8	8.0	14.4	6.4	7.2	2.4	2.4	0.8	0.8	0	0	0	125
Shetland (6 species)	56.9	31.9	11.2	0	0	0	0	0	0	0	0	0	0	0	0	116

they are not found at all under certain conditions and it is very relevant to an understanding of the general ecology of the group to see from what kinds of habitats they appear to be excluded. Table 1 shows the general occurrences of Ephemeroptera among sites examined during the three surveys described above. It can be seen that there is a considerable difference between Great Britain and Tayside on the one hand, and Shetland on the other.

Only six species were recorded from Shetland, and most of the sites studied (56.9%) contained no Ephemeroptera. The majority of the others (31.9%) included one species, though at a number of sites (11.2%) two species occurred. None of the sites ever included more than this. In Tayside and Great Britain as a whole, however, many more species were recorded (28 and 38 respectively) and relatively few sites contained no Ephemeroptera (11.2 and 19.8% respectively). In Tayside 34.4% and in Great Britain 20.2% of the sites contained five or more species, and in both areas sites with ten or more species occurred.

An analysis of the characteristics of sites in Tayside and Great Britain which have no Ephemeroptera shows that the majority fall into one of the following four categories: (1) Small water bodies of of any type which appear to be temporary and thus liable to periodic desiccation. (2) Estuarine reaches of running waters or standing waters sufficiently near the coast to have saline water. (3) Nutrient-poor upland waters in very exposed areas. (4) Nutrient-rich lowland waters subject to eutrophication and pollution. Very few natural waters outside these categories do not have Ephemeroptera, except in island areas like the Shetlands, where, apparently the extreme exposure to wind renders many habitats unavailable to this poorly flying group.

Most of the sites with large numbers of Ephemeroptera are in medium or low altitude areas of base rich geology away from dense human populations or intensive agriculture. Characteristically a wide variety of microhabitats are available at each site, a majority of which are running waters.

The 47 species of Ephemeroptera recorded from Great Britain are given in the check list included in Table 2. The numerical data in this table are based on the results from most of the sites in the Tayside and Great Britain surveys discussed above (a total of 425 sites - 216 standing and 209 running waters) together with (for the sake of completeness) data from the literature for a few species which were not recorded during the present surveys. It can be seen that there is a wide range of occurrences, from common species like *Baetis rhodani*, *Ephemerella ignita* and *Caenis horaria* which were found at about 20% or more of the sites examined, to rare species such as *Baetis digitatus* and *Arthroplea congener* which did not occur at any of the 425 sites included in the table and have only ever

Table 2. A check list of the British Ephemeroptera and their occurrence in standing (S) or running (R) water. This is expressed as a percentage of the total number of sites (N) at which each species was found. The data refer to collections from Great Britain (including Tayside), together with a few records for the less common species taken from the literature and included in the total number of sites*.

Family		Genus	Species	S	R	N
SIPHLONURIDAE	1.	*Siphlonurus*	*armatus* Eaton	50	50	4*
	2.		*lacustris* Eaton	93	7	15
	3.		*linnaeanus* (Eaton)	60	40	5*
	4.	*Ameletus*	*inopinatus* Eaton	29	71	17
BAETIDAE	5.	*Baetis*	*fuscatus* (L.)	0	100	7
	6.		*scambus* Eaton	4	96	53
	7.		*vernus* Curtis	7	93	42
	8.		*buceratus* Eaton	0	100	11
	9.		*rhodani* (Pictet)	6	94	109
	10.		*atrebatinus* Eaton	0	100	4
	11.		*muticus* (L.)	0	100	66
	12.		*niger* (L.)	6	94	18
	13.		*digitatus* Bengtsson	0	100	1*
	14.	*Centroptilum*	*luteolum* (Muller)	35	65	79
	15.		*pennulatum* Eaton	0	100	8
	16.	*Cloeon*	*dipterum* (L.)	79	21	66
	17.		*simile* Eaton	89	11	54
	18.	*Procloeon*	*bifidum* Bengtsson	11	89	18
HEPTAGENIIDAE	19.	*Rhithrogena*	*semicolorata* (Curtis)	1	99	80
	20.		*germanica* Esben-Petersen	0	100	3*
	21.	*Heptagenia*	*sulphurea* (Muller)	0	100	17
	22.		*longicauda* (Stephens)	0	100	6*
	23.		*fuscogrisea* (Retzius)	50	50	2*
	24.		*lateralis* (Curtis)	3	97	34
	25.	*Arthroplea*	*congener* Bengtsson	0	100	1*
	26.	*Ecdyonurus*	*venosus* (Fabricius)	6	94	32
	27.		*torrentis* Kimmins	6	94	17
	28.		*dispar* (Curtis)	29	71	45
	29.		*insignis* (Eaton)	0	100	3*
LEPTOPHLEBIIDAE	30.	*Leptophlebia*	*marginata* (L.)	94	6	35
	31.		*vespertina* (L.)	94	6	35
	32.	*Paraleptophlebia*	*submarginata* (Stephens)	10	90	10
	33.		*cincta* (Retzius)	0	100	5*
	34.		*tumida* Bengtsson	0	100	5*
	35.	*Habrophlebia*	*fusca* (Curtis)	0	100	10
EPHEMERELLIDAE	36.	*Ephemerella*	*ignita* (Poda)	7	93	84
	37.		*notata* Eaton	0	100	8
POTAMANTHIDAE	38.	*Potamanthus*	*luteus* (L.)	0	100	4*
EPHEMERIDAE	39.	*Ephemera*	*vulgata* (L.)	0	100	3*
	40.		*danica* Muller	27	73	15
	41.		*lineata* Eaton	0	100	5*
CAENIDAE	42.	*Brachycercus*	*harrisella* Curtis	0	100	3*
	43.	*Caenis*	*macrura* Stephens	0	100	6*
	44.		*moesta* Bengtsson	56	44	72
	45.		*robusta* Eaton	50	50	6
	46.		*horaria* (L.)	74	26	81
	47.		*rivulorum* Eaton	0	100	27

been recorded once each in Great Britain - the former from the River Frome (Macan 1979) and the latter at Stanmore, Middlesex (Kimmins 1954).

Table 2 shows that there are no species which occur exclusively in standing waters, though there are a few which are found mainly in this habitat (e.g. *Leptophlebia marginata, L. vespertina* and *Cloeon simile*). On the other hand, many species appear to occur only in running water (e.g. *Baetis muticus, Caenis rivulorm* and *Heptagenia sulphurea*) or mainly so (*Baetis scambus, B. rhodani* and *Ephemerella ignita*) and indeed well over half of the British species can be described as showing a distinct preference for the lotic environment.

Ephemeroptera Associations and Ecodata

In order to describe the associations among different species in communities, Fager (1957) has suggested an index which relates the probabilities of the joint occurrence of any two species to the sum of their total occurrences This index does not consider negative associations and is a useful preliminary step to the delimitation of communities. An association analysis of the affinities among the 38 species of Ephemeroptera occurring at sites surveyed during the Great Britain survey described above has been carried out and the results are shown in Figure 1.

It can be seen from this figure that there are several strong groupings among the network of associations produced from the analysis. Two of these are large: one centred around *Baetis scambus/ Ephemerella ignita/B. rhodani*, the other around *Baetis buceratus/ B. niger*. Table 2 shows that both these groupings are dominated by running water species. There are two smaller groupings evident in Figure 1: one associated with *Leptophlebia vespertina/ L. marginata*, the other with *Cloeon simile/Caenis horaria*. From Table 2 it can be seen that these groupings are dominated by standing water species. The species linking these standing and running water groupings (e.g. *Ecdyonurus dispar* and *Caenis robusta*) are commonly found in both standing and running waters (Table 2).

Using the extensive background information available from the large number of sites surveyed during this study it is possible to define fairly clearly the conditions in which each species occurs Presumably these can be considered as associated with their tolerance limits for at least some factors. As an example, the values for altitude of the common Ephemeroptera occurring in Tayside are shown in Figure 2. It can be seen that the sites at which *Ameletus inopinatus* occurs have a mean altitude of 517 m, but that this species can occur at a considerable range around this. At the other end of the scale, *Ephemerella notata* occurs at sites which have a mean altitude of only 40 m and a range around this of 31 m.

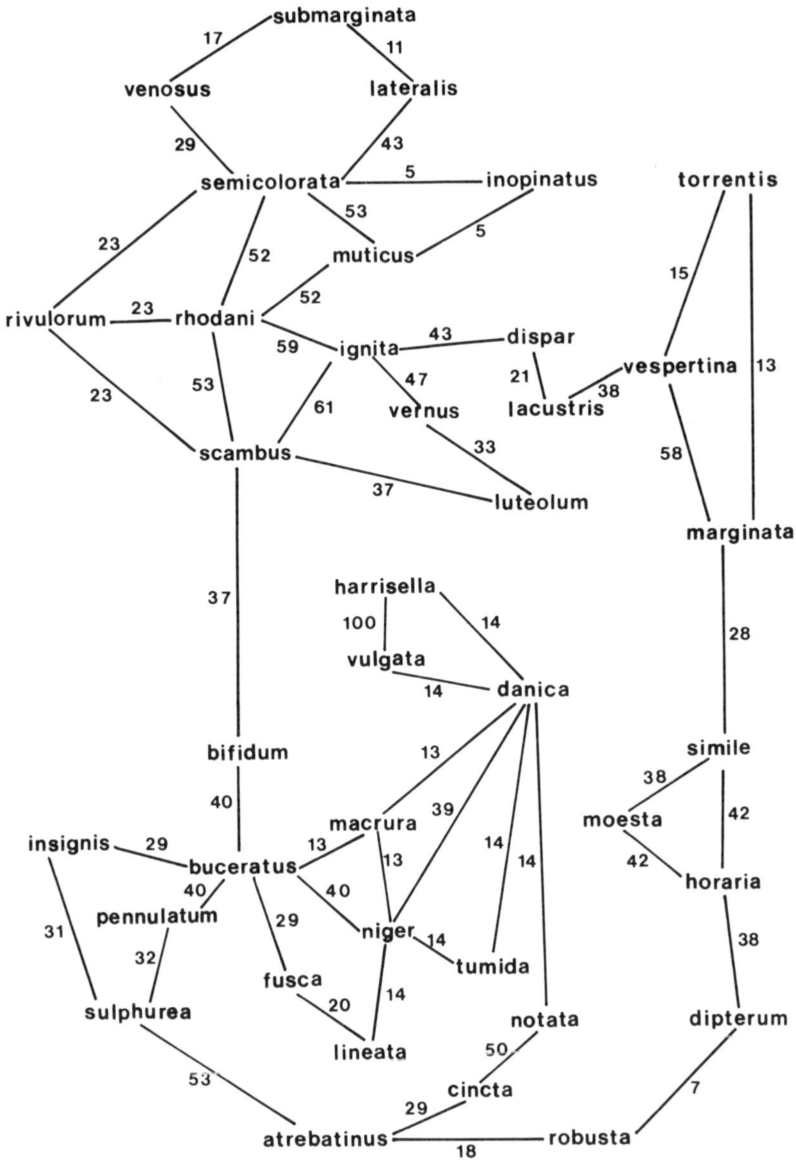

Figure 1. An association network for Ephemeroptera recorded during a survey of 300 sites (150 standing and 150 running waters) in Great Britain. The association values (indicated by the figures between species) were obtained as recommended by Fager (1957) and the network has been produced by linking each species with the two others with which it is most highly associated.

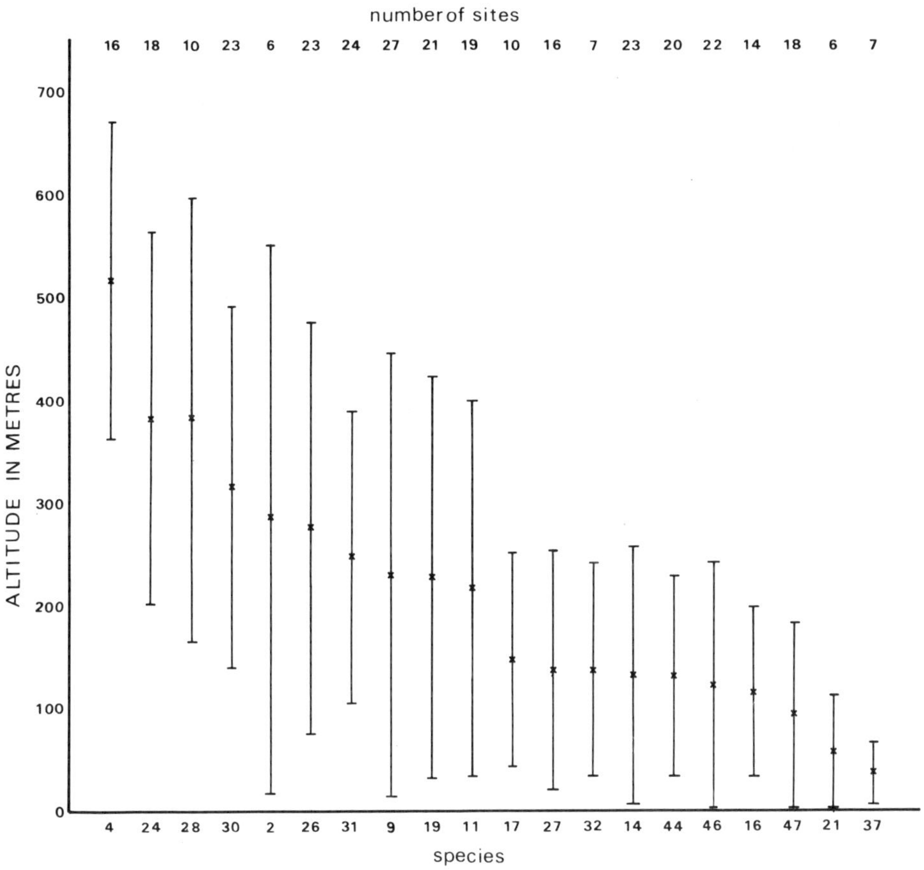

Figure 2. The means and standard deviations of the altitudes of sites in the Tayside Region at which common species of Ephemeroptera occurred. Only those occurring at six or more sites are figured. The numbers at the top of the figure refer to the number of sites; the numbers at the bottom to the species listed in Table 2.

If the altitude values for the species shown in Figure 2 are related to the species associations given in Figure 1, it can be seen that the running water *scambus/ignita/rhodani* and standing water *vespertina/marginata* groupings are dominated by high altitude species. In contrast, the running water *buceratus/niger* and standing water *simile/horaria* groupings are dominated by low altitude species. Thus it seems that the ecodata concerned are relevant to defining the nature of the communities of Ephemeroptera involved and that further more sophisticated, multivariate analyses are likely to be profitable.

Species Ecodata and Econiches

It will be clear from the methodology and the results described so far that a large data bank of information on individual Ephemeroptera and their habitats in Great Britain is now available. In this short paper it is only possible to indicate by example how these ecodata may be used in autecological studies. A single species has been selected for this purpose: *Cloeon simile*. It has already been shown that this is primarily a standing water species (Table 2), associated over Great Britain as a whole with *Caenis horaria* and *Caenis moesta* (Fig. 1) and occurring at low to medium altitudes (Fig. 2).

The distribution of *C. simile* at sites surveyed within Great Britain (excluding the Tayside and Shetland surveys) is indicated in Figure 3. It is clearly a widespread species. Even in Shetland where Ephemeroptera are relatively uncommon it occurred at 16 out of the 66 standing water sites studied: it was never found in running water there. The characteristics of the sites at which it occurred are indicated in Table 3 and these are contrasted there with the means of all the standing waters surveyed. The differences between the two sets of means are assumed to indicate something of the preferences of the species concerned, though naturally the casual relationships are not apparent.

At all the lochs on Shetland at least three samples (usually more) of zoobenthos from the dominant microhabitats were collected. The general results from the 16 lochs in which *C. simile* occurred are shown in Table 4. It is clearly a littoral species and occurred in none of the profundal samples collected. Within the littoral zone it occurred in a variety of microhabitats and apparently showed no special preference for one particular type of substrate. However, when the data for abundance are examined (Fig. 4) it is very clear that high densities occurred only in shallow water in areas of fine silt.

DISCUSSION

One of the important and original aspects of the surveys described in this paper, especially those of Shetland and Tayside is that they have attempted a more or less complete coverage of all the types of water body occurring in the areas concerned. Thus ditches, temporary pools, and high altitude waters (normally ignored in ecological surveys in Great Britain) have been included. The data therefore provide a firm basis for the assessment of the habitat of any one species. The large total number of sites which have now been surveyed is also relevant here.

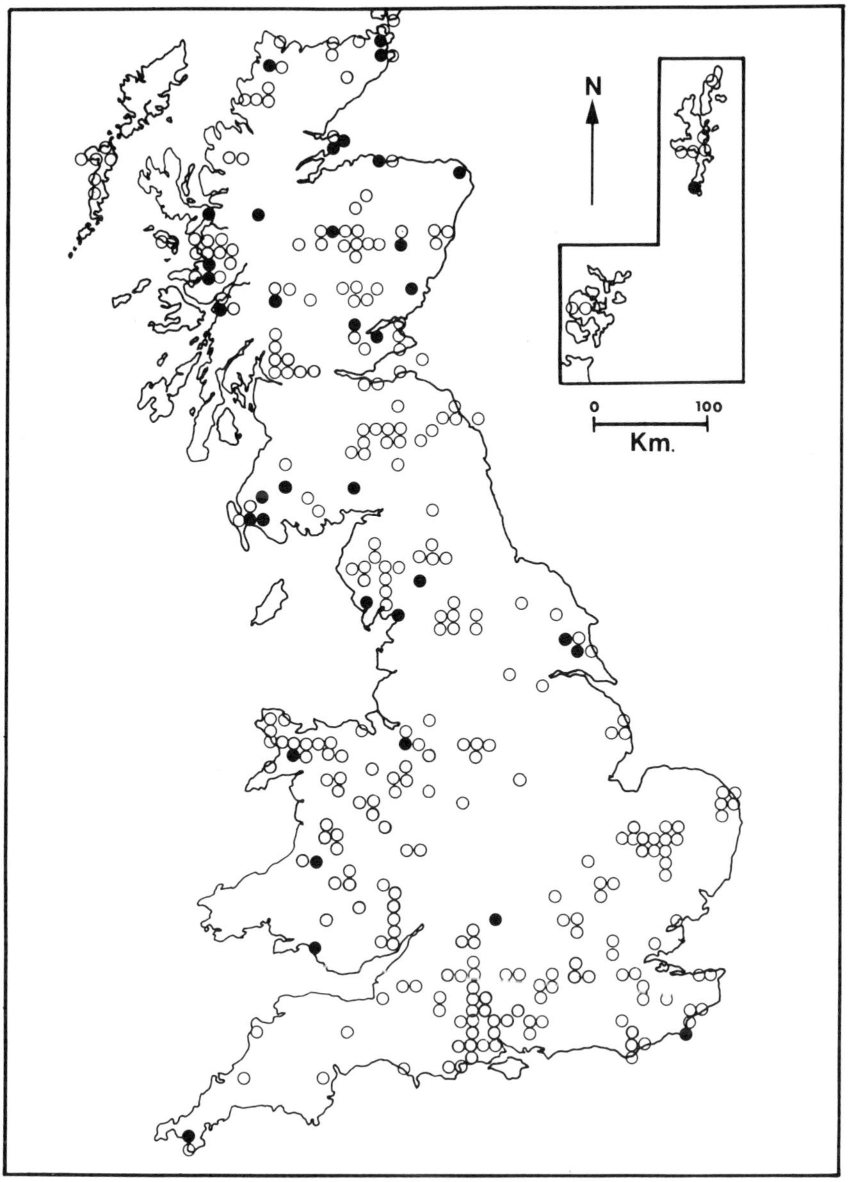

Figure 3. The occurrence of *Cloeon simile* at freshwater sites surveyed in Great Britain. Positive records are indicated by closed symbols.

Table 3. Sample ecodata for *Cloeon similie* on Shetland. This species does not occur in running waters there so the data refer to standing waters only.

	Feature	Background		+ve Occurrences	
		Mean	S.D.	Mean	S.D.
1.	National Grid Reference	4184/11773	0103/0243	4328/11599	0076/0161
2.	Surface Area (ha)	8.9	15.9	17.5	27.6
3.	Altitude (m)	60.2	46.1	39.0	28.2
4.	Catch. Altitude (m)	135.5	111.8	121.6	80.4
5.	% Arable	7.8	21.4	5.6	14.1
6.	% Forest	0	0	0	0
7.	% Peat	36.6	42.6	21.8	30.9
8.	% Water	3.6	6.7	6.9	8.6
9.	Nos. People	1.3	3.6	8.3	28.4
10.	Area (ha)	112.3	178.5	227.4	435.3
11.	Rainfall (cm)	110.4	10.7	103.1	11.3
12.	Distance from Sea (km)	0.95	0.76	0.80	0.81
13.	% Weed Cover	36.2	32.3	38.9	29.2
14.	Mean Depth (m)	1.12	1.47	1.92	1.89
15.	Maximum Depth (m)	3.12	4.29	4.16	5.34
16.	pH	6.6	1.6	6.8	1.0
17.	Calcium (mg/l)	8.1	8.4	7.8	5.1
18.	Alkalinity (mg/l)	26.4	26.2	23.4	19.4
19.	Phosphorus (mg/l)	0.027	0.083	0.070	0.040
20.	% Shore with Trees	0	0	0	0
21.	Water Colour (Hazen Units)	53.1	62.7	29.0	20.4
22.	Suspended Solids (mg/l)	0.025	0.032	0.012	0.014
23.	Sodium (mg/l)	46.1	15.6	50.5	18.9
24.	Conductivity (μS)	295.9	121.1	304.4	130.9

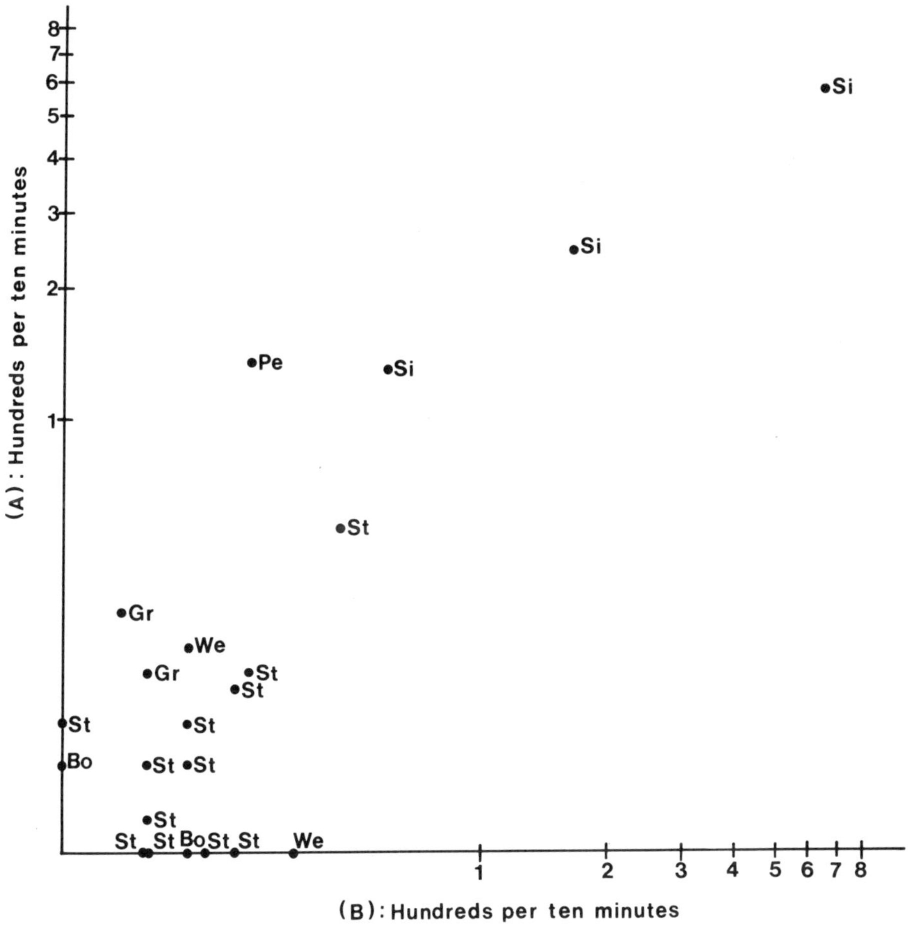

Figure 4. Plots of the numbers of *Cloeon simile* (per 10 minute collection) in pairs of samples (A and B) taken from various substrates in lochs on Shetland (Si: silt; Pe: peat; Gr: gravel; St: stones; Bo: boulders; We: weed). Only those pairs in which *C. simile* occurred in at least one sample are plotted.

Though not mentioned in the results presented above, it must be remembered that data on other aspects of the biota of these sites are also available. These include information on algae, higher aquatic plants, zooplankton and the other zoobenthos. Thus for example, an association diagram for the common Shetland zoobenthos, similar to that for the British Ephemeroptera (Fig. 1) is now available (Maitland 1979). In it, the association of the Shetland

Ephemeroptera with the other zoobenthos is clearly seen. *Baetis rhodani* is a characteristic memeber of the running water community there, and *Cloeon simile* and *Caenis horaria* of the standing water one

This brief account indicates the strategy and tactics of the synoptic work at present in progress. The amount of data potentially available offers scope for a wide variety of analyses in many different fields. It is already evident that considerable interest has been generated by the project and that demands from other scientists and user organisations are likely to increase.

It is the eventual aim of the study that all parts of the data bank which are relevant to synoptic analyses will be computerised. So far, only selected portions of the data have been stored and analysed by computer, mainly with specific objectives in mind. Currently, the bulk of the data are stored by conventional manual filing systems, largely on a site basis using standard pro formas, or increasingly, edge notched cards. This system has proved invaluable for a number of purposes and will be maintained even after computerisation. Its prime use is the speed and ease with which full site information may be retrieved for examination.

Beyond the value of access to information on individual sites lies the goal of a full synopsis of British fresh waters. A major part of this will be comprehensive classification system covering the full range of aquatic habitats in Great Britain. The evidence so far, from both running waters (e.g. Maitland 1966) and standing waters, indicates that the framework for such a classification will be an arbitrary one - its limits bound by the range of fresh waters

Table 4. The occurrence of *Cloeon simile* in samples taken from different microhabitats in those Shetland lochs in which this species occurred.

		Profundal	Littoral			
		Silt	Silt/ Peat	Sand/ Gravel	Stones/ Boulders	Weed/ Silt
No. of samples		16	10	8	38	4
No. with *C. simile*		0	8	4	22	3
% with *C. simile*		0	80	50	58	75
C. simile per 10 mins:	mean	0	199.0	6.5	8.2	14.0
:	S.D.	0	236.8	9.6	11.9	11.4

occurring in Great Britain, its classes defined partly by natural boundaries and partly by convenient ones proposed by limnologists. This is because there appears to be a more or less complete transition from one extreme type of water to another with intermediate types occurring throughout the range.

Synecological studies of animal and plant communities appear to confirm that a critical definition of discrete communities is not possible, and rather that a continuum exists along which there are important, probably stable, nodes which can be described in detail. Apart from the scientific interest and value of synecological studies, the accurate description and quantification of the natural communities has a number of applied values, particularly in relation to impact studies.

The data derived from the studies in synoptic limnology described offer a promising field for autecological research. The definition of natural species niches and apparent tolerances to a wide variety of environmental parameters is a relatively straightforward exercise from the comprehensive data available, and such analyses lead naturally to comparisons of the ecological requirements of different species. Again, such studies have both pure and applied values - e.g. in the case of a species proposed for introduction to any area it is invaluable to have detailed figures for its natural range of habitat.

Finally, the predictive value of synoptic limnology must be emphasised, especially in the fields of conservation and resource management. One of the realistic developments of the synecological and autecological studies described above is that it should eventually be possible to predict, for example, the type of community found in any water whose geographic position and main physical characters are defined. Likewise the probability of a particular animal or plant species thriving at a new site or an existing site under new conditions should be estimated. Naturally, as in all biological situations, such predictions could never be completely accurate, nevertheless they must represent one of the most promising methods available to ecologists.

ACKNOWLEDGMENTS

I am grateful to many of my colleagues for their help in field and laboratory, particularly Mr. K. East and Mr. K.H. Morris who are closely involved in the work on zoobenthos. Most of the data obtained during the Nature Conservation Review (Ratcliffe 1977) were collected by Mr. N.C. Morgan and Mr. R.H. Britton and I thank them for allowing me to use unpublished results. The manuscript was criticised by Ms. B.D. Smith and typed by Mrs. M. Wilson. The figures were drawn by Ms. G.M. Dennis.

RESUME

Quarante-sept espèces d'éphéméroptères se rencontrent en Grande-Bretagne, ce qui est beaucoup moins que dans l'ensemble de l'Europe. Au cours d'études synoptiques d'écosystèmes d'eau douce de l'Angleterre, on a analysé plus de mille plans d'eau. On a recueilli des échantillons de faune zoobenthique à tous les endroits et plusieurs contenaient des éphéméroptères. On procéda ensuite à l'analyse de données sélectionnées pour faire une description quantitative des habitats des éphéméroptères britanniques ainsi que de la nature des populations et des écosystèmes au sein desquelles ils vivent ou non. Quelques espèces sont si rares qu'il n'est pas possible de décrire leur habitat. D'autres espèces plus communes se recontrent dans toutes sortes d'eaux, d'autres encoure dans un seul type particulier. Les facteurs déterminants paraissent liés au courant, à la température, à la composition chimique, à l'exposition de l'eau et à d'autres variables complexes, certains de types biotiques. L'association ainsi que d'autres analyses multiples montrent que, malgré l'existence de transitions, certaines populations d'éphéméroptères peuvent être définies. L'auteur parle de la valeur de ses recherches dans le domaine des sciences pures et des sciences appliquées.

ZUSSAMENFASSUNG

In Großorittannien gibt es 47 Ephemeropterenarten. Das sind viel weniger als in Europa insgesamt. Während synoptischer Studien von Süßwasser Ökosystemen in England wurder über tausend verschieden Gewässer überprüft. An allen Stellan wurden zoobehtische Proben genommen und viele davon enthielten Ephemeropteren. Eine Auswahl von Daten wurde analysiert, um quantitativ die Heimat der britischen Ephemeropteren und die Art der Kommunitäten zu beschreiben, in welchen sie sich befinden oder nicht auftreten. Eien klein Anzahl von Arten ist so selten, daß es nicht möglich wäre, ihre Beheimatung darzustellen. Von den gewöhnlicheren Arten treten einige in verschiedenartigen Gewässern auf, andere nur in einem bestimmten Typ von Wasser. Die definierenden Faktoren scheinen in Beziehung zu stehen zu Wasserfluß, vorherrschender Temperatur, Chemie, Lage und anderen komplexen Variablen, von denen einige biotischer Art sind. Assoziation und andere multiple Analysen zeigen, daß, obwohl es Ubergänge gibt, einige Ephemeropterengemeinschaften definiert werden können. Der praktische unde theoretische Wert der Untersuchung wird in der vorliegenden Arbeit erörtert.

REFERENCES

Brittain, J.E. 1972. The life cycles of *Leptophlebia vespertina* (L.) and *L. marginata* (L.) (Ephemeroptera) in Llyn Dinas, North Wales. *Freshwat. Biol.* 2: 271-277.

Brown, D.S. 1961. The food of the larvae of *Cloeon dipterum* L. and *Baetis rhodani* (Pictet) (Insecta, Ephemeroptera). *J. Anim. Ecol.* 30: 55-75.

Elliott, J.M. 1967. The life histories and drifting of the Plecoptera and Ephemeroptera in a Dartmoor stream. *J. Anim. Ecol.* 36: 343-362.

Fager, E.W. 1957. Determination and analysis of recurrent groups. *Ecology* 38: 586-595.

Kimmins, D.E. 1954. A revised key to the adults of the British species of Ephemeroptera with notes on their ecology. *Freshwat. Biol. Assoc. Sci. Publ.* 15: 1-71.

Macan, T.T. 1949. The taxonomy of the nymphs of the British species of the genus *Ecydonurus*. *Entomol. Mon. Mag.* 85: 64-70.

Macan, T.T. 1950. Descriptions of some nymphs of the British species of the genus *Baetis* (Ephem.). *Trans. Soc. Br. Entomol.* 10: 143-166.

Macan, T.T. 1951. The taxonomy of the British species of Siphlonuridae (Ephem.). *Hydrobiologia* 3: 84-92.

Macan, T.T. 1961. A key to the numphs of the British species of Ephemeroptera with notes on their ecology. *Freshwat. Biol. Assoc. Sci. Publ.* 20: 1-63.

Macan, T.T. 1979. A key to the nymphs of the British species of Ephemeroptera with notes on their ecology. *Freshwat. Biol. Assoc. Sci. Publ.* 20: 1-79.

Maitland, P.S. 1965. The distribution, life cycle, and predators of *Ephemerella ignita* (Poda) in the River Endrick, Scotland. *Oikos.* 16: 48-57.

Maitland, P.S. 1966. The fauna of the River Endrick. Blackie, Glasgow.

Maitland, P.S. 1977. A coded checklist of animals occurring in fresh water in the British Isles. Institute of Terrestrial Ecology, Cambridge.

Maitland, P.S. 1979. Synoptic limnology: the analysis and classification of British freshwater ecosystems. Institute of Terrestrial Ecology, Cambridge.

Minshall, G.W. and R.A. Kuehne. 1969. An ecological study of invertebrates of the Duddon, an English mountain stream. *Arch. Hydrobiol.* 66: 169-191.

Ratcliffe, D.A. 1977. A nature conservation review. University Press, Cambridge.

Smith, I.R. and A.A. Lyle. 1979. The extent and distribution of fresh waters in Great Britain. Institute of Terrestrial Ecology, Cambridge.

Wise, E.J. 1976. Studies on the Ephemeroptera of a Northumbrian river system. I. Serial distribution and relative abundance. *Freshwat. Biol.* 6: 363-373.

LA ZOOGEOGRAPHIE, L'ECOLOGIE ET LA PROTECTION DES EPHEMEROPTERES EN POLOGNE, ET LEUR UTILISATION EN TANT QU'INDICATEURS DE LA PURETE DES EAUX COURANTES

Ryszard Sowa

Université Jagellonne, Institut de la Biologie d'Environnement, Laboratoire d'Hydrobiologie
2a, rue Oleandry, 30-063 Cracovie 19, Pologne

RESUME

L'état de la connaissance zoogéographique et écologique des éphémères de la Pologne est présenté. La structure spatiale et temporale des groupements dans les différentes zones le long des rivières carpathiques et de la Vistule est envisagée, et les régularités zoogéographiques et écologiques concernant la zonation des éphémères sont discutées ensemble avec les problèmes de la protection de leurs groupements. Une méthode pratique est proposée pour la reconnaissance de l'état naturel des groupements des éphémères dans les eaux courantes en Pologne, permettant aussi l'estimation en termes neutres du degré d'influence anthropogenique.

La faune des éphémères de la Pologne comporte 113 espèces, appartenant aux 32 genres et 17 familles. On peut croire que c'est un nombre hautement représentatif malgré que certaines régions du pays ne soient pas encore bien prospectées. Les Carpathes polonaises, la Grande Pologne et le centre du pays en général appartiennent aux terrains dont on a la majorité des informations faunistiques et écologiques. Le reste du pays et surtout les Sudètes, la Poméranie et le Nord-est de la Pologne demandent encore des très intensives recherches. Les espèces et la structure de leurs groupements dans divers torrents et rivières de la haute et moyenne montagne et dans des régions submontagneuses (Sowa 1975, 1975a, 1979) ainsi que dans de plus grandes rivières de plaine (Keffermüller 1960, 1972, 1978; Jazdzewska 1979, Wójcik 1963) sont relativement bien connus mais, par contre, on ne connait pas bien la répartition et le comportement des espèces et de le leurs groupements dans plusieurs petits et moyens cours d'eau submontagneux et de plaines, ainsi que dans des eaux stagnantes et surtout des lacs.

S'il s'agit de leur distribution maximale, on peut diviser les éphémères de la Pologne en quelques groupes (Sowa 1975). Dans le centre et au Nord du pays vivent les espèces ayant la répartition plus ou moins large: nord-centro-européennes, européennes, eurasiatiques et même holarctiques. Dans le Sud du pays et notamment dans les Carpathes existent les espèces de distribution plus restreinte: alpino-carpathique ou sud-centro-européenne. Les Baetidae et Heptageniidae sont ici riches en taxons et les espèces du genre *Rhithrogena* et *Epeorus* ainsi que plusieurs espèces d'*Ecdyonurus* atteignent en ce territoire la frontière nord-est de leur distribution maximale. Sauf un petit nombre d'espèces, localisées dans un ou l'autre cours d'eau, les éphémères de la Pologne sont communes dans les régions montagneuses, submontagneuses de plaines et peuvent être retrouvées dans leurs propres milieux à condition que ceux-ci se trouvent en état naturel. La situation est bien pire si l'on veut tenir compte des travaux écologiques plus approfondis surtout sur l'importance des éphémeres dans la transformation de la matière et de l'énergie dans les eaux courantes et sur l'autécologie des espèces plus nombreuses. A part des travaux mentionnés par Sowa (1975, 1975a) on peut citer ici les études plus anciens de Kawecka (1977) sur la nourriture de 43 espèces de la rivière Raba et celles de Kamler (1971) et Cianciara (1979) sur la bioénergétique et le developpement de *Cloeon dipterum**. L'utilité de telles recherches en Pologne est liée à sa position géographique: les résultats peuvent être rapportés aux terrains plus vastes d'Europe continentale, surtout dans le cas des espèces de plaine. Aussi les recherches elles mêmes sont-elles ici encore possibles parce que plusieurs cours d'eau se trouvent en situation naturelle, étant peuplés par des espèces qui dans plusieurs régions européennes n'existent plus. Pourtant la réserve des eaux naturelles diminue graduellement sous l'influence de la pollution organique et industrielle, de l'amenagement, de la construction des barrages et de l'essor du tourisme (cas des rivières de montagne et submontagneuses). Les plus influencées sont les rivières de plaines, du centre et de l'Ouest du pays, des rivières carpathiques et de celles des Sudètes. Par conséquence l'aire de répartition de plusieurs espèces, même nombreuses, devient de plus en plus réduite et elles peuvent même disparaître dans l'avenir. Aux unités les plus menacées appartiennent: *Behningia ulmeri*, *Oligoneuriella keffermuellerae*, *O. mikulskii*, *O. pallida*, *Oligoneurisca borysthenica*, *Ephoron virgo*, *Palingenia longicauda* (existe-t-elle encore?), *Ametropus fragilis* et *Centroptilum shadini*, donc celles qui habitent de grandes rivières. On peut

---

* La nomenclature des espèces citées est celle présentée par Puthz (1978).

estimer qu'en somme 14% des espèces citées de Pologne sont en situation grave et 17% en situation assez dangereuse. Il s'agit des formes rares ou limitées à certains cours d'eau particulièrs ou enfin des formes à répartition vaste mais à la fois restreinte à certaines zones le long des torrents et des rivières. La protection de ces espèces devient donc la question d'urgence. La protection directe de l'une ou l'autre espèce a peu de chance de réussir et il est bien évident qu'elle doit être effectuée au niveau de toute biocoenose donc entre autres de tout groupement des éphémères. Pour réaliser une telle protection les biologistes doivent savoir répondre à la question de savoir en quel état se trouve la biocoenose dans un lieu et un temps donné. Est-ce qu'elle est naturelle ou modifiée, et en quel degré? Les biologistes doivent donc connaître au moins les changements spatiaux et temporels dans la structure des unités biocenotiques le long des cours d'eau. Etant donné la variabilité géographique des groupements le jugement doit se baser sur les études dans la même région ou des régions voisines.

Les tableaux 1 à 5 présentent la distribution saisonnière des stades larvaires des éphémères et leur abondance estimée en trois degrés, dans les différentes zones de la distribution des groupements le long des eaux courantes de la Pologne, en allant de l'amont vers aval. Les données ont été élaborées d'après les observations faites pendant plusieurs années dans beaucoup de stations. Elles concernent les rivières principales des Carpathes (Fig. 1) dont les affluents s'écoulent par des vallées profondes (les torrents principaux et latéraux) et dont les sources se trouvent à une certaine hauteur, dans la haute ou moyenne montagne (Sowa 1975, 1975a). Ces données sont ici révisées et synthétisées pour chacune des zones. Certaines zones ont été divisées, s'il s'agit de l'abondance des espèces, en parties: supérieure, moyenne et inférieure. Une telle distinction des groupements semble assez artificielle, puisque les changements le long des cours d'eau font une continuité (un "cline" écologique) mais, dans un but pratique, la manipulation paraît justifiée. Il faut admettre d'autre part que les zones de la distribution longitudinale des éphémères correspondent bien à la zonation générale des biocoenoses présentée par Illies et Botosaneanu (1963). Ainsi la zone 1 est conforme à l'hypocrenon, la zone 2 a l'épirhithron, la zone 3 au metarhithron, la zone 4 à l'hyporhithron, et la zone 5 à l'épipotamon. Les groupements vivent donc dans les eaux courantes de l'hypocrenal jusqu'à l'épipotamal (Fittkau 1976).

En comparant la structure et le comportement des groupements dans des zones successives, nous pouvons remarquer quelques régularités zoogéographiques et écologiques. La plus grande diversité est à remarquer dans la zone 4, et le plus grand acroissement du nombre des espèces (par rapport à la zone précédente) - dans la zone 2. En zone 4 dissparaissant les espèces de *Rhithrogena* et d'*Epeorus*, et en zone 5 la majorité des espèces d'*Ecdyonurus*. Il y a

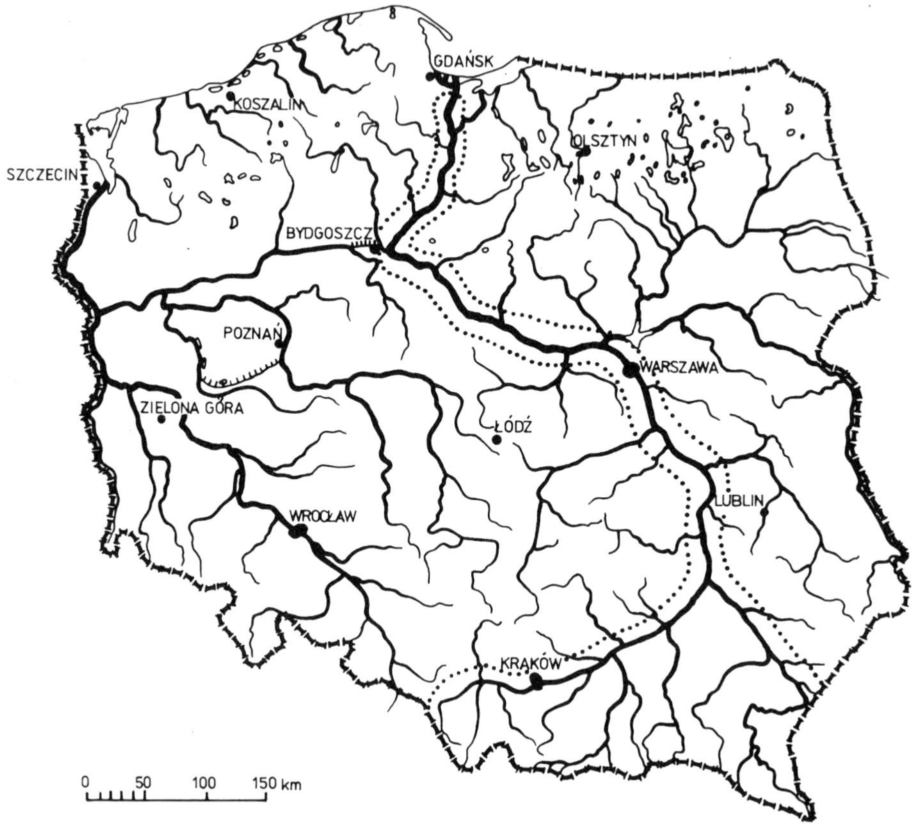

Figure 1. Le reseau hydrographique de la Pologne. Les points limitent le fragment du reseau étudié plus en détail.

peu d'espèces limitées a l'une ou a l'autre zone et en majorité elles vivent dans deux, trois ou même quatre zones successives. Les zones sont délimitées surtout par les changements nets dans la structure des espèces dominantes. Les espèces plus étendues en distribution longitudinale, appartiennent au trois groupes. Le premier englobe des espèces qui abondent dans ce type des eaux courantes dont nous parlons ici. Le deuxième est composé des espèces plutôt ubiquistes, qui abondent aussi dans les autres écosystèmes aquatiques, et le troisième est constitué par des espèces peu nombreuses ou sporadiques, qui habitent des eaux voisines du type différent où elles sont nombreuses. Ce sont

Tableau 1. La présence saisonnière et l'abondance des larves des Éphéméroptères dans la premiere zone de la distribution de leurs groupements le long des principaux cours d'eau carpathiques. 3 - espèce nombreuse, 2- assez nombreuse, 1 - peu nombreuse ou sporadique.

espèce	mois	XII	I	II	III	IV	V	VI	VII	VIII	IX	X	XI	
Baetis alpinus		---	---	---	---	---	---	---	---	---	---	---	---	3
Rhithrogena loyolaea		---	---	---	---	---	---	---	---	---	---	---	---	2
Ameletus inopinatus		---	---	---	---	---	---	---	---	---	---	---	---	2
Ecdyonurus subalpinus		---	---	---	---	---	---	---	---	---	---	---	---	2
Baetis rhodani		---	---	---	---	---	---	---	---	---	---	---	---	2
Rhithrogena iridina		---	---	---	---	---	---	---	---	---	---	---	---	2
- gorganica		---	---	---	---	---	---	---		---	---	---	---	2
Baetis vernus						---	---	---	---	---	---	---		1
somme des espèces:		7	7	7	7	8	8	8	7	8	8	8	7	

Tableau 2. La présence saisonnière et l'abondance des larves des Éphéméroptères dans la deuxieme zone de la distribution de leurs groupements le long des principaux cours d'eau carpathiques. 3 - espèce nombreuse, 2 - assez nombreuse, 1 - peu nombreuse ou sporadique, s - partie supérierue de la zone, i - partie inférieure.

espèces	mois: XII I II III IV V VI VII VIII IX X XI	s	i
Baetis alpinus	--	3	3
Rhithrogena iridina	--	3	3
Baetis melanonyx	----------------------	3	3
Ameletus inopinatus	--	3	2
Rhithrogena loyolaea	--	3	1
Baetis muticus	--	1	3
Ecdyonurus venosus	--	1	3
Epeorus sylvicola	--	1	3
Baetis rhodani	--	1	3
Habroleptoides modesta	--------------------------- -------------	1	3
Ecdyonurus subalpinus	--	2	1
Baetis vernus	--------------------------	2	1
Rhithrogena hybrida	------------------------------------ --------	1	2
- gorganica	--------------------------- ----------	1	2
Ephemerella krieghoffi	------------------------------- --------	1	1
Ecdyonurus lateralis	--	0	2
- torrentis	--------------------------- ---------	0	2
Baetis lutheri	--	0	2
- scambus	-------------------------	0	2
Centroptilum luteolum	--	0	2
Ephemera danica	--	0	2
Habrophlebia lauta	--	0	2
Rhithrogena ferruginea	--	0	1
Ecdyonurus quadrilineatus	-------------------------------- -------	0	1
- carpathicus	--	0	1
- submontanus	---------------------	0	1
Paraleptophlebia submarginata	------------------------------- ----------	0	1
Ephemerella major	--------------------------- ----------	0	1
- ignita	-------------------------	0	1
Caenis rivulorum	------------------------------- --------	0	1
somme des espèces:	25 25 25 25 28 30 30 26 23 29 28 25		

p.ex.: *Baetis vernus, B. niger,* les espèces de *Siphlonurus* et la plupart des Leptophlebiidae. Les changements de la distribution saisonnière des espèce dans les zones successives montrent des différences caractéristiques. Dans la zone 1 le groupement est pratiquement le même pendant toute l'année. Dans les zones situées plus bas la composition des espèces pendant les mois successifs devient de plus en plus différente, comme résultat de la différenciation de leurs cycles vitaux. A partir de la zone 2 c'est en hiver qu'on trouve le moins d'espèce et l'optimum de leur présence a lieu au printemps et surtout en avril et en mai (la zone 4 et 5) ou en mai et en juin (la zone 2 et 3). A ce moment on trouve dans chacune des zones le nombre maximal ou presque maximal d'espèces. La participation des espèces hivernales diminue, dans des zones successives a l'avantage des espèces estivales, mais une équilibration ne s'effectue qu'en zone 5. De l'amont vers l'aval augmente aussi le nombre d'espèce polyvoltines, tant hivernales qu'estivales. La participation de celles-ci est surtout importante dans la zone 5. Les tendances seront plus nettes si nous prenons en considération les espèces plus nombreuses.

Pour le cours moyen et inférieur de la Vistule nous avons seulement les données fragmentaires. Celles-ci ainsi que les données des autres rivières voisines (la Warta, la Pilica, le Bug) nous permettent pourtant de reconstruire le peuplement naturel de ce grand fragment de la Vistule. Le groupement diffère nettement de celui du San inférieur, on peut donc croire qu'il se place dans la zone 6, celle de métapotamal. Par rapport a la zone précédente disparaissent ici les espèces suivantes: *Baetis gracilis, B. niger,B. rhodani, B. vardarensis, Choroterpes picteti Habroleptoides modesta, Oligoneuriella rhenana, Ecdyonurus aurantiacus, E. insignis, E. macani,* et probablement aussi *E. dispar.* Ces pertes sont compensées par l'entrée des espèces nouvelles: *Siphlonurus alternatus, Baetis tracheatus, Baetopus wartensis, Centroptilum shadini, Oligoneuriella keffermuellerae, Oligoneurisca borysthenica, Behningia ulmeri* et *Palingenia longicauda.* Le groupement montre donc presque le même nombre d'espèces. Bien que nous n'ayons pas des données plus précises sur les cycles de développement de quelques espèces il semble qu'en somme la distribution saisonnière du groupement envisage les mêmes tendances que celui de la zone 5. Le groupement se retrouve plus dans les cours d'eau que le groupement de la zone 5, puisqu'il peut exister aussi dans les rivières qui ne prennent pas leur source dans la montagne.

La distribution saisonnière de la présence des espèces dans les zones particulières montre la situation optimale. Pour des localités et des termes choisis de la récolte, les nombres d'espèces trouvées peuvent être plus bas, même jusqu'à 20%. Pour l'interprétation des résultats il faut prendre cela en considération. Le groupement de l'une ou l'autre zone est en état naturel s'il

Tableau 3. La présence saisonnière et l'abondance des larves des Éphéméroptères dans la troisième zone de la distribution de leurs groupements le long des principaux cours d'eau carpathiques. 3 - espèce nombreuse, 2 - assez nombreuse, 1 - peu nombreuse our sporadique, s - parite supérieure de la zone, m - partie moyenne, i - parite inférieure.

espèces	mois: XII I II III IV V VI VII VIII IX X XI	s	m	i
Rhithrogena ferruginea	--	3	3	3
Baetis lutheri	--	3	3	3
- rhodani	--	3	3	3
- muticus	--	3	3	3
Habroleptoides modesta	----------------------- -----------------	3	3	3
Ecdyonurus lateralis	--	3	3	3
Baetis scambus	--------------------	2	3	3
Ecdyonurus torrentis	--------------------------- -----------	2	3	3
Epeorus sylvicola	--	3	2	2
Baetis alpinus	--	3	2	1
- melanonyx	--------------------	3	2	1
Ephemerella ignita	------------------------	2	2	2
Ecdyonurus dispar	--	2	2	2
Ephemerella major	--	2	2	2
Habrophlebia lauta	--	2	2	2
Rhithrogena hercynia	------------------------- ---------------	2	2	2
Caenis rivulorum	------------------------------ ------	1	2	2
Siphlonurus lacustris	------------------	2	2	2
Centroptilum luteolum	--	2	1	1
Ephemera danica	--	2	1	1
Ecdyonurus venosus	------------------------------ -----------	2	1	1
Baetis fuscatus	---------------------------	1	2	2
- sinaicus	------------------	0	2	2
Rhithrogena semicolorata	-------------------------- ---------	1	1	2
Ecdyonurus submontanus	------------------	1	1	1
Ephemerella krieghoffi	------------------------- ---------	1	1	1
Ecdyonurus starmachi	--	1	1	1
Baetis vernus	----------------------	1	1	1
Paraleptophlebia submarginata	------------------------- ---------	1	1	1
Ecdyonurus carpathicus	--	1	1	0
Caenis beskidensis	--------------------	1	1	0
Rhithrogena germanica	---------------- ---------------------	0	1	1
Baetis beskidensis	---------------------	0	1	1
Caenis macrura	--	0	1	1
Rhithrogena iridina	--	2	0	0
Baetis vardarensis	--	0	0	2
Caenis pseudorivulorum	--------------------	0	0	1
Oligoneuriella rhenana	-----------------	0	0	1
Procloeon bifidum	----------------------	0	0	1
Ephemera lineata	--	0	0	1
Ecdyonurus insignis	------------------	0	0	1
somme des espèces:	27 27 27 28 40 40 40 36 34 33 28 27			

Tableau 4. La présence saisonnière et l'abondance des larves des Éphéméroptères dans la quatrième zone de la distribution de leurs groupements le long des principaux cours d'eau carpathiques. 3 - espèce nombreuse, 2 - assez nombreuse, 1 - peu nombreuse ou sporadique, s - partie superiéure de la zone, m - partie moyenne, i - partie inferiéure.

espèces	mois: XII I II III IV V VI VII VIII IX X XI	s	m	i
Baetis vardarensis	--	3	3	3
- fuscatus	-----------------------	3	3	3
Ephemerella ignita	--------------------	3	3	3
Rhithrogena semicolorata	-------------------- ---------	3	3	2
Baetis rhodani	--	3	3	2
Ephemerella major	----------------------- ---------	3	3	2
Ecdyonurus dispar	--	3	3	2
Rhithrogena diaphana	--------------	2	3	3
Oligoneuriella rhenana	----------------	2	3	3
Potamanthus luteus	--	2	3	3
Caenis pseudorivulorum	-------------------	2	3	3
Habroleptoides modesta	------------------- -------------	3	2	2
Ephemera lineata	--	2	3	2
Baetis muticus	--	3	2	1
Ecdyonurus aurantiacus	-----------------------	1	2	3
Heptagenia sulphurea	--	1	2	3
Procloeon bifidum	-------------------------	1	2	3
Rhithrogena germanica	------------- -----------------	2	2	2
Caenis macrura	--	2	2	2
Ecdyonurus torrentis	----------------------- ---------	2	2	1
Centroptilum luteolum	--	2	2	1
Baetis beskidensis	-----------------------	2	2	1
Ecdyonurus insignis	----------------------	2	2	1
- macani	------------------- ---------	1	2	2
Baetis buceratus	--	1	2	2
- lutheri	--	3	1	0
- scambus	-----------------------	2	1	1
Centroptilum pennulatum	-----------------------	1	2	1
- pulchrum	-----------------------	0	2	2
Epeorus sylvicola	--	2	1	0
Heptagenia coerulans	---------------------------- ------	0	1	2
- longicauda	--	0	1	2
Baetis gracilis	----------------------	0	1	2
Choroterpes picteti	-----------------	0	1	2
Baetis niger	--	1	1	1
- vernus	-------------------	1	1	1
Caenis horaria	--	1	1	1
Paraleptophlebia submarginata	------------------- -------------	1	1	1
Ephoron virgo	--------------------	0	1	2
Ecdyonurus lateralis	--	1	1	0
Baetis sinaicus	---------------------	1	1	0
Ephemerella krieghoffi	------------------- -------	1	1	0
Siphlonurus lacustris	---------------	1	1	0
Habrophlebia lauta	--	1	1	0
Pseudocloeon inexpectatum	-------------------	0	1	1
Ephemerella notata	--------------------- -----	0	1	1
Rhithrogena ferruginea	------------------- ---------	2	0	0
Ecdyonurus submontanus	--------------------	1	0	0
Caenis rivulorum	----------------------- ---------	1	0	0
Ephemera danica	--	1	0	0
Ecdyonurus starmachi	--	1	0	0
Baetis calcaratus	? ? ? ? ? -------------------- ? ?	0	0	1
Ecdyonurus fascioculatus	? ? ? ? ? ----------------- ? ? ?	0	0	1
- affinis	--	0	0	1
Paraleptophlebia cincta	-------------------------- --------	0	0	1
Procloeon ornatum	---------------------	0	0	1
Heptagenia flava	--	0	0	1
somme des espèces:	34 34 34 44 56 56 51 47 47 49 37 34			

posséde l'ensemble des espèces nombreuses et assez nombreuses, et s'il contient aussi un certain nombre d'espèces peu nombreuses ou sporadiques. La détérioration de l'état naturel du groupement concerne d'habitude la réduction du nombre d'espèces en général et la diminution d'abondance de celles qui survivent. A la base des tableaux 1 à 5 il est également possible d'estimer d'autres changements dans les groupements. A l'aide de ces tableaux on peut enfin préciser le nombre minimum des échantillonages nécessaires pour retrouver toutes les espèces du groupement donné. Pour un seul échantillonage la période optimale est le mois de mai pour chacune des zones. L'analyse doit se baser sur un nombre considérable d'échantillons quantitatifs ou demi-quantitatifs tank du facies lotique d'une station que du facies lentique, tous les substrats étant pris en considération. Les échantillons doivent être grands, d'une surface de 5 dm^2 ou du volume de 2 dm^3 du substrat. Pour le terme choisi et la station donnée il est nécessaire de prendre au moins 10 prélèvements dans la zone 1 et 2, et 20 prélèvements dans la zone 3 à 6.

Le degré de la menace potentielle des groupements n'est pas le même dans les zones différentes. Les moins menacés sont des groupements de la zone 1 et 2 tant par leur grande répétition dans les cours d'eau que par leur position dans des terrains élevés et forestiers, faiblement peuplés par l'homme. Le groupement de la zone 3 est déjà souvent soumis au changement, ou même entièrement détruit dans certains cours d'eau. Mais la répétition de cette zone dans le réseau aquatique étudié est encore assez considérable, de l'ordre de quelques dizaines, et plusieurs unités se trouvent en état naturel. La répétition de la zone 4 ne dépasse pas quelques rivières carpathiques et leurs affluents les plus grands. Dans toutes ces rivières les groupements des éphémères sont très menacés ou plus au moins détruits par l'activité de l'homme et demandent le contrôle et la protection immédiats. Le groupement de la zone 5 existe dans le San inférieur seulement (jadis, on pouvait le trouver aussi dans la Vistule supérieure). Le San subit la pollution dans quelques secteurs et la surveillance de éphémères de cette zone est surtout importante et urgente. La zone 6 se répète dans quelques grandes rivières de plaine, mais le groupement correspondent est limité à certains secteurs seulement. Aussi ce groupement demande-t-il la protection urgente. En cas de destruction totale et simultanée des groupements de la zone 4 à 6 dans un cours d'eau donné leur pleine réstitution prochaine par récolonisation serait très difficile ou même impossible. Dans la zone 6 il s'agira des espèces citées dans la page 142, dans la zone 4 et 5 des taxons suivants: *Baetis vardarensis*, *B. calcaratus*, *B. gracilis*, *Pseudocloeon inexpectatum*, *Centroptilum nana*, *C. pulchrum*, *Oligoneuriella mikulskii*, *O. pallida*, *O. rhenana*, *Isonychia ignota*, *Ecdyonurus aurantiacus*, *E. fascioculatus*, *E. affinis*, *Heptagenia coerulans*, *Choroterpes picteti*, *Brachycercus minutus*, *B. pallidus*. Pour toutes ces espèces les trois zones constituent des biotopes caractéristiques

Tableau 5. La présence saisonnière et l'abondance des larves des Éphéméroptères dans la cinquieme zone de la distribution de leurs groupements le long des principaux cours d'eau carpathiques. 3 - espèce nombreuse, 2 - assez nombreuse, i - peu nombreuse ou sporadique, s - partie supérieure de la zone, m - partie moyenne, i - partie inférieure.

espèces	mois: XII	I	II	III	IV	V	VI	VII	VIII	IX	X	XI	s	m	i
Caenis pseudorivulorum					–	–	–	–	–	–			3	3	3
Baetis fuscatus					–	–	–	–	–	–	–		3	3	3
- buceratus	–	–	–	–	–	–	–	–	–	–	–	–	3	3	3
Ecdyonurus aurantiacus				–	–	–	–	–	–	–			3	3	3
Heptagenia flava	–	–	–	–	–	–	–	–	–	–	–	–	2	3	3
Oligoneuriella mikulskii				–	–	–	–	–	–				2	3	3
Ephoron virgo				–	–	–	–	–	–	–			2	3	3
Baetis tricolor	–	–	–	–	–	–	–	–	–	–	–	–	2	3	3
Heptagenia sulphurea	–	–	–	–	–	–	–	–	–	–	–	–	3	3	1
Oligoneuriella rhenana				–	–	–	–	–					3	3	1
Ephemerella mesoleuca					–	–	–	–					1	3	3
Heptagenia coerulans	–	–	–	–	–	–	–	–	–		–	–	3	2	2
Potamanthus luteus	–	–	–	–	–	–	–	–	–	–	–	–	3	2	2
Ephemerella ignita				–	–	–	–	–					2	2	2
Ecdyonurus dispar	–	–	–	–	–	–	–	–	–	–	–	–	2	2	2
Caenis macrura	–	–	–	–	–	–	–	–	–	–	–	–	2	2	2
Baetis calcaratus	?	?	?	–	–	–	–	–	–	–	?	?	2	2	1
Isonychia ignota				–	–	–	–	–	–				1	2	2
Centroptilum pulchrum				–	–	–	–	–	–	–			1	2	2
Brachycercus harrisella				–	–	–	–	–	–				1	2	2
Procloeon ornatum				–	–	–	–	–	–	–	–		1	2	2
Ecdyonurus affinis	–	–	–	–	–	–	–	–	–	–	–	–	1	2	1
Siphlonurus aestivalis				–	–	–	–	–					1	2	1
Ephemera lineata	–	–	–	–	–	–	–	–	–	–	–	–	2	1	1
Baetis gracilis				–	–	–	–	–	–	–			2	1	1
- rhodani	–	–	–	–	–	–	–			–	–	–	2	2	0
Oligoneuriella pallida					–	–	–	–	–				0	2	2
Brachycercus minutus	?	?	?	–	–	–	–	–	–	?	?	?	0	2	2
- pallidus	?	?	?	–	–	–	–	–	–	?	?	?	0	2	2
Ecdyonurus macani	–	–	–	–	–	–				–	–	–	2	1	0
Baetis vardarensis	–	–	–	–	–	–	–	–	–	–	–	–	2	1	0
Choroterpes picteti					–	–	–	–	–				1	1	1
Heptagenia longicauda	–	–	–	–	–	–	–	–	–	–	–	–	1	1	1
Ecdyonurus insignis				–	–	–	–	–	–				1	1	1
Paraleptophlebia cincta	–	–	–	–	–	–	–			–	–	–	1	1	1
- submarginata	–	–	–	–	–	–			–	–	–	–	1	1	1
Centroptilum luteolum	–	–	–	–	–	–	–	–	–	–	–	–	1	1	1
Procloeon bifidum				–	–	–	–	–	–	–			1	1	1
Pseudocloeon inexpectatum				–	–	–	–	–	–				1	1	1
Baetopus tenellus	?	?	?	–	–	–	–	–	–	?	?	?	1	1	1
Cloeon simile	–	–	–	–	–	–	–	–	–	–	–	–	1	1	1
- cognatum	–	–	–	–	–	–	–	–	–	–	–	–	1	1	1
- dipterum	–	–	–	–	–	–	–	–	–	–	–	–	1	1	1
Caenis horaria	–	–	–	–	–	–	–	–	–	–	–	–	1	1	1
Baetis niger	–	–	–	–	–	–	–	–	–	–	–	–	1	1	0
Habroleptoides modesta	–	–	–	–	–	–			–	–	–	–	1	1	0
Ephemerella major	–	–	–	–	–	–			–	–	–	–	1	0	0
Ametropus fragilis	–	–	–	–	–	–	–			–	–	–	0	0	2
Heptagenia fuscogrisea	–	–	–	–	–	–					–	–	0	0	1
Centroptilum nana				–	–	–	–	–	–				0	0	1
Leptophlebia marginata	–	–	–	–	–	–				–	–	–	0	0	1
Caenis luctuosa	–	–	–	–	–	–	–	–	–	–	–	–	0	0	1
somme des espèces:	28	28	28	47	52	52	46	43	42	38	30	28			

Dans l'estimation biologiques des changements anthropogéniques des eaux courantes et surtout dans l'estimation du degré de leur pollution organique, certains éphémères, comme les autres invertebrés, forment le groupe des indicateurs de la valeur indicatrice allant de la xénosaprobie à l'alphamésosaprobie (Zelinka et Marvan 1961). La liste des espèces utilisées a augmenté ces dernieres années (Braasch et Jacob 1976, Janeva 1979, Russev 1979). Bien que l'utilisation des éphémères ayant des valeurs indicatrices ait une certaine importance, il faut remarquer que ces valeurs sont souvent attribuées aux espèces d'une facon, semble-t-il, trop universelle, D'autre part il est bien évident que la présence fréquente des éphémères dans un cours d'eau témoigne d'habitude de son état naturel plutôt que de sa pollution. Grâce à la sensibilité générale dans la réaction envers tous les changements du milieu les groupements des éphémères peuvent aussi être utilisés avec succès pour la constatation de l'état naturel ou faiblement changé par l'homme des eaux courantes. L'évaluation doit se baser sur l'analyse du groupement entier c'est-à-dire de sa diversité et de son abondance et les résultats doivent être exprimés en termes neutres. La condition primordiale c'est la reconnaissance des groupements naturels et de leur zonation au territoire géographique donné. La technique du travail est la même que celle présentée sur les pages 148 et 150. On peut proposer l'interprétation suivante des résultats: 100 à 75% du nombre des espèces par rapport à l'état naturel, avec l'abondance constante ou faiblement changée - situation bonne; 74 à 50% des espèces avec l'abondance des dominants moyennement réduite ou changée - situation assez bonne; 49 à 25% des espèces avec l'abondance des dominants nettement réduite ou changée - situation douteuse, état de menace; au-dessous de 25% des espèces avec l'abondance des dominants très réduite ou nulle - situation mauvaise, état de pollution sensible. La méthode d'estimation est facile à l'appliquer dans les zones 3 à 6. Dans la zone 1 et partiellement dans la zone 2 les résultats obtenus peuvent montrer la situation meilleure que ceux qu'on doit aux autres méthodes (p.ex. chimiques). La pénétration exacte de tous les substrats dans la zone 6 et 5 est plus difficile que dans les autres zones.

ABSTRACT

The current status of zoogeographic and ecological knowledge of the Ephemeroptera of Poland is presented. The spacial and temporal structure of the communities of the various zones along both the Carpathian rivers and the River Vistula is presented. The zoogeographical and ecological characteristics concerning the zonation of mayflies are discussed, together with the problems involved in conserving their communities. A practical method is proposed for recognizing the natural relationships of the mayfly communities in the running water of Poland. The method also allows estimation, in unbiased terms, of the amount of human influence.

ZUSSAMENFASSUNG

In der vorliegenden Arbeit soll der zoogeographische und ökologische Kenntnisstand über Ephemeropteran in Polen dargelegt werden. Dabei wird die räumliche und temporale Struktur der Gruppierungen in den verschiedenen Zonen entlang der karpathischen Flüsse und der Weichel erörtert. Die Arbeit befaßt sich weiterhin mit den zoogeographischen Gleichmäßigkeiten, die für die Verteilung der Ephemeropteren auf verschiedene Zonen von Bedeutung sind und geht auch auf Probleme bezüglich der Protektion dieser Gruppierungen ein. Ferner wird eine praktische Methode In Vorschlag gebracht, die zur Erkenntnis des Naturzustandes der Ephemerengruppierungen in den fließenden Gewässern Polens nützlich ist, und die zugleich eine vorsichtige Schätzung erlaubt, in wie weit anthropogenische Einflüsse eine Rolle spielen.

OUVRAGES CITES

Braasch, D., et U. Jacob. 1976. Die Verwendung von Ephemeropteren (Insecta) der DDR als Indikatoren für die Wassergüte. *Entomol. Nachr.* 20: 101-111.

Cianciara, S. 1979. Some study on the biology and bioenergetics of *Cloeon dipterum* L. Ephemeroptera (preliminary data) p. 175-192. *In:* Proc. Sec. Int. Conf. Ephem. PWN, Kraków.

Fittkau, E.J. 1976. Kinal und Konon, Lebensraum und Lebensgemeinschaft der Oberflächendrift am Beispiel amazonischer Fliessgewässer. *Biogeographica* 7: 101-113.

Illies, J., et L. Botosaneanu. 1963. Problèmes et méthodes de la classification et de la zonation écologique des eaux courantes considerées surtout du point de vue faunistique. *Int. Ver. theor. angew. Limnol. Mitt.* 12: 1-57.

Janeva, I. 1979. Einige Vertreter der Gattung *Baetis* (Ephemeroptera) als limnosaprobe Bioindikatoren. p. 139-143. *In:* Proc. Sec. Inter. Conf. Ephem., PWN, Kraków.

Jazdzewska, T. 1979. Premiers résultats des recherches sur la faune des Ephéméroptères de la rivière Pilica. p. 133-137. *In:* Proc. Sec. Inter. Conf. Ephem., PWN, Kraków.

Kamler, E. 1971. Reactions of two species of aquatic insects to the changes of temperature and oxygen concentration. *Pol. Arch. Hydrobiol.* 18: 303-323.

Kawecka, B. 1977. The food of dominant species of bottom fauna larvae in the River Raba (Southern Poland). *Acta Hydrobiol.* 19: 191-213.

Keffermüller, M. 1960. Investigations on the fauna of Ephemeroptera in Great Poland. *P. Kom. Biol. Pozn. Tow. Przyk. Nauk,* 19: 1-59.

Keffermüller, M. 1972. Investigations on the mayflies (Ephemeroptera) fauna in Wielkopolska. V. *Pol. Pismo Entomol.* 42: 527-533.

Keffermuller, M. 1978. Investigations on the mayflies (Ephemeroptera) fauna in Great Poland. VI. *Bad. Fizjogr. Pol. Zach., Ser. C.* 31: 95-103.

Puthz, V. 1978. Ephemeroptera. 256-263. J. Illies, ed. In Limnofauna Europaea, Fischer Berlin.

Russev, B.K. 1979. Die Anpassungsfähigkeit der Ephemeropteren an die Verunreinigung der Gewässer und die Möglichkeit ihrer Ausnützung als limnosaprobe Bioindikatoren. p. 145-149. *In:* Proc. Sec. Inter. Conf. Ephem., PWN, Krakow.

Sowa, R. 1975. Ecology and biogeography of mayflies (Ephemeroptera) of running waters in the Polish part of the Carpathians. 1. Distribution and quantitative analysis. *Acta Hydrobiol.* 17: 223-297.

Sowa, R. 1975a. Ecology and biogeography of mayflies(Ephemeroptera) of running waters in the Polish part of the Carpathians. 2. Life cycles. *Acta Hydrobiol.* 17: 319-353.

Sowa, R. 1979. Le développement des Ephéméroptères de la rivière Dunajec aux environs de Pieniny. p. 125-131. *In:* Proc. Sec. Int. Conf. Ephem., PWN, Kraków.

Wójcik, S. 1963. Fauna jetek (Ephemeroptera) Wisly pod Tczewem. *Zesz. Nauk. Uniw. Adama Mickiewiza, Biol.* 4: 102-120.

Zelinka, M., et P. Marvan. 1961. Zur Präzisierung der biologischen Klassifikation der Reinheit fliessender Gewässer. *Arch. Hydrobiol.* 57: 389-407.

THE DISTRIBUTION OF EPHEMEROPTERA IN NORTHERN CANADA

Donald G. Cobb and John F. Flannagan

Department of Fisheries and Oceans
Freshwater Institute
501 University Crescent
Winnipeg, Manitoba, Canada
R3T 2N6

ABSTRACT

Mayflies from six drainage systems in North-Central Canada were examined. The results, together with published distribution records from other systems, indicated that drainage systems arising in the north and flowing to the ocean have a lower diversity of mayflies than drainage systems arising further south (where there is a diverse fauna) and flowing north to the ocean. This suggests that most mayfly dispersal occurs via water, and is most likely by larvae. An arctic species, *Baetis lapponicus* (Bengtsson), previously reported from Northern Europe, Northeastern Russia and from Baffin Island (Canada) is recorded on mainland Canada for the first time. Its distribution indicates that its dispersal was transatlantic.

INTRODUCTION

Until recently, few species of Ephemeroptera have been reported from Canada north of latitude 60°N. McDunnough (1936) reported *Baetis foemina*, a new species, and *B. lapponicus* (Bengtsson), a north European species (Müller-Liebenau 1969), from Lake Harbor on Baffin Island. Lehmkuhl (1973) described a new species, *Baetis bundyae* from the edge of arctic tundra ponds near Rankin Inlet, Northwest Territories (N.W.T.). Brunskill *et al.* (1973) and Wiens *et al.* (1975) have provided an extensive list of genera and species for the Mackenzie and Porcupine River drainage systems of the N.W.T. and the Yukon Territories (Y.T.). Moore (1977) collected several species of mayflies from the N.W.T.

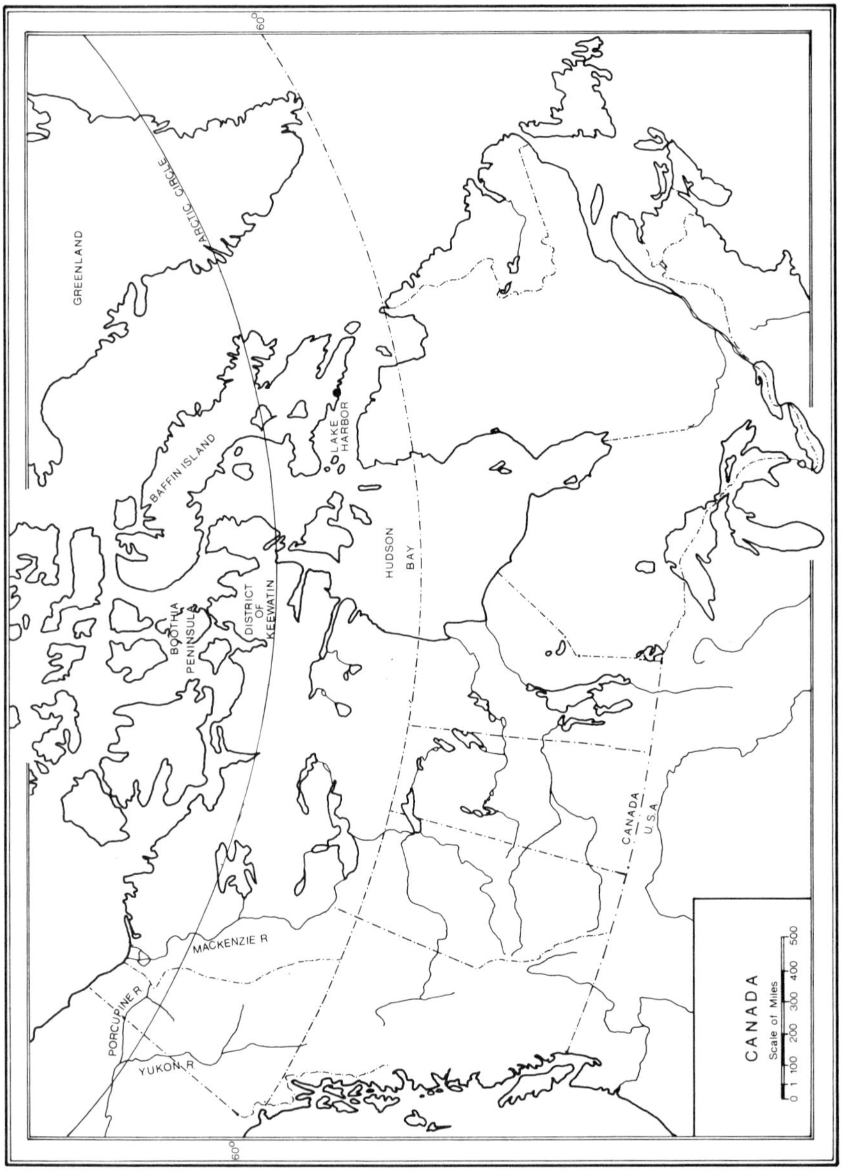

Figure 1. Location of drainage areas in Canada.

Mayfly nymphs were identified from Surber samples taken by Lawrence *et al.* (1977) from 118 streams in six drainage systems in the District of Keewatin and Boothia Peninsula, N.W.T. A species distribution list of mayflies of north-central Canada was compiled using the above information. Possible explanations of dispersal and origins of some species are presented and new distributions of several species are noted.

THE DRAINAGE SYSTEMS

The six drainage systems of the District of Keewatin and Boothia Peninsula (Table 1, Figs. 1 and 2) all arise locally and flow to Hudson Bay, the Gulf of Boothia, or Franklin Strait. The area containing these systems is predominately glaciated Canadian Shield, mostly devoid of trees with an altitude < 600 m above sea level (Lawrence *et al.* 1977). Drainage is poorly organized except for a few well defined rivers and tributaries. Climate is typically low and mid-arctic with short cool summers (mean maximum temperature in July is 17°C in the south and 10°C in the north). Since most streams and rivers in the drainage systems are not named, only the drainage system will be referred to.

The Mackenzie River (Figs. 1 and 3) starts at the outflow of Great Slave Lake, but its drainage system includes the Peace, Athabasca and Liard Rivers, together draining more than 1,000,000 km^2. In this drainage system there is a wide diversity of river and stream types ranging from the large Mackenzie and Liard Rivers with their steep banks of alluvial sediments to the smaller humic, shallow, boulder-cobble tributaries such as the Harris and Trail Rivers (Brunskill *et al.* 1973). The Mackenzie Delta (Fig. 3) covers 12,000 km^2 and consists of the West, the East, and the Main Channels and a myriad of lakes. Thus a wide variety of habitats exist within these systems.

The Porcupine River, Y.T. (Fig. 3), flows west to the Yukon River which drains to the Pacific Ocean. Alpine creeks, foothill streams and floodplain rivers provide habitats for a diverse fauna of mayflies. This area was not glaciated (Matthews 1979) and may have been a glacial refugium.

DISTRIBUTION OF MAYFLIES

Fifty-five taxa were recorded north of latitude 60°N. The richest fauna (44 taxa) was recorded in the Mackenzie River drainage; the Porcupine River system was less diverse (15 taxa) (Table 2). Fauna of the eastern N.W.T. was poor by comparison.

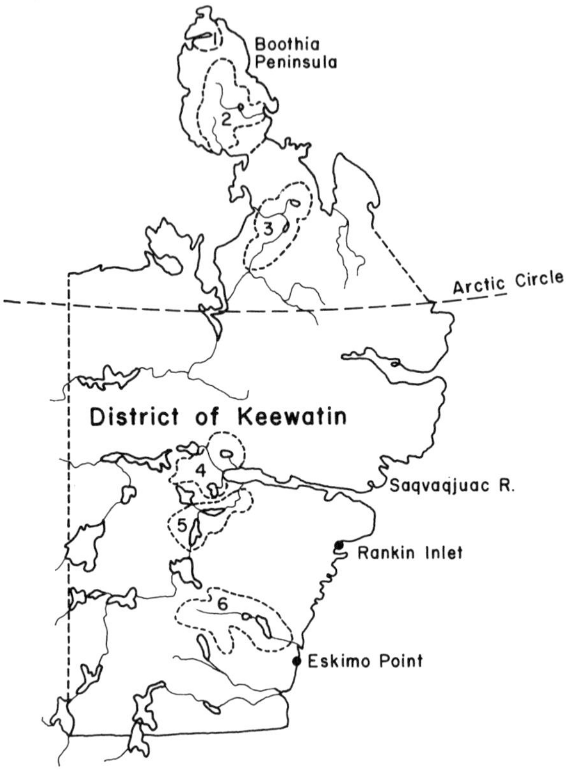

Figure 2. Drainage areas of the District of Keewatin and Boothia Peninsula (from Lawrence *et al.* 1977).

1. Amitoryouak Lake Drainage System.
2. Sanagak Lake Drainage System.
3. Murchison Lake Drainage System.
4. Pitz Lake Drainage System.
5. Thirty Mile Lake Drainage System.
6. Maguse River Drainage System.

Two, and possibly three mayfly species are common to the Mackenzie River drainage system and the Keewatin-Boothia area. *Baetis foemina* McDunnough (McDunnough 1936) occurs on Baffin Island, and in the Flat River (Moore 1977) a tributary of the Mackenzie River *Metretopus borealis* Eaton, a holarctic species which occurs across northern Canada (Edmunds *et al.* 1976), is found in the Pitz Lake and Thirty Mile Lake drainages and in the Harris River, a tributary of the Mackenzie River (Wiens *et al.* 1975). *Leptophlebia* sp. which

Table 1. Description of Keewatin-Boothia Drainage Systems. (Compiled from Lawrence *et al.* 1977).

	Geographic Location	Area Drained km^2	Date Sampled	Stream Substrate	Vegetation
Amitoryouak Lake Drainage Area	71°16'N to 71°55	910	Aug. 22-23 1976	Rock outcroppings & gravel.	Lichens, cushion plants
Sanagak Lake Drainage Area	69°30'N to 70°45'N	3,755	Aug. 16-26 1976	Rock, gravel & sand.	Sparce. Moss - sedge.
Murchison Lake Drainage Area	67°32'N to 68°46'N	7,760	Aug. 27 - Sep. 5 1976	Sand, silt & cobble.	Scrub willow.
Pitz Lake Drainage Area	63°40'N to 65°00'N	7,030	Jul. 18 - Aug. 15 1976	Boulder, cobble, gravel, sand & fragmented rock.	Scrub willow.
Thirty Mile Lake Drainage Area	62°45'N to 64°00'N	8,575	Jul. 5-17 1976	Cobble, boulder, gravel & sand.	Scrub willow & sedge.
Maguse River Drainage Area	61°12'N to 62°26'N	10,600	Jun. 8-22 1976	Clean gravel, cobble & boulder.	Scrub willow & sedge.

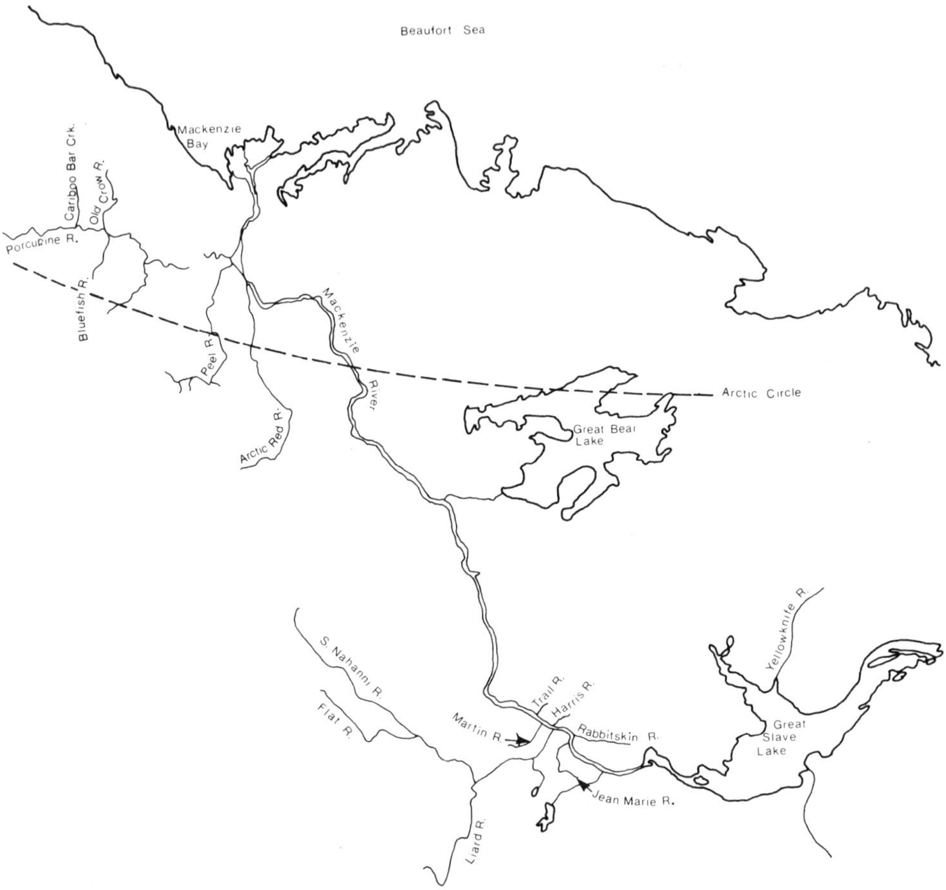

Figure 3. Mackenzie-Porcupine Drainage Systems (from Brunskill *et al.* 1973).

keys to either *L. cupida* Say or *L. nebulosa* (Walker) is also found in both areas, but the nymphs of these two species are not distinguishable (Burks 1953).

Baetis lapponicus (Bengtsson) previously reported by McDunnough (1936) from Lake Harbor on Baffin Island is recorded from mainland Canada for the first time. It was the only species of mayfly found in the streams of the Sanagak Lake drainage system of the Boothia Peninsula. This record establishes the most northerly location from which mayflies have been recorded in Canada (70°12'N). Distribution of *Baetis flavistriga* McDunnough, *B. parvus* Dodds and *B. pygmaeus* (Hagen) are extended to the District of Keewatin. *Baetis pygmaeus*

(Hagen) and *Heptagenia* sp. were found in one Surber sample from a stream in the Maguse River drainage area. *Leptophlebia* sp. was collected at an inlet of a stream to a lake in the same area. Further sampling of this area may reveal a different species composition from the more northern drainage systems.

DISCUSSION

Baetis lapponicus (Bengtsson) and *B. macani bundyae* Lehmkuhl both occur only in northern Europe and in the District of Keewatin-Boothia Peninsula. Mayflies, being the oldest winged insects, have existed since Carboniferous and Permian times (Lehmkuhl 1979), so they were present before continental drift separated Europe, Asia and North America. Therefore it appears that these species dispersed between Europe and Canada via a north Atlantic route.

Several factors could be responsible for the large difference in diversity of mayflies between the Mackenzie River and Keewatin-Boothia drainage systems. The climate of the Mackenzie River system is less severe than that of the Keewatin-Boothia area. The deciduous vegetation in many areas of the Mackenzie drainage is an important energy source, while vegetation in the Keewatin-Boothia area is sparce (Table 1). Despite these differences, the streams and rivers of the Keewatin-Boothia area have a large diversity of habitats and thus would be expected to have a greater diversity of mayflies than was found (Thienemann 1954).

Freshwater fish are presently distributed according to post-glacial drainages (McPhail and Lindsey 1970). Stoneflies have followed the cold meltwaters of receding glaciers resulting in their present post-glacial distribution (Dosdall and Lehmkuhl 1979, Flannagan 1978, Brinck 1949). Neotropical species of mayflies dispersed through the changing drainage patterns caused by glaciation (Ide 1955). Therefore cold stenothermic mayfly species probably recolonized their present day northern locations in a similar fashion. Thus, it would appear that little or no recolonization, or additional colonization by southern species, has occurred in northern drainage systems not connected by a river to southern systems. This would explain the low diversity of species in the Keewatin-Boothia systems. This indicates that in the north of Canada dispersal of mayflies probably occurs in the nymphal stage. Long dispersal flights of adults recorded in the southern latitude (Müller-Liebenau 1971) may be impossible in the north because of adverse climatic conditions.

CONCLUSION

The eastern arctic systems in this study are probably more vulnerable to environmental perturbations than those in the western

Table 2. Distribution list.

	Mackenzie Delta (Channels and Lakes)*	Mackenzie Mainstream and Tributaries**	Porcupine River Drainage System*	Amitoryouak Lake Drainage System	Sanagak Lake Drainage System	Murchison Lake Drainage System	Pitz Lake Drainage System	Thirty Mile Lake Drainage System	Maguse River Drainage System	Rankin Inlet***	Lake Harbor (Baffin Island)****	Saqvaqjuac River
Siphlonuridae												
Ameletus prob. *sparsatus* McDunnough		+										
Ameletus sp.		+	+									
Parameletus sp.		+										
Siphlonurus cf. *alternatus* (Say)		+										
S. occidentalis inflatus McDunnough	+											
S. spectabilis Traver		+										
Siphlonurus spp.	+	+	+									
Metretopidae												
Metretopus borealis Eaton		+					+	+				
Siphloplecton interlineatum (Walsh)		+										
Siphloplecton sp.		+	+									
Baetidae												
Baetis cf. *bicaudatus* Dodds		+										
B. macani bundyae Lehmkuhl						+	+	+		+		+
B. flavistriga McDunnough						+	+	+				
B. foemina McDunnough		+									+	
B. lapponicus (Bengtsson)					+	+	+	+			+	
B. parvus Dodds						+						
B. propinquus (Walsh)			+									
B. pygmaeus (Hagen)									+			
B. prob. *vagans* McDunnough		+										
Baetis spp.	+	+	+									
Callibaetis sp.			+									
Centroptilum sp.		+	+									
Cloeon sp.	+	+	+									
Pseudocloeon sp.		+	+									
Heptageniidae												
Heptageniinae												
Cinygmula cf. *tarda* McDunnough		+										
Cinygmula sp.			+									
Epeorus deceptivus (McDunnough)		+										
Epeorus sp.			+									
Heptagenia cf. *flavescens* (Walsh)		+										
H. hebe McDunnough and/or *maculipennis* Walsh		+										
H. julia Traver		+										
H. pulla (Clemens)		+										
Heptagenia spp.	+	+	+						+			
Rhithrogena undulata (Banks)		+										
Rhithrogena sp.	+	+	+									
Stenonema tripunctatum tripunctatum (Banks)		+										
S. vicarium (Walker)		+										
Stenonema spp.		+										
Arthropleinae												
Arthroplea bipunctata McDunnough		+										

Table 2. Distribution list (cont'd.)

	Mackenzie Delta (Channels and Lakes)*	Mackenzie Mainstream and Tributaries**	Porcupine River Drainage System*	Amitoryouak Lake Drainage System	Sanagak Lake Drainage System	Murchison Lake Drainage System	Pitz Lake Drainage System	Thirty Mile Lake Drainage System	Maguse River Drainage System	Rankin Inlet***	Lake Harbor (Baffin Island)****	Saqvaqjuac River
Leptophlebiidae												
Leptophlebia nebulosa (Walker)		+										
Leptophlebia sp.	+	+							+			
Paraleptophlebia guttata (McDunnough)		+										
Paraleptophlebia sp.	+	+										
Ephemerellidae												
Ephemerella (Dannella) simplex McDunnough		+										
E. (Ephemerella) aurivillii (Bengtsson)		+										
E. (E.) excrucians Walsh		+										
E. (Eurylophella) bicolor Clemens		+										
E. (E.) sp.							+	+				
Ephemerella spp.	+	+	+									
Tricorythidae												
Tricorythodes sp.		+										
Caenidae												
Brachycercus sp.		+										
Caenis sp.	+	+	+									
Baetiscidae												
Baetisca obesa (Say)		+										
Ephemeridae												
Ephemera simulans Walker		+										

* Material of Wiens *et al.* 1975

** Material of Wiens *et al.* 1975, Moore 1977

*** Lehmkuhl 1973

**** McDunnough 1936

arctic. Elimination of mayfly species from northern areas not connected by a water route to southern sources of recolonization may be permanent.

ACKNOWLEDGMENTS

We would like to thank the Environmental Impact Section, Freshwater Institute for providing the mayflies from the Keewatin-Boothia area. We appreciate the promptness with which Dr. D. Morihara verified nymphal identifications. Thanks to Dr. D. Rosenberg for suggesting improvements to the manuscript. Ms. G. Decterow and Ms. M. Ziprick typed the manuscript and Ms. D. Taite prepared the figures.

RESUME

Ce document étudie les éphéméroptères provenant de six systèmes de drainage situés dans le centre-nord du Canada. Les résultats de l'analyse, ainsi que les rapports publiés sur la répartition des insectes provenant d'autres systèmes, révèlent que les systèmes de drainage s'écoulant du nord en direction de la mer contiennent une moins grande variété d'éphéméroptère que les systèmes de drainage qui prennent naissance plus au sud (où la faune est plus variée) montant vers le nord jusqu'à l'océan. Cela permet de supposer que le gros de la dispersion des éphéméroptères s'effectue par voir d'eau et sans doute par les larves. Une espèce arctique, la *Baetis lapponicus* (Bengtsson) dont la présence avait déjà été signalée en Europe septentrionale, au nord-est de la Russie et à l'île Baffin (Canada), vient d'être observée pour la première fois sur le continent canadien. Sa répartition indique qu'elle y est parvenue d'outre-Atlantique.

ZUSSAMENFASSUNG

In Nord-Zentralkanada wurden Eintagsfliegen von sechs Entwässerungssystemen studiert. Die Ergebnisse in Verbindung mit veröffentlichten Verteilungsberichten ergaben folgendes: Entwässerungssysteme, die im Norden entstehen und in den Ozean fließen, weisen eine geringere Mannigfaltigkeit von Eintagsfliegen auf, als Entwässerungssysteme, deren Ursprung sich weiter im Süden befindet (wo eine abwechslungsreiche Fauna besteht) und die nach Norden in den Ozean fließen. Es liegt hahe, daß die Verbreitung der Eintagsfliegen hauptsächlich im Wasser und zwar höchst wahrscheinlich durch Larven geschient. Eine arktische Art, *Baetis lapponicus* (Bengtsson), zuvor von Nordeuropa, Nord-Ostrußland, und Baffin Island (Kanada) gemeldet, wird nun in Kanada zum ersten Mal auf dem Festland verzeichnet. Die Verteilung zeigt, daß ihre Verbreitung auf transatlantischem Wege vor sich ging.

REFERENCES

Brinck, P. 1949. Studies on Swedish stoneflies. *Opusc. Entomol. Suppl.* 11: 1-250.

Burks, B.D. 1953. The mayflies or Ephemeroptera of Illinois. *Ill. Nat. Hist. Surv. Bull.* 26(1): 1-216.

Brunskill, G.J., D.M. Rosenberg, N.B. Snow, G.L. Vascotto and R. Wagemann. 1973. Ecological studies of aquatic systems in the Mackenzie-Porcupine drainages in relation to proposed pipeline and highway developments. Vol. 1. *Canada Task Force N. Oil. Dev. Environ. Social Comm. Rep.* 73-40: 1-131.

Dosdall, L. and D.M. Lehmkuhl. 1979. Stoneflies (Plecoptera) of Saskatchewan. *Quaest. Entomol.* 15: 3-116.

Edmunds, G.F., Jr., S.L. Jensen and L. Berner. 1976. The mayflies of North and Central America. Univ. Minnesota Press, Minneapolis.

Flannagan, J.F. 1978. The winter stoneflies *Allocapnia granulata* (Capniidae), *Taeniopteryx nivalis* and *T. parvula* (Taeniopterygidae) in southern Manitoba. *Can. Entomol.* 110: 111-112.

Ide, F.P. 1955. Two species of mayflies representing southern groups occurring at Winnipeg, Manitoba (Ephemeroptera). *Ann. Entomol. Soc. Am.* 48: 15-16.

Lawrence, M., S. Davies, G. Collins, F. Hnytka, K. Kroeker and R. Sie. 1977. A survey of aquatic resources of the District of Keewatin and Boothia Peninsula. *Interim Report for the Arctic Islands Pipeline Project. Department of Fisheries and the Environment. Fisheries and Marine Service. Western Region.* Vols. 1-8.

Lehmkuhl, D.M. 1973. A new species of *Baetis* (Ephemeroptera) from ponds in the Canadian Arctic, with biological notes. *Can. Entomol.* 105: 343-346.

Lehmkuhl, D.M. 1979. Ephemeroptera. p. 305-308. *In*: Canada and its insect fauna. H.V. Danks (ed.). *Mem. Entomol. Soc. Can.* 108.

McDunnough, J. 1936. A new arctic Baetid (Ephemeroptera). *Can. Entomol.* 68: 32-34.

McPhail, J.D. and C.C. Lindsey. 1970. Freshwater fishes of northwestern Canada and Alaska. *Bull. Fish. Res. Board Can.* 173: 1-381.

Matthews, J.V., Jr. 1979. Tertiary and quaternary environments: historical background for an analysis of the Canadian insect fauna. p. 31-86. *In:* Canada and its insect fauna. H.V. Danks (ed.). *Mem. Entomol. Soc. Can.* 108

Moore, I.A. 1977. Species descriptions of some Plecoptera, Ephemeroptera, Trichoptera and Diptera from three areas of the Northwest Territories. *Can. Environ. Protect. Serv. MS. Rep.* NW-77-9: 1-57.

Müller-Liebenau, I. 1969. Revision der europaüschen Arten der Gattung *Baetis* Leach, 1815 (Insecta, Ephemeroptera). *Gewässer Abwässer* 48/49: 1-214.

Müller-Liebenau, I. 1971. Ephemeroptera (Insecta) von den Kanarishen Inselin. *Gewässer Abwässer* 50/51: 7-40.

Thienemann, A. 1954. Ein drittes biozönotisches Grundprinzip. *Arch. Hydrobiol.* 49: 421-422.

Wiens, A.P., D.M. Rosenberg and N.B. Snow. 1975. Species list of aquatic plants and animals collected from the Mackenzie and Porcupine River watersheds from 1971-1973. *Can. Fish. Mar. Serv. Tech. Rep.* 557: 1-39.

THE OCCURRENCE AND BIOLOGY OF *SIPHLONISCA AERODROMIA* NEEDHAM (EPHEMEROPTERA: SIPHLONURIDAE) IN MAINE, U.S.A.

K. Elizabeth Gibbs

Department of Entomology
University of Maine at Orono
Orono, Maine 04469

ABSTRACT

Siphlonisca aerodromia Needham has been collected from three locations in Penobscot and Aroostook Counties in Maine. Nymphs were found in areas of *Carex* sp. flooded as a result of stream overflow following snow melt. They were collected between April 21 and June 5 in 1978 and between April 26 and May 24 in 1979. Adult emergence occurred in early June in 1978 and late May in 1979.

Observations on the feeding habits of the nymphs and analysis of their gut content indicated that they were predaceous and fed on other mayfly nymphs, especially those of *Siphlonurus* sp.

RESUME

Le *Siphlonisca aerodromia* Needham a été recueilli à trois endroits dans les comtés de Penobscot et Aroostook dans le Maine. Les nymphes furent décourvertes dans des champs de *Carex* inondés par un cours d'eau au moment de la fonte des neiges. Les échantillons furent recueillis entre le 21 avril et le 5 juin 1978 et entre le 26 avril et le 24 mai 1979. Les adultes sont apparus au début de juin en 1978 et à la fin de mai en 1979.

Des observations effectuées sur les habitudes alimentaires des nymphes et l'analyse des matières contenues dans leur intestin indiquent que ce sont des prédateurs et qu'elles se nourrissent d'autres nymphes éphéméroptères, en particulier de celles de *Siphlonurus* sp.

ZUSSAMENFASSUNG

Siphlonisca aerodromia Needham wurde an drei Stellan in Penobscott und Aroostock (Maine) gesammelt. Nymphen wurden in *Carex* sp. in Gebieten gefunden, die durch die Schneeschmelze von Flüssen überflutet waren. Die Sammlung fand zwischen dem 21 April und dem 5 Juni 1978 und zwischen dem 26 April und 24 Mai 1979 statt. Ausgewachsene Fliegen kamen 1978 Anfang Juni zum Vorschein und 1979 gegen Ende Mai.

Beobachtungen, die den Nahrungserwerb der Nymphen betrafen und eine Analyse ihres Darminhaltes zeigten, daß sie Raubtiere wafen und sich von anderen Eintagsfliegennymphen, besonders *Siphlonurus* sp., ernährten.

ABUNDANCE AND ALTITUDINAL DISTRIBUTION OF EPHEMEROPTERA IN A ROCKY MOUNTAIN STREAM

James V. Ward and Lewis Berner*

Department of Zoology and Entomology
Colorado State University
Fort Collins, Colorado 80523, U.S.A.

ABSTRACT

The nymphs of 29 species of Ephemeroptera representing six families were collected from rubble riffles in a Colorado stream system, from alpine tundra to the plains. Density ranged from 44 nymphs/m^2 at the headwater site to 2,031 nymphs/m^2 in the Foothills Limnological Zone. Heptageniidae comprised the majority of the mayflies at upstream sites and were numerically important at all but the plains location. Baetidae attained maximum diversity and relative abundance (62% of total mayflies) in the plains stream. Seven species of *Ephemerella* exhibited overlapping distribution patterns from 3109 m a.s.l. to the plains.

INTRODUCTION

The Colorado portion of the Cordilleran chain contains numerous peaks rising over 4,000 m above sea level (a.s.l.). The altitudinal gradient offers excellent opportunities to investigate factors which control the diversity, composition, and abundance of stream organisms. Illies and Botosaneanu (1963), and more recently Hynes (1970) and Hawkes (1975), have reviewed results of studies of longitudinal faunal changes in stream systems.

Remarkably little previous work on stream zonation has been conducted in Colorado. Dodds and Hisaw's (1925) pioneering study of South Boulder Creek employed only qualitative sampling techniques, was limited to the late June to early September period, and did not

* Department of Zoology, University of Florida, Gainesville, Florida.

include physico-chemical data. Other studies have been limited in sampling period, altitudinal gradient, or taxonomic scope (Knight and Gaufin 1966, Elgmork and Saether 1970, Mecom 1972, Allan 1975).

St. Vrain Creek in the Front Range of Colorado flows freely from a glacier-fed source to the plains over an altitudinal gradient of nearly 1900 m. The ensuing data on the abundance and distribution of lotic Ephemeroptera as a function of altitude were collected during a comprehensive study of stream zonation which will be reported elsewhere.

STUDY AREA AND METHODS

Middle St. Vrain Creek begins near the St. Vrain Glaciers, which occupy cirques on the east side of the Continental Divide in northern Colorado. The stream originates in alpine tundra (3414 m a.s.l.) and drops 1870 m to the plains over a distance of 54 km. The lakes and terrestrial vegetation of the region have been described in detail (Pennak 1958, Marr 1967). Bottom fauna-substrate relationships were examined at a single location on North St. Vrain Creek, a major tributary of the study stream (Ward 1975).

Benthic samples were collected from rubble riffles at eleven sites (Fig. 1). Six Surber samples (929 cm^2) were taken from rubble riffles at each site, three with a regular mesh net (720 μm) and three with fine mesh (240 μm). Only the results of the regular mesh samples are reported herein.

From the Montane Limnological Zone (2633 m) to the plains, sampling was conducted monthly from June 1975-May 1976. Upper sites, which were reached by foot trail, were only accessible during summer due to the deep snowpack. Sites 4 and 5 were accessible by June, Sites 2 and 3 by July, but sampling at Site 1 was not possible until August due to an especially heavy snowpack the preceding winter. Because the open season is so short at higher elevations, it is unlikely that any common species were missed despite the restricted sampling period at upper sites.

Because of the difficulty of assigning species names to mayfly nymphs without associated adults for confirmation, question marks are used liberally to indicate uncertainty. Some identifications are presumptive because of the proximity of St. Vrain Creek to South Boulder Creek, from which several species were described by Dodds (1923), and the great likelihood of the streams having similar faunas. Edmunds' (1952) keys (with emendation, G.F. Edmunds, Jr., personal communication) were used to distinguish *Baetis* nymphs, although the confused state of *Baetis* taxonomy in North America is fully recognized.

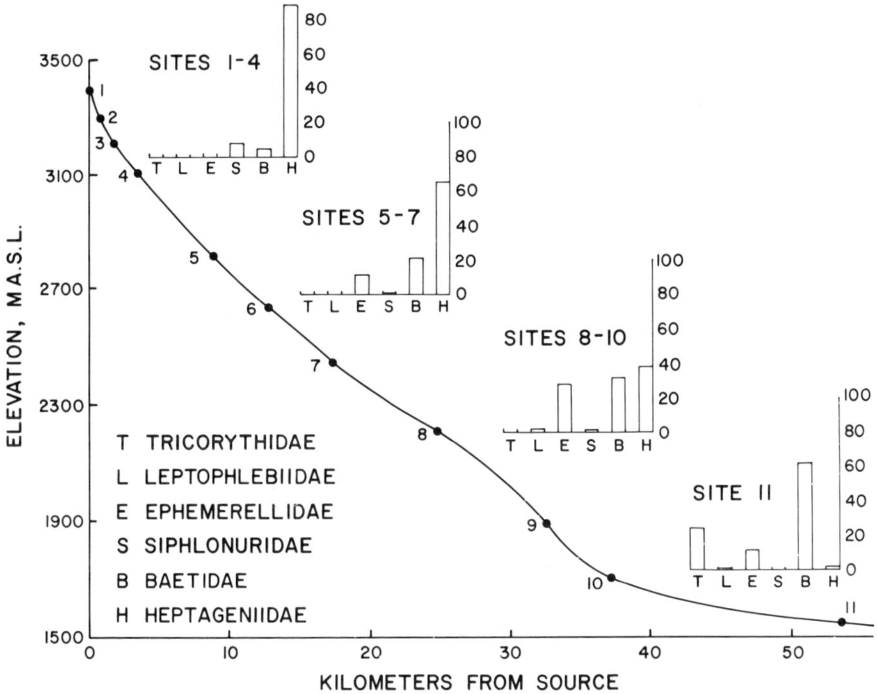

Figure 1. The relative contribution (mean % composition by density) of each family to total mayfly nymphs as a function of altitude in St. Vrain Creek, Colorado. Site locations are indicated by Arabic numerals on the longitudinal stream profile.

Detailed data on physico-chemical parameters, periphton, and macrophytes were collected but will only be referred to briefly.

RESULTS AND DISCUSSION

The nymphs of 29 species of Ephemeroptera from six families were collected during the study. A few additional species not found as immatures in riffles (e.g., *Siphlonurus occidentalis* Eaton) were collected in sweep nets. The number of species exhibited a general increase downstream. Only three species were found at Site 1 in the alpine tundra, whereas 18 species were collected at the plains location. The mean density ranged from 44 nymphs/m^2 at Site 1 to 2,031 nymphs/m^2 at Site 8 in the foothills. Density at the plains location (Site 11) averaged 1,060 nymphs/m^2.

The extent to which various families contributed to the total mayfly fauna varied as a function of altitude (Fig. 1; the percentage representation by family is shown for sites at different elevations). Heptageniids exhibited a progressive decrease downstream and were the predominant mayflies at all but the plains location. Siphlonurids were numerically important only at high elevations, whereas leptophlebiids (which were never abundant) were collected only at the lower sampling sites. The family Baetidae exhibited a progressive downstream increase in relative abundance, comprising 62% of total mayflies at Site 11. Ephemerellids reached a maximum relative abundance (28% of total mayflies) at Sites 8-10 in the foothills. The family Tricorythidae only occurred at the plains location where *Tricorythodes minutus* Traver comprised 24% of the mayfly fauna.

The nymphal species distributions observed during this study are presented in Figure 2. The altitudinal range appears greater in a few instances than shown for a given family in Figure 1 due to a small number of individuals which did not show up on the scale. A few species (e.g., *Baetis bicaudatus* Dodds) occurred over nearly

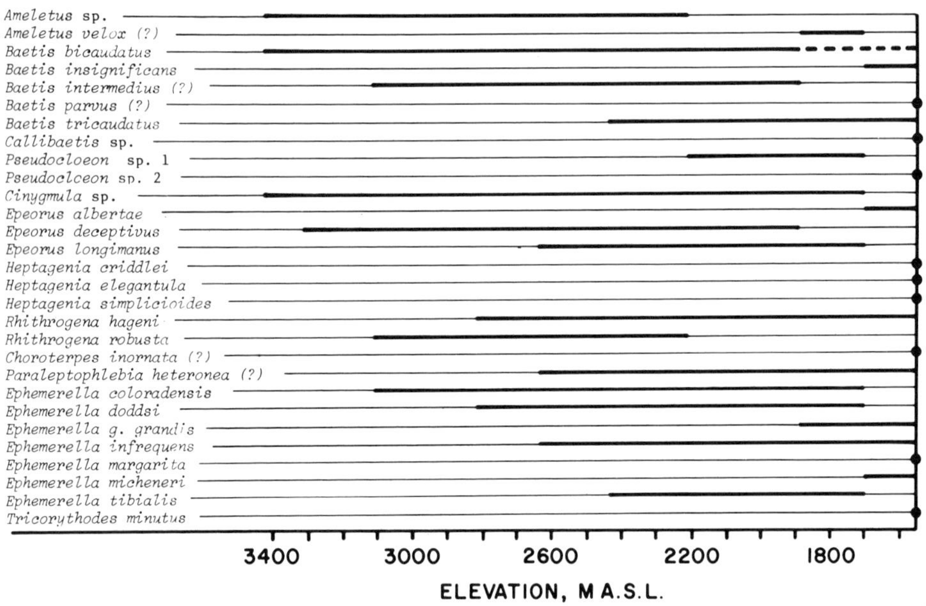

Figure 2. Altitudinal distribution (dark horizontal lines and solid circles) of Ephemeroptera nymphs in St. Vrain Creek, Colorado.

the entire altitudinal gradient. Nine species were found at only one location (solid circles on Figure 2); all of these were from Site 11 on the plains. Three additional species were collected only from the plains and the lowermost foothills site. No species was restricted to the headwaters. The three species occuring at Site 1 were all much more abundant at downstream locations and were found over all or nearly all of the altitudinal gradient. Although a few species were restricted to middle reaches, the general pattern appears to be a selective elimination of species toward higher elevations with only three species able to tolerate conditions in the alpine tundra stream.

Dodds (1923) collected mayflies over the same altitudinal range in a drainage system immediately to the south of St. Vrain Creek. Dodds listed 26 species of nymphs and adults, including specimens from lakes and ponds as well as from South Boulder Creek and its tributaries. Synonomies reduce that number to 24 species, only 13 of which were collected as nymphs from lotic waters, compared to 29 species of nymphs from St. Vrain Creek. The restricted sampling period (late June to early September) of the South Boulder study probably accounts for the smaller number of species, especially at lower elevations. Many of the species collected during the present study, but not reported by Dodds, were restricted to the plains site or to the plains and lowest foothills location (Fig. 2). Some of these species were represented by only a few individuals, whereas others (e.g., *Tricorythodes minutus*) were abundant, although restricted in distribution.

Elgmork and Saether (1970) conducted an intensive 4-day study in mid-July of the portion of North Boulder Creek above the tree line in which they reported that *Cinygmula mimus* (Eaton) occurred at higher elevations (3550 m) than any mayfly. *Baetis bicaudatus* Dodds and a few specimens of *Ephemerella coloradensis* Dodds were collected up to 3400 m. At Site 1 of the present study, three species of mayflies were collected, *Baetis bicaudatus*, *Ameletus* sp., and *Cinygmula* sp. *C. ramaleyi* "was the most abundant species in streams between 8,000 and 11,000 feet" in the South Boulder Creek Basin (Dodds 1923). Adult *Cinygmula par* (Eaton) were collected at Site 1 on St. Vrain Creek; the gills of the nymphs, however, contained a small fibrillar portion unlike *C. par*. Dodds reported *C. par* imagos at locations where *C. ramaleyi* nymphs were collected. In the present study, *Cinygmula* sp. nymphs were found at all except the plains locations with greatest density (958 individuals/m^2) at Site 3.

The genus *Ephemerella* was represented by seven species. *E. coloradensis* occupied the widest range of altitude and extended the highest (Fig. 2), although not approaching the 3400 m upper limit reported by Elgmork and Saether (1970). *E. margarita* Needham was restricted to the plains.

Six species of *Ephemerella* occurred sympatrically at Site 10. *E. coloradensis* and *E. micheneri* Traver were represented by marginal populations of a few individuals at the lower and upper limits, respectively, of their altitudinal distribution. Three of the four remaining species, *E. g. grandis* Eaton, *E. infrequens* McD., and *E. doddsi* Needham, exhibited slow seasonal cycles (Hynes 1970). *E. tibialis* McD. showed a fast seasonal cycle. Analysis of production and life histories as a function of altitude will be presented in a later paper.

The distribution of stream organisms has been attributed to a variety of factors (Hynes 1970), many of which vary as a function of altitude and thus may be responsible, directly or indirectly, for the zonation pattern of Ephemeroptera. Although temperature and altitude determine oxygen solubility, it is unlikely that oxygen was limiting at any site since all values were near saturation. Variations in current and substrate were reduced by restricting sampling to rubble riffles.

Several physico-chemical parameters varied significantly with altitude. Stream size and associated changes in habitat diversity may be important since Site 1 is a first-order stream, whereas the plains stream is fifth-order. The pH increased from near neutrality in the headwaters to 8.1 in the plains stream. Total dissolved salts which averaged less than 10.0 mg/L at Site 1 increased to nearly 200 mg/L at Site 11. Suspended matter values and hardness also exhibited increases downstream. Many of these physico-chemical parameters varied only slightly between sampling sites in the mountains, but exhibited great increases from Site 10 to Site 11.

Mean summer temperatures increased from 2.3°C to 20.8°C over the altitudinal gradient. The importance of temperature in determining the distribution of stream fauna has been emphasized by several authors (e.g., Dodds and Hisaw 1925, Ide 1935, Kamler 1965, Hynes 1970), and most certainly plays a major role in determining the altitudinal ranges of species. The thermal pattern, as well as mean temperatures, must be considered (Lehmkuhl 1972, Ward and Stanford 1979).

Aquatic angiosperms were not encountered at mountain stream sites. The dense beds at the plains locations may provide microhabitats and current refugia not present at other sites.

Allochthonous inputs and *in situ* autochthonous production vary in quality and quantity over the gradient of altitude, thus providing a range of trophic conditions.

Other biotic factors, such as predation and competition, vary with altitude, although few studies have dealt with the influence of these phenomena in determining the distribution of Ephemeroptera or other stream invertebrates. The presence of congeners may, for example, influence the niche breadth and therefore the altitudinal range of a species.

CONCLUSIONS

Dodds and Hisaw concluded that "...there is one group of species confined to the high altitudes, another to low, and a third including those which have an extended range from low to high." In St. Vrain Creek, the general pattern appears to be the addition of species in the downstream direction without loss of those present at higher elevations (Fig. 2). However, a sharp faunal break occurs between Site 10 in the lower foothills and Site 11 on the plains coincident with the abrupt change in physico-chemical parameters. Seven species present at Site 10 were not collected at Site 11 during the year's study. In addition, nine species not found at mountain stream sites were collected at Site 11. Only three species were restricted to the middle range of elevation. In these cases, congeners occupy higher or lower elevations, or both.

The gradient of environmental conditions over short distances in the Front Range of the Colorado Rocky Mountains provides an ideal setting not only for investigating factors controlling distribution patterns and life cycle phenomena, but also for testing models of competitive interactions and segregation mechanisms.

ACKNOWLEDGMENTS

The authors wish to thank Dr. R.W. Pennak, University of Colorado, for reading the manuscript and Dr. G.F. Edmunds, Jr., University of Utah, for taxonomic assistance.

RESUME

Les nymphes de 29 especes d'éphéméroptères représentant familles furent recueillies dans des basses eaux rocailleuses du bassin d'un cours d'eau du Colorado s'étendant de la toundra alpine jusqu'aux plaines. La densité de population s'échelonnait de 44 nymphes/m^2 à l'origine à 2,031 nymphes/m^2 dans la zone limnologique des avant-monts. Vers l'amont, les heptagéniidés formaient la majorité des éphéméroptères et elles étaient nombreuses à tous les endroits sauf dans la plaine. Dans le cours d'eau de la plaine c'étaient les bäetis qui étaient les plus variées et relativement les plus abondantes (62% du nombre total d'éphéméroptères). On a enfin découvert qu'il y avait chevauchement quant à la répartition de sept espèces d'*Ephemerella* de 3109 mètresmer jusqu'aux plaines.

ZUSSAMENFASSUNG

Nymphen von 29 Ephemeropterenarten aus sechs Familien wurden aus dem Geröll der Stromschnellen eines Flußnetzes in Colorado gesammelt , das sich von alpiner Tundra bis in die Ebene erstreckt. Die Densität schwankte zwischen 44 Nymphen/m^2 am Oberlauf und 2,031 Nymphen/m^2 in der limnologischen Zone des Vorgebirges. Heptageniiden stellten den Hauptanteil der Eintagsfliegen au stromaufwärts gelegenen Stellen dar und waren zahlenmäßig von Bedeutung an allen, außer den in der Ebene gelegenen Plätzen. Baetiden erreichten im Flußlauf der Ebene maximale Vielfalt und verhältnismaßig größe Zahlen. (62% der gesamten Eintagsfliegen). Sieben Arten von *Ephemerella* wiesen sich überschneidende Verteilungspattern auf, bei Gebieten von 3,109 m ü. M. bis in die Ebene.

REFERENCES

Allan, J.D. 1975. The distributional ecology and diversity of benthic insects in Cement Creek, Colorado. *Ecology* 56: 1040-1053.

Dodds, G.S. 1923. Mayflies from Colorado: Description of certain species and notes on others. *Trans. Am. Entomol. Soc.* 49: 93-114.

Dodds, G.S. and F.L. Hisaw. 1925. Ecological studies on aquatic insects. IV. Altitudinal range and zonation of mayflies, stoneflies and caddisflies in the Colorado Rockies. *Ecology* 6: 380-390.

Edmunds, G.F., Jr. 1952. Studies on the Ephemeroptera. Part II. The taxonomy and biology of the mayflies of Utah. Unpubl. Ph.D. Thesis, Univ. Massachusetts.

Elgmork, K. and O.R. Saether. 1970. Distribution of invertebrates in a high mountain brook in the Colorado Rocky Mountains. *Univ. Colo. Stud. Ser. Biol.* 31: 1-114.

Hawkes, H.A. 1975. River zonation and classification, p. 312-374. *In*: B.A. Whitton (ed.), River Ecology, Blackwell Scientific, Oxford.

Hynes, H.B.N. 1970. The ecology of running waters. Univ. Toronto Press, Toronto.

Ide, F.P. 1935. The effect of temperature on the distribution of the mayfly fauna of a stream. *Publ. Ont. Fish. Res. Lab.* 50: 1-76.

Illies, J. and L. Botosaneanu. 1963. Probèmes et méthodes de la classification et de la zonation écologique des eaux courantes, considerées surtout du point de vue faunistique. *Int. Ver. theor. angew. Limnol. Mitt.* 12: 1-57.

Kamler, E. 1965. Thermal conditions in mountain waters and their influence on the distribution of Plecoptera and Ephemeroptera larvae. *Ekol. Pol. Ser.* A. 13: 377-414.

Knight, A.W. and A.R. Gaufin. 1966. Altitudinal distribution of stoneflies (Plecoptera) in a Rocky Mountain drainage system. *J. Kan. Entomol. Soc.* 39: 668-675.

Lehmkuhl, D.M. 1972. Change in thermal regime as a cause of reduction of benthic fauna downstream of a reservoir. *J. Fish. Res. Board Can.* 29: 1329-1332.

Marr, J.W. 1967. Ecosystems of the East Slope of the Front Range in Colorado. *Univ. Colo. Stud. Ser. Biol.* 8: 1-134.

Mecom, J.O. 1972. Productivity and distribution of Trichoptera larvae in a Colorado mountain stream. *Hydrobiologia* 40: 151-176.

Pennak, R.W. 1958. Regional lake typology in Northern Colorado, U.S.A. *Int. Verein. theor. angew. Limnol. Verh.* 13: 264-283.

Ward, J.V. 1975. Bottom fauna-substrate relationships in a northern Colorado trout stream: 1945 and 1974. *Ecology* 56: 1429-1434.

Ward, J.V. and J.A. Stanford. 1979. Ecological factors controlling stream zoobenthos with emphasis on thermal modification of regulated streams. *In*: J.V. Ward and J.A. Stanford (eds.), The Ecology of Regulated Streams. Plenum, New York (in press).

MAYFLY STRATEGIES IN A NORWEGIAN SUBALPINE LAKE

John E. Brittain

Zoological Museum
University of Oslo
Sars gate 1
Oslo 5, Norway

ABSTRACT

Mayfly ecology was studied in the lake, Øvre Heimdalsvatn, 1090 m a.s.l. in the Jotunheimen Mountains. Life cycle, winter growth, biomass, nutrition, predation, emergence and fecundity are considered. The four main species have adopted different strategies - *Baetis macani* has opted for low mortality and rapid summer growth, *Siphlonurus lacustris* for large size and high biomass, *Leptophlebia vespertina* for high density and *L. marginata* for winter growth. Each strategy has its disadvantages, but they enable coexistence in spite of the environmental restrictions imposed by a long period of lake ice cover and a short summer.

INTRODUCTION

The ecology of mayflies (Ephemeroptera) was studied in the subalpine lake, Øvre Heimdalsvatn as part of multidisciplinary ecosystem study under the auspices of IBP/PF. Some data have been published earlier on Ephemeroptera of this lake (Brittain 1974, 1978, 1979; Brittain and Lien 1978, Brittain and Lillehammer 1978,and Larsson *et al.* 1978). The aim of the present paper is to consider the life cycle strategies employed by the major ephemeropteran species in Ø. Heimdalsvatn by synthesising data from the above studies supplemented by new information on fecundity.

There are four major species in the lake: *Leptophlebia marginata* (L.), *L. vespertina* (L.), *Siphlonurus lacustris* Eaton and *Baetis macani* Kimmins. *Siphlonurus aestivalis* Eaton, *Ameletus inopinatus*

Eaton and *Baetis rhodani* (Pictet) have also been recorded. However, they have not been considered here as they only constitute 1-4% of the emerging mayflies (Brittain 1978).

HABITAT DESCRIPTION

Øvre Heimdalsvatn is siutated at 1090 m a.s.l. in the Jotunheimen Mountains of central southern Norway. It has a surface area of 0.78 km^2 a maximum depth of 13 m and is ice-covered from October until early June. During June there is a relatively rapid rise in temperature and lake temperatures generally remain between 10 and 14°C during the latter part of June, and in July and August. During this period the shallow waters of the lake, where Ephemeroptera are most abundant, can reach 20°C.

METHODS

Field sampling of nymphs and adults, together with a description of sampling sites and laboratory methods treated in this paper are described in detail in Brittain and Lillehammer (1978). Methods used in the predation analysis are given in Brittain and Lien (1978).

Fecundity, defined as the total number of eggs per female, was estimated from subimago material using a Sedgewick-Rafter cell (Clifford and Boerger 1974). One quarter of the total field was counted, except for *B. macani* where all eggs were counted. Subimago material was taken within two days of peak emergence for the species in question. Densities and fecundities are expressed in terms of total lake area in m^2, in order to compare values with other ecosystem components. However, the Ephemeroptera are more or less restricted to the shallow waters of Øvre Heimdalsvatn which only account for about 18% of the total lake area (Aarefjord *et al.* 1978). Therefore, when comparing values with other studies, especially those carried out in lotic habitats, this fact should be taken into account.

RESULTS AND DISCUSSION

All four major species are univoltine, but their strategies differ markedly (Fig. 1, Table 1). There is a progression from *L. marginata* in which nymphs are present for about 10 months of the year and in which nearly 60% of total nymphal growth is accomplished during the period of ice cover, to *B. macani* where nymphal growth is restricted to 2-3 months during the summer. *L. vespertina* has a similar life cycle to *L. marginata*, although emerging later and having only about 30% of its growth under the ice. *S. lacustris*, despite its more or less delayed egg hatching, is primarily a summer species, most of its nymphal growth being accomplished between ice break and

Table 1. Selected parameters of the ephemeropteran populations in Øvre Heimdalsvatn.

	L. marginata	*L. vespertina*	*S. lacustris*	*B. macani*
% of growth under ice	58	30	23	0
Consumed by trout ($nos/m^2/year$)	2.97	11.39	0.52	0.22
Av. indiv. wt at maturity (mg dw)	5.2	2.1	7.3	1.3
Nymphal biomass prior to emergence ($mg\ dw/m^2$)	4.34	6.94	7.86	1.03
C.V. of body length in pre-emergence popln.	15.1	14.1	27.7	17.1
Emergence peak 1971	3 Jul.	29 Jul.	9 Aug.	12 Aug.
1972	29 Jun.	23 Jul.	1 Aug.	10 Aug.

emergence in August.

These differences in nymphal growth strategy are reflected in emergence periods (Table 1). There is a succession in emergence of all the ephemeropterans in Øvre Heimdalsvatn (Brittain 1978). This succession is maintained from year to year despite changes in lake temperature (Brittain 1979).

The biometric variation in the pre-emergence nymphal populations can provide further information on strategies (Ulfstrand 1975). The delayed hatching in *S. lacustris* is clearly reflected in the high coefficient of variation in body length (Table 1). Maintenance of a coherent cohort in *B. macani* is perhaps not surprising as it has such a short period of nymphal growth and ice break-up appears to be the factor associated with egg hatching in this species (Brittain 1975). In the two *Leptophlebia* species the reduction in growth rate during the winter may operate in a regulatory manner.

The differences in nymphal growth periods are also reflected in their nutrition. All four species are herbivores. However, while detritus dominated in the two *Leptophlebia* species, vascular plant material in *S. lacustris* and diatoms in *B. macani* were almost as important as detritus.

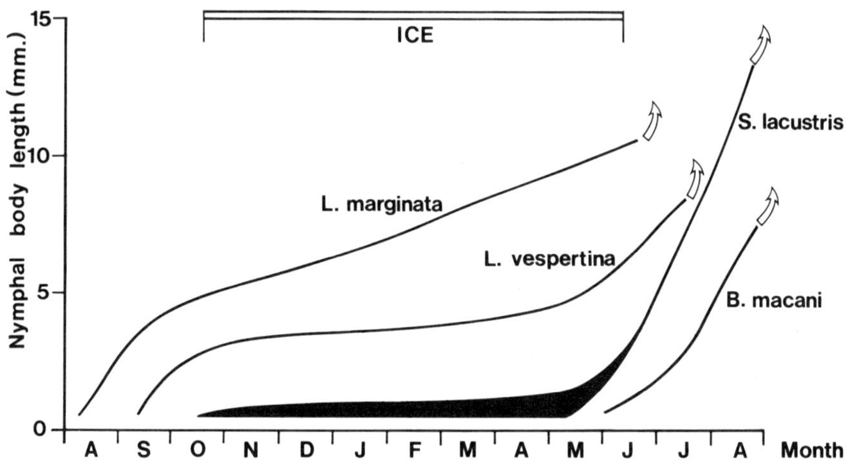

Figure 1. Nymphal growth of the major mayfly species in Øvre Heimdalsvatn. Emergence of each species and the period of lake ice cover are indicated.

Size is an important factor in ecological segregation. In Ø. Heimdalsvatn the four main species differ in their size at maturity and also throughout their nymphal lives (Fig. 1). *S. lacustris* is by far the largest of the four with a mean individual dry weight at maturity of 7.3 mg. In contrast nymphs of *B. macani* weigh only 1.3 mg. Such differences in nymphal size are important when considering biomass. For instance the nymphal biomasses of *S. lacustris* and *B. macani* prior to emergence were 7.86 and 1.03 mg/m^2 dry weight respectively (Table 1). By virtue of its large size, *S. lacustris* has a similar biomass to *L. vespertina*, the most abundant ephemeropteran in Ø. Heimdalsvatn.

The number of eggs required to produce an adult insect is also a measure of success and gives an indication as to whether the species inhabits a favourable environment (Clifford and Boerger 1974). The four main species in Ø. Heimdalsvatn differ in their fecundity and the subsequent survival of the eggs and nymphs (Tables 2 and 3). Clifford and Boerger (1974) and Benech (1972) showed that adult size is a major factor determining fecundity. Therefore it is not surprising to find that *S. lacustris* has the highest and *B. macani* the lowest fecundity of the four species. Degrange's (1960) data for *S. lacustris* and *L. vespertina* are similar to the values from Ø. Heimdalsvatn, although the range is greater in the present material. The fecundity of *B. macani* is exceptionally low, even when its small size is taken into account. The other three species produce about 120-130 eggs per mm body length, while *B. macani* produces only 30 eggs per mm body length.

However, if the number of eggs required to produce one adult is considered, the picture changes radically. Mortality from egg to imago is far greater in *S. lacustris* than in *B. macani* with the two *Leptophlebia* species occupying an intermediate position (Table 3).

Table 2. Fecundity and body length in mayfly subimagos (10 of each species examined) from Øvre Heimdalsvatn.

	Fecundity		Av. body length (mm)	No. eggs/mm body length
	Range	Av. no.		
L. marginata	600-2020	1380	10.7	129.0
L. vespertina	640-1832	1079	8.7	124.0
S. lacustris	624-2792	1740	13.2	131.8
B. macani	129-258	201	6.8	29.6

The oviposition behaviour in *Baetis* in which the female lays her eggs on the substratum rather than scattering them over the water surface may considerably reduce egg mortality. Also, *B. macani* by its presence as a nymph for only a short period largely avoids trout predation (Table 1). It is also able to utilise the peak biomass of benthic and epiphytic algae while the two *Leptophlebia* species are in the egg stage. Unlike the other three species *B. macani* is not at or near its altitudinal limit. The *Leptophlebia* species are rare above the subalpine vegetation belt and *S. lacustris* usually occurs in low numbers in the low alpine (Brittain 1974). In contrast *B. macani* can occur in high densities in both low and mid-alpine localities (Brittain 1974, 1975). Nevertheless, if one compares these values with Clifford and Boerger's data (1974) all four major species in Ø. Heimdalsvatn can be considered to be living in a more or less favourable environment.

Thus, in summary, the four main species in the subalpine lake Øvre Heimdalsvatn, appear to have adopted different approaches: *B. macani* has opted for low mortality and rapid summer growth, *S. lacustris* for large size and high biomass, *L. vespertina* for high density and *L. marginata* for winter growth. Each strategy carries with it disadvantages, but the adoption of differing strategies enables these four species to coexist in spite of the environmental restrictions imposed by a long period of ice cover and a short summer.

Table 3. Egg and adult mayfly densities in Øvre Heimdalsvatn.

	Av. no. eggs per female	Total no. females emerging $/m^2/year$	Total no. eggs $/m^2/year$	Total no. males and females emerging $/m^2/year$	No. eggs required to produce 1 adult
L. marginata	1830	0.380	524.4	0.833	630
L. vespertina	1079	1.382	1491.2	3.232	461
S. lacustris	1740	0.576	1002.2	1.077	931
B. macani	201	0.409	82.2	0.797	103

RESUME

L'écologie des éphéméroptères a été etudiée dans le lac Øvre Heimdalsvatn, situé a 1090 m d'altitude. Le cycle saisonnier, la croissance en hiver, la biomasse, la nourriture, la prédation, l' emergence et la fécondité ont tous été considerés. Les quatres espèces les plus importantes ont adopté des stratégies individuelles: *Baetis macani* a choisi la mortalité basse et la croissance rapide en été, *Siphlonurus lacustris* la grande taille et la biomasse importante, *Leptophlebia vespertina* la grande densité et *L. marginata* la croissance en hiver. Chacune de ces stratégies ont des désavantages, tout en permettant la coexistence malgré les restrictions du milieu imposées par de longues periodes de gel et des étés courts.

ZUSSAMENFASSUNG

Die Ökologie der Eintagsfliegen im See Övre Heimdalsvatn 1090 m ü.m. im Gebirge Jotunheimen, wurde studiert. Lebenslauf so wohl als Winterwachstum, Biomasse, Ernährung, Predation, Auftauchen und Fecunditet wurden in Betracht gezogen. Vier Arte von Eintagsfliegen haben verschiedene Strategien entwickelt um zu überleben. *Baetis macani* hat niedrige Mortalitet und raschen Sommerwachstum, *Siphlonurus lacustris* hat Grösse und viel Biomasse, *Leptophlebia vespertina* hat hohe Densitet und *L. marginata* hohen Winterwachstum. Jede dieser Strategien hat ihre Nachteile, aber sie erlauben Koexistence trotz der Restrictionen der Umgebung; lange Eisperioden und kurzer Sommer.

REFERENCES

Aarefjord, F., J.E. Brittain and A. Lillehammer. 1978. The benthos of Øvre Heimdalsvatn. *Holarct. Ecol.* 1: 219-220.

Benech, V. 1972. La fécondité de *Baetis rhodani* Pictet. *Freshwat. Biol.* 2: 337-354.

Brittain, J.E. 1974. Studies on the lentic Ephemeroptera and Plecoptera of southern Norway. *Norsk ent. Tidsskr.* 21: 135-154.

Brittain, J.E. 1975. The life cycle of *Baetis macani* Kimmins (Ephemerida) in a Norwegian mountain biotope. *Ent. scand.* 6: 240-244.

Brittain, J.E. 1978. The Ephemeroptera of Øvre Heimdalsvatn. *Holarct. Ecol.* 1: 239-254.

Brittain, J.E. 1979. Emergence of Ephemeroptera from Øvre Heimdalsvatn, a Norwegian subalpine lake. p. 115-123. *In:* K. Pasternak and R. Sowa ed. Proceedings of the Second International Conference on Ephemeroptera, August 23-26, 1975. Panstwowe Wydawnictwo Naukowe, Warszawa.

Brittain, J.E. and L. Lein. 1978. Seasonal and interspecific variation in consumption of Ephemeroptera by brown trout in a subalpine lake. *Int. Ver. theor. angew. Limnol. Verh.* 20: 728-733.

Brittain, J.E. and A. Lillehammer. 1978. The fauna of the exposed zone of Øvre Heimdalsvatn: Methods, sampling stations and general results. *Holarct. Ecol.* 1: 221-228.

Clifford, H.F. and H. Boerger. 1974. Fecundity of mayflies (Ephemeroptera), with special reference to a brown-water stream of Alberta, Canada. *Can. Entomol.* 106: 1111-1119.

Degrange, C. 1960. Recherches sur la réproduction des Ephémeroptères. *Trav. Lab. Hydrobiol. Piscicult. Grenoble* 51: 7-193.

Larsson, P., J.E. Brittain, L. Lien, A. Lillehammer and K. Tangen. 1978. The lake ecosystem of Øvre Heimdalsvatn. *Holarct. Ecol.* 1: 304-320.

Ulfstrand, S. 1975. Diversity and some parameters of Ephemeroptera and Plecoptera communities in subarctic running waters. *Arch. Hydrobiol.* 76: 499-520.

SOME ASPECTS OF THE BIOLOGY OF *EPHEMERA DANICA* MÜLL (EPHEMERIDAE: EPHEMEROPTERA) IN IRISH WATERS

Kenneth F. Whelan

The Inland Fisheries Trust Incorporated
Balnagowan, Mobhi Boreen
Glasnevin, Dublin 9
Ireland

ABSTRACT

From the anglers point of view the mayfly *Ephemera danica* is one of the most important insects found in Irish waters. During the past decade it has shown marked fluctuations in abundance and it some waters it has practically disappeared. Laboratory reared nymphs emerged to adults after two years while a batch of nymphs, reared from the egg, took only eleven months to develop to adults in a small limestone lake. It is concluded that the life cycle of *Ephemera danica* is flexible, achieving development in a shorter time in places where food is abundant and habitat conditions are ideal. The principal factors affecting the abundance of *Ephemera danica* in Irish lakes are discussed.

INTRODUCTION

The mayfly *Ephemera danica* is the only burrowing species of Ephemeroptera found in Irish waters. It shows a preference for habitats characterised by high alkalinity (100-240 mg/l Ca CO_3) and high pH values (7.9-8.5). The central plain of Ireland is saucer shaped, lying on a geological base of Carboniferous limestone (Fig. 1). The majority of rivers and lakes occupying this area contain populations of *Ephemera danica*. These waters also hold good stocks of large, fast-growing, brown trout (*Salmo trutta* L.) which can reach 2 kg in weight in five years of growth. With an abundance of other invertebrate food to feed on, it is only during hatches of *Ephemera danica* that these large trout can be readily taken while fly fishing. The mayfly season (mid-May to mid-June) is considered one of the most important annual angling events by both resident and visiting trout anglers.

During the past decade, the hatches of *Ephemera danica* have shown marked fluctuations in abundance and in some instances they have almost disappeared. Since the mayfly is of such importance to the larger Irish trout fisheries, the Inland Fisheries Trust, a semi-state body charged with the development and management of these waters, decided to initiate research into the basic biology of *Ephemera danica* in Irish waters. These studies form the basis of the present paper.

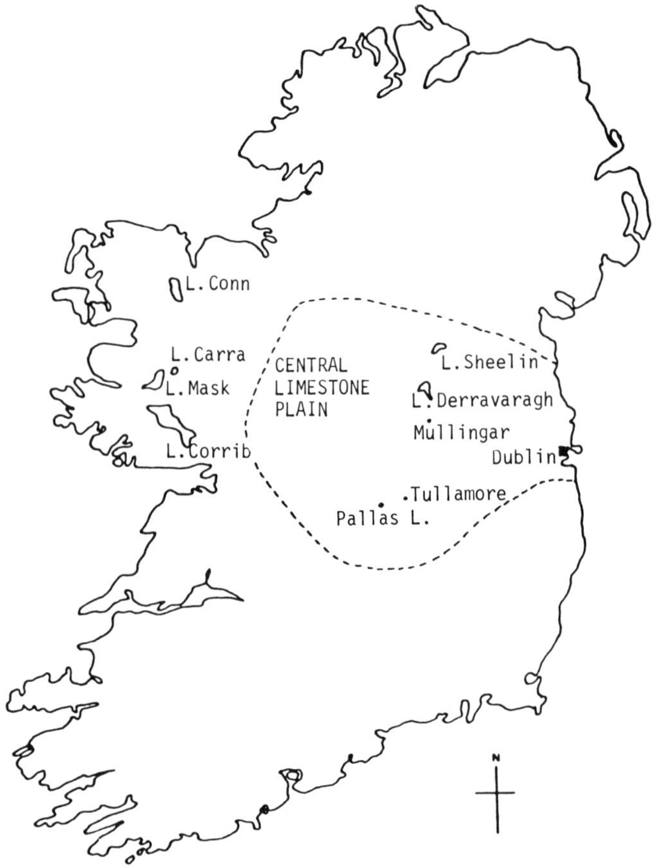

Figure 1. General map of Ireland showing central limestone plain and waters mentioned in text.

ANALYSIS OF QUALITATIVE COLLECTIONS

In the period March 1974 to June 1976 the field staff of the Inland Fisheries Trust collected in excess of 5,000 specimens of *Ephemera danica* from waters throughout Ireland. The majority of mayflies collected were in the nymphal stages; however, some 500 adult mayflies were also included. Length frequency analysis of these insects showed some interesting trends.

In the case of the adult *Ephemera danica* there was a sexually dimorphic growth pattern present. The males averaged from 14 to 18 mm in length and were 4 to 6 mm smaller than their female counterparts. The specimens from the western lakes were smaller than those collected from midland waters. River adults were significantly larger than those taken in still water.

The length frequency analysis of the nymphs showed that in many of the waters examined, for each month sampled, two distinct size groups were present (Fig. 2). The size discrepancy was too large to be explained in terms of the sexually dimorphic growth pattern mentioned earlier. It was proposed that two distinct age groups of *Ephemera danica* nymphs were present. A set of field experiments were designed to test this hypothesis.

FIELD STUDIES

A one square metre section of Lough Sheelin, a large, 1862 hectare, limestone lake, was cordoned off using a metal frame and tygon sheeting. Metal plates were attached to the legs of the frame and these were sunk into the lake bottom. When in place these metal plates protruded approximately 20 centimetres above the substratum. A tight-fitting wooden cover, containing a trap door, completed the structure. The cover served a dual purpose in that it ensured the capture of all emerging adults and also prevented adults from laying eggs in the experimental area.

The trap was in place from May 1975 until November 1976 when a series of severe winter storms irreparably damaged the frame and covering. Samples taken from areas lying adjacent to the trap showed that two size groups of nymphs were present.

In May/June 1975 108 adults emerged from the enclosed area. The wooden cover was removed in July 1975 and it was not replaced until early May 1976. Twenty five adults emerged from the trap during the 1976 emergence period. This experiment showed that at least a percentage of the mayfly population spent in excess of one year as nymphs.

To verify the length of the *Ephemera danica* life cycle the

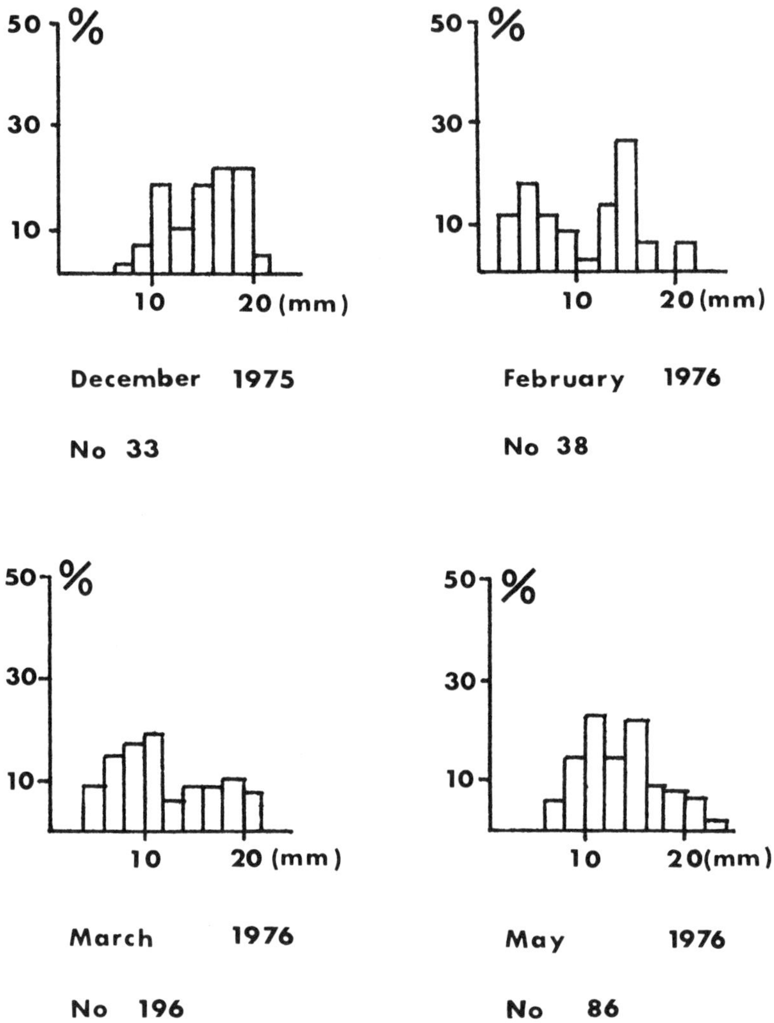

Figure 2. Cumulative length frequency histograms of mayfly nymphs collected from the western lakes (Loughs Mask, Carra and Conn) in Decmeber 1975 and February, March and May 1976.

insects were reared from the egg to the adult stage under experimental conditions. A technique based on that used by Lunn (Hills 1934) was developed for collecting fertilized ova.

COLLECTION AND SUBSEQUENT DEVELOPMENT OF FERTILIZED OVA

Having completed metamorphosis from the subimago (Green Drake) to the imago (Gnat), the male spinners of *Ephemera danica* gather in large clouds and swarm. As they do so, they display a most characteristic rise and fall motion. The females join the swarm as they are ready to mate.

Each female that enters the swarm is immediately persued by a group of males. One of these approaches the female from underneath and grasps her with his long forelegs in the region of her thorax. The male then tilts his abdomen upwards and partly forwards over its back and clasps the eight or ninth segment of the female abdomen with his abdominal forceps or claspers (Whelan 1975). If only one male is attached to the female copulation is effected while the pair continue flying. However, it is more usual for three to four male spinners to grasp onto the female or to the original copulating male. The attendant males force the female to the ground where copulation is completed. It is at this stage that fertilized females may be gathered.

When a suitable number have been collected they are transported to the laboratory, placed on containers of water and their abdomens gently stimulated until they are seen to emit a steady, cone-shaped, stream of eggs.

Estimates for the fecundity of *Ephemera* have ranged between 2,300 and 6,500 eggs per female (Clifford and Boerger 1974, Halford 1889, Hills 1934, Peart 1916, Percival and Whitehead 1926). Detailed examination of adult females collected during the course of the project has shown that fecundity values in Irish specimens of *Ephemera danica* range between 1,700 and 3,000 eggs per female or between 94 and 136 egg/mm body length. These estimates are based on total egg counts from eleven selected individuals.

The ova of *Ephemera danica* are adhesive and it has been found that plastic containers offer a very suitable substratum for them. After several hours, a transparent egg mat is formed. The ova are now firmly attached to the containers and it is necessary to dissect sections of the mat if specimens are required for microscopic examination.

After eight days of incubation the first outlines of an S-shaped larva appear in the ova. By the fourteenth day, a wedge-shaped larva is present. These larvae show signs of segmentation, and can be made to move by light or heat stimuli. The eggs hatch after eighteen days at 15°C. Newly hatched nymphs are characterised by five pigment spots on the head and a mottled brown, wedge-shaped oesophagus.

ADULTS REARED FROM THE EGG

Aquaria which contained suitable burrowing substrata and also a variety of aquatic plants were prepared and set up in the grounds of the laboratory. They were, therefore, subject to normal climatic variation.

First instar *Ephemera danica* nymphs were placed into the aquaria on the 10th June 1976. A table outlining their subsequent growth is shown (Table 1).

Table 1. Mean size of laboratory reared mayfly nymphs.

Date	Mean Size of Mayfly Nymphs	Number of Nymphs Measured
29.06.76	1.5 mm	40
09.08.76	3.0 mm	40
15.09.76	4.5 mm	25
01.11.76	5.7 mm	14
07.06.77	12.2 mm	12

No attempt was made to artificially feed these nymphs, however, as previously stated, the aquaria in which they were held contained natural plant detritus. It was noticed when sub-sampling the nymphs present, that many were concentrated around the roots of the *Potamogeton* sp. present in the aquaria.

From the five thousand first instar nymphs stocked five hatched as adult insects in June 1978. There were three males (mean length 14.9 mm) and two females (mean length 18.5 mm) present. The nymphs of these adults had spent a full two years as larvae.

MASS REARING OF *EPHEMERA DANICA*

Following the success attained in artificially rearing

and propagating mayfly eggs during 1976, experiments were carried out in late May and early June of 1977 to ascertain the feasibility of mass rearing *Ephemera danica* to the first nymphal instar stage. Between May 23 and 30 some 230 fertilized female adult mayflies were collected. These were transported to the Inland Fisheries Trust fish farm at Mullingar and spawned into plastic containers. Approximately ten females were placed on each container and moved about the water's surface to ensure even dispersion of the eggs. The containers were placed in troughs in the trout hatchery. A constant flow of water insured sufficient oxygenation and it was hoped to leave the boxes undisturbed for at least ten to twelve days.

Thirty of the female mayflies collected were spawned in the laboratory and the eggs from these were held in an aerated plastic aquarium.

Fourteen days after fertilization, June 10, the eggs held at the laboratory hatched. The eggs at the fish farm hatched between June 12 and June 20, eighteen to twenty days after fertilization. The hatchery water was an average 3.5°C cooler than the laboratory water during the incubation period.

The half a million first instar *Ephemera danica* nymphs resulting from this experiment were subsequently stocked into Pallas Lake, a small 10 hectare limestone lake. Although *Ephemera danica* had never been recorded from Pallas Lake a biological survey showed that the resident fauna was similar in composition and abundance to many of the larger midland limestone lakes, which had established populations of *Ephemera danica*.

Between May 19 and May 21, 1978 small hatches of adult *Ephemera danica* were recorded from Pallas Lake. These mayflies, which were stocked as first instar nymphs in June 1977, had hatched after a nymphal life of only eleven months.

DISCUSSION

Results from research carried out prior to 1978 indicated that the life cycle of *Ephemera danica* was completed in two years. The occurrence of small nymphs during the adult hatching period and the results obtained from rearing experiments in 1975/78 seemed to rule out the possibility of *Ephemera danica* completing its growth cycle in one year. The hatch of *Ephemera danica* recorded from Pallas Lake in June 1978 showed that under favourable environmental conditions adults can emerge after spending only eleven months as nymphs.

It is often assumed that the life cycle of an insect is a species characteristic whereas in fact many life cycles are extremely flexible, achieving development in a shorter time in places where food is abundant and environmental conditions are ideal. One species of dragonfly may complete its development in one, two or three years (Macan, personal communication).

Many of the water keepers on the English chalk streams during the nineteenth and early twentieth centuries were ardent entomologists and it is to them credit must be given for the initial research into the life cycle of *Ephemera danica* (Halford 1889, Hills 1934, Peart 1916, Percival and Whitehead 1926). They maintained that the life cycle was completed in two years. However, their only proof of this was the occurrence of juvenile nymphs during the adult hatching period. Previous research has shown that river mayfly adults are consistently larger than their lake counterparts. This size differential may be due to a faster nymphal growth rate in the river environment.

It has been recently demonstrated that the progeny of river mayflies, given virgin growing conditions, can hatch after one year. Damerham Fisheries are small, artifically constructed, "put and take" ponds lying adjacent to the River Allen, which is itself a tributary of the English Hampshire Avon. The lakes were excavated in midwinter and by the following June gnats, from the adjacent stream, were seen ovipositing on the newly constructed ponds. One year later there was a substantial hatch of *Ephemera danica*, providing an unexpected bonus to the fishery owners. A mayfly hatch, although not as large as the original emergence, has become a regular feature on these lakes (Walker 1978).

Peart (1916) also records that ponds constructed near a stream containing *Ephemera danica* produced a hatch of adult mayfly after they had been only one year in existence. He discounts the possibility that the larval stages of *Ephemera danica* had gained access to the ponds.

The life cycle of the large North American species of mayfly, *Hexagenia limbata* has also caused much discussion and controversy. Both one and two year life cycles have been documented for this insect. Some authors claimed that it had a one year cycle in the Southern States of America and a two year cycle in the Northern States and Canada. (Craven and Brown 1969, Neave 1932 and Swanson 1967). A recent intensive study has shown that both one and two year cycles may occur in any given population of these insects. Temperature is the principal factor regulating their growth. The majority of the mayflies take two years to complete their development while the nymphs hatching from eggs laid in the early part of the adult emergence period completed their development in one year (Rutter and Wissing 1975). One factor which emerges from much of the research to date on the various species of mayflies is the correlation between the rate of nymphal growth and water temperature (Brittain 1972; Rutter and Wissing 1975).

Analysis has shown that two separate regimes of temperatures are present in the Irish midland and western lakes. The western lakes display a higher mean winter temperature. They become warmer at a slower rate than the midland lakes in spring but they ultimately reach a higher mean summer temperature. The drop in temperature in the autumn is more pronounced in the midland lakes.

The temperature regime present in the western lakes facilitates the early hatching of eggs laid in the summer or early autumn. Initial growth continues for a longer period and adults may emerge after twelve to eighteen months. This leads to a situation where a staggered mayfly emergence is present, with peaks in May and August. From the results obtained in the case of the American mayfly, *Hexagenia limbata* it is reasonable to assume that the time of egg laying and first instar nymph hatching regulates the length of the life cycle and so the time of adult emergence. In the midland lakes the temperature regime limits adult emergence to a confined period from mid-May to mid-June.

Natural fluctuations in the intensity of *Ephemera danica* hatches have been recorded from many waters in Britain and Ireland during the nineteenth and twentieth centuries. The mayfly disappeared on sections of the river Test between 1893 and 1917. Poor weather conditions during the period 1893-1906 were considered responsible for this drop in numbers (Hill 1934). Minor fluctuations in abundance have also occurred in the past ten years.

Past annual reports of the Inland Fisheries Trust record occasions when adverse weather conditions effected mayfly populations. Unusually cold weather in the May/June period of 1962 prevented the imagines from laying their eggs. Mayfly populations were diminished in 1963 and 1964, but normal mayfly hatches were recorded from 1965 to 1972. The fluctuations in abundance of both mayflies and Trichoptera experienced at this time promoted the authors of the 1962 and 1963 Inland Fisheries Trust reports to write an explanatory note on cycles in insects population abundance (I.F.T. 1962 and 1963). Weather conditions were assumed to regulate either directly or indirectly the intensity of insect hatches during any given year.

The next major natural catastrophe occurred in 1972 when gale force winds drowned millions of emerging adult mayflies and prevented those that survived from making their egg-laying flight to the lake margins (Table 2).

On Lough Sheelin the mayfly has failed to appear in any significant numbers since that time. Other waters, such as Lough Derravaragh, were also badly hit by the storm but in recent years moderate hatches of mayflies have occurred.

Table 2. Wind velocities in the Irish midlands May 1972.

Date	Mean Wind Speed (m/sec)	Highest Gust (m/sec)	Beaufort Number	Description
22.5.72	5.2	16	7	Near Gale
23.5.72	6.4	20	8	Gale
24.5.72	7.6	24	9	Strong Gale
25.5.72	8.1	28	10	Storm
26.5.72	10.7	25	10	Storm
27.5.72	8.6	20	8	Gale
28.5.72	6.8	19	8	Gale
29.5.72	8.7	23	9	Strong Gale
20.5.72	8.7	20	8	Gale

The reason for the disappearance of *Ephemera danica* from Lough Sheelin may be attributed to three main causes, the storm of 1972, cultural eutrophication, and increased predation from both brown trout and perch populations.

Increased cultural eutrophication became evident in Lough Sheelin during the spring of 1971 when blue green algal blooms and high chlorophyll *a* levels were recorded. The situation improved somewhat during 1972 but in January 1973 intensive algal blooms reappeared.

Britt (1955a) observed that the populations of *Hexagenia* in Lake Erie were decimated during the early fifties as a result of intense cultural eutrophication. The low numbers of mayfly nymphs resulting from the 1972 egg laying season on Lough Sheelin were subjected to changing environmental conditions which favoured the more pollution resistant organisms such as Chironomidae and Oligochaeta.

Another factor which must be taken into account when discussing the fate of *Ephemera danica* in Lough Sheelin is the increase in the population density of both perch (*Perca fluviatilis* L.) and brown trout (*Salmo trutta* L.) that occurred in the early seventies. This followed many years of predator control which resulted in a dramatic reduction of the pike (*Esox lucius* L.) stocks in 1970-71. Stocking of large numbers of artificially reared fish farm trout began in 1970, reaching a peak in 1973 when 35,000 summerlings, 80,000 fingerlings, 128,800 spring yearlings and 61,000 autumn yearlings were placed in the lake. The numbers of perch present in Lough Sheelin increased dramatically from 1974 to 1976. The weight of perch captured during predator control operations was 1,112 kg in 1974, 8,330 kg in 1975 and 8,338 kg in 1976. The perch feeds on a mixed diet of invertebrates and fish, whereas the pike, after its first year, is mainly a fish feeder. The increased population density of both perch and brown trout must have resulted in a greater cropping of the invertebrate populations present in the lake. For a species such as *Ephemera danica* whose survival was already under pressure from changing environmental conditions, the unprecedented rise in predation levels must have further decreased the possibility of its reappearance in significant numbers.

The results of the project to date indicate that *Ephemera danica* populations are exceptionally sensitive to environmental change. Natural fluctuations in abundance occur and where these coincide with habitat or species dominance changes, they can result in the disappearance of the species.

In Lough Sheelin at the present time (1979) levels of nutrient input are still excessively high. The lake is subject to continuing cultural eutrophication and so the chances of the mayfly re-establishing itself as an angling species are presently poor. However, the technique developed for the artificial rearing of *Ephemera danica* may enable its eventual re-introduction into areas where suitable habitats are available.

ACKNOWLEDGMENTS

My thanks are due to my colleagues in the Inland Fisheries Trust for the practical help and encouragement they have given to me during the course of my research. I also wish to express sincere appreciation to Professor C.F. Humphries and Dr. E.J. Wise of the Zoology Department, University College, Dublin, whose knowledge of the Ephemeroptera did so much to stimulate my interest in the group. Brian Hayes worked with me as a summer bursar during 1978 and ably assisted me with the fecundity aspects of this study. Travel costs to the conference were met jointly by the Inland Fisheries Trust and also by the Zoology Department, University of Dublin. To both of these organisations I offer my sincere thanks.

RESUME

Du point de vue des pêcheurs à la ligne, l'éphéméroptére, *Ephemera danica*, est un des insectes les plus importants que l'on trouve en Irlande. Pendant les dix dernieres années, l'abondance de l'insecte a beaucoup varié et dans certaines eaux il a presque disparu. Des nymphes élevées au laboratoire ont atteint le stade adulte en deux ans. Il n'a fallu qu'onze mois pour le développement complet d'un certain nombre de nymphs élevées depous 'oeuf et placées dans un petit lac a pH éleve. On en conclut que le cycle de vie d'*Ephemera danica* est flexible et que l'insecte atteint son développement en un temps plus court quand la nourriture est abondante et les conditions d'habitat idéales. Le travail se termine avec une discussion des prineipaux facteurs qui influencent l'abondance d'*Ephemere danica* dans les eaux irlandaises.

ZUSSAMENFASSUNG

Vom Standpunkt des Anglers aus gesehen ist die Eintagsfliege *Ephemera danica* eines der wichtigsten Insekten, die sich in irischen Gewässern befinden. Während des letzten Jahrzehnts zeigte sie markante Abundanzfluktuierungen, und in einigen Gewässern ist sie praktisch ganz ausgestorben. Nymphen, die in einem Labor aufgezogen wurden, traten nach zwei Jahren als ausgewachsene Tiere zutage, während ein Schub von Nymphen die aus Eiern aufgezogen worden waren, nur elf Monate brauchte um sich in einem kleinen Kalksteinsee zu ausgewachsenen Tieren zu entwickeln. Daraus läßt sich schließen, daß der Lebenszyklus von *Ephemera danica* flexibel ist. Ihre Entwicklung vollzieht sich schneller in Plätzen, die reichlich Nahrung enthalten und ideale Lebensbedingungen bieten. In der vorliegenden Arbeit werden die wichtigsten Faktoren erörtert, die die Abundanz von *Ephemera danica* in irischen Seen beeinflussen.

REFERENCES

Britt, N.W. 1955a. Stratification in western Lake Erie in summer of 1953; effects on the *Hexagenia* (Ephemeroptera) population. *Ecology* 36: 239-244.

Brittain, J.E. 1972. The life cycle of *Leptophlebia vespertina* (L.) and *L. marginata* (L.) (Ephemeroptera) in Llyn Dinas, North Wales. *Freshw. Biol.* 2: 271-277.

Clifford, H.F. and H. Boerger. 1974. Fecundity of mayflies (Ephemeroptera) with special reference to mayflies of a brown-water stream in Alberta, Canada. *Can. Entomol.* 106: 1111-1119.

Craven, R.E. and B.E. Brown. 1969. Ecology of *Hexagenia* naiads. (Insecta-Ephemeridae) in an Oklahoma reservoir. *Am. Midl. Nat.* 82: 346-356.

Halford, F.M. 1889. Dry fly fishing, theory and practice. Barry Shurlock and Co., Reading.

Hills, J.W. 1934. River Keeper, The Life of William James Lunn. Geoffrey Bles, London.

Inland Fisheries Trust. 1962 and 1963. Annual Report. I.F.T., Glasnevin, Dublin 9.

Neave, F. 1932. A study of the Mayflies (*Hexagenia*) of Lake Winnipeg. *Contrib. Can. Biol. Fish.*, N.S. 7: 179-201.

Peart, A.R. 1916. Notes on the introduction and propagation of the Ephemeridae. *Salmon Trout Mag.* 14: 29-42.

Percival, E. and H. Whitehead. 1926. Observations on the biology of *Ephemera danica*, Mull. *Proc. Leeds Phil. Lit. Soc.* 1: 136-148.

Rutter, R.P. and T.E. Wissing. 1975. Distribution, abundance and age structure of a population of the burrowing mayfly *Hexagenia limbata* in an Ohio pond. *Ohio J. Sci.* 75: 7-13.

Swanson, G.A. 1967. Factors influencing the distribution and abundance of *Hexagenia* nymphs (Ephemeroptera) in a Missouri River reservoir. *Ecology* 48: 216-225.

Walker, R. 1978. The truth about mayfly nymphs. *Trout Salmon* 276: 53-54.

Whelan, K.F. 1975. Five Ephemeroptera - their life cycles and importance. *Proc. Biol. Soc. Univ. Coll. Dublin.* 1: 4-16.

REGIME ALIMENTAIRE DES LARVES DE DEUX GENRES D'EPHEMEROPTERES : *RAPTOBAETOPUS* MÜLLER-LIEBENAU, 1978 ET *PROSOPISTOMA* LATREILLE, 1833

Josette Fontaine

Département de Biologie animale et Zoologie,
Université Lyon I, 43, Bd du 11 Novembre 1918
69621 - Villeurbanne, France

RESUME

L'étude des larves des Ephéméroptères montre que, parmi une majorité d'espèces herbivores détritivores, certains genres sont carnivores et possèdent des mandibules sans région molaire ou avec une région molaire hérissée de pointes ou d'épines. Dans le Rhône, grand fleuve du Sud-Est de la France, deux espèces de ce type furent trouvées : *Raptobaetopus tenellus* (Albarda) qui mange larves de Chironomidae et oligochètes; et *Prosopistoma foliaceum* Fourcroy qui semble ingérer détritus et algues, tandis que parmi les espèces malgaches du genre, une est carnivore et deux autres détritivores herbivores. L'examen des pièces buccales ne suffit pas pour déduire le régime des larves des Ephéméroptères.

INTRODUCTION

Raptobaetopus et *Prosopistoma* sont deux genres d'Ephéméroptères rencontrés dans le Rhône. Ce grand fleuve francais possède une riche faune d'eau courante, et même très courante (Lafon 1953, Perrin et Roux 1978) qui comprend de nombreuses espèces d'Ephéméroptères. Parmi celles-ci, beaucoup sont constantes et se retrouvent d'année en année en fonction de leur cycle biologique. L'espèce dominante est *Heptagenia sulphurea* (Müller) suivie, en nombre moindre, par *Baetis fuscatus*.

A côté de ces espèces constantes, figurent deux espèces accidentelles : *Raptobaetopus tenellus* (Albarda) et *Prosopistoma foliaceum* Fourcroy. La première de ces espèces est rare puisque, malgré des prélèvements réguliers, seules 7 larves ont été capturées le 26/10/77 dans un substrat artificiel du type décrit par Roux *et al* (1976), déposé dans le fleuve à une quinzaine de kilomètres en amont de Lyon. Aucune nouvelle larve n'a été récoltée depuis. La rareté des captures de larves de *Raptobaetopus tenellus* n'est pas propre au Rhône, mais semble être la règle pour cette espèce (Sowa 1975, Soldan 1978)*. Des larves de *Prosopistoma foliaceum* ont été capturées dans la faune littorale du Rhône à Lyon: 20 larves en 1951 (Lafon 1952) et près de 400 larves en 1953 (Fontaine 1955). Aucune larve n'a été retrouvée depuis, mais le piégeage lumineux de nuit a permis la capture de 24 imagos mâles en 1958. Aucun adulte n'a été récolté depuis malgré des piégeages réguliers.

A l'originalité que montrent ces deux espèces par leur présence discontinue dans le temps, s'ajoutent leurs habitudes nutritionnelles.

En effet, la majorité des larves des Ephéméroptères se nourissent de détritus ou d'algues, en râclant souvent les pierres sur lesquelles elles se déplacent (Bengtsson 1924, Trama 1958, Katchalova 1960, Brown 1960, 1961). D'autres larves, en nombre bien moindre présentent un comportement prédateur (Edmunds 1957, 1960; Müller-Liebenau 1978).

Alors que les larves de *Raptobaetopus* se rangent dans la catégorie des prédateurs, celles de *Prosopistoma* sont soit détritivores herbivores, soit carnivores selon les espèces considérées.

REGIME ALIMENTAIRE DE *RAPTOBAETOPUS* MÜLLER-LIEBENAU

Müller-Liebenau, dans son important travail de 1978, crée le genre *Raptobaetopus* pour une espèce de Malaisie et fait état du régime carnivore des larves de *Raptobaetopus tenellus*: quatre larves de Chironomides sont trouvées dans le contenu stomacal d'une des larves examinées. Lestage (1919) avait pressenti ces habitudes nutritionnelles en se fondant sur l'aspect particulier de la zone molaire de la mandibule, profondément transformée par la présence de fortes denticulations.

L'examen du contenu stomacal de deux des larves capturées dans le Rhône nous permet de confirmer ce régime:

- chez une larve de 4.3 mm de longueur (cerques non compris) sont dénombrées trois larves de Chironomides (1 Orthocladiinae et

* Une étude morphologique et écologique de cette espèce est actuellement en cours : J. Fontaine et J.F. Perrin "Structure et fonctionnement des écosystèmes du Haut-Rhône francais. XIII. *Raptobaetopus tenellus* (Albarda), nouvelle espèce pour la faune francaise (Ephemeroptera : Baetidae ?)."

et 2 Tanytarsiens) (Planche I, 4).

- chez l'autre larve de 5 mm de longueur, sont trouvées six de Chironomidae (5 Orthocladiinae et 1 Tanytarsien) et de nombreuses soies d'Oligochètes Naïdidae, trop nombreuses et groupées pour ne pas admettre l'ingestion d'un ou plusieurs de ces Vers (Planche I, 1, 2, 3).

Il est intéressant de noter que, pour certaines proies nouvellement ingérées et par suite peu altérées par la digestion, celles-ci se retrouvent à peu près intactes dans le contenu stomacal (Planche I, 1, 3, 4). Il n'est donc pas interdit de penser que malgré la transformation profonde du lobe molaire en une sorte de deuxième lobe incisif, les larves de *Raptobaetopus* ne se servent pas de leurs pièces buccales pour "couper" ou "broyer" leurs proies, mais qu'elles avalent des animaux entiers. Les pièces buccales pourraient alors servir à la préhension et à la rétention des proies.

REGIME ALIMENTAIRE DE *PROSOPISTOMA* LATREILLE

Le Rhône possède l'espèce *Prosopistoma foliaceum*. Cette espèce fut d'abord considérée comme carnivore par Trägardh (1911), cet auteur revenant ensuite sur ses premières observations en 1924 (in Bengtsson 1924).

Au moment des captures nombreuses de larves, j'ai examiné uniquement le contenu stomacal de trois larves sans y rencontrer autre chose que des débris variés, minéraux et végétaux, en très petite quantité. Ceci ne me permet donc pas de conclure de façon formelle à l'absence de comportement prédateur chez ces larves, car le nombre d'exemplaires examinés est insuffisant. En effet, à l'époque, tel n'était pas mon problème car je cherchais à obtenir l'imago mâle encore inconnu alors et il n'était pas question de sacrifier de nombreuses larves nouvellement récoltées : toutes les larves capturées étaient placées en élevage.

Par contre, l'examen des contenus stomacaux des larves de *Prosopistoma* malgaches récoltées par la mission Starmühlner, les missions du Muséum de Paris ou H. Bertrand apportent beaucoup de renseignements sur le comportement alimentaire du genre *Prosopistoma*. Parmi les cinq espèces recensées à Madagascar,* une est prédatrice,

* Une étude systématique du matériel récolté est actuellement terminée et sera sous presse prochainement; J. Fontaine "Les espèces malgaches du genre *Prosopistoma* Latreille (Ephemeroptera: Prosopistomatidae).

deux sont herbivores détritivores et deux autres, récoltées en un seul exemplaire, possédaient un tube digestif vide.

La larve de l'espèce carnivore photographiée Planche II (1), d'une longueur de 7.5 mm (sans les cerques) venait d'ingérer une larve entière de *Baetis*, absolument intacte de 2 mm de long (cerques non compris) !! On peut voir, à travers la carapace du Prosopistome, les yeux de la larve de *Baetis* (flèche). La paroi distendue et transparente de l'estomac laissee apparaître nettement le Baetidae, lorsque tête et carapace du Prosopistome sont enlevées (Planche II, 2). Outre cette grosse larve de *Baetis*, le contenu stomacal de ce Prosopistome renfermait:

- une tête et les pièces buccales d'une autre larve de *Baetis*, d'espèce différente;
- toutes les pièces buccales et les pattes d'une larve de Tricorythidae (Pl. II, 5, 7, 8, 9). En examinant la taille des pièces buccales et en comparant celle-ci avec les dimensions d'autres larves de la même espèce, on peut évaluer la taille du Tricorythidae ingéré aux environs de 1.65 mm;
- une tête et les soies postérieures d'une larve de Chironomidae;
- une branchie de Leptophlebiidae.

L'examen du contenu stomacal d'autres larves de la même espèce donne les résultats suivants:

Prosopistome de 6.6 mm:
- 2 larves de Plécoptères (Pl. II, 3);
- 1 larve d'Ephéméroptère (Baetidae ?);
- les pièces buccales d'une larve de *Baetis* (Pl. II, 4).

Prosopistome de 4.3 mm:
- 1 larvule d'Ephéméroptère (Baetidae ?);
- 1 larve de Chironomidae (Pl. II, 6);
- 1 larve de *Baetis*.

Prosopistome de 4.2 mm:
- segments abdominaux, cerques et pattes d'une larve de *Baetis*;

Planche I
1. Larve de Chironomide ingérée par une larve de *Raptobaetopus tenellus* de 5 mm long (x 120)
2. Soies d'Oligochète Naïdidae de la même larve (x 300).
3. Larve de Chironomide de la même larve (x 120).
4. Larves de Chironomides trouvées dans le contenu stomacal d'un autre *Raptobaetopus* de 4.3 mm (x 45).

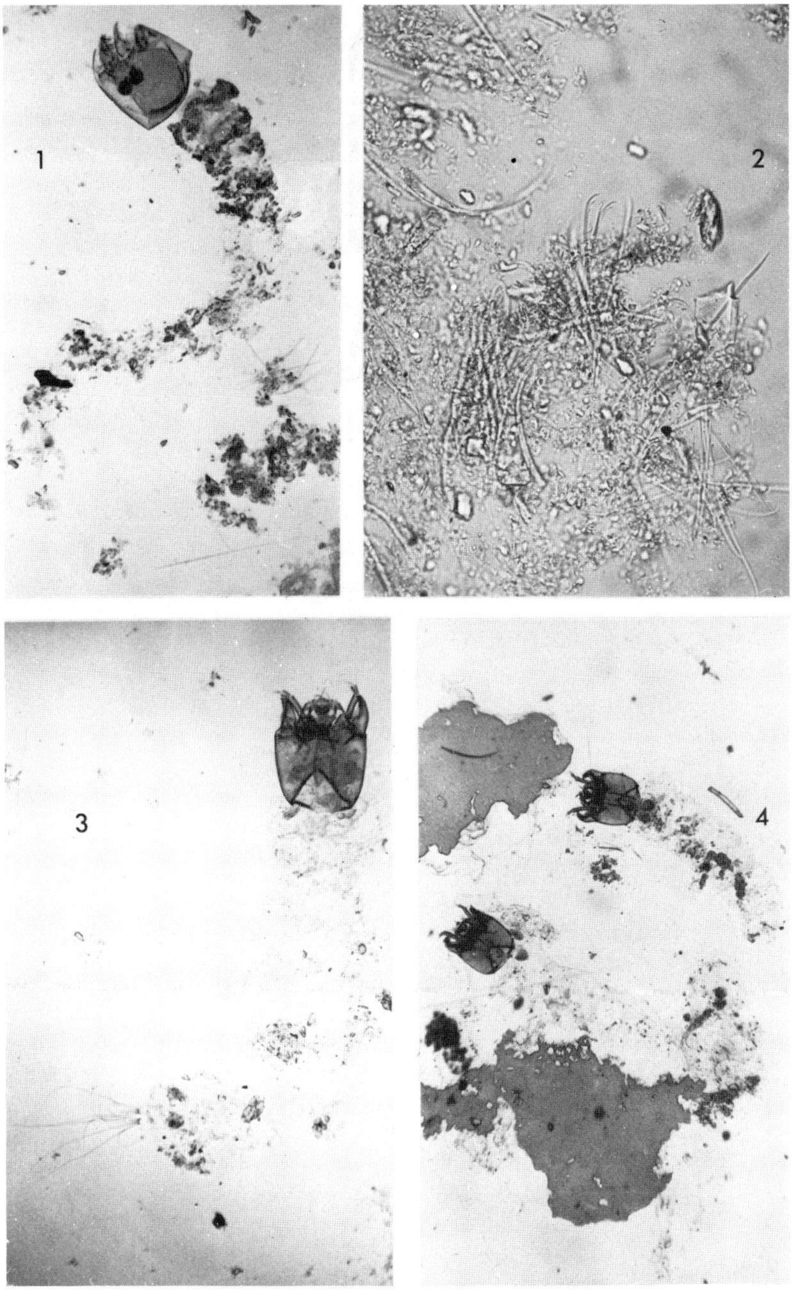

Planche I

- tête et soies postérieures d'une larve de Chironomidae.

Comme pour *Raptobaetopus*, on constate que les proies nouvellement ingérées par les larves de *Prosopistoma* prédatrices sont absorbées entières, sans être cisaillées ou broyées. Il est intéressant de noter, chez cette espèce, l'éventail très large des proies : larves de Diptères Chironomidae, de Plécoptères, d'Ephéméroptères (Baetidae, Tricorythidae, ...).

L'examen du contenu stomacal des deux espèces de *Prosopistoma* malgaches détritivores herbivores montre :

- pour une espèce, du sable et de nombreuses Diatomées ovalaires (Planche II, 11),
- pour l'autre espèce, les mêmes Diatomées mêlées à du sable, à des fragments d'autres Diatomées plus allongées et à des débris d'algues (Planche II, 10).

L'examen du contenu stomacal des larves des diverses espèces de *Prosopistoma* malgaches est très intéressant car toutes ces larves sont morphologiquement très voisines les unes des autres, spécialement par leurs pièces buccales : par exemple, les mandibules de toutes les espéces ne possèdent qu'un lobe incisif et pas de lobe molaire. L'examen du matériel malgache confirme donc parfaitement la remarque de Peters (1967) de la page 210 de son travail : "Only by careful dissection and use of high power dissecting and compound

Planche II

1. Larve de *Prosopistoma* malgache carnivore, de 7.5 mm; la flèche indique les yeux d'une larve de *Baetis* nouvellement ingérée.
2. Même larve qu'en 1; tête et carapace mésothoracique sont enlevées.
3. Partie antérieure d'une larve de Plécoptère ingérée par un Prosopistome carnivore de 6.6 mm. (x 85).
4. Pièces buccales d'une larve de *Baetis* capturée par le même individu que 3 (x 200).
5. Mandibule d'une larve de Tricorythidae mangée par la larve 1 (x 90).
6. Tête d'une larve de Diptère Chironomidae absorbée par un Prosopistome carnivore de 4.3 mm (x 75).
7. Maxille de Tricorythidae ingérée par 1 (x 85).
8. L'autre mandibule de Tricorythidae de 1 (x 90).
9. Partie du labium et pattes du même Tricorythidae (x 70).
10. Fragment du contenu stomacal d'une larve d'une espèce malgache de *Prosopistoma* détritivore herbivore (x 600).
11. Contenu du tube digestif de l'autre espèce détritivore herbivore (x 825).

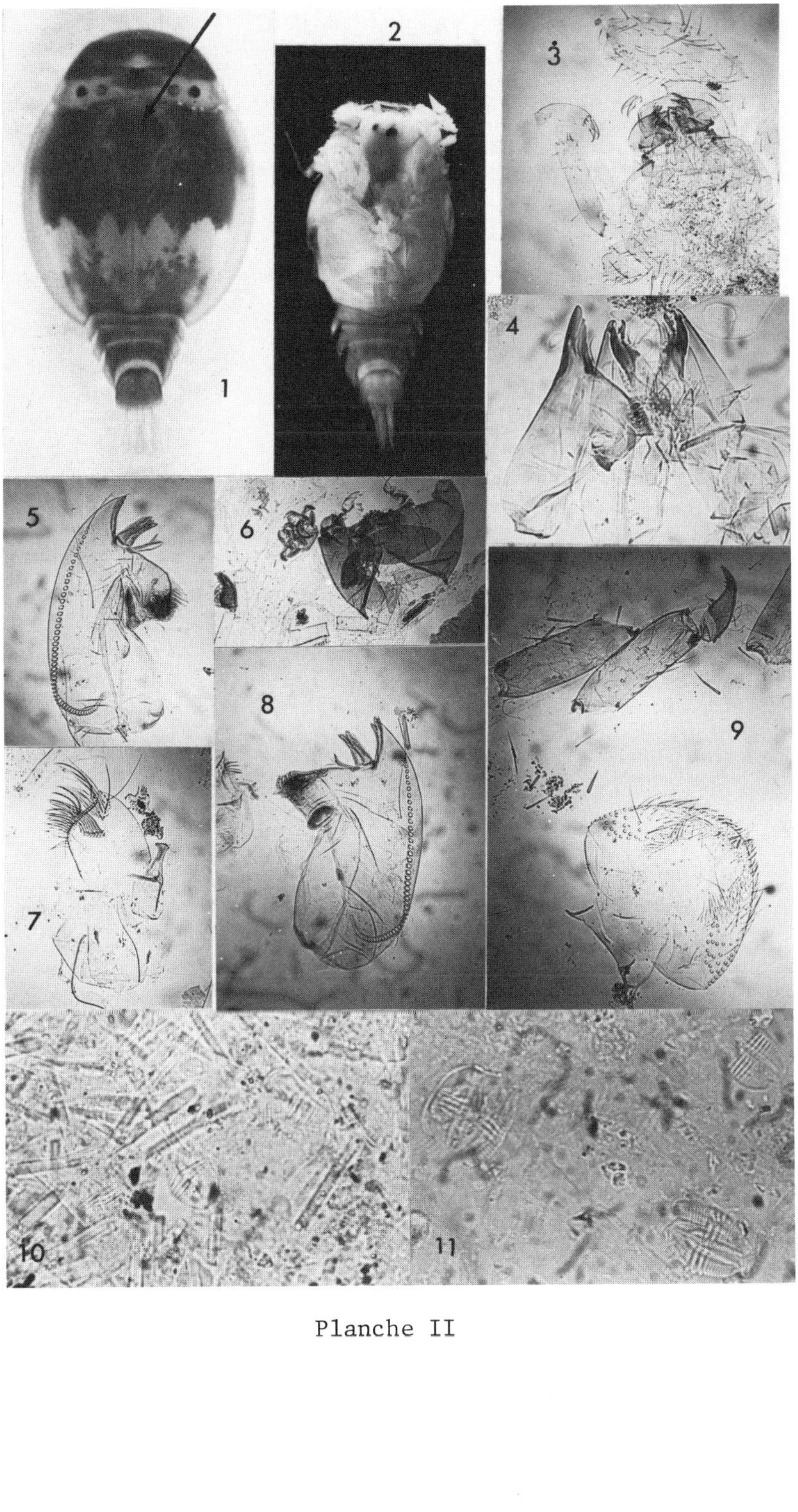

Planche II

microscopes can most species of *Prosopistoma* be identified in the nymphs". Malgré ces analogies morphologiques, le régime alimentaire varie considérablement d'une espèce à l'autre.

CONCLUSION

Le présent travail confirme le régime prédateur de *Raptobaetopus* Müller-Liebenau, en ajoutant à ses proies déjà décrites (larves de Chironomides) les Oligochètes Naïdidae.

Il apporte la preuve irréfutable à la fois du comportement prédateur de certaines espèces de *Prosopistoma* et du régime détritivore herbivore d'autres espèces. Ceci est intéressant car il ne s'agit plus d'un genre carnivore mais de variations de la nutrition à l'intérieur du genre, c'est-à-dire au niveau des espèces, malgré un conformation des pièces buccales très analogue d'une espèce à l'autre. Il est donc certain que la forme des pièces buccales d'une larve ne peut absolument plus être considérée comme un indice sûr de son régime.

Il est nécessaire également de reconsidérer la place des Ephéméroptères dans un écosystème aquatique ; consommateurs primaires ou secondaires? Les Ephéméroptères, selon les genres, étaient placés dans l'une ou l'autre catégorie. A l'avenir, il sera nécessaire de faire une détermination rigoureuse jusqu'à l'espèce puisque c'est à ce niveau que peuvent se situer les variations importantes du régime alimentaire.

ABSTRACT

Studies on the food-habits of Ephemeroptera nymphs show that among predominantly herbivorous or scavenger larvae, some genera are carnivorous, with mandibles without a molar area or with a molar area bearing strong spines or denticles. In the large river Rhône, in Southeastern France, two species with such mandibles are found : *Raptobaetopus tenellus* (Albarda) which eats Chironomid larvae and Oligochaeta; and *Prosopistoma foliaceum* Fourcroy which seems to eat detritus and algae. Among the species of *Prosopistoma* from Madagascar, one is carnivorous and two others are scavengers. Carnivorous habits cannot be inferred only from the observation of the mouthparts.

ZUSSAMENFASSUNG

Untersuchungen der Freßgewohnheiten von Ephemeropterennymphen zeigen, daß unter vorwiegend pflanzen- und detritusfressenden Larven einige Gattungen auch fleischfressend sind. Diese haben Kinnladen ohne Molargebiet oder mit einen Molargebiet, das mit Zähnchen oder Stacheln bedeckt ist. In der Rhône, einen großen Fluß Süd-Ostfrankreichs, wurden zwei Arten mit solchen Kinnladen gefunden: *Raptobaetus tenellus* (Albarda), der Chironomidenlarven und Oligochaeten frißt, und *Prosopistoma foliaceum* die anscheinend von Detritus und Algen lebt. Dagegen gibt es unter den Arten der *Prosopistoma* Gattung eine, die fleischfressend und zwei andere, die detritusfressend sind. Die Gewohnheiten der Fleischfresser können aber nicht nur von Untersuchungen der Mundteile abgeleitet werden.

TRAVAUX CITES

Bengtsson, S. 1924. La nutrition des larves de Ephémères. *Ann. Biol. Lacust.* 13: 215-217.

Brown, D.S. 1960. The ingestion and digestion of algae by *Chloëon dipterum* L. (Ephemeroptera). *Hydrobiologia* 16: 81-96.

Brown, D.S. 1961. The food of the larvae of *Chloëon dipterum* L. and *Baetis rhodani* (Pictet) (Insecta, Ephemeroptera). *J. Anim. Ecol.* 30: 55-75).

Edmunds, G.F., Jr. 1957. The predaceous mayfly nymphs of North America. *Proc. Utah. Acad. Sci. Arts Lett.* 34: 23-24.

Edmunds, G.F., Jr. 1960. The food habits of the nymphs of the mayfly *Siphlonurus occidentalis. Proc. Utah Acad. Sci. Arts Lett.* 37: 73-74.

Fontaine, J. 1955. Les formes ailées de *Prosopistoma foliaceum* Fourcroy (Ephéméroptère). *Bull. Soc. Linn. Lyon* 24: 60-65.

Katchalova, O. 1960. (en russe) *Latv. PSR Zinat. Akad. Vestis* 6: 147-152.

Lafon, J. 1952. Note sur *Prosopistoma foliaceum* Fourc. (Ephéméroptère). *Bull. Soc. Zool. Fr.* 77: 425-436.

Lafon, J. 1953. Recherches sur la faune aquatique littorale du Rhône à Lyon. *Bull. Soc. Linn. Lyon* 22(2): 36-46.

Lestage, J.A. 1919. Contribution à l'étude des larves des Ephémères paléarctiques (série 2). *Ann. Biol. Lacust.* 9: 79-182.

Müller-Liebenau, I. 1978. *Raptobaetopus,* eine neue carnivore Ephemeropteren - Gattung aus Malaysia (Insecta, Ephemeroptera : Baetidae). *Arch. Hydrobiol.* 82: 465-481

Perrin, J.F., et A.L. Roux. 1978. Structure et fonctionnement des écosystèmes du Haut-Rhône français. 6. La macrofaune benthique du fleuve. *Int. Ver. theor. angew. Limnol. Verh.* 20: 1494-1502.

Peters, W.L. 1967. New species of *Prosopistoma* from the Oriental region (Prosopistomatidae ; Ephemeroptera). *Tijdschr. Entomol.* 110: 207-222.

Roux, A.L., H. Tachet, et M. Neyron. 1976. Structure et fonctionnement des écosystèmes du Haut-Rhône français. III. Une technique simple et peu onéreuse pour l'étude des macroinvertébrés

benthiques des grands fleuves. *Bull. Ecol.* 7: 493-496.
Soldán, T. 1978. Mayflies (Ephemeroptera) new to the fauna of Czechoslovakia found in 1972-1977. *Acta entomol. bohemoslov.* 75: 319-329.
Sowa, R. 1975. Ecology and biogeography of mayflies (Ephemeroptera) of running waters in the Polish part of the Carpathians. 1. Distribution and quantitative analysis. *Acta Hydrobiol.* 17: 223-297.
Trägårdh, I. 1911. Om *Prosopistoma foliaceum* Fourc. en för Sverige ny Ephemerid. *Entomol. Tidskr.* 32: 91-104.
Trama, F.B. 1958. The transformation of energy by an aquatic herbivore *Stenonema pulchellum* (Ephemeroptera). *Diss. Abstr.* 18: 2275.

LIFE HISTORY, DEVELOPMENTAL PROCESSES, AND ENERGETICS OF THE BURROWING MAYFLY *DOLANIA AMERICANA*

R.S. Harvey, R.L. Vannote* and B.W. Sweeney*

Savannah River Laboratory
E.I. du Pont de Nemours and Company
Aiken, South Carolina 29801

ABSTRACT

Dolania americana has a two year life history in Upper Three Runs Creek at Savannah River Plant near Aiken, SC. Eggs deposited in early June hatch the following spring; larvae require 12-14 months to mature. Adult metamorphosis is synchronous, with > 95% of individuals emerging during the first week. Females produce an average of 77 eggs per individual; adult size and fecundity are correlated positively. Adult size of both males and females appears to decrease during the emergence period. Caloric measurements on various tissues and developmental stages ranged from 4.63 to 6.55 cal/mg. Eggs had the highest caloric content (avg 6.18 cal/mg); subimaginal molt skins had the lowest (4.63 cal/mg). Caloric content per unit weight of larval tissue increased with larval size. Chemical analysis of eggs revealed a high protein content (50.3%) relative to lipid (25.5) and carbohydrate (9.0) fractions. Weight specific respiration rates for larvae were inversely related to larval size, but correlated positively with water temperature between 6 and 23°C. Larval growth appeared continuous, although seasonal differences in the amount of growth were observed. Net production efficiency for larvae ranged from 5.3 to 29.2% and varied seasonally, averaging 8.1, 16.6, 21.7 and 19.4% for winter, spring, summer and fall, respectively.

---

* Stroud Water Research Center, Academy of Natural Sciences of Philadelphia, R.D. 1, Avondale, Pennsylvania 19311

INTRODUCTION

Dolania americana Edmunds and Traver is a burrowing mayfly known only from three coastal plain streams in the southeastern United States (Peters and Peters 1977). In Upper Three Runs, a spring-fed blackwater stream in South Carolina (33°23'N; 81°37'W), *D. americana* is the dominant mayfly species inhabiting the sandy substrate. Larvae are very distinct morphologically, being highly adapted for burrowing in sandy substrate (Edmunds and Traver 1959). *D. americana* are predators and feed mainly on chironomid larvae.

Aside from brief field notes accompanying the original taxonomic description of *D. americana* (Edmunds and Traver 1959), only the adult habits and emergence patterns have been reported (Peters and Peters 1977). This paper describes quantitatively: (1) the bioenergetics and developmental dynamics of the larval stage; (2) adult emergence and fecundity; and (3) biochemical characteristics of the eggs. These data are used to delineate both the life history of *D. americana* and the overall importance of this species to energy flow in Upper Three Runs.

METHODS

Larvae were taken at random from all potential habitats throughout the year. Larval collections were made at about two week intervals from September 1972 through August 1974 and also from June 1978 through September 1978. Small larvae (dry weight < 2.0 mg), which were most abundant in fine to coarse (0.1-1.0 mm) sand near the stream edge, were obtained by hand picking through sand collected with a dip net. Large larvae (dry weight > 2.0 mg) were usually collected with a modified Needham scraper (5.0 mm mesh) from coarse sand and fine gravel (11-34 mm) adjacent to the thalweg. Figure 1, derived from transect core samples from Upper Three Runs, shows that *D. americana* are more likely to occur where the sand size is small. Sample size ranged from ≥ 50 when larvae were small to ca. 20 for large larvae.

Larval respiration was measured with a Differential Submarine Respirometer (Gilson 1963). Larvae were collected from the stream, placed immediately in test vessels containing 1.5 cm of washed stream sand and 7 mL of filtered (0.45 µm) stream water, and acclimatised to test conditions for two hours. Test temperatures were with 0.5°C of ambient stream water at the time of collection. One larva was placed in each vessel and oxygen uptake was measured at 30 min intervals for three hours. Carbon dioxide which evolved during respiration was absorbed by a 20% KOH solution. Following the three hour experiment at ambient (AMB) stream temperature, the respirometer water bath was gradually increased 5.0°C over a 48 hour period. Respiration was then measured at 30 min intervals for three hours at AMB + 5.0°C temperature. Larvae used in metabolic studies were

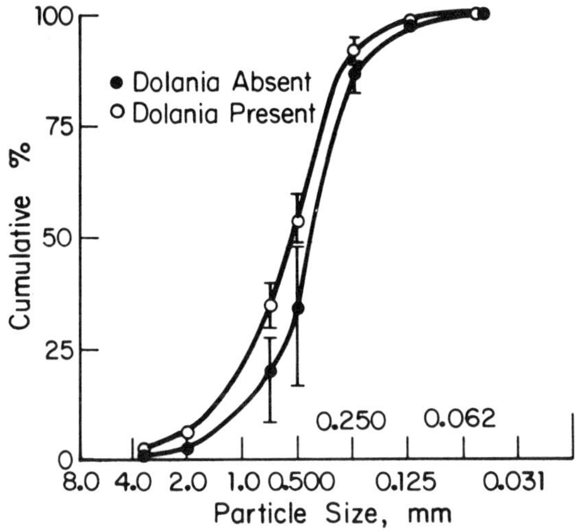

Figure 1. Size distribution of sand from stream transect samples.

killed in ETOH,* sexed, dried at 100°C for 24 hours, and weighed to the nearest 0.1 mg.

Adult emergence was monitored by: (1) light trapping at the stream edge; (2) surface drift samples of terminal instar larval molt skins; and (3) drift collections of spent male and female adults. Drift nets (1 mm mesh) were 1 m wide and filtered the top 15 cm of the stream. Two drift nets were deployed: one about 3 m from the stream edge (net 1) and the other at mid-channel (net 2). Drift was sampled daily from 0400 to 0700 h during peak emergence.

Caloric values for eggs, larvae, subimagos and adults were estimated using a Phillipson microbomb calorimeter. The lipid, protein, and carbohydrate content of eggs were estimated by the following methods: (1) lipid-weight loss following a 24 hour hot (50°C) chloroform:methanol (2:1) extraction solvent in a Soxyhlet apparatus; (2) protein-modified Lowry *et al.* (1951) technique (Price 1965); (3) carbohydrate-a phenol-sulfuric acid extraction (Dubois *et al.* 1956).

* The amount of weight loss from ETOH was inversely related to size and ranged from 6.8 to 21.7%. Percent weight loss (y) can be predicted from initial dry weight (x) by:
$y = 24.4 - 0.845x$, correlation coefficient (r^2) = 0.82.

Energy budgets were constructed according to the following equation:

$$A = G + M + R$$

where

A = assimilation = energy consumed in food minus the energy lost by egestion, excretion, and secretion
G = total change in energy value of body materials
M = energy accumulated in molt skins (exuviae)
R = energy metabolically used or released in all ways for all purposes

Growth was measured by changes in biomass and converted to calories using different conversion factors which varied according to size and sex of larvae (see results below). Relative growth rates (mg/mg/day) for each sampling period and season were calculated using an exponential model:

$$\text{Growth Rate} = (\ln W_f - \ln W_i) / t$$

where

W_f = weight at the end of the time period
W_i = weight at the beginning of the time period
t = duration of time period

Molt skin losses during the larval period were estimated by assuming that: (1) each male and female larva molts 13 times (T. Fink, personal communication) while growing from 0.1 mg to about 11 and 25 mg respectively; (2) the first larval molt occurs at 0.15 mg dry weight for both sexes; (3) each larva grows exponentially; (4) larval weight for each successive molt (m + 1) can be predicted from the weight of the previous molt (m) by the following equations: males: $Wt_{(m+1)} = Wt_{(m)}\, e^{0.336}$, females: $Wt_{(m+1)} = Wt_{(m)}\, e^{0.383}$, where 0.336 and 0.383 are the average instantaneous growth rates for the entire larval growth period of the fastest growing male and femal respectively; and (5) each molt skin weights about 15% of the larva's dry weight at the time of molting. Molt skin losses were converted to calories separately for males and females by using a caloric equivalent of 4.935 cal/mg dry weight.

Respiration costs were estimated for a given size animal and stream temperature using empirically derived regression equations. Total respiration for a given period was obtained by summing daily oxygen consumption for an animal of median weight. Oxygen consumption (μLO_2) was converted to calories by using the oxycaloric equivalent of 4.825×10^{-3} cal per μLO_2 (Brody 1945).

The energy budget reported is for the fastest growing male and female larvae in the subpopulation of Upper Three Runs. For each larval collection, we assumed that the average of the three largest larvae approximated the weight of the fastest growing female and the average larval weight for the entire collection represented the fastest growing male. These assumptions appear valid for *D. americana* because: (1) large (> 3.5 mg) larvae are easily sexed and observations on winter, and spring collections (when most larvae generally exceed 3.5 mg) bear out our assumptions; and (2) rearing data for other mayfly species show that the largest and average sized larvae for each collection made early in the larval growth period are always the largest female and male respectively (Sweeney 1978, Sweeney and Schnack 1977).

The energy budget only includes periods of positive growth. For example, if the mean larval weight decreases or remains the same in consecutive samples then the time interval for calculating growth, respiration, and molt loss is lengthened sequentially until a sample is obtained that exhibits an increase in mean larval weight. This avoids entering negative values for growth into the budget.

RESULTS

Life History

It appears that *D. americana* has a two year life cycle in Upper Three Runs. Adult emergence and egg deposition in 1978 began on June 1 and continued through June 19. Bottom samples collected on June 2 with fine (500 μm) mesh nets revealed many small (0.1-0.5 mg) larvae (Fig. 2B). It is hypothesized that these larvae were not first instar, but rather had hatched in the early spring of 1978 because: (1) dry weight measurements of eggs indicate that the first instar larvae weigh at most 0.1 mg; (2) larvae weighing 0.5 mg were present on the date of first adult emergence; and (3) instantaneous growth rates of small (< 1.0 mg) larvae in June or July (when stream temperatures are highly favorable for growth; Fig. 2A) suggest that at least five weeks would be needed for larvae to grow from 0.1 to 0.5 mg.

The above hypothesis was tested by dissecting eggs in June 1978 from newly emerged females, fertilizing them with sperm stripped from adult males, and incubating the eggs in glass jars (5.5 cm OD, 6.5 cm deep) containing 100 mL of filtered (0.45 μm) stream water. Jars were kept in a constant temperature water bath at the Stroud Waste Research Center that was adjusted weekly to trace approximately the average annual thermal regime of Upper Three Runs. Mature embryos were observed by November, but eggs did not begin hatching until April 4, 1979. Egg hatch continued through May 25, with first instar larvae being about 0.06 mg dry weight.

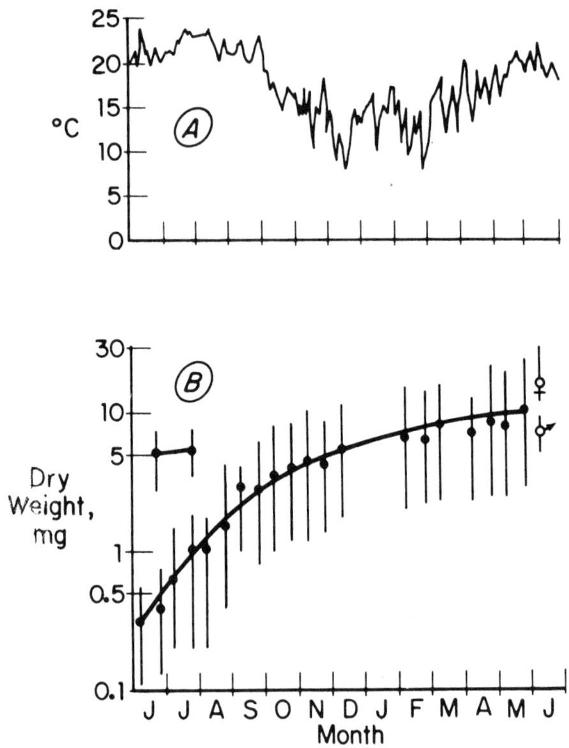

Figure 2. (A) Daily temperature recordings for Upper Three Runs from June 1973-June 1974. (B) Larval growth of *Dolania americana* in Upper Three Runs. Solid dots = mean weight of each collection; vertical lines indicate the range. Male and female signs designate the mean weight and range of male and female subimagos, respectively. Line drawn through sample means is described by: $y = 0.027x - 0.244$ where y is average weight and x is number of days since June 1.

Collections made after June indicate that larval weight increase exponentially during the warm months and growth ends by the following May. Maximum larval weights at emergence were 12 and 34 mg for males and females respectively. Adult emergence was synchronous with most adults on wing during the early part of June. A few intermediate size (range: 2.5-6.0 mg) larvae failed to emerge with the main cohort in June (see Fig. 2B). These remaining larvae showed no indication of adult maturation (e.g. wing bud development, darkened cuticle, etc.) and disappeared from the stream by early August. These larvae apparently contribute little if any to population recruitment in Upper Three Runs because: (1) drift net samples indicate that they probably do not emerge, and (2) if they did emerge, female size would be below the threshold for egg production (i.e. ca. 9 mg).

Calorimetry of Life History Stages

The caloric content of *D. americana* varies substantially with its developmental state (Table 1). Eggs have the highest caloric content (ca. 6.18 cal/mg) while newly hatched larvae have the fewest calories (ca. 5.33 cal/mg). Caloric content per unit weight increases with larval growth due probably to accumulation of stored lipid materials. Mature female larvae contain more lipid material per unit weight than mature male larvae and thus have a higher energy content. Stored energy in mature female larvae is represented largely by mature eggs whose lipid content per unit weight is apparently higher than any other developmental stage.

The energy content of larval molt skins, both male and female, is only about 15-20% lower than for the larvae themselves. This suggests that molting may represent a large energy loss especially since *D. americana* molts about 13 times during larval growth (T. Fink, personal communication) and each molt skin is about 10-15% of the molting larva's dry weight.

Table 1. Caloric content (calories per mg ash free dry weight) of various developmental stages and tissue types for *Dolania americana*.

Dolania americana				
Sample Type	n	Mean	S.D.	Range
Small larvae (♂&♀)	2	5.33	0.22	5.17-5.49
Intermediate larvae (♂&♀)	1	5.80	-	-
Mature ♂ larvae	3	5.68	0.29	5.38-6.08
♂larval molt skin	3	5.09	0.32	4.90-5.48
♂Subimago	3	5.86	0.14	5.72-6.00
♂Subimago molt skin	1	4.63	-	-
Mature ♀ larvae	3	6.09	0.12	5.97-6.26
♀ larval molt skin	5	5.16	0.23	4.91-5.52
♀ Subimago (pre-oviposition)	3	6.03	0.30	5.77-6.36
♀ Subimago (post-oviposition)	2	5.55	0.01	5.54-5.55
Eggs	5	6.18	0.21	5.90-6.43

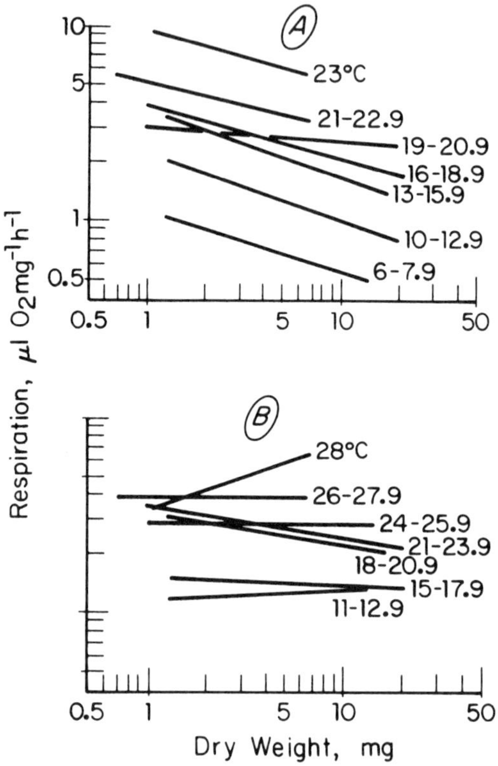

Figure 3. Weight specific respiration rate for *Dolania americana* larvae. Larvae were collected, placed in respiration vessels at ambient stream temperatures, and oxygen uptake measured (Figure 2A); larvae were then gradually acclimated over a 48 hour period to ambient +5°C before taking further measurements (Figure 2B). Two or more experiments within a given temperature range were combined for the regression analysis.

Larval Respiration

Respiration rates were correlated positively with temperature (between 6° and 23°C), but inversely related to body size for any specific temperature (Fig. 3A). Regression coefficients of fitted equations did not vary significantly over the range of test temperatures (Analysis of Covariance: $F_{6,879} = 2.03^*$, $p > 0.05$; Table 2). This suggests that the metabolism of small and large larvae responds similarly to changes in temperature. Sex did not appear to alter respiration rates significantly when males and females of similar weight were compared ($p > 0.05$).

For increasing stream temperatures +5°C above ambient, weight specific respiration increased for large animals but decreased for small animals (Figure 3B). This reflects a general tendency for regression coefficients to shift from negative values at ambient temperatures to zero (or even positive values) at ambient plus 5°C. We do not interpret these results as a form of overcompensation or acclimation by small larvae, but rather a type of metabolic collapse associated with stress, particularly among small individuals. The stress involved may be related to the short-term increase in temperature or other factors associated with the experimental design (e.g. food deprivation, build-up of waste materials in test vessels, or prolonged exposure to artificial test conditions).

Table 2. Regression equations (log Y = b log X + log a) for weight specific metabolism ($\mu l\ O_2\ mg^{-1}\ h^{-1}$) of *Dolania americana* at various temperature intervals.

Temperature °C	$b_{Y \cdot X}$	log a	Degrees of freedom	Level of significance
6-7.9	-0.316	0.074	1,022	0.002
10-12.9	-0.371	0.380	1,093	< 0.001
13-15.9	-0.346	0.592	1,229	< 0.001
16-18.9	-0.266	0.590	1,269	< 0.001
19-20.9	-0.087	0.499	1,085	0.112
21-22.9	-0.238	0.721	1,158	< 0.001
23-23.9	-0.284	1.004	1,030	0.024

* Although this F value borders on significance at the 0.05 level, most of the variance in the analysis can be attributed to the 19-20.9°C data. Elimination of the 19-20.9°C data, which do not yield a significant regression, results in more convincing statistics showing that the regression coefficients do not vary significantly (Analysis of Covariance: $F_{5,795} = 0.85$, $p > 0.5$).

Table 3. Partial energy budget for the fastest growing male and female larvae of *Dolania americana* in Upper Three Runs.*

MALE LARVAE

	°C		mg		Total calories per median size larva				%	$mg\ mg^{-1}\ d^{-1}$
Date	Average Temp.	No. Days	Average Weight (s.d.)	Median Weight	Growth	Respiration	Exuvia	Assimi-lation	Net Production Efficiency	Instantaneous Growth Rate
June 7			0.314 (0.101)							
	20.8	15		0.354	0.419	2.127	0.291	2.837	24.9	0.015
June 22			0.395 (0.169)							
	21.1	15		0.529	1.436	5.625	0.414	7.475	24.7	0.034
July 7			0.663 (0.324)							
	21.7	15		0.870	2.205	8.221	0.577	11.003	25.2	0.032
July 22			1.077 (0.441)							
	23.0	31		1.334	2.770	33.598	1.999	38.367	12.4	0.013
Aug. 22			1.591 (0.988)							
	20.8	31		2.155	6.212	28.031	1.687	35.980	22.0	0.017
Sep. 22			2.721 (1.563)							
June 7 - Sep. 22	21.6	107		1.517	13.042	77.602	4.968	95.612	18.8	0.020
Sep. 22			2.721 (1.563)							
	20.4	15		3.180	4.817	15.740	2.294	22.851	31.1	0.019
Oct. 7			3.641 (1.822)							
	17.3	15		3.890	2.509	18.341	0.000	20.850	12.0	0.008
Oct. 22			4.139 (1.855)							
	15.4	16		4.321	1.656	17.688	3.072	22.419	21.1	0.005
Nov. 7			4.504 (2.463)							
	14.4	30		5.129	6.308	39.543	0.000	45.851	13.7	0.008
Dec. 7			5.755 (2.979)							
Sep. 22 - Dec. 7	16.4	76		4.238	15.293	91.312	5.366	111.971	18.4	0.009
Dec. 7			5.755 (2.979)							
	12.9	62		6.234	5.024	79.015	4.441	88.425	10.7	0.003
Feb. 7			6.893 (3.501)							
	12.1	28		7.610	6.179	27.783	0.000	33.957	18.2	0.006
Mar. 7			8.328 (4.289)							
Dec. 7 - Mar. 7	12.7	90		7.038	11.203	106.793	4.441	122.434	12.7	0.004
Mar. 7			8.328 (4.289)							
	16.1	46		8.493	1.172	94.761	6.292	102.225	7.3	0.001
Apr. 22			8.659 (6.155)							
	18.4	30		9.913	11.895	81.677	8.261	101.833	19.7	0.008
May 22			11.168 (6.784)							
Mar. 7 - May 22		76		9.748	13.067	176.438	14.554	204.058	18.3	0.004

* Net production efficiency (%) was obtained by dividing the sum of growth and exuvia energy by total assimilation for a given period. See text for description of other parameters.

Table 3. (Continued) Partial energy budget for the fastest growing male and female larvae of *Dolania americana* in Upper Three Runs.*

FEMALE LARVAE

	°C		mg		Total calories per median size larva				%	$mg\ mg^{-1}\ d^{-1}$
Date	Average Temp.	No. Days	Average Weight (s.d.)	Median Weight	Growth	Respiration	Exuvia	Assimi- lation	Net Production Efficiency	Instantaneous Growth Rate
June 7			0.574 (0.018)							
	20.8	15		0.677	1.084	3.844	0.503	5.431	29.2	0.020
June 22			0.781 (0.079)							
	21.1	15		1.155	4.038	10.205	0.740	14.983	31.9	0.044
July 7			1.530 (0.090)							
	21.7	15		1.739	2.184	13.937	1.110	17.231	19.1	0.016
July 22			1.948 (0.112)							
	23.0	31		3.124	12.862	61.269	3.997	78.129	21.5	0.025
Aug. 22			4.300 (0.360)							
	20.8	31		5.333	11.508	59.802	3.405	74.715	19.9	0.012
Sep. 22			6.366 (0.550)							
June 7 - Sep. 22		107			31.676	149.057	9.755	190.488	21.7	0.022
Sep. 22			6.366 (0.550)							
	20.4	15		7.366	10.938	33.951	0.000	44.889	24.3	0.018
Oct. 7			8.366 (1.101)							
	17.3	15		8.499	1.468	32.533	4.885	38.886	16.3	0.002
Oct. 22			8.633 (1.106)							
	15.4	16		9.616	10.855	31.814	0.000	42.669	25.4	0.012
Nov. 7			10.600 (1.757)							
	14.4	30		10.933	3.664	64.867	7.402	75.933	14.5	0.002
Dec. 7			11.266 (1.193)							
Sep. 22 - Dec. 7	16.4	76			26.925	163.165	12.287	202.377	19.4	0.007
Dec. 7			11.266 (1.193)							
	12.9	62		13.649	2.640	131.984	10.363	144.987	8.9	< 0.001
Feb. 7			16.033 (1.877)							
	12.1	28		16.249	2.540	44.901	0.000	47.441	5.3	< 0.001
Mar. 7			16.466 (2.369)							
Dec. 7 - Mar. 7	12.7	90			5.180	176.885	10.363	192.428	8.1	0.004
Mar. 7			16.466 (2.369)							
	16.1	46		19.999	41.907	175.486	0.000	217.393	19.3	0.007
Apr. 22			23.533 (2.357)							
	18.4	30		24.364	9.762	172.289	17.910	199.961	13.8	0.002
May 22			25.196 (0.600)							
Mar. 7 - May 22		76			51.669	347.775	17.910	417.354	16.5	0.006

* Net production efficiency (%) was obtained by dividing the sum of growth and exuvia energy by total assimilation for a given period. See text for description of other parameters.

Larval Growth

Larval growth appeared continuous for *D. americana*, although seasonal differences in the amount of growth occurred (Table 3). The magnitude and efficiency of growth were greatest in non-winter months when stream temperatures exceeded 15°C. Females generally grew more efficiently and faster than males during all seasons. The daily caloric requirement per female larva was usually about twice the amount estimated for male larvae. Instantaneous growth rates appeared to be inversely related to body size and positively correlated with stream temperatures for both sexes. Energy lost from respiration greatly exceeded that used in net production (growth plus exuviae) for most time periods. Net production efficiency was highest for small animals growing at temperatures above 20°C. Maximum net production (cal/day) per individual occurs in spring for both sexes. The data do not show when maximum net production occurs for the subpopulation as a unit in Upper Three Runs because larval density estimates were not made.

ADULT EMERGENCE

Drift collections of larval molt skins, which are shed at metamorphosis suggest that 98% of the individuals emerged synchronously in a seven day period (Table 4). Adult emergence in Upper Three Runs began each day at about 0530 h (sunrise) and ended about 0615 h. Observations support Peters and Peters (1977) contention that *D. americana* are sensitive to small changes in light intensity occurring just before sunrise. During the peak emergence period, nine flow-through microcosm streams were stocked with mature male and female larvae. Adult emergence among microcosms was synchronous and could be initiated earlier in the morning than normal by artificially lighting the microcosms for one hour early in the morning, then darkening them again, and re-exposing the trays to natural sunrise (Table 5).

The sex ratio at emergence was skewed towards males according to data from drifting larval molt skins, but towards females for drifting adults (Table 4). The skewed sex ratio for drifting adults has also been reported for other *D. americana* subpopulations (Peters and Peters 1977). A predominance of females among drifting adults may be an artifact because adult males fail to return to the river after mating due to intense predation by bats, birds, etc. while patrolling for females. The skewed sex ratio of drifting cast molt skins could result if female skins sink after floating a shorter distance than males due to their heavier weight (avg. = 2.787 vs. 1.804 mg). We suggest, therefore, that skewed sex rations based on drifting adults and larval skins may be artifacts, especially since larval collections made prior to emergence (late May) indicate a 1:1 sex ratio.

Table 4. Number and sexual composition of cast larval molt skins and dead adults collected in drift nets on the Upper Three Runs. Net #1 = near right bank; Net #2 = mid-channel.

Date	Larval Skins					Adults			
	Net #1		Net #2		Net #2	Net #1		Net #2	
	Male	Female	Male	Female	Cumulative %	Male	Female	Male	Female
6-3-78	18	4	28	15	1.5	-	-	-	-
6-4-78	47	45	279	187	17.8	-	-	51	60
6-5-78	1	-	30	-	18.9	-	-	1	-
6-6-78	290	240	329	264	39.7	-	-	45	61
6-7-78	(discontinued)		5	8	40.2	(discontinued)		-	-
6-8-78			821	626	91.1			160	199
6-9-78			121	93	98.6			12	18
6-10-78			5	4	98.9			-	-
6-11-78			7	11	99.5			-	-
6-12-78			2	-	99.6			-	-
6-13-78			4	2	99.8			1	-
6-14-78			3	1	99.9			-	-
6-15-78			-	-				-	-
6-16-78			-	-				-	-
6-19-78			1	-	100.0			-	-
Totals	356	289	1635	1211		-	-	269	338

Table 5. Influence of artificial and natural lighting on the onset of diel emergence for *Dolania americana*

	Laboratory Trays		Upper Three Runs	
Date	Lights	First Emergence	Sunrise	First Emergence
6-3-78	0400	0401	0518	0534
6-4-78	0330	0333	0518	0540
6-5-78	0416	0417	0518	0536
6-6-78	0230 (0530)	0233 (0533)	0518	0542

The mean weight of drifting larval skins appears to decrease for both sexes during the emergence period (Fig. 4). Since larval skin weight is correlated positively with the weight of the larva (and thus emerging adult), these data suggest that adult size of both sexes decrease through the emergence period. Thus, the relative contribution of the last emerging females to population recruitment may be decreased considerably by: (1) reduced size and fecundity; and (2) decreased probability of mating successfully.

Adult Fecundity

The number of eggs per female averaged about 77 and ranged from 44 to 148 (Table 6). There was no significant difference between mature larvae and subimagos when comparing the average number of eggs, weight per egg, and total egg weight per individual. The quantity of nonreproductive tissue appeared higher in female larvae than adults (10.9 vs. 7.201 mg); but when the average weight of the cast larval skin (2.787 mg) is subtracted from the larval weight, the differences are negligible (7.299 vs. 7.201 mg). About 56% of the total dry weight of an emerging female is eggs. After ovipositing, adult females weigh about the same as adult males.

A highly significant linear relationship was observed for fecundity as a function of female dry weight (Fig. 5). Since there were no significant differences between the fecundity dry weight regressions for subimagos and larvae (Analysis of Covariance: $F_{1,37} = 0.73$, $p > 0.05$), the two data sets were combined into a single regression. Although *D. americana* produces fewer eggs per female than has been reported for any other mayfly, indivudual eggs are at least ten times larger and heavier than reported for any other

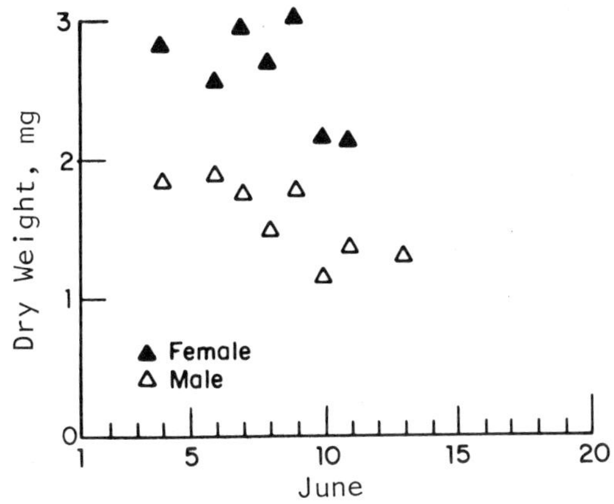

Figure 4. Average weight of male (△) and female (▲) larval molt skins collected from Upper Three Runs in drift nets during the adult emergence period. Molt skins were counted and weighed collectively for each date; the variance of each mean is unknown.

Table 6. Average dry weight and fecundity of *Dolania americana* females collected from Upper Three Runs from June 3 through June 6, 1978.

	Subimagos*	Larvae+
	Avg. ± S.D.	Avg. ± S.D.
Nonreproductive tissue wt (mg)	7.201±1.628	10.086±2.489
Egg wt (mg)	9.182±2.580	9.581±3.304
Total female wt (mg)	16.390±3.238	19.671±5.521
Egg number per female	75±20	79±25
Wt per egg (mg)	0.122±0.009	0.120±0.011

* sample size = 18
\+ sample size = 22

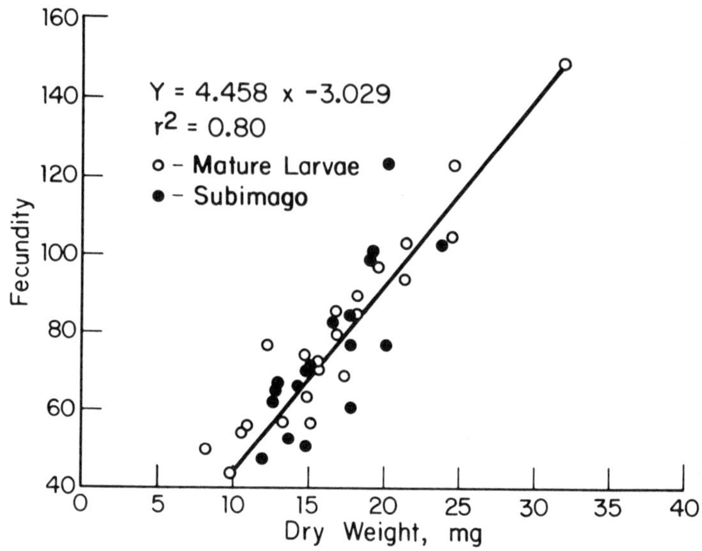

Figure 5. Fecundity of *Dolania americana* as a function of individual dry weight. Open circles = mature larvae; closed circles = subimago.

species. The adaptive significance of a low fecundity-large egg reproductive strategy is unclear, but may be related more to the species trophic habit (i.e. predator) than to the shifting, sandy nature of the habitat.

Egg Quality

Biochemical analysis revealed that eggs were mainly protein (65.5%) and lipid (25.2%) by weight, with small amounts of carbohydrate (9.0%; Table 7). Although similar data on other aquatic species are not available for comparison, the results agree well with the limited published data for terrestrial insects (Singh and Sinha 1977). Assuming the non-lipid dry component is used primarily for synthesis rather than respiration and the average dry weight of one egg is 0.12 mg (see Table 6), the maximum dry weight of first instar larvae would be 0.08 mg.

Table 7. Chemical composition of *D. americana* eggs.

Nutrient Chemical	n	Mean %	Range
Carbohydrate	5	9.0	7.6-12.8
Lipid	3	25.2	24.7-25.6
Protein	6	65.5	58.7-69.1

DISCUSSION

Although a two year life cycle has been suggested for other species of burrowing mayflies (Bartholomae and Meier 1977, Hynes 1970), the two years are spent mainly in the larval stage. In contrast, *D. americana* remains in the egg for most of the first year and as a larva the second. The ecological significance of a ten month embryonic period is unclear, particularly since the substrate of Upper Three Runs is loose, shifting sand. The egg of *D. americana*, however, appears adapted to endure a prolonged embryonic period because: (1) eggs contain a large supply of nutrient chemicals needed for embryonic development; (2) the chorion is thick and seems very resistant to mechanical breaking or puncturing; and (3) the intact chorion is not readily infected by aquatic fungi. The large size of the egg may also be highly adapted because it yields a large first instar larva. This assumes that larval mortality is inversely related to size and increased larval size expands the size range of potentially exploitable prey.

For most insect species, egg quantity and quality are generally considered to be important life history parameters that are related directly to the probability of the average animal to survive during the period between egg deposition and larval maturation. Our analyses of the size and number of eggs for *D. americana* suggests that individual mortality for eggs and larvae is probably very low relative to other mayfly species. Since the degree of egg mortality would seem to be correlated positively with the duration of the egg stage, one might expect strong selection against a prolonged egg diapause. Our results however, on the bioenergetics of larval development indicate that if there was no diapause and larvae hatched in June (as opposed to April), they would never achieve sufficient growth by the following June to emerge with more than a few eggs, if they could emerge at all. The intermediate sized larvae which fail to emerge with the remaining subpopulation of Upper Three Runs may result from a few individuals each generation failing to undergo an embryonic diapause and hatching prematurely.

Although annual or seasonal production of *D. americana* was not estimated in Upper Three Runs, the resource requirements of individual larvae can be obtained from the energy budget. We estimate the minimum daily caloric intake (i.e. assimilation/day) of a male and female larva growing during the summer - fall - winter - spring period to be 0.88 - 1.46 - 1.35 - 2.68 and 1.77 - 2.65 - 2.13 - 5.49 cal/day, respectively. Assuming *D. americana* eat mainly chironomid larvae that average about 0.15 mg and have a caloric equivalent of about 5.0 cal/mg, the number of prey consumed per day would range from 1.2 to 3.5 for males 2.3 to 7.3 for females depending on the season and stage of larval development.* Although the daily caloric intake per animal increases during larval development (except in winter), resource exploitation by the subpopulation as a whole may actually decrease or remain in a quasi-equilibrium throughout the year. To approach an equilibrium, larval mortality would have to be about 75% during the growing season. Higher mortality, which seems reasonable, would decrease the amount of resources used by large larvae relative to earlier stages.

RESUME

Le cycle évolutif de la *Dolania americana* dans le Upper Three Runs Creek de la Savannah River Plant près d'Aiken, en South Carolina, est de deux ans. Les oeufs pondus au début de juin éclosent le printemps suivant; les larves mettent de douze à quatorze mois à atteindre à leur plein développement. La métamorphose des larves à l'état adulte se produit en même temps, l'éclosion de plus de 95% des éphémères ayant lieu au cours de la première semaine. Les femelles pondent en moyenne 77 ouefs chacune; il y a corrélation entre la taille de l'adulte et sa fécondité. Les mâles et les femelles adultes semblent rapetisser pendant la période de métamorphose. La teneur en calories des divers tissus durant toute la durée du développement s'établit entre 4.63 à 6.55 cal/mg. Ce sont les oeufs qui ont la plus forte teneur en calories (moyenne 6.18 cal/mg) et la mue au stade subimaginal la plus faible (4.63 cal/mg). La teneur en calories par unité de poids du tissue larvaire augmente proportionnellment à la taille de la larve. L'analyse chimique des oeufs révèle une forte teneur en protéines (50.3%) contre 25.5% de lipides et 9% de glucides. Le taux de respiration relatif au poids des larves et inversement proportionnel à leur taille, mais directement proportionnel à la température de l'eau qui se situe entre 6 et 23°C. La croissance des larves semble être continue, bien que le rythme de croissance diffère de saison en saison. La capacité totale de production des larves se situe entre 5.3 et 29.2% avec les

* These values may be quite conservative because in this analysis we did not include the energy expended in the actual search and capture of prey.

variantes saisonnières s'établissant en moyenne à 8.1, 16.6, 21.7 et 19.4% en hiver, au printemps, en été et à l'automne respectivement.

ZUSSAMENFASSUNG

Dolania americana hat eine zweijährige Lebensgeschicte im Upper Three Runs Creek der Savannah River Plant nahe bei Aiken, South Carolina. Anfang Juni abgelegte Eier werden im folgenden Frühjahr ausgebrütet. Larven brauchen 12 bis 14 Monate, um zu reifen. Die Metamorphose der ausgewachesenen Tiere ist synchron, wobei 95% der einzelnen Fliegen während der ersten Woche zum Vorschein kommen. Die Weibchen produzieren durchschnittlich je 77 Eier. Größe und Fekundität der reifen Tiere sind positiv aufeinander bezogen. Ausgewachsene Tiere beider Geschlechter scheinen während der Periode des Auftauchens an Größe abzunehmen. Kalorienmessungen an verschiedenen Geweben und Entwicklungsstadien schwankten von 4.63 cal/mg bis 6.55 cal/mg. Eier hatten den höchsten Kaloriengehalt (durchschnittlich 6.18 cal/mg). Subimaginale Häutungsschalen hatten den niedrigsten Kaloriengehalt. Der Kaloriengehalt pro Einheit von Larvengewebe nahm mit der Larvengröße zu. Eine chemikalische Analyse von Eiern ergab einen hohen Gehalt an Protein (50.3%) im Verhältnis zu Lipoid (25.5%) und Kohlehydrat Bruchteilen (9.0%). Die gewichtsspezifischen Respirationsraten standen in entgegengesetztem Verhältnis zur Larvengröße, waren jedoch positiv auf die Wassertemperaturen zwischen 6 Grad und 23 Grad C. bezogen. Das Wachstum erschien stetig, obwohl Unterschiede, die durch die Jahreszeit bedingt waren, in der Wachstumsrate beobachtet wurden. Die Netto Produktionswirksamkeit für die Larven reichte von 5.3% bis 29.2% und variierte jahreszeitlich, mit Durchschnittswerten von 8.1% für Winter, 16.6% für Frühling 21.7% für Sommer und 19.4% für Herbst.

REFERENCES

Bartholomae, P.G. and P.G. Meier. 1977. Notes on the life history of *Potomanthus myops* in Southeastern Michigan (Ephemeroptera: Potamanthidae). *Great Lakes Entomol.* 10: 227-237.

Brody, S. 1945. Bioenergetics and growth. Reinhold, New York.

Dubois, M., K. Gilles, J. Hamilton, P. Rebers and F. Smith. 1956. Colorimetric method for determination of sugars and related substances. *Anal. Chem.* 28: 350-356.

Edmunds, G.F. and J.R. Traver. 1959. The classification of the Ephemeroptera I. Ephemeroidea, Behningiidae. *Ann. Entomol. Soc. Am.* 52: 43-51.

Gilson, W.E. 1963. Differential respirometer of simplified and improved design. *Science (Wash. D.C.)* 141: 531-532.

Hynes, H.B.N. 1970. The ecology of running waters. University of Toronto Press, Toronto.

Lowry, O.H., N.J. Rosebrough, A.L. Farr and R.J. Randall. 1951. Protein measurement with the folin phenol reagent. *J. Biol. Chem.* 193: 265-275.

Peters, W.L. and J.G. Peters. 1977. Adult life and emergence of *Dolania americana* in Northwestern Florida (Ephemeroptera: Behningiidae). *Int. Rev. gesamten. Hydrobiol.* 62: 409-438.

Price, C.A. 1965. A membrane method for determination of total protein in dilute algal suspensions. *Anal. Biochem.* 12: 213-218.

Sweeney, B.W. 1978. Bioenergetic and developmental response of a mayfly to thermal variation. *Limnol. Oceanogr.* 23: 461-477.

Sweeney, B.W. and J.A. Schnack. 1977. Egg development, growth, and metabolism of *Sigara alternata* (Say) (Hemiptera: Corixidae) in fluctuating thermal environments. *Ecology* 58: 265-277.

Singh, N. B. and S.P. Sinha. 1977. Carbohydrate, lipid and protein in the developmental stages of *Sitothilus oryzae (S. granarius)* Coleoptera, Curculionidae. *J. Entomol. Soc. Am.* 70: 107-111.

This paper was prepared in connection with work under Contract No. AT(07-2)-1 with the U.S. Department of Energy.

THE NUTRITIONAL DYNAMICS OF THE BURROWING MAYFLY, *HEXAGENIA LIMBATA* *

Melvin C. Zimmerman** and Thomas E. Wissing

Department of Zoology
Miami University
Oxford, Ohio 45056

ABSTRACT

Aspects of the nutritional dynamics of the burrowing mayfly, *Hexagenia limbata*, were examined. Freshly deposited detritus and surface sediments were analyzed (organic, carbon, nitrogen, caloric contents) and compared with gut material from two size-classes of naiads (< 15 and ≥ 15 mm). Both the chemical analysis and ATP-biomass carbon estimates suggest a higher nutritive potential for detritus. In both size-classes the ingested material consisted primarily of fine inorganic particles, and of the organic gut material, over 90% consisted of detrital aggregates. Analysis of the structure of the mouthparts and observations of feeding behavior were used to develop a description of the feeding process.

INTRODUCTION

To assess the importance of an organism in the utilization and cycling of nutrients, information is needed on its food habits and the quality and quantity of food ingested on a daily and seasonal basis. Although there are qualitative data available on the food materials ingested by *Hexagenia* (Neave 1932, Morgan and Grierson 1932, Hunt 1953, Fremling 1967, Walker 1970), there has been no real effort made to analyze feeding in relation to seasonal variation in the food materials available to the organism. Naiads of *Hexagenia*

* Supported in part by a grant from Sigma Xi.

** Department of Biology, Lycoming College, Williamsport, Pennsylvania 17701 *limbata* can be classified as deposit feeders or gatherer-collectors, using the terminology of Cummins (1973). Collectors appear to utilize fine particulate organic matter (FPOM) that is primarily surface-colonized by bacteria (Anderson and Cummins 1979).

The material presented here is part of a larger study (Zimmerman 1977) of the life history and nutritional dynamics of *H. limbata*. Our approach to describing the nutritional dynamics of this species was to examine the quantity and quality of potential foods (i.e., surface sediments and freshly deposited detritus) available to naiads seasonally and to compare these characteristics to those of material actually ingested by nymphs and stored in the foregut. A description of the feeding apparatus and behavior is also given. The effects of temperature on the feeding intensity of *H. limbata* have been reported elsewhere (Zimmerman and Wissing 1978).

STUDY AREA

Brandenburg Pond is located in southwestern Ohio. It is roughly circular, with a diameter of 88 m, total surface area of 0.61 ha, and a maximum depth of 2.9 m. Rutter and Wissing (1975) and Zimmerman *et al.* (1975) have reported on the physical, chemical and biological characteristics of the pond. Sediment composition and distribution in the pond are shown in Figure 1 (Zimmerman *et al.* 1975). The mean particle-size diameter of zone 1 (2.8 Ø; medium sand) is larger than zone 4 (4.3 Ø; very fine sand). Zones 2 and 3 are intermediate in mean particle-size diameter (3.8 Ø and 3.6 Ø, respectively; fine sand). Mean separation with Duncan's New Multiple Range Test demonstrated that zones 1 and 4 are most dissimilar in all particle-size categories (Zimmerman 1977).

Hexagenia limbata naiads were abundant (often > 400 nymphs/m^2) in zones 1 and 2 of the pond and sparse (often < 80 nymphs/m^2) in zone 4 (Zimmerman *et al.* 1975). Late summer density estimates, taken from samples along two transects (Fig. 1), indicated that the mean population size remained relatively stable (approximately 351 nymphs/m^2) during 1971-1976. Nymphs of various sizes can usually be collected from the pond as no short periods of synchronous emergence and mating exist. Instead, an emergence period extending from April through September was observed.

METHODS AND MATERIALS

Characteristics of Potential Food

To examine seasonal changes and to reduce variability due to sample site, surface sediments and detritus were collected monthly (November 1974 - November 1975) at two sites along the 1.2 and

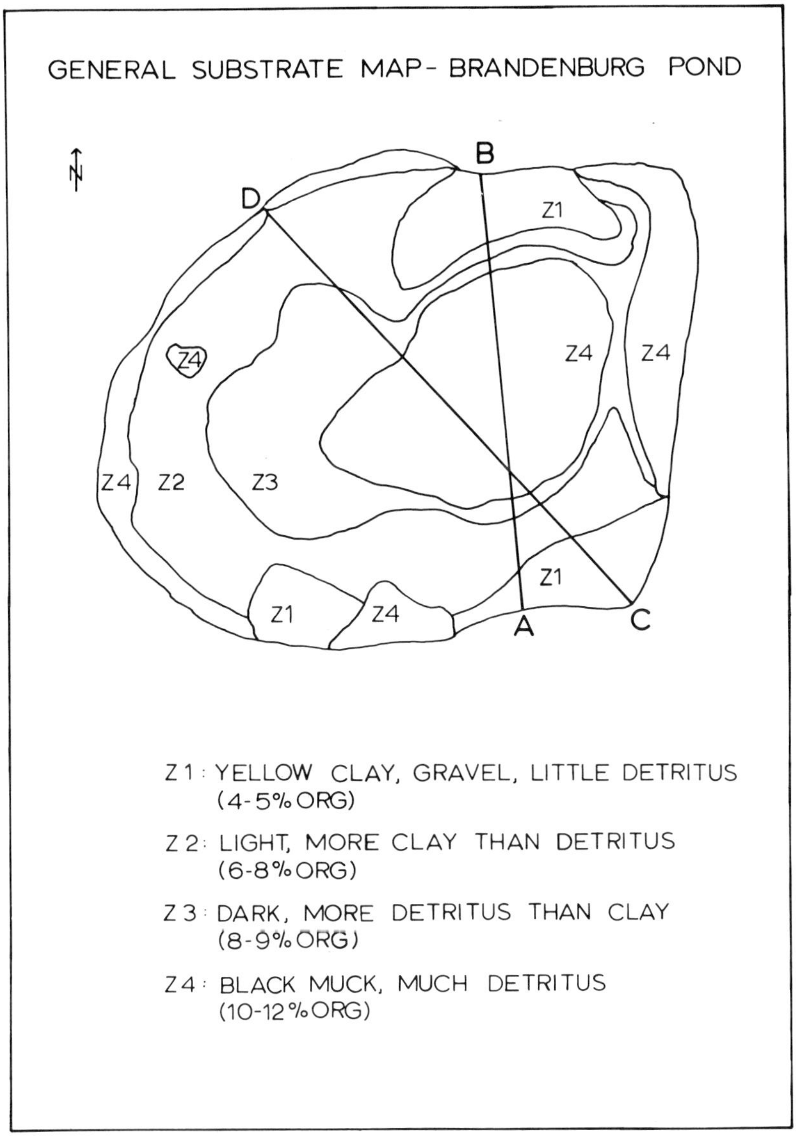

Figure 1. Types and distribution of sediments in Brandenburg Pond.

1.8 m contours in the southeast area of the pond (zones 1 and 2, respectively; see Fig. 1). Minimally disturbed surface sediments were collected with a K.B. gravity-type core sampler. Two replicate samples taken at each site were composed of the top 2 cm of sediment from four cores. All samples were transported to the laboratory on ice, oven-dried (60°C) to constant weight, and stored for later analysis. Suspended detritus, which settles to the bottom and hence becomes available as a potential food resource, was collected in paired acrylic plastic cylinders (9.5 cm d, 9 cm high) attached to bricks such that the effective collecting area (70.8 cm^2) was approximately 20 cm above the bottom. The traps were suspended for three, 5-day periods monthly. At the end of each period the total contents of a trap were emptied into plastic containers, oven-dried (60°C) and weighed. The two sediment and detritus samples collected at each depth were analyzed separately for their caloric (cal/g ash-free dry weight), nitrogen (as % dry weight), organic (as % dry weight), and carbon (as % dry weight) contents. Caloric contents were determined with the modified iodate sulfuric acid-wet oxidation method of Karzinkin and Tarkovskaya (1964), as described in Hughes (1969). Total organic nitrogen was estimated from the micro-Kjeldahl method of the Association of Official Agricultural Chemists (A.O.A.C. 1970). Organic contents were determined from comparison of weighed samples before and after combustion in a muffle furnace (550°C for 3 h). Before the final weight was taken, the ashed material was wetted with double-distilled water to reintroduce the water of hydration of clay and other minerals (Weber 1973) and dried. Organic carbon contents of sediment and detritus samples were estimated by wet oxidation with chromic acid (Walkley and Black 1934), as described by Morgans (1956). During October, November and December of 1976, and April and May of 1977, total microbial ATP was acid-extracted (0.6 N H_2SO_4) from fresh and lyophilized sediments and from fresh detritus. The ATP extraction methods were modified (see Zimmerman 1977) from Lee *et al.* (1971a, b), Karl and LaRock (1975), and Perry (1974). An Aminco Chem Glow photometer (American Instrument Co.) coupled to a strip chart recorder was used to measure light intensity following the injection of 0.1 mL of a reconstituted lyophilized luciferan-luciferase firefly lantern extract (Sigma Chemical Co.) into 0.5 mL of an extracted ATP sample.

Ingested Foods

The gut contents of 25-80 naiads of two size-classes (< 15 (small) and ≥ 15 mm (large)) were removed monthly during November 1974 - October 1975 for determinations of their organic, nitrogen and caloric contents using the procedures outlined above. All nymphs were collected in areas corresponding to zones 1 and 2 of the pond (Fig. 1).

The biotic composition and range of detritus particle sizes in the gut contents of the two size-classes of naiads were determined monthly. Preliminary observations during November 1974 - April 1975 indicated that preserved (70% ETOH) gut contents from the entire digestive tracts did not lend themselves to detailed enumeration. Subsequently, the anterior one-third of the gut contents of five of the small and three of the large size-class of nymphs were dissected monthly from May 1975 - December 1975, placed in distilled water, and dispersed with forceps. The dispersed gut contents were washed onto a gridded Millipore filter (0.45 μm) (Cummins 1973). Two such filters were prepared for each size-class and examined microscopically at 100, 200 and 450X. Determinations of biotic composition were made on the basis of 20 randomly chosen ocular grid squares (each square = 0.12 mm^2 at 200X). Particles were enumerated as a percentage of total present and were placed in the categories of organic detritus, diatoms, filamentous algae, and animal fragments. In addition, estimates of the areas (maximum length x maximum width; at 100X) of 40 detrital particles were made from each filter.

From September 1976 - March 1977, anterior gut contents of nymphs were observed periodically with a scanning electron microscope (SEM) (Coates and Welter 106). The contents were prepared for observation with the technique of Paerl and Shimp (1973). Gut contents were filtered on Nucleopore filters (0.4 μm), fixed (20% v/v gluteraldehyde solution buffered in 0.1 M sodium cacodylate), dehydrated in a graded ETOH series, transferred to 100% methanol. critical-point-dried (liquid CO_2 as transition fluid), mounted on aluminum stubs, sputter-coated with palladium-gold (40:60), and examined at 15 kV accelerating voltage.

Feeding Apparatus and Behavior

Naiad feeding behavior was observed in the field (from observations near shore) and laboratory. In the laboratory, nymphs were placed in 3 cm-wide aquaria filled with sediment and pond water. In addition, movements of mouthparts of nymphs placed with their ventral sides up in hanging drop slides (1.75 mm deep) were observed with a binocular microscope.

Mouthparts of naiads of various sizes (range = 10-25 mm) were observed with the SEM at an accelerating voltage of 1 kV. Whole nymphal heads or heads plus forelegs were separated from the thorax, preserved in 70% ETOH, and prepared for viewing as described above.

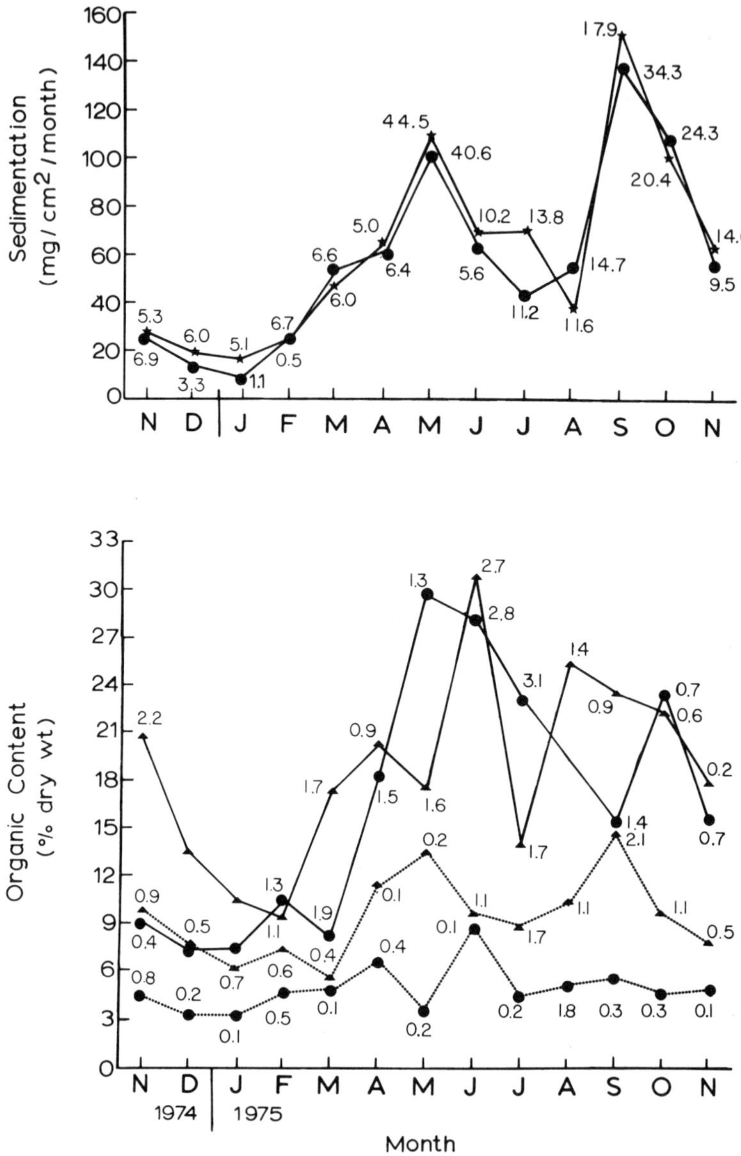

Figure 2. Sedimentation rates (mg-dry wt/cm^2/month) and organic contents (% dry wt) of detritus and surface sediments from Brandenburg Pond. Numbers represent the standard errors of the means. Sedimentation: * 1.2 m; ● 1.8 m. Organic content, detritus: ●---● 1.2 m; ▲---▲ 1.8 m; surface sediments: ●---● 1.2 m, ▲---▲ 1.8 m.

RESULTS AND DISCUSSION

Potential Food Materials

Mean monthly sedimentation rates (mg-dry wt/cm^2/month) and organic contents (% dry weight) of detritus and surface sediments are shown in Figure 2. The rates of sedimentation revealed a distinct seasonal pattern of low sedimentation during winter and high sedimentation during spring, summer and autumn. The monthly median rates of sedimentation at the two sites were similar. The highest rates of sedimentation were recorded in May and September, during and following a period of mixing, respectively. Distinct seasonal trends in the organic contents of detritus and sediments were not observed. However, during all but two months (November and December, 1974) the organic contents of detritus were always higher than those of surface sediments. The relationship between resuspended bottom sediments and water turbulence has been discussed by a number of investigators (Davis 1968, Steele and Baird 1972, Pennington 1974, Gasith 1975, and Jones 1976). The low organic content of the detritus indicates that during all months a large fraction of the material collected in the traps is resuspended sediment. As algal blooms develop very infrequently in this system (the pond may be considered oligotrophic), tripton of high organic content is never present in high concentrations.

Mean monthly nitrogen, carbon, and caloric contents of sedimentary detritus and surface sediments are shown in Figures 3 and 4. No distinct trends between sites or among seasons were observed for detritus nitrogen and carbon contents (Fig. 3). The caloric contents of detritus, at both depths, were higher in the May-September period (i.e., the range at 1.8 m = 4374-3366 cal/g ash-free dry wt) than in the November-March period (range = 3598-2919 cal/g ash-free dry wt). There appeared to be two peaks (April and August) in monthly nitrogen, carbon and caloric contents of the sediments (Fig. 4). During the winter the sediment nitrogen content was low or not detectable. Most studies of sedimentation have shown that material settling on the bottom has higher carbon, nitrogen and caloric contents than that of sediments in the same area. This has been attributed to the degree of oxidation of materials in the sediments (Pamatmat and Bhagwat 1973).

Annual grand mean values for the organic, caloric, carbon and nitrogen contents (Table 1) of detritus were always higher than those for sediments at each depth. Thus, it appeared that freshly deposited detritus has more nutritive potential for the naiads of *H. limbata*. However, Mann (1972) has noted that many animals that feed on sedimentary deposits are more nutritionally dependent on the microbiota than on the nonliving organic debris. Newell (1965) suggested that bacterial carbon and nitrogen are present in quantities sufficient to account for a large proportion of sediment carbon and

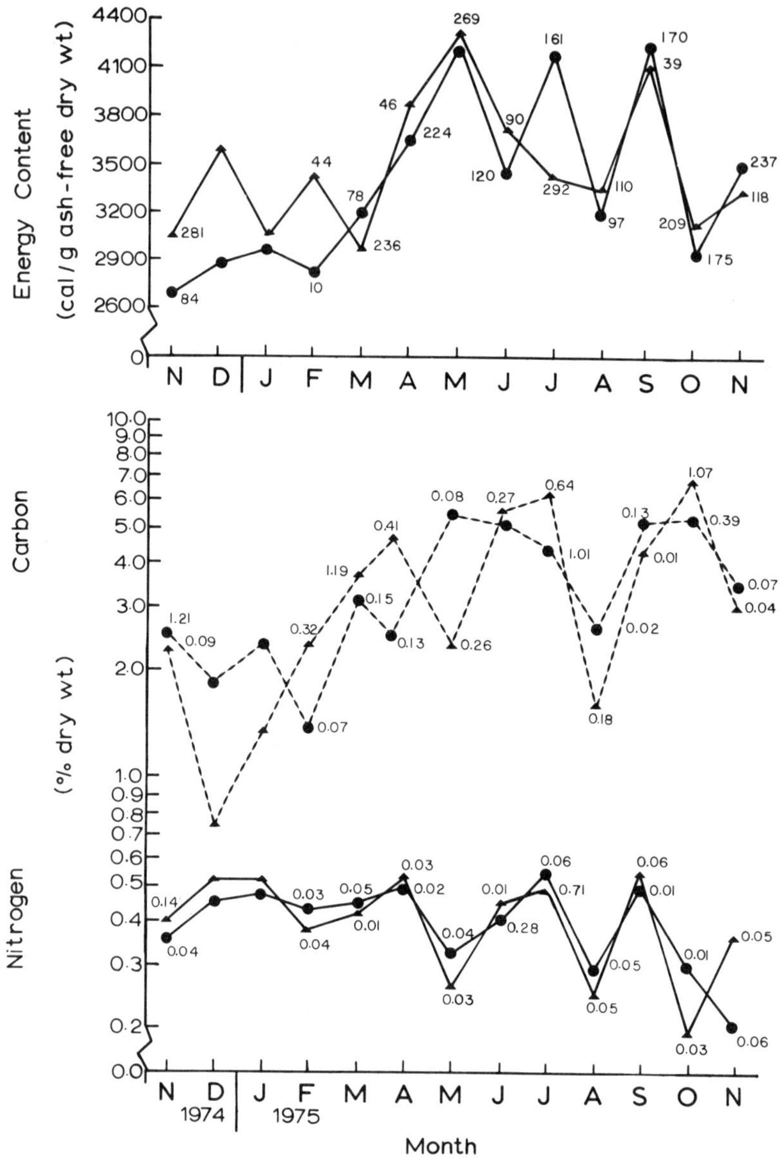

Figure 3. Percentage nitrogen and carbon (dry wt) and caloric contents (cal/g ash-free dry wt) of sedimentary detritus in Brandenburg Pond. Numbers represent the standard errors of the means. ● 1.2 m; ▲ 1.8 m.

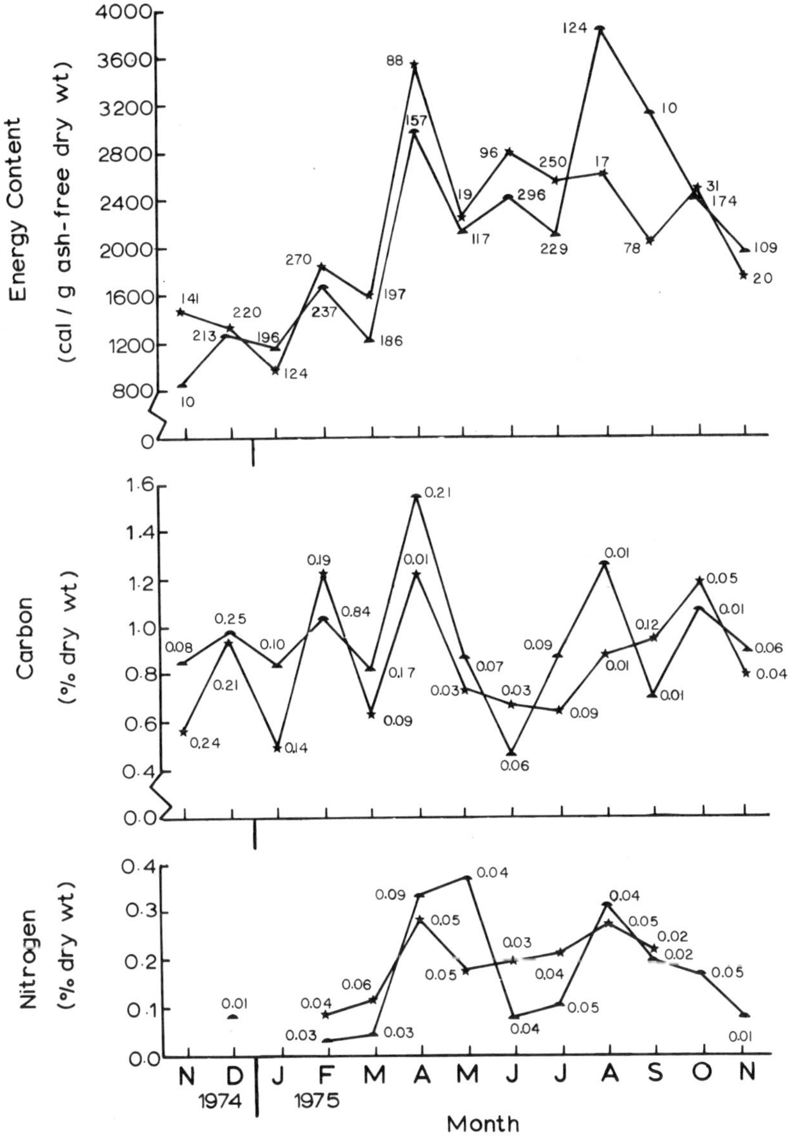

Figure 4. Percentage nitrogen and carbon (dry wt) and caloric contents (cal/g ash-free dry wt) of surface sediments from Brandenburg Pond. Numbers represent the standard errors of the means. * 1.2 m; • 1.8 m.

Table 1. Annual grand mean values for organic, carbon (C), nitrogen (N), (as percentage dry weights), caloric contents (cal/g ash-free dry wt) and C:N ratios of detritus and surface sediments from Brandenburg Pond. Numbers in parentheses represent standard errors of the means.

	Characteristic					
	Depth (m)	Organic (% dry wt)	Caloric (cal/g ash-free wt)	Carbon (% dry wt)	Nitrogen (% dry wt)	C:N
Detritus						
	1.2	16.5 (2.2)	3390 (154)	3.61 (0.42)	0.40 (0.03)	9.1:1
	1.8	18.7 (1.7)	3505 (125)	3.50 (0.53)	0.41 (0.03)	8.5:1
Sediments						
	1.2	5.0 (0.4)	2092 (195)	0.86 (0.07)	0.12 (0.03)	7.2:1
	1.8	9.3 (0.7)	2079 (239)	0.95 (0.07)	0.13 (0.04)	7.3:1

nitrogen, and Dale (1974) observed high correlations between carbon and nitrogen concentrations and bacterial numbers in sediments. He noted further that bacteria alone do not probably account for the total carbon and nitrogen in sediments. It seems reasonable that the primary source of carbon and nitrogen in sediments would include the total microbial community, consisting of bacteria, protozoans, fungi and epipelic algae. The potential nutritive contribution by the microbial community in sediments and detritus from Brandenburg Pond is shown by their low C:N ratios (see Table 1) and by preliminary values for ATP-biomass carbon (Table 2). Lee *et al.* (1971a) and Perry (1974) have reported values of sediment ATP ranging from 0.34-9.4 μg ATP/g dry wt and 11.24-50.5 μg ATP/g dry wt for nine Wisconsin lakes and Marion Lake, respectively. In this study, the highest values for fresh surface sediment were observed in April and May (range = 9.9-14.7 μg ATP/g dry wt).

The organic carbon estimates (as % dry wt) reflected the organic carbon present in living and dead material (Fig. 3 and 4). The biomass carbon values were based only on living material (Table 2). Comparison of the two sets of values provide information on what fraction of the sediment or detritus was tied up in live microbes. If it is assumed that the 1974-1975 carbon values (Figs. 3 and 4 were representative of any year, then most of the organic carbon present at any time is in microbial biomass. For example, the December and April organic carbon estimates for detritus and sediments at 1.2 m were 1.7 and 0.93%, and 2.5 and 1.2%, respectively. The corresponding ATP-biomass carbon estimates (converted to % dry wt) for these months were 0.9 and 2.5%.

Analysis of Ingested Foods

Changes in the organic, nitrogen and caloric contents of gut material from two size-classes of naiads are shown in Figure 5. Values of the gut material for both size-classes were similar during each interval. In general, the highest values for each group were observed during March-September. This trend was especially evident in data on the caloric content of gut material (e.g., for large nymphs, March-September range was 5588 cal/g ash-free dry wt, while the November-February range was 4329-4286 cal/g ash-free dry wt). The organic and caloric contents of gut material were always higher than values for sedimentary detritus and surface sediments (Fig. 3 and 4). However, the nitrogen contents of gut material, though higher than values for surface sediments, were similar to values for sedimentary detritus. For example, for large nymphs the July nitrogen content of ingested material was 0.63% of dry weight, while the sediment and detritus nitrogen values at 1.2 m were 0.21 and 0.54% of dry weight, respectively. These results suggest that *H. limbata* naiads utilize the freshly-deposited detritus as food. However, the contributions of gut secretions and the effects of

Table 2. Surface sediment and detritus ATP-biomass carbon values for Brandenburg Pond, 1976-1977. Biomass carbon calculated by multiplying the ATP measurements by 250 (Holm-Hansen, 1973).

Year	Depth (m)	No. of Samples	Biomass ATP ($\bar{x}$ ± SE μg/g dry wt)	Biomass Carbon ($\bar{x}$ ± SE μg/g dry wt)
1976				
Lyophilized sediments				
October	1.2	3	3.5±0.2	875±38
November	1.2	3	4.2±0.1	1042±22
December	1.2	3	3.9±0.1	967±36
1977				
Fresh sediments				
April	1.2	3	9.9±0.1	2487±13
	1.8	3	12.6±0.1	3147±29
May	1.2	3	16.6±0.4	4141±92
	1.8	3	14.7±0.3	3645±73
Detritus				
May	1.8	2	15.8±0.3	3945±70

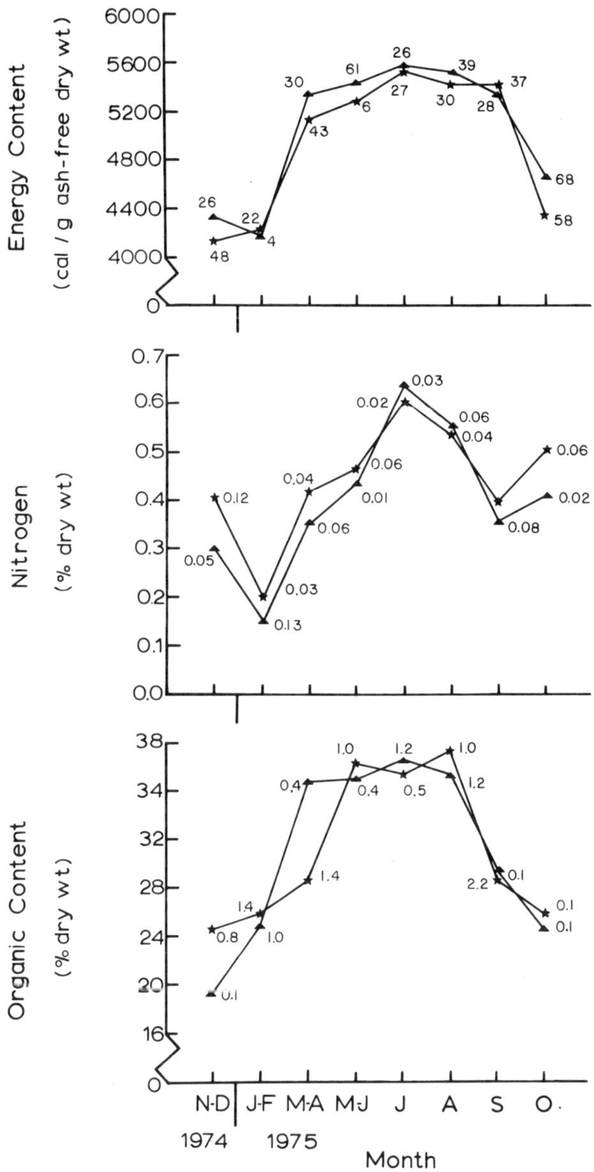

Figure 5. Percentage organic and nitrogen (dry wt) and caloric contents (cal/g ash-free dry wt) of gut contents from two size groups (< 15 and ≥ 15 mm) of *H. limbata* naiads. Numbers represent the standard errors of the means. * < 15 mm naiads; ▲ ≥ 15 mm naiads.

digestive processes on the organic, nitrogen and caloric values of gut materials are not known.

Results of the visual analysis of the gut contents of the two size-classes of nymphs are shown in Table 3. In both groups, detrital aggregates accounted for over 90% of the organic fraction of the gut material on a monthly basis. Diatoms were present (< 5% of total gut contents) in the foreguts of nymphs in all months except July. Filamentous algae and plant remains were of minor importance in the diet, and no animal fragments were encountered. Shapas and Hilsenhoff (1976) obtained similar results (92% detritus and 8% diatoms) for ten *H. limbata* nymphs collected in early spring. Though his data were only qualitative, Hunt (1953) reported that no significant differences were discernible in the amounts or composition of material ingested by *H. limbata* of different sizes or during different seasons. Although no animal fragments were observed in the gut contents, Rutter (1972) noted the presence of an oligochaete (*Sparganophilus* sp.) protruding from the mouthparts of a late instar naiad taken from Brandenburg Pond. Cummins (1973) has noted that the food items digested more rapidly, and hence seldom observed in gut contents, may actually have the highest nutritive value. In aquatic insects, these would probably include soft-bodied Diptera and Oligochaeta. Recently Moore (1977) has concluded that the importance of algae in the diet of aquatic insects is highly variable and dependent on the species, age and habitat. Brown (1960) observed that many of the algae removed from the hindguts of the detritivore *Chloeon dipterum* were viable. However, Cummins (1973) has emphasized that organic compound secreted or leaked by the algae also may be of nutritional importance to the consumer organism. Furthermore, Monk (1976) has observed weak cellulase activity in a number of aquatic insects, including ephemeropterans.

Mean size (μm^2) and range of detrital aggregates in the foreguts of the two size groups of naiads are also shown in Table 3. The mean sizes (19,375 μm^2) of detrital particles in the guts of small nymphs were significantly different ($p \leq 0.05$; Student's *t*-test) from those (38,239 μm^2) in large numphs. These clumps of particles can be seen in the contents from freshly-dissected guts and are not artifacts of filtration. It is possible that naiads mechanically compact these particles while straining debris through the molar surfaces and that the size differences are due to the variability in the size of the molar ridges between the large and small size-classes (see description of molar surfaces below). Treatment of the particles with acetone, which causes a white precipitate on contact with mucopolysaccharides (= mucus), revealed that these aggregates may also be held together by this material. Bacteria are known to secret sticky acidic polysaccharides (Corpe 1970, Floodgate 1972) and have been linked to formation of detrital aggregates in freshwater systems (Paerl 1973, 1974, 1975; Rodina 1963). Recently, Canfield and Backman (1978) reported high

Table 3. Summary of the food habits of two size-classes (<15 and ≤ 15 mm) of *H. limbata* naiads. Analyses are based on food material removed from the anterior one-third of the digestive tract.

	Month							
	M	J	J	A	S	O	N	D
Naiads ≥ 15 mm:[1]								
Detritus remains								
% of total[2]	95.8	97.7	96.9	96.5	98.2	97.2	96.8	95.6
mean size (μm^2)[3]	36,750	47,050	39,070	38,800	40,250	36,405	45,285	37,052
size range of particles (μm^2)	3,000 to 159,600	3,000 to 161,000	2,800 to 144,000	1,200 to 160,000	2,000 to 156,000	600 to 159,600	650 to 165,000	575 to 158,000
Diatoms								
% of total	3.1	1.3	0	3.2	1.2	2.5	3.2	4.4
Filamentous algae								
% of total	1.1	1.0	3.1	0	0.6	0	0	0
Plant remains								
% of total	0	0	0	0.3	0	0.3	0	0
Animal remains								
% of total	0	0	0	0	0	0	0	0
Naiads < 15 mm:								
Detritus remains								
% of total	97.3	96.0	97.0	99.4	97.3	95.9	97.2	97.2
mean size (μm^2)	18,670	17,200	18,588	19,875	18,990	18,622	20,847	18,932
size range of particles (μm^2)	800 to 72,000	600 to 87,800	700 to 73,800	1,200 to 86,800	400 to 120,900	1,300 to 60,000	600 to 85,100	950 to 68,130
Diatoms								
% of total	2.4	1.0	0	0.6	2.5	3.8	2.5	2.5
Filamentous algae								
% of total	0.3	3.0	3.0	0	0.2	0.3	0.3	0
Plant remains								
% of total	0	0	0	0	0	0	0	0
Animals remains								
% of total	0	0	0	0	0	0	0	0

[1] Size range of nymphs ≥ 15 mm was 15-25 mm; size range of nymphs < 15 mm was 8-14 mm.

[2] Percent detritus, diatoms, filamentous algae, animal, and plant remains based on total particles present and not size or weight. Estimates for detritus do not include inorganic particles.

[3] Size based on maximum length x maximum width of organic detritus and does not include inorganic particles.

concentrations (range = 4-274 million/L) of detrital aggregates (size range = 4-200 μm dia.) in the seston of nine Iowa lakes and reservoirs.

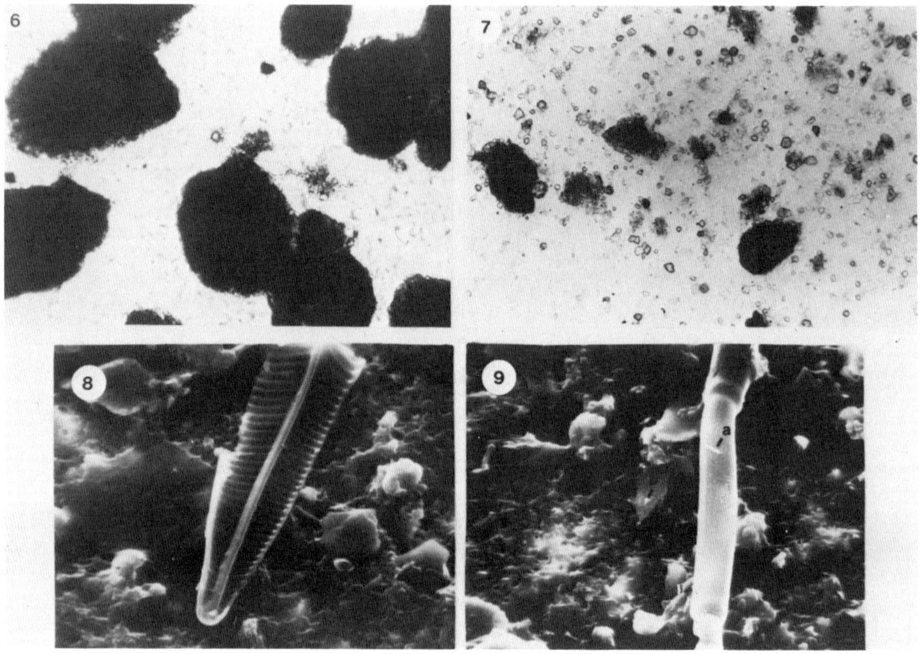

Plate 1. Figs. 6 and 7. Light micrographs of Millipore-filtered gut contents of *H. limbata* naiads. 13, large detrital aggregates at 100X; 14, a region of the filter showing smaller detrital aggregates and particles at 100X.

Figs. 8 and 9. Scanning electron photomicrographs of Nucleopore-filtered gut contents of *H. limbata* naiads. 8, fractured diatom frustule at 2700X; 9, filamentous algae at 2500X (a = colony of (?) bacteria attached to filamentous algae; b = (?) bacterium attached to inorganic particle). Accelerating voltage at 16 kV.

(Figs. 6 - 9 reduced 50% for reproduction).

Light and SEM micrographs of gut contents of *H. limbata* are shown in Plate 1. The light micrographs (at 100X) depict the size range of detrital particles (Figs. 6 and 7). The number and size range of inorganic particles (especially sand and clay) observed were large. Walker (1970) enumerated the inorganic and organic particles of *H. munda* naiads and found that 90% of the gut material was composed of sand and clay particles. Examination of the gut contents with the SEM supported the observation that the most abundant component of the gut contents is fine inorganic particulate material (many particles <1 μm; Figs. 8 and 9). Diatoms, pollen grains, filamentous algae, fungal hyphae, and bacteria were observed with the SEM. Many of these were associated with the detrital aggregates. Over 90% of the diatom frustules in the gut contents of naiads were fragmented (Fig. 8). While some appeared to be colonized by microbes, and ingested in that state, others seemed to have been fragmented by the animal during feeding. The grinding action of the molar surface of naiad mandibles, in combination with inorganic particles, may fragment the silicate frustules. Richardson and Gaufin (1971) concluded that sand particles in the foregut of the stonefly nymph, *Pteronarcys californica*, are used for grinding food and are regurgitated after this process. Regurgitation of sand particles does not seem to occur in deposit feeders. In fact, Cummins (1973) has suggested that these materials have nutritional value, owing to adsorbed organic material and associated bacteria.

Feeding Behaviour and Apparatus

It was difficult to observe the mechanism by which fine detritus is swept into the preoral cavity, processed by the molars, and swallowed. The following account of the feeding process is derived from microscopic analysis of the structure (Figs. 10-25) and movements of the mouthparts, laboratory and field observations of feeding behaviour, and from scattered reports in the literature (Morgan 1913), Needham 1920, Needham *et al.* 1935, Spieth 1941, and most recently McCafferty 1971). Most of the literature descriptions involve *Hexagenia recurvata*, now described as *Litobrancha recurvata* (McCafferty 1971, Edmunds *et al.* 1976). The most detailed description of the morphology, function and articulations of the mouthparts of mayflies was reported by Brown (1965) who studied two baetid forms (*Chloeon dipterum* and *Baetis rhodani*).

Food material is brought to the mouthparts of *H. limbata* in two ways. Nymphs occasionally approach the mouth of their burrows and dislodge surrounding material with their forelegs. Food material is also brought into contact with the mouthparts as the gills beat, thereby forcing water and debris into the burrow. On numerous occasions, I observed "puffs" of debris entering and leaving the burrows. The fine detritus debris is presumably collected by the labial palps, galea-laciniae, and canines, and swept into the preoral cavity, while the hypopharynx is retracted, by the movements of the galea-laciniae. The food is retained in the ventral part of the preoral cavity by setae and spines situated on the labrum (inner surface and outer edge, Fig. 11) and tips of the hypopharyngeal lobes (Fig. 25). Debris is forced towards the mouth by the mandibles, as they close, and by the pressure exerted through the protraction of the hypopharynx. A relatively dry bolus of food is prepared for swallowing by the straining and crushing action of the molar surfaces. Support for these observations can be derived from a detailed examination of the structure of the mouthparts.

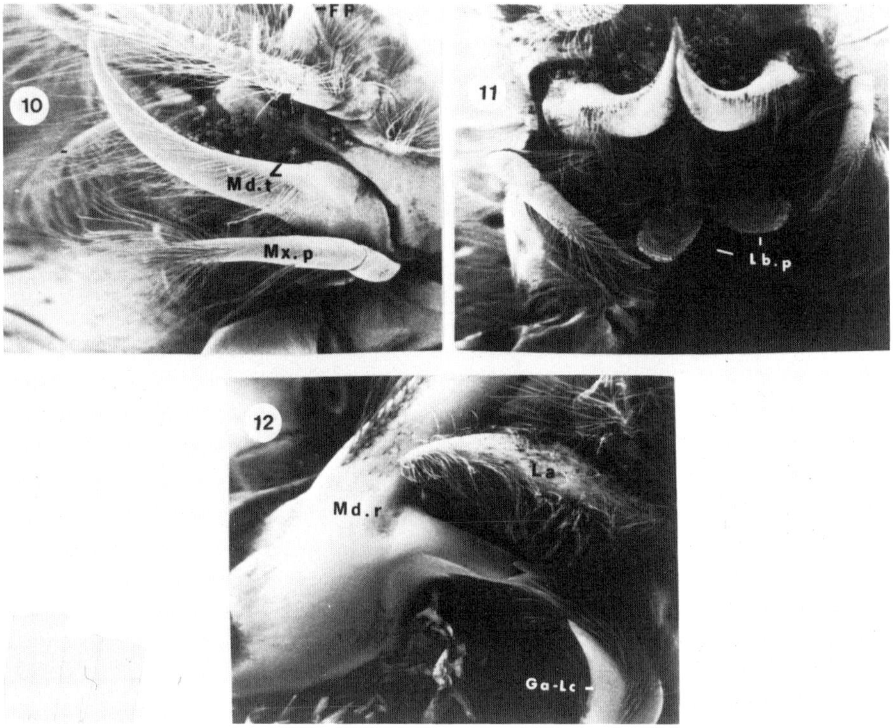

Plate 2. Figs. 10-12. Scanning electron photomicrographs of feeding appendages of *H. limbata* naiads. 10, side view of head at 68.2X; oblique view of head at 53.2X; 12, ventral view of head with labium and right maxilla removed exposing right mandible at 61.0X. Abbreviations: FP = frontal process; Ga-Lc = galea-lacinia of maxilla; La = labrum; Md. t = mandibular tusk; Md. r = right mandible; and Mx. p = maxillary palp. (Figs. 10 - 12 reduced 50% for reproduction).

The most noticeable diagnostic features of the head of *H. limbata*, as well as other *Hexagenia* species, are the long, curved tusks (mandibular tusks) extending beyond the head, and the dome-shaped prominence (frontal process) on the front of the head (Fig. 10) The maxillary and labial palps also extend beyond the head. All

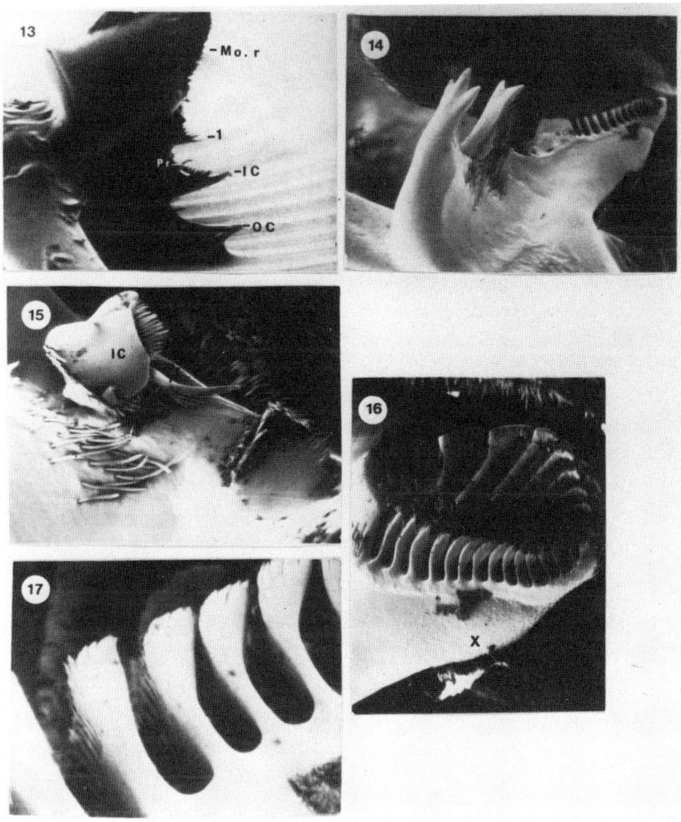

Plate 3. Figs. 13-17. Scanning electron photomicrographs of various aspects of the right mandible of *H. limbata* naiads. 13, profile of dissected right mandible at 87.7X; 14, ventral view of dissected mandible exposing canines and molar surfaces at 123.4X (note inner canine has two teeth); 15, median view of inner canine showing position of prostheca at 418.7X; 16, right molar surface at 286.1X (note, X = anterior surface); 17, close-up of anterior ridges of molar surface at 1100X. Abbreviations: IC = inner canine; Mo. r = right molar; OC = outer canine; and Pr = prostheca. Note, no 1 in Figures 13 and 15 are spines which lie between inner canine and molar surface. (Figs. 13 - 14 reduced 50% for reproduction).

mouthparts have an assortment of setae, spines and bristles which are directed toward the oral opening and apparently aid in food handling (Figs. 10 and 11). The proximal portions of the mandibles, maxillae, and labium are enclosed anteriorly by the clypeo-labral lobe, and the posterior surface of the lobe forms the anterior wall of the preoral cavity. In addition to the mandibular tusks, each mandible is composed of two canines (composed of two or three teeth), a prostheca, a number of spines and a molar surface (Figs. 13-21). The canines are on the median side of the tusk near the base, and, when the mandibles are in normal position (Fig. 12), they extend downward and probably function in grasping food. The food is passed from the canines to the grinding surface of the molars. Asymmetry in the structure of the right (Figs. 13-17) and left (Figs. 18-21) mandibles is produced by the arrangement and structure of the canines prostheca and molar surfaces. The apex of the outer canine of the right mandible contains three teeth, while the inner canine contains two teeth (Fig. 14). The dorsal surface of each canine is cup-shaped and numerous spines line the outer edge. This is more easily seen in the dorsal view of the canines from the left mandible (Fig. 18). At the lateral base of the inner canine lies the bristled prostheca (also called lacinia mobilis or mandibular palp), which is freely movable, but lacking muscles (Fig. 15). Its significance in feeding has not been determined (Needham *et al.* 1935, Berner 1950, Brown 1965). Between the prostheca and the right molar surface is a thick cluster of spines (Figs. 13 and 15). The anterior end of the right molar surface bears 14 ridges, which are blunt and toothed. These traverse the entire molar surface and terminate as blunt projections on the posterior side (Figs. 16 and 17).

Both canines of the left mandible contain three teeth, and the perimeter of their cup-shaped dorsal surface is composed of two alternating rows of toothed spines (Fig. 18). The left prostheca is stouter than the right and bears several short, horny spines (compare Fig. 19 with Fig. 15). The anterior molar surface of the left mandible is formed by 12 blunt ridges which traverse the body of the molar and terminate as sharp projections (Figs. 20 and 21). There is no tuft of spines between the prostheca and molar surface of the left mandible.

When the mandibles are closed, the canines of the left mandible lie anterior to (overlap) those of the right, and the ridges of the molar surfaces overlap (Fig. 12). Brown (1965) noted that the spines and teeth are sharpest and most numerous on the mandibles of larvae that have recently undergone ecdysis. These structures become worn with use. A worn molar surface from a right mandible is shown in Figure 22. The wear exerted by the sediments on the molar surfaces during feeding may account in part for the large number of instars (approximately 40) that may occur in this species (N.W. Britt, personal communication).

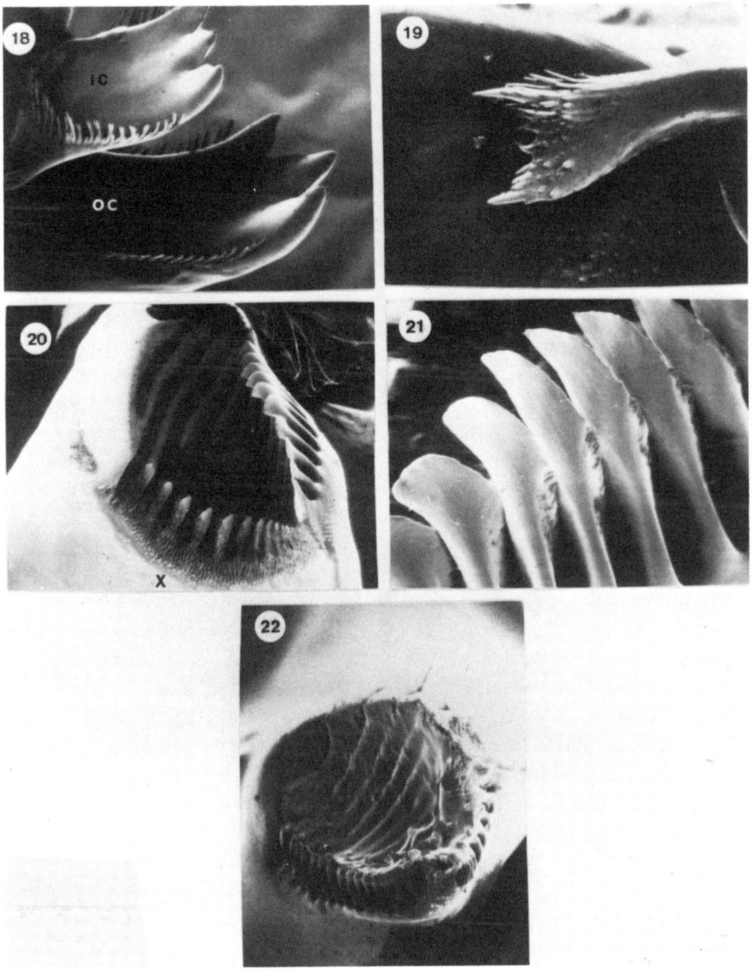

Plate 4. Figs. 18-22. Scanning electron photomicrographs of various aspects of the left mandible of *H. limbata* naiads. 18, lateral view of canines at 370.9X (note, inner canine has three teeth); 19, prostheca of left mandible at 840.1X; 20, left molar surface at 269.1X (note, X = anterior surface); 21, close-up of ridges on posterior molar surface at 827.6X; worn molar surface from the right mandible at 256.8X. Abbreviations: IC = inner canine; and OC = outer canine. (Figs. 18 - 22 reduced 50% for reproduction).

The positions of the maxillary and labial palps of *H. limbata* are shown in Figure 23. The palpi of the maxillae consist of three segments and terminate in a blunt, apical spine. The galea-laciniae of the maxillae are acute apically, with one large bidenticulate apical spine and two small subapical spines (not easily seen in Fig. 23). Though not distinct in Figure 25, the hypopharynx is composed of two superlinguae and a median sublingua.

The labium is formed by two palpi, each consisting of two segments and two inner paraglossae (Fig. 23). The dorsal surface of the terminal segment of a palp bears numerous spines (Fig. 24). The only differences in the morphology of the feeding appendages of large (none > 25 mm observed) and small (none < 10 mm observed)

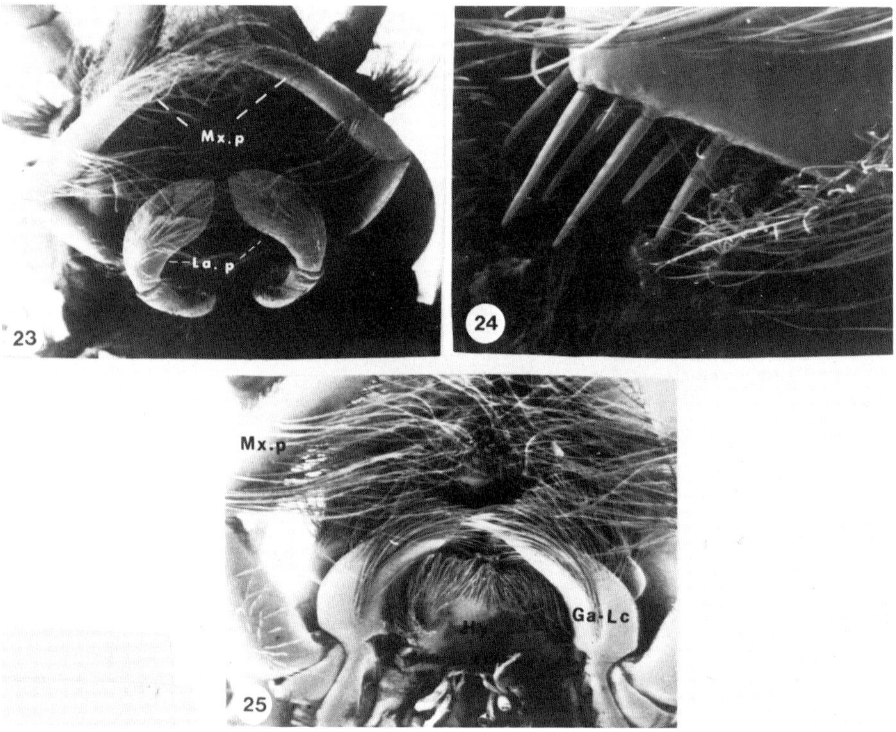

Plate 5. Figs. 23-25. Scanning electron photomicrographs of various aspects of ventral view of naiad head. 23, ventral view at 45.5X; 24, apical view of terminal segment of labial palp at 436.0X; 25, ventral view of head with labium removed exposing maxillae and hypopharynx at 61.9X. Abbreviations: Ga-Lc = galea-lacinia; Hy = hypopharynx; La. p. = labial palp; Mx. p. = maxillary palp. (Figs. 23 - 25 reduced 50% for reproduction).

naiads were in the sizes of structures, which in turn are dependent on length (or age) of the organism.

In summary, it appears that all size-classes of *H. limbata* feed in a similar manner and ingest material of high caloric, organic, nitrogen and carbon contents (i.e., the most freshly-deposited detritus). This material also appears to have a high microbial content. Other data indicate that naiads feed continuously, and at most temperatures (range = 10-25°C) ingest over 100% of their dry body weight per day (Zimmerman and Wissing 1978). An estimate of how much sediment is processed by the *H. limbata* population in the pond may be made by multiplying the densities of nymphs by their ingestion rates calculated from equations given in Zimmerman and Wissing (1978). When this is done for nymphs with mean lengths of 8, 12, 18 and 22 mm (for zone 1 of pond, using densities reported in Zimmerman *et al.* 1975; assume temp.= 20°C), the estimates of the quantities of sediment ingested are 1116, 1416, 1893 and 587 mg/m^2/day, respectively. For all size-classes combined, this value would be 5.01 g/m^2/day. Such information indicates that a deposit feeder such as *H. limbata* can contribute significantly to detritus-processing in small aquatic ecosystems.

RESUME

On a examiné les aspects des dynamiques nutritives de l'éphémère fouisseuse, *Hexagenia limbata*. On a analysé le détritus nouvellement déposé et les sédiments superficiels (les volumes organiques, du carbone, de l'azote, et calorique) et comparés avec la matière dans l'intestin antérieur des larves de deux classes de taille (< 15 et ≥ 15 mm). Et l'analyse chimique et les devis du ATP-biomasse carbone suggèrent un potentiel nutritif plus haut du détritus. Dans les deux classes de taille, la matière ingerée se compose surtout de fines particules inorganiques, et plus de 90% de la matière organique dans l'intestin antérieur se compose de agrégats de détritus. On a utilisé l'analyse de la structure des pièces buccales et les observations du comportement alimentaire pour développer une description du procés alimentaire.

ZUSSAMENFASSUNG

Diese Arbeit behandelt Aspekte der Ernährungsdynamik der sich eingrabenden Eintagsfliege, *Hexagenia limbata*. Frisch abgeschiedener Detritus und Oberflächensedimente wurden analysiert (organischer Gehalt, Wärmegehalt, Kohlestoff - und Stickstoffgehalt) und verglichen mit Darmmaterial von Najaden zweier Größenklassen. (< 15 und ≥ 15 mm). Beides, die chemische Analyse und ATP-Biomasse-Kohlenstoff Schätzungen lassen vermuten, daß Detritus ein höheres Ernähungspotential besitzt. Bei beiden Größenklassen enthielt das verschlungene

Material hauptsächlich feine, anorganische Partikel. Des organische Darmmaterial bestand zu mehr als 90% aus Detritalaggregaten. Eine Analyse der Struktur der Mundteile und Beobachtungen bezüglich des Ernährungsverhaltens wurden verwendet, um daraus eine Beschreibung des Nahrungsaufnahme Prozesses zu entwickeln.

REFERENCES

Anderson, N.H. and K.W. Cummins. 1979. Influence of diet on the life histories of aquatic insects. *J. Fish. Res. Board Can.* 36: 335-342.

Association of Official Analytical Chemists. 1970. Official methods of analysis. 11th ed. The Association, Washington, D.C.

Berner, L. 1950. The mayflies of Florida. Univ. Florida Press, Gainesville.

Brown, D.S. 1960. The ingestion and digestion of algae by *Chloeon dipterum* L. (Ephemeroptera). *Hydrobiologia* 16: 81-96.

Brown, D.S. 1965. The morphology and function of the mouthparts of *Chloeon dipterum* L. and *Baetis rhodani* (Pictet). *Proc. Zool. Soc. Lond.* 136: 147-176.

Canfield, D.E. and R.W. Bachman. 1978. Detrital aggregates in some Iowa lakes and reservoirs. *Hydrobiologia* 57: 275-279.

Corpe, W.A. 1970. Attachment of marine bacteria to solid surfaces. p. 73-85. *In:* Manly, R. ed. Adhesion in biological systems. Academic Press, New York.

Cummins, K.W. 1973. Trophic relations of aquatic insects. *Annu. Rev. Entomol.* 18: 183-206.

Dale, N.G. 1974. Bacteria in intertidal sediments: factors related to their distribution. *Limnol. Oceanogr.* 19: 509-518.

Davis, M.B. 1968. Pollen grains in lake sediment, redeposition caused by seasonal water circulation. *Science (Wash., D.C.)* 162: 796-799.

Edmunds, G.F., S.L. Jensen and L. Berner. 1976. The mayflies of North and Central America. Univ. Minnesota Press, Minneapolis.

Floodgate, G.D. 1972. The mechanisms of bacterial attachment to detritus in aquatic systems. *Mem. Ist. Ital. Idrobiol.* 29 (Suppl.): 309-323.

Fremling, C.R. 1967. Methods for mass rearing *Hexagenia* mayflies (Ephemeroptera: Ephemeridae). *Trans. Am. Fish. Soc.* 96: 407-410.

Gasith, A. 1975. Tripton sedimentation in eutrophic lakes, simple correction for resuspended matter. *Int. Ver. theor. angew. Limnol. Verh.* 19: 116-122.

Holm-Hansen, O. 1973. The use of ATP determinations in ecological studies. *Bull. Ecol. Res. Comm. (Stockholm)* 17: 215-222.

Hughes, R.N. 1969. Appraisal of the iodate sulphuric acid-wet oxidation procedure for caloric content of marine sediments. *J. Fish. Res. Board Can.* 26: 1959-1964.

Hunt, B.P. 1953. The life history and economic importance of a burrowing mayfly, *Hexagenia limbata*, in southern Michigan Lakes. *Bull. Inst. Res., Mich. Dep. Conserv.* 4: 1-151.

Jones, J.G. 1976. The microbiology and decomposition of seston in open water and experimental enclosures in a productive lake. *J. Ecol.* 64: 241-278.

Karl, D.M. and P.A. LaRock. 1975. Adenosine triphosphate measurements in soil and marine sediments. *J. Fish. Res. Board Can.* 32: 599-607.

Karzinkin, G.S. and O.I. Tarkovskaya. 1964. Determination of caloric values of small samples. p. 122-124. in Pavlovskii, E.N., ed. Techniques for the investigation of fish physiology. Oldbourne Press, London.

Lee, C.C., R.F. Harris, J.D. Williams, D.E. Armstrong and J.K. Syers. 1971a. Adenosine triphosphate in lake sediments 1. Determination. *Soil Sci. Soc. Am. Proc.* 35: 82-86.

Lee, C.C., R.F. Harris, J.D. Williams, D.E. Armstrong and J.K. Syers. 1971b. Adenosine triphosphate in lake sediments 2. Origin and significance. *Soil Sci. Soc. Am. Proc.* 35: 86-91.

Mann, K.H. 1972. Detritus and its role in aquatic ecosystems: introductory remarks. *Mem. Ist. Ital. Idrobiol.* 29 (Suppl.): 13-16.

McCafferty, W.P. 1971. New genus of mayflies from Eastern North America (Ephemeroptera: Ephemeridae). I. *J. N.Y. Entomol. Soc.* 79: 45-51.

Monk, D.C. 1976. The distribution of cellulase in freshwater invertebrates of different feeding habits. *Freshw. Biol.* 6: 471-475.

Moore, J.W. 1977. Some factors effecting algal consumption in subarctic Ephemeroptera, Plecoptera and Simuliidae. *Oecologia (Berl.)* 27: 261-273.

Morgan, A.H. 1913. A contribution to the biology of mayflies. *Ann. Entomol. Soc. Am.* 6: 371-413.

Morgan, A.H. and M.C. Grierson. 1932. The functions of the gills in burrowing mayflies (*Hexagenia recurvata*) during the winter and early spring. *Physiol. Zool.* 5: 230-245.

Morgano, J.F.C. 1956. Notes on the analysis of shallow water soft substrata. *J. Anim. Ecol.* 25: 367-487.

Neave, F. 1932. A study of the mayflies (*Hexagenia*) of Lake Winnipeg. *Contrib. Can. Biol. Fish.* 15 (ser. A): 179-201.

Needham, J.G. 1920. Burrowing mayflies of our larger lakes and streams. *Bull. U.S. Bur. Fish.* 36: 269-292.

Needham, J.G., J.R. Traver and Y.C. Hsu. 1935. The biology of mayflies. Comstock Pub. Co., Ithaca, NY.

Newell, R. 1965. The role of detritus in the nutrition of two marine deposit feeders, the prosobranch *Hydrobia ulvae* and the bivalve *Macoma balthica*. *Proc. Zool. Soc. Lond.* 144: 25-45.

Paerl, H.W. 1973. Detritus in Lake Taho: structural modification by attached microflora. *Science (Wash., D.C.)* 180: 496-498.

Paerl, H.W. 1974. Bacterial uptake of dissolved organic matter in relation to detrital aggregation in marine and freshwater systems. *Limnol. Oceanogr.* 19: 966-972.

Paerl, H.W. 1975. Microbial attachment to particles in marine and freshwater ecosystems. *Microb. Ecol.* 2: 73-83.

Paerl, H.W. and S.L. Shimp. 1973. Preparation of filtered plankton and detritus for study with scanning electron microscopy. *Limnol. Oceanogr.* 18: 802-805.

Pamatmat, M.M. and A.M. Bhagwat. 1973. Anaerobic metabolism in Lake Washington sediments. *Limnol. Oceanogr.* 18: 611-627.

Pennington, W. 1974. Seston and sediment formation in five lake district lakes. *J. Ecol.* 62: 215-251.

Perry, E.A. 1974. The biomass and activity of bacteria in the sediment of Marion Lake, British Columbia. Thesis, Univ. of British Columbia.

Richardson, J.W. and R.A. Gaufin. 1971. Food habits of some western stonefly nymphs. *Trans. Am. Entomol. Soc.* 97: 91-121.

Rodina, A.G. 1963. Microbiology of detritus. *Limnol. Oceanogr.* 8: 388-393.

Rutter, R.P. 1972. Distribution, abundance and rates of respiration and food turnover of the burrowing mayfly, *Hexagenia limbata*, in a pond ecosystem. Thesis, Miami Univ., Oxford, OH.

Rutter and T.E. Wissing. 1975. Distribution, abundance and age structure of a population of the burrowing mayfly, *Hexagenia limbata*, in an Ohio pond. *Ohio J. Sci.* 75: 7-13.

Shapas, T.J. and W.L. Hilsenhoff. 1976. Feeding habits of Wisconsin's predominant lotic Plecoptera, Ephemeroptera and Trichoptera. *Great Lakes Entomol.* 9: 175-188.

Spieth, H.T. 1941. Taxonomic studies on the Ephemeroptera. II. The genus *Hexagenia*. *Am. Midl. Nat.* 26: 233-280.

Steele, J.H. and I.E. Baird. 1972. Sedimentation of organic matter in a Scottish sea loch. *Mem. Inst. Ital. Idrobiol.* 29 (suppl.): 73-88.

Walker, W.H. 1970. A study of the ecology and distribution of burrowing mayfly nymphs (Ephemeroptera: *Hexagenia munda* Eaton) in Stone Mountain Lake, Thesis, Emory Univ.

Walkley, A. and I.A. Black. 1934. An examination of the Deytjareff method for determining soil organic matter and a proposed modification of the chromic acid titration method. *Soil Sci.* 37: 29-38.

Weber, C. 1973. Recent developments in measurement of the response of plankton and periphyton to changes in their environments. p. 119-138. in: Glass, G. ed. Bioassay techniques and environmental chemistry. *Ann Arbor Sci. Publ.* Ann Arbor, MI.

Zimmerman, M.C. 1977. Aspects of the life history, feeding behavior and nutritional dynamics of the burrowing mayfly, *Hexagenia limbata*, in an Ohio pond. Dissertation, Miami Univ., Oxford, OH.

Zimmerman, M.C. and T.E. Wissing. 1978. Effects of temperature on gut-loading and gut-clearing times of the burrowing mayfly, *Hexagenia limbata*. *Freshw. Biol.* 8: 269-277.

Zimmerman, M.C., T.E. Wissing and R.P. Rutter. 1975. Bioenergetics of the burrowing mayfly, *Hexagenia limbata*, in a pond ecosystem. *Int. Ver. theor. angew Limnol. Verh.* 19: 3039-3049.

OVERWINTERING STRATEGY OF *CLOEON DIPTERUM* (L.) LARVAE

Björn Nagell

Institute of Zoophysiology
University of Uppsala
Uppsala S-752 36 Sweden

ABSTRACT

In Sweden, larvae of *Cloeon dipterum* L. are very abundant in ponds that are anoxic for three to four months each winter. Adaptation to this winter environment was investigated. Some comparative studies were carried out on larvae from an English non-anoxic winter environment.

Investigations on the anaerobic metabolism of the Swedish larvae indicate that glycogen is stored in the autumn and used during the winter anaerobiosis. The end product of the anaerobic catabolism of glycogen was lactate, but there were also some indications of a higher energy yielding metabolic pathway, including the simultaneous catabolism of carbohydrate and amino acids.

INTRODUCTION

In Sweden the larvae of *Cloeon dipterum* L. live in high numbers in small ponds, especially where there are no fish since they are very susceptible to fish predation. These ponds are characterized by being small, shallow, receiving allochthonous plant material and having little or no flow-through. These ponds are usually anoxic for three to four months each winter. Two Swedish fish species can survive anoxia, the crucian carp (*Carassius carassius* L.) and the carp (*Cyprinus carpio* L.). However, these fish species only occur in ponds of this kind, in central Sweden, where they have been introduced by man. Thus *Cloeon* largely avoids fish predation - although to do so they must withstand adverse winter conditions.

The susceptibility of the larvae to fish predation was clearly demonstrated when their abundance was compared in four very similar and closely situated ponds. In one of the ponds containing a few crucian carp their abundance was only about one twentieth of that of the other three.

WINTER ENVIRONMENT

The winter environment of a typical pond of this type in central Sweden is usually as follows: in November or at the beginning of December the pond freezes over. This produces pronounced temperature stratification with 0°C at the surface and 4°C at the bottom. Within a few days, due to this stratification and to the oxygen consumption of the mud, the bottom water becomes anoxic. Ten to fourteen days after freeze-up the whole water mass is usually anoxic and contains H_2S in low concentrations. The redox potential of the water is negative and decreases towards the bottom. After about six weeks gas-filled plant material floats up beneath the ice, mainly in eutrophic ponds, and increases the H_2S concentrations there. Towards the end of the winter, usually at the end of March, the maximum concentrations of undissociated H_2S are $\simeq$ 0.5-1.0 mg/L During anoxia the water is grayish in colour due to FeS in suspension.

To all appearances the whole water mass is anoxic except during thaw periods when oxygen rich melt water can percolate through the ice. Due to its low density this oxygen rich water remains in a thin, one or two cm thick, layer just beneath the ice. However, shortly after the end of the thaw period anoxic conditions return. Further details of the winter environment are given by Nagell and Brittain (1977).

OVERWINTERING STRATEGY

Chronologically the overwintering strategy is as follows: the first response, indicative of approaching winter, is that the larvae successively, over a 10 to 20 day period, cease feeding. This happens when the temperature of the ambient water is close to or below 4°C. The larvae start feeding again if the temperature rises to 5°C. Experimentally it has been shown that the factor that ultimately controls the cessation in feeding is the low temperature and not, as might be expected, the decreasing photoperiod during the autumn (Nagell in prep.).

The cessation of feeding and the development of resistence to anoxia, measured as survival after a four day exposure to anoxia at 0.1°C, were investigated with respect to the time of acclimation to 0.1°C. The results showed that these two features were closely related. After ten days of acclimation about 90% of the larvae had ceased feeding and 10% survived exposure to anoxia. After 20 days 100% had ceased feeding and about 70% survived anoxia. With a further ten days of acclimation almost 100% survived (Nagell in prep.). Another group of larvae in otherwise identical conditions but acclimated to 10°C for six weeks did not shown any resistance to anoxia (exposed for 40 hours at 0.1°C) (Nagell and Fagerström 1978). Thus the acclimation to low temperature induces resistance to anoxia. Fully developed, resistance corresponds to a LT_{50} value of about 130 days and some larvae survived as long as 155 days (Nagell 1977a).

In oxygenated water the larvae are negatively phototactic but when the water becomes anoxic they become positively phototactic, moving up to the underside of the ice (Nagell 1977b). This location greatly aids their survival, giving the larvae access to oxygen-rich melt water during thaw periods. In a pond it was shown experimentally that this is very important for survival during the winter. Sixty percent of larvae with access to the percolating water survived for 138 days (December to mid-April), compared with 18% of larvae screened off from this water by a plastic film but otherwise given identical conditions (Nagell 1979). Evidently during overwintering the larvae are facultative anaerobes. In anoxia the larvae crawl and swim when disturbed, although considerably more sluggishly than in aerated water.

In order to study adaptation to winter conditions in a region where the period of ice-cover is so short that no anoxia occurs, larvae from the English Lake District were investigated. It was found that these larvae did not cease feeding as quickly as the Swedish larvae. When acclimated to 0.1°C it took 42 days before close to 100% ceased feeding. The English larvae continued to feed unaffectedly at temperatures as low as 2.5 to 1.1°C. It was evident that the English larvae were not adapted to rapidly changing conditions. Continuation of feeding is probably of value for them since risk of anoxia is very low and eating at low temperatures permits the larvae to grow during a greater part of the winter. This probably shortens the development time in the spring, thereby increasing the possibility of two generations per year. This happens only exceptionally in central Sweden, but is not uncommon in Britain (Macan 1979).

In spite of a six week long acclimation to 0.1°C and a decreasing photoperiod the English larvae were not found to be very resistant to anoxia. Of 144 larvae tested at 0.1°C, 134 did not survive the first 24 hours. However, one larva survived three days, three four to five days, two seven to eight days and one as long as 37 days. Thus there are individuals with a certain resistance. This will be important in the case of an unusually cold winter with a period of anoxia. The presence of this character within the population may be taken as an indication that winter anoxia occurs occasionally in the pond.

Recently the species *Cloeon dipterum* L. has been systematically revised (Sowa 1975). According to Dr. Sowa, who kindly examined some experimental larvae and imagines from both localities, the Swedish material belonged to *C. inscriptum* Bengtsson and the English to *C. cognatum* Stephens, two very closely related species. The differences found in the present investigation might be taken as a further reason for their separation.

ANAEROBIC METABOLISM

Investigations are in progress to elucidate the anaerobic pathways in the larvae. It has been found that glycogen is stored during the autumn and used during winter anaerobiosis. From September to December, glycogen increased more than four times reducing slightly below the September value (11.2 mg/g wet wt.) during the overwintering. However under aerobiosis, triglycerides were used and the glycogen reserve was only reduced by 11.5%. This is somewhat surprising as carbohydrates are usually consumed first during starvation (Wigglesworth 1965). However, the larvae are in a diapause and there is probably a mechanism to retard the consumption of glycogen in order to save it for use during anoxia.

After 73 days of anoxia at 0.1° C, lactate concentration in the haemolymph had increased three fold. This indicates that the end product of glycogen degradation is lactate, a metabolic pathway that is not very effective in yielding ATP. Thus many facultatively anaerobic invertebrates probably do not catabolize carbohydrates solely in this simple way. Hochachka *et al.* (1973) have suggested more effective pathways such as simultaneous catabolism of carbohydrate and amino acids, by which the ATP yield can be increased above that of the lactic acid pathway without seriously reducing the cell. In *Cloeon* larvae there are some indications that such a pathway is also present. In the haemolymph of the 73 day anoxia larvae there was an increase in alanin (55%) and a decrease in glutamate (65%) and aspartate (15%) which coincide with some of the suggestions made by Hochachka *et al.* (1973).

ACKNOWLEDGMENTS

To the Freshwater Biological Association, Windemere, for supplyi supplying *C. dipterum* larvae and to Dr. J. Brittain for improving the English.

RESUME

Les larves du *Cloeon dipterum* foisonnent dans les mares suédoises qui sont très pauvres en oxygène pendant trois ou quatre mois chaque hiver. Des recherches furent entreprises sur la façon dont l'insecte s'adapte à ce milieu, en prenant comme point de comparaison des études effectuées sur des larves en Angleterre dans un milieu hivernal qui n'est pas privé d'oxygène. Poursuivant l'étude du métabolisme anaérobique des larves suédoises, on constate qu'elles emmagasinent glycogène pendant l'automne et l'utilisent pendant l'anáerobiose d'hiver. Le produit final du catabolisme anaérobique du glycogène est le lactate et l'on a pu constater également à certains indices l'existence d'un métabolisme produisant une plus grande quantité d'énergie, y compris le catabolisme simultané de glucides et d'amino-acides.

ZUSSAMENFASSUNG

In Schweden treten Larven von *Cloeon dipterum* sehr zahlreich in Teichen auf, die im Winter drei bis vier Monate keinen Sauerstoff enthalten. Die Anpassung an diese winterliche Umwelt wurde in der vorliegenden Arbeit untersucht. Auch einige komparative Studien mit englischen Larven aus einer Winterlandschaft, die sauerstoffhaltig ist, wurde vorgenommen. Untersuchungen am anaerobischen Metabolismus der schwedischen Larven weisen nach, daß im Winter Glykogen gespeichert und während der Winteranaerobiose genutzt wird. Als Endprodukt des anaerobischen Katabolismus ergab sich Laktat, jedoch waren auch einige Anzeichen eines stärker energiespendenden Prozesses zu bemerken, der den gleichzeitigen Katabolismus von Kohlehydraten und Aminosäuren umfaßt.

REFERENCES

Hochachka, W., J. Fields and T. Mustafa. 1973. Animal life without oxygen: basic biochemical mechanisms. *Am. Zool.* 13: 543-555.

Macan, T.T. 1979. A key to the nymphs of the British species of Ephemeroptera with notes on their ecology, 3rd edition. *Freshw. Biol. Assoc. Sci. Publ.* 20: 1-80.

Nagell, B. 1977a. Survival of *Cloeon dipterum* (Ephemeroptera) larvae under anoxic conditions in winter. *Oikos* 29: 161-165.

Nagell, B. 1977b. Phototactic and thermotactic responses facilitating survival of *Cloeon dipterum* (Ephemeroptera) larvae under winter anoxia. *Oikos* 29: 342-347.

Nagell, B. 1979. Overwintering of larvae of *Cloeon dipterum* (L.) in an ice covered and anoxic pond. p. 193-198. *in:* K. Pasternak & R. Sowa ed. Proceedings of the Second International Conference on Ephemeroptera, August 23-26, 1975. Panstwowe Wydawnictwo Naukowe, Warszawa.

Nagell, B. and J. Brittain. 1977. Winter anoxia - a general feature of ponds in cold temperate regions. *Int. Rev. gesamten Hydrobiol.* 62: 821-824.

Nagell, B. and T. Fagerström. 1978. Adaptations and resistance to anoxia in *Cloeon dipterum* (Ephemeroptera) and *Nemoura cinerea* (Plecoptera). *Oikos* 30: 95-99.

Sowa, R. 1975. What is *Cloeon dipterum* (Linnaeus, 1761)? *Entomol. Scand.* 6: 215-223.

Wigglesworth, V.B. 1965. The principles of insect physiology. 6th ed. Methuen, London.

STAGES AND PHYSIOLOGICAL PERIODS IN THE DEVELOPMENT OF *CLOEON DIPTERUM* (L.) (BAETIDAE)

Stanislawa Cianciara

Department of Ecological Bioenergetics
Institute of Ecology
Polish Academy of Sciences
Dziekanów Leśny
05-105 Lomianki, Poland

ABSTRACT

The postlarvullar period in the development of *C. dipterum* covers eight aquatic stages, which differ in morphology, biometry and respiration value. Body size, number of moults, and the duration of development may vary, and may depend on environmental conditions. With the constant increase of dry weight in relation to body length, and the constant increase of body caloric value there occur two periods which vary in physiological processes: period I (L_I - L_V stages), is characterized by the accelerated increase of wet weight and body hydration and period II, (L_{VI} - N stages), is characterized by increased oxygen consumption and a decrease in body hydration.

INTRODUCTION

Cloeon dipterum, an extremely common species, frequently exhibiting mass occurrence, may have an important ecological role. Unfortunately, in the literature, discrepancies both in terminology and methods used to evaluate degree of development, result in physiological and bioenergetical data which are often not comparable. The same problem also occurs in other species of Ephemeroptera.

The criteria for evaluating the degree of development of an insect should be based on easily distinguishable morphological features and should divide the development of the species into equal periods. There are two important periods of morphogenesis in the

Table 1. Stages in development of Ephemeroptera (terminology and methods of dividing).

<table>
<tr><td colspan="16">Periods in morphogenesis with forming of:</td><td rowspan="2">Authors</td></tr>
<tr><td colspan="4">Gills</td><td colspan="12">Wing-pads</td></tr>
<tr><td colspan="16">LARVAE</td><td>Weber 1933</td></tr>
<tr><td colspan="16">NYMPHS</td><td>Macan and Worthington 1972</td></tr>
<tr><td colspan="6">N_{I}</td><td colspan="7">N_{II}</td><td colspan="3">N_{III}</td><td>Macan 1961</td></tr>
<tr><td colspan="3">N_{I}</td><td colspan="8">N_{II}</td><td colspan="4">N_{III}</td><td>N_{IV}</td><td>Clifford 1970</td></tr>
<tr><td>N_{I}</td><td colspan="2">N_{II}</td><td colspan="5">N_{III}</td><td colspan="7">N_{IV}</td><td>N_{V}</td><td>Lehmkuhl 1970, Hall et al. 1975</td></tr>
<tr><td colspan="3">LARVAE</td><td colspan="13">NYMPHS</td><td>Bertrand 1954</td></tr>
<tr><td colspan="3">LARVULLAE</td><td colspan="6">LARVAE</td><td colspan="7">NYMPHS</td><td>Mikulski 1936, Schmidt 1951</td></tr>
<tr><td colspan="4">(L_{o}) LARVAE</td><td colspan="3">N_{I}</td><td colspan="2">N_{II}</td><td colspan="6">N_{III}</td><td>N_{IV}</td><td>Pleskot 1958, Hilmy 1962</td></tr>
<tr><td colspan="2">L_{o}</td><td colspan="2">L</td><td colspan="3">N_{I}</td><td colspan="2">N_{II}</td><td colspan="6">N_{III}</td><td>N_{IV}</td><td>Bretschko 1965, Levanidova 1968,
Levanidova and Rubanienkova 1965</td></tr>
<tr><td colspan="2">L_{o}</td><td colspan="2">L_{I}</td><td>L_{II}</td><td colspan="3">L_{III}</td><td>L_{IV}</td><td>L_{V}</td><td colspan="2">L_{VI}</td><td colspan="2">L_{VII}</td><td colspan="2">N</td><td>Cianciara 1979a</td></tr>
</table>

← time of development →

development of the Ephemeroptera from an egg to a subimago: the period of formation of the respiratory abdominal organs (gills), and the period of formation of the wing-pads. Changes in these morphological features constitute the basis for distinguishable stages.

DEVELOPMENTAL STAGES

The criterion proposed for evaluating the degree of development and the basis for differentiating stages is the size of the mesothoracic wing-pads in relation to the segments of the thorax and the abdomen (Cianciara 1979a). Larvullae (L_O), larvae (L) and nymphs (N) are distinguished in a similar way to Mikulski (1936), Schmidt (1951), Pleskot (1958) and others (see Table 1). Larvullae cover all the developmental forms up to the moment when gills were fully formed, the term having wider connotations than in the classification of Pleskot (1958). The larvae (L_I - L_{VII}) covered the whole period of formation and growth of the wing-pads, and the nymphs (N) have wing-pads of maximum length. The term "larvae" is broadened, while the term "nymphs" is restricted in relation to the other authors except Weber (1933)(see Table 1). Many authors use the term "nymphs" for all the aquatic forms in the development of Ephemeroptera, starting with hatching from an egg (Macan 1961, Clifford 1970, Lehmkuhl 1970, Hall *et al.* 1975), or when the gills are fully formed (Bertrand 1954), or for the entire period of growth of the wing-pads (Pleskot 1958, Hilmy 1962), or when their length exceeds that of the metathorax (Mikulski 1936, Schmidt 1951). Most authors distinguish subsequent stages on the basis of distinct morphological features, but the younger stages are too "frequent" compared with the older ones which comprise longer physiological periods of development. Presuming that the assumed time axis, divided into stages on the basis of laboratory culture (Cianciara 1979b), represents, at least approximately, the physiological possibilities of the species studied in the natural environment, the division of time for particular stages gives the most equal sections (Table 1).

A number of experiments were carried out (Cianciara 1979a, b, c, d) in order to provide evidence supporting these divisions of developmental stages.

NUMBER OF DEVELOPMENTAL STAGES

In holometabolous insects the number of stages is determined by a constant and small number of moults. Contradictory data on the number of moults in hemimetabola, including the Ephemeroptera, exist in the literature. Some authors (e.g. Pleskot 1958, Pescador and Peters 1974) give a constant and definite number of moults for a given species, in others (e.g. Mikulski 1936, Schmidt 1951, Clifford

1970) the number changes and may even reach 50 moults. The latter was confirmed by culture (Cianciara 1979b) where, 27 to 40 moults were observed, depending on the type of food. The number of moults may also be influenced by other environmental factors, e.g. temperature (Schmidt 1951), and light (Khoo 1964). In the same species, the number of moults is greater in the winter generation than in the summer one (Schmidt 1951, Bretschko 1965). In addition, injuries and resulting regeneration may increase the number of moults with out influencing the rate of development (Schmidt 1951). It is impossible to define the moult number on the basis of any metrical parameters, since, under environmental pressure the larger nymphs may be physiologically younger than the smaller ones (Clifford 1970). The size of the larva need not bear any relationship to the number of moults (Schmidt 1951, Khoo 1964), these being dependent not just on growth but being, primarily, the method of excretion of the waste products of metabolism, collected in the exuvium.

DURATION OF DEVELOPMENT

The duration of total development and of development of each stage may also change, and may be influenced by environmental factors. *C. dipterum* usually has two generation per year (Landa 1968, Schmidt 1951, Bertrand 1954, Bretschko 1965, Cianciara 1979c), but only one in unfavourable climatic conditions (Macan 1962, Brittain 1972). Thus, the total development may last from two months (for the summer generation) to ten months (for the winter one) or even to 12 months. In natural conditions, environmental changes (i.e., temperature, day length) influence changes in the rate of the development, and subsequently the duration of particular stages. In the winter generation the stages L_{II} - L_{IV} last for about 150 days in the period of cessation of growth (Bretschko 1965, Cianciara 1979c). Under relatively stable laboratory conditions, (Cianciara 1979b) cultures showed a steady rate of growth, but type of food greatly influenced the duration of development (from about 5 to 7 months).

LIFE CYCLE ANALYSIS

When analysing the life cycles or the physiological processes of Ephemeroptera, many authors identify age groups on basis of body size (Trama 1957, Brown 1961, Lehmkuhl and Anderson 1970, Brittain 1972, 1974; Zimmerman *et al.* 1975) or body weight (Winberg 1956, Ivanova 1958, Pattee 1968, Kamler 1970, Alimov 1971 and others). However, the above parameters may be modified due to geographical location, food supplies or other variables. Independent of the type of food, the cultures (Cianciara 1979b) showed a uniform growth and approximately the same body sizes (both in early stages and in the final ones), though at different times and with a different number of moults. In natural conditions, however, it was observed (Cianciar

1979c) that the summer generation was represented by individuals smaller than those in the winter one. Seasonal changes in body size of *C. dipterum* was also observed by Bertrand (1954), Macan (1961) and Bretschko (1965). The typical growth pattern may also be influenced by diapause or periodical cessation of growth (Britt 1962), or by delayed egg hatching (Clifford 1970). The same body size (and weight) may then correspond to different age and different degrees of advancement in development of the insects (Clifford 1970).

Thus, the number of moults, duration of development, body size, and weight of mayflies are changeable and cannot be the starting point for defining the degree of their development.

BIOMETRIC AND BIOENERGETIC RELATIONSHIPS

If, however, stages in the development of Ephemeroptera exist, the fact should be confirmed by biometrical features and physiological properties of the insects. To test this, a number of biometrical and bioenergetical measurements were made for all the identified stages, and the course of changes in these parameters during the development of *C. dipterum* was studied (Cianciara 1979d). It was found that changes of the linear body dimensions (body length, head capsule width) were relatively uniform and in proportion to the growth of a given stage (Fig. 1). A uniform increase in the caloric value of the body was also observed. The increase in the wet and dry body weights (Fig. 2) followed a typical S-shaped curve (with flying forms also considered). The pattern of change of the parameters is not dissimilar to the patterns observed in other animals (Winberg 1968). No stepwise changes were observed, which would confirm the occurrence of stages in the development of *C. dipterum*. The only distinct differences between the parameters concerned the sexual differences in nymphs. It was found that in *C. dipterum* there were constant positive relationships between dry body weight and body length and between caloric equivalent and dry body weight (Fig. 3). The increase in the wet body weight in relation to the linear dimensions, however, was more rapid in the young larval stages (L_I - L_V) than in the older forms (L_{VI} - N). Between the stages L_V and L_{VI} the change is associated with an initial increase and then with a decrease in body hydration (Fig. 4A). Significant differences were also observed in intensity of oxygen consumption (Fig. 4B): initially it decreased (from maximum in stage L_I to minimum in stage L_{VI}) whereas in older stages it increased. These changes were observed in *C. dipterum* by Ivanova (1958) and in three other species of mayflies by Hilmy (1962). Significant changes between a group of younger stages (L_I - L_V) and a group of older ones (L_{VI} - N) were noted, oxygen consumption being dependent on body weight (Fig. 4C). The data confirms the occurrence of two periods with different intensity of physiological processes in the development of *C. dipterum*. The initial period of larval development (stages L_I - L_V) is characterized

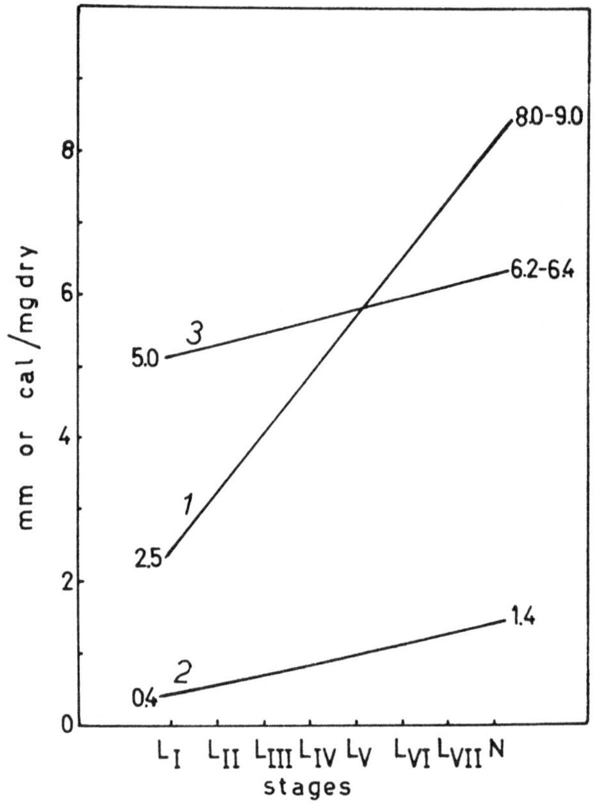

Figure 1. Parameters in the development of *C. dipterum* showing linear growth: 1 - body length (mm), 2 - width of head capsule (mm), 3 - caloric value (cal/mg dry wt); L - larvae, N - nymphs; I - VII - number of stages.

by a rapid rate of increase in body weight (and increase in body hydration). In the second period (stages L_{VI} - N) the increase in wet body weight is slower and body hydration decreases, but the processes of metabolism occur with greater intensity which is reflected by higher oxygen consumption. Most probably it is connected with a rapid development of gonads, first in females, then in males (Soldan, personal communication).

CONCLUSION

Thus the biometrical and bioenergetical data revealed two periods in the development of aquatic forms of *C. dipterum* (i.e. two

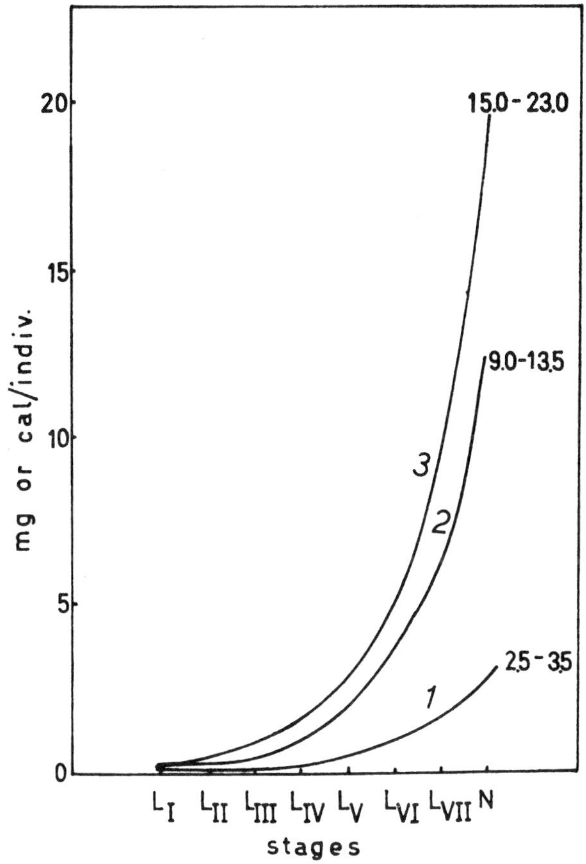

Figure 2. Parameters in the development of *C. dipterum* showing parabolic growth: 1 - dry weight (mg), 2 - wet weight (mg), 3 - caloric equivalent (cal/indiv.). Symbols as in Fig. 1.

physiological stages are confirmed). Since the two periods cover very long stages in the development, it seems more useful, for experimental reasons, to use the earlier accepted division (Table 1). This appears to be substantiated by statistical analysis of measured parameters (Cianciara 1979d). Highly significant statistical differences were confirmed between stages from L_I to N. However, there were no such differences (or not significant ones) between primarily (Cianciara 1979a) distinguished nymphs N_I and N_{II}. In the literature (Pleskot 1958, Hilmy 1962, Bretschko 1965 and others - Table 1), the final nymphal forms (N_{II}) are usually identified in mayfly development as certain, evidently different, stages (ripe nymphs). There is

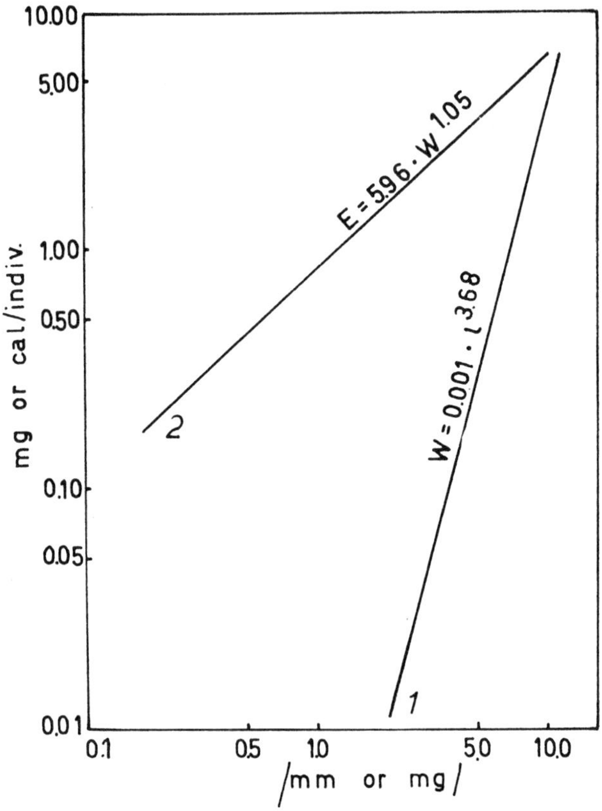

Figure 3. Constant dependence in stages L_I - N in development of *C. dipterum:* 1 - the relationship between dry body weight (mg) and body length (mm), 2 - the relationship between caloric equivalent (cal/indiv.) and dry weight (mg). Symbols as in Fig. 1.

no evidence whatsoever in the analysis of results of biometry and respiration for this division. Thus, in spite of distinct morphological differences, N_{II} nymphs cannot constitute a separate stage from N_I nymphs, since they are only the final instar before metamorphosis and not the stage. In summary, from the above data it can be concluded that in the postlarvullar period of the development of *C. dipterum* eight stages can be distinguished (Table 1): seven larval ones (L_I - L_{VII}) and one nymphal stage (N).

Since the development of *C. dipterum* shows a gradual metamorphosis, as do all hemimetabolous insects, the identified stages cover

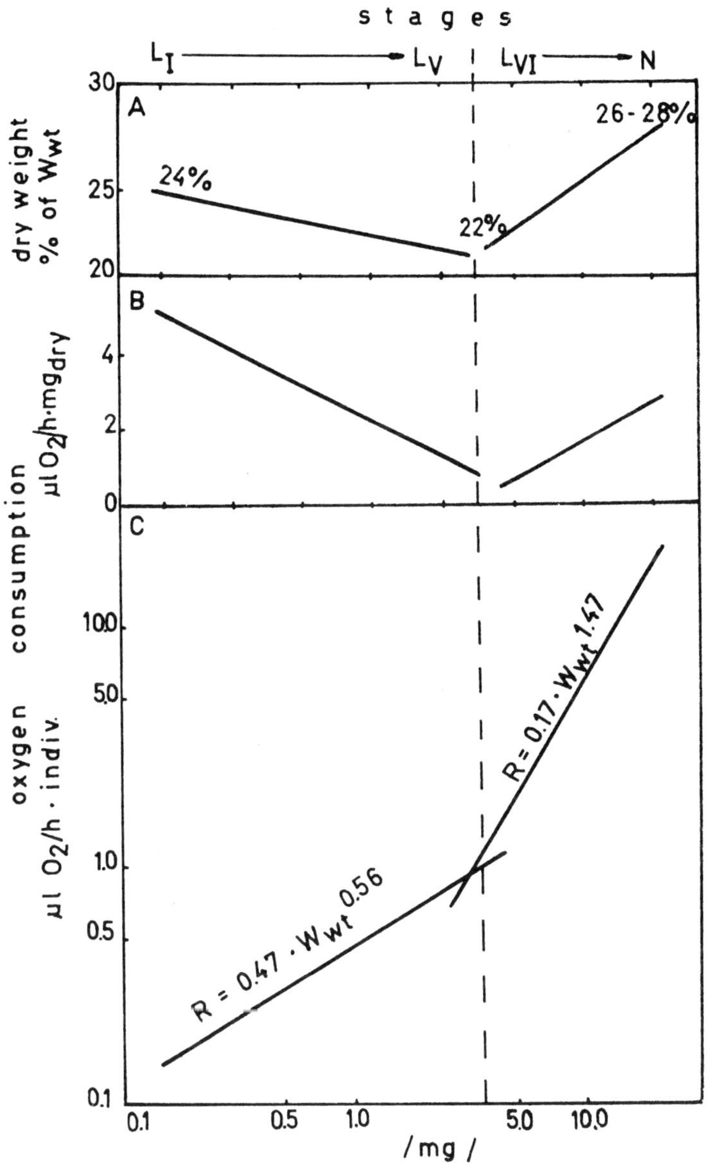

Figure 4. Parameters and their relationships in the development of *C. dipterum*, showing the existence of two periods: I - stages L_I - L_V, II - stages L_{VI} - N; A - dry body weight (percent of wet body weight), B - oxygen consumption (μL O_2/h/mg dry wt), 3 - the relationship between oxygen consumption (μL O_2/h/indiv.) and wet body weight (mg). Symbols as in Fig. 1.

the developmental phase determined by morphological criteria where particular larvae or nymphs may be ascribed definite biometrical and bioenergetical features. Though the division into the stages is arbitrary and not univocal as in holometabola, such separation may be helpful in experimental work.

RESUME

Le *Cloeon dipterum* passe par huit étapes aquatiques de développement, différant les unes des autres par leur morphologie, leur biométrie et leur capacité de respiration. La taille du corps, la quantité de mues et le temps que l'espèce met à se développer varient en fonction des conditions du milieu. En maintenant constant le taux d'augmentation du poids à l'état sec par rapport à la taille du corps et à l'augmentation constante de son pouvoir calorifique, il se produit deux périodes aux processus physiologiques distincts. La période I, comportant les étapes L_I à L_V, se caractérise par une augmentation accélerée du poids à l'état mouillé (concomitante à l' hydratation de l'organisme) et la période II, comportant les étapes L_{VI} à N, se caractérise par une augmentation de la consommation d'oxygène (concomitante à la déshydratation de l'organisme).

ZUSSAMENFASSUNG

Die Entwicklung von *Cloeon dipterum* erstreckt sich über acht aquatische Stadien, die sich unterschieden in Bezug auf Morphologie, Biometrie und Respirationswerte. Körpergröße, Zahl der Häutungen und Entwicklungsdauer der Arten können variieren und von Bedingungen der Umgebung abhängen. Innerhalb der konstanten Vermehrungsrate des Trockengewichts im Verhältnis zur Körperlänge und der konstanten Zunahme der vom Körper erzeugten Wärmewerte treten zwei Perioden auf, die in physiolgoischen Prozessen variieren. Periode I enthält die Stadien L_I bis L_V und ist charakterisiert durch beschleunigte Zunahme sowohl an Naßgewicht, als auch an Körperhydration. Periode II erstreckt sich über die Stadien L_{VI} bis N und zeichnet sich sowohl durch vermehrten Oxygenverbrauch, als auch verminderte Körperhydration aus.

REFERENCES

Alimov, A.F. 1971. Oxygen consumption by larvae of some aquatic insects. Izd. "Nauka", Leningrad, 10: 47-50. (In Russian).

Bertrand, H. 1954. Les Ephemeropteres. p. 215-266. *In:* Les insectes aquatiques d'Europe. *Encylopedie Entomologique* 30 (1). Lechevalier, Paris.

Bretschko, G. 1965. Zur Larvalentwicklung von *Cloeon dipterum*, *Cloeon simile*, *Centroptilum luteolum* und *Baetis rhodani*. *Z. wiss. Zool.* 172(1/2): 17-36.

Britt, N.W. 1962. Biology of two species of Lake Erie mayflies, *Ephoron album* (Say) and *Ephemera simulans* (Walker). *Bull. Ohio Biol. Surv.* 1(5): 1-70.

Brittain, J.E. 1972. The life cyles of *Leptophlebia vespertina* (L) and *L. marginata* (L) (Ephemeroptera) in Llyn Dinas, North Wales. *Freshwater Biol.* 2: 271-277.

Brittain, J.E. 1974. Studies on the lentic Ephemeroptera and Plecoptera of southern Norway. *Nor. Entomol. Tidsskr.* 21: 135-154.

Brown, D.S. 1961. The life-cycle of *Cloeon dipterum* (L.) (Ephemeroptera: Baetidae). *Entomologist* 94: 114-120.

Cianciara, S. 1979a. Some study on the biology and bioenergetics of *Cloeon dipterum* (L.) (Ephemeroptera). (In prep.)

Cianciara, S. 1979b. Food preference of *Cloeon dipterum* (L.) larvae and dependence of their development and growth on the type of food. *Pol. Arch. Hydrobiol.* 26(4). (In press).

Cianciara, S. 1979c. Life cycles of *Cloeon dipterum* (L.) in natural environment. *Pol. Arch. Hydrobiol.* 26(3). (In press).

Cianciara, S. 1979d. Biometrical and bioenergetic characteristics of development of *Cloeon dipterum* (L.). *Pol. Arch. Hydrobiol.* 26(4). (In press).

Clifford, H.F. 1970. Analysis of a northern mayfly (Ephemeroptera) population with special reference to allometry of size. *Can. J. Zool.* 48: 305-316.

Hall, R.J., L. Berner and E.F. Cook. 1975. Observations on the biology of *Tricorythodes stratus* McDunnough (Ephemeroptera: Tricorythidae). *Proc. Entomol. Soc. Wash.* 77(1): 34-49.

Hilmy, A.M. 1962. Experimente zur Atmungsphysiologie von Ephemeropteren Larven. *XI Internationaler Kongress fur Entomologie, Wien* 3: 254-258, 1960.

Ivanova, S.S. 1958. Feeding of larvae of some mayflies. *Trudy Mosk. Tekhnol. Inst. Ryb. Prom. Khoz.* 9: 102-120. (In Russian).

Kamler, E. 1970. The main parameters regulating the level of energy expenditure in aquatic animals. *Pol. Arch. Hydrobiol.* 17(1/2): 201-216.

Khoo, S.G. 1964. Studies on the biology of *Capnia bifrons* (Newmann) and notes on the diapause in nymphs of this species. *Gewass. Abwass.* 34/35: 23-30.

Landa, V. 1968. Developmental cycles of central European Ephemeroptera and their interrelations. *Acta. Entomol. Bohemoslov.* 65: 276-284.

Lehmkuhl, D.M. 1970. The life cycle of *Rhithrogena morrisoni* (Banks) in western Oregon. (Ephemeroptera: Heptageniidae). *Pan-Pac. Entomol.* 46(2): 124-127.

Lehmkuhl, D.M. and N.H. Anderson. 1970. Observations on the biology of *Cinygmula reticulata* (McDunnough) in Oregon (Ephemeroptera: Heptageniidae). *Pan-Pac. Entomol.* 46(4): 268-274.

Levanidova, I.M. 1968. Benthic organisms of the Amur river system. *Izv. Tikhookean. Nauchn-Issled. Inst. Ryb. Khoz. Okeanogr.* 64: 181-289. (In Russian).
Levanidova, I.M. and L.S. Rubanienkova. 1965. On the methods of studying life cycles of amphibiotic insects. *Zool. Zh.* 44: 34-45. (In Russian, English summary).
Macan, T.T. 1961. A key to the nymphs of the British species of Ephemeroptera. *Freshwater Biol. Assoc. Sci. Publ.* 20: 1-64.
Macan, T.T. 1962. Ecology of aquatic insects. *Annu. Rev. Entomol.* 7: 261-288.
Macan, T.T. and E.B. Worthington. 1972. Life in lakes and rivers. (Paperback revised edition). Collins, London. (Fontana New Naturalist).
Mikulski, J.S. 1936. Mayflies (Ephemeroptera). *Fauna slodkowodna Polski.* 15: 1-168. (In Polish).
Pattee, E. 1968. Relation entre poids frais et poids sec chez quelques invertébrés d'eau douce. *Hydrobiologia* 31: 417-420.
Pescador, M.L. and W.L. Peters. 1974. The life history and ecology of *Baetisca rogersi* Berner (Ephemeroptera: Baetiscidae). *Bull. Fl State Mus., Biol. Sci.* 17(3): 151-209.
Pleskot, G. 1958. Die Periodizität einiger Ephemeropteren der Schwechat. *Wässer Abwässer* 1958: 188-219.
Schmidt, H. 1951. Amputation und Regeneration von Schwanzfaden und Abdominalsegmenten bei Larven der Ephemeridenart *Cloeon dipterum* L. und ihr Einfluss auf die Hautungsintervalle. *Zool. Jahrb. Abt. allg. Zool. Physiol. Tiere* 62: 395-428.
Trama, F.B. 1957. The transformation of energy by an aquatic herbivore *Stenonema pulchellum* (Ephemeroptera). Ph.D. Dissertation, Univ. Mich.
Weber, H. 1933. Lehbruch der Entomologie. Fischer, Jena.
Winberg, G.G. 1956. On dependence of metabolic rate on body dimension in arthropods. *Uch. Zap. Beloruss. Gos. Univ.* 26: 243-254. (In Russian).
Winberg, G.G. 1968. The methods for the estimation of production of aquatic animals. *Izd. "Vyšejšaja Škole" Minsk.* (In Russian).
Zimmerman, M.C., T.E. Wissing and R.P. Rutter. 1975. Bioenergetics of burrowing mayfly, *Hexagenia limbata*, in a pond ecosystem. *Int. Ver. theor. angew. Limnol. Verh.* 19: 3039-3049.

PREDATION, CLIMATE, AND EMERGENCE AND MATING OF MAYFLIES

George F. Edmunds, Jr. and Christine H. Edmunds

University of Utah
Salt Lake City
Utah, 84112 U.S.A.

ABSTRACT

The abbreviated adult life of Ephemeroptera is an adaptation to minimize exposure time to predators. In eight to ten independent specialized lineages adult life is reduced to two hours or less. Predation by Odonata and birds is intensive in the lowland tropics and most unspecialized longer-lived forms emerge as subimagos in the first two hours of darkness, transform to imagos before dawn and mate and oviposit by mid-morning. In temperate regions cool night-time temperatures often preclude the possibility of the tropical pattern; furthermore lessened predation allows other temporal patterns. Temperate mayflies thus have a variety of emergence and swarming times. Short-lived specialized genera are subject to fewer restraints on emergence and swarming times and tropical and temperate forms are similar. Seasonal emergence and coordinated mass emergence are mechanisms for satiating predators. Remote nuptial flight probably evolved as a mechanism of escape from predation; it is assumed that swarm markers are essential to allow remote flight. It is probable that swarm markers are more common in tropical than in temperate mayflies. There is some evidence that ancestral emergence and swarming habits may persist when mayflies disperse to new areas.

Bishop (1973) in his study of the Gombak River of the Malay Peninsula noted that adult aquatic insects were uncommon. Use of a sweep net for sampling adults proved to be almost useless. He attributed the small numbers of adults to heavy predation, especially by numerous insectivorous birds. In two months in Malaysia we did not observe the emergence of a single subimago and we saw very few mayfly imagos in nuptial flight. The abundance and high species diversity in many Malaysian streams suggests that the adults are much more common than these observations indicate. It was obvious that subimaginal and imaginal behaviour patterns in these lowland tropics must be strikingly different from the temperate patterns and apparently different also from tropical highland areas.

About half-way through the Malaysian studies we made a major effort to understand the activity cycles of the mayflies. As the patterns of activity became clear, we compared them with other experiences with tropical and temperate mayflies. The comparisons led us to some interesting contrasts between tropical and temperate forms. It is obvious that some of the factors discussed form a continuum and that the comparison is of extremes of a continuum. This paper contains considerable speculation, but we are hopeful that these broad comparisons will point the way to areas where data is needed.

Apparently many of the activity patterns and adaptations of adult mayflies (subimagos and imagos) have formed in response to selection pressure from predators. Mayfly subimagos are slow and clumsy fliers and are highly vulnerable to predation. Flying imagos are much less vulnerable but resting subimagos or imagos, being Paleoptera, are unable to fold the wings and hide in leaf litter, crevices or other protected areas. The brevity of their winged lives is itself an adaptation to reduce exposure to predation. We agree with Bishop (1973) that birds are serious daytime predators in Malaysia. We believe, however, that the most significant day-time predators are Odonata, and birds are secondary. Many mayflies are killed also in spider webs, but the webs are probably equally effective, day or night. The only significant night-time predators are bats. Both Odonata and bats are generally more abundant and diverse in the tropics. It is abundantly clear that predation on mayfly subimagos and imagos is several times as great in the day as it is at night.

In Malaysia it is probable that nearly 100% of the vulnerable subimagos emerge in almost total darkness. Apparently the lowest predation rate of subimagos in the temperate regions also would be during hours of darkness. The selection pressure which seems to counteract selection for night-time emergence appears to be cool climate that slows transformation from the subimago to the imago. In the lowland tropics, nights are warm and most imagos which emerge in the first 1½ hours of darkness transform to the imago stage before 0300 h the next morning. At higher elevations, higher latitudes and in dry climates the contrast between daytime and night-time temperatures is greater, and this coupled with the lower average temperatures at higher latitudes and altitudes prolongs the temperature-dependent subimaginal stage. In the tropics it appears that most mayflies mate and lay eggs before noon, and most do so by 1030 h following their emergence the night before. Tjonneland (1970) states that in Uganda most Ephemeroptera (and Trichoptera) swarm between sunset and sunrise. If this statement is based on the arrival of imagos at light and the lack of observations on swarming at other times, it agrees with our observations in Malaysia. However, because it is not possible to see the swarms in the dark and because of swarming away from the stream the problem of when swarming occurs is rather complex.

The risk of predation is much lower per unit time in temperate regions and emergence of the subimago during full daylight occurs in many species. The most common time of emergence for most temperate species is from late afternoon through the first hour of darkness. In warm temperate regions with suitable night-time temperatures subimagos of various species may emerge during the night. The favored times of nuptial flight in tropical species are very different than in temperate species. The time of nuptial flight appears to be the result of selection to reduce the time of daylight exposure. With the favorable temperatures the mayflies have become imagos by daylight and swarming takes place from before dawn through the morning hours. Dusk appears to be the most common swarming time in temperate species. Table 1 compares my estimate of the most to least frequent times of swarming in tropical and temperate species.

Table 1.

NUPTIAL FLIGHT TIME

Decreasing Frequency ↓	TROPICS	TEMPERATE
	Morning	Dusk
	First dark hours	Afternoon
	Dawn	Afternoon/Morning
	Midday	Morning
	Dusk	Midday
	Afternoon	Dawn
		First dark hours

It is interesting to speculate that ancestral emergence and swarming habits may persist when mayflies disperse to new areas. For example, the emergence and swarming of *Tricorythodes* (Hall *et al.* 1975) and *Traverella* (Edmunds 1948) are basically tropical patterns and both appear to be extensions of South American genera into North America. In some cases, however, tropical forms in temperate regions presumably have shifted to temperate behaviour. *Dactylobaetis*, a genus of presumed South American origin, swarms in the evening in Idaho. The persistence of ancestral swarming patterns may account for the swarming of some Ephemerellidae and Heptageniidae in Sabah (Borneo) in early afternoon, but the area where such swarming was observed is at 1050 meters elevation. No afternoon or dusk swarming was observed at low elevations.

Eight to ten times the Ephemeroptera have evolved very short adult lives, in which the subimago stage lasts only for a few minutes, and the imagos live 2 hours or less; in some cases the female mates and dies as a subimago. This is true of all the Oligoneuriinae (and probably Chromarcyinae) (Oligoneuriidae), possibly in some Leptophlebiidae (*Isca)*, the known Behningiidae, some or all Euthyplociidae (*Proboscidoplocia*, *Polyplocia*), apparently all Polymitarcyidae, some or all Palingeniinae, some Tricorythidae (*Tricorythus* s. l., *Teloganella*), (*Teloganella* is a tricorythid, Edmunds in MS.), all Caenidae and in *Prosopistoma* (Prosopistomatidae). These short-lived mayflies are not subject to many of the mating flight time restraints imposed on species with longer lives, especially those with long-lived subimagos. Mating swarms of short-lived mayflies are crepuscular (e.g. *Ephoron)*, after dark (Euthyplociidae), pre-dawn or dawn (Behningiidae, *Prosopistoma*, some Caenidae) or in morning sunlight (some Caenidae). Swarming times of short-lived forms vary within a genus in the same general area; some species swarm in the morning while others are crepuscular, e.g.

Figure 1. A generalized comparison of emergence and moulting to imago stages in specialized short adult life (<2 hours) and longer adult life (>6 hours) unspecialized species of Ephemeroptera in tropical and temperate regions. Specialized genera have the potential of being similar in the two regions, but nuptial flight in tropical regions is most frequently in the first hour of darkness. Unspecialized genera in the tropical lowlands tend to be strongly restrained to a short daylight life, although exceptions are known. In unspecialized genera of the temperate region, emergence of subimagos is scattered, and length of subimaginal and imaginal life is variable; the diagram indicates a common generalized pattern but does not show the range of variation.

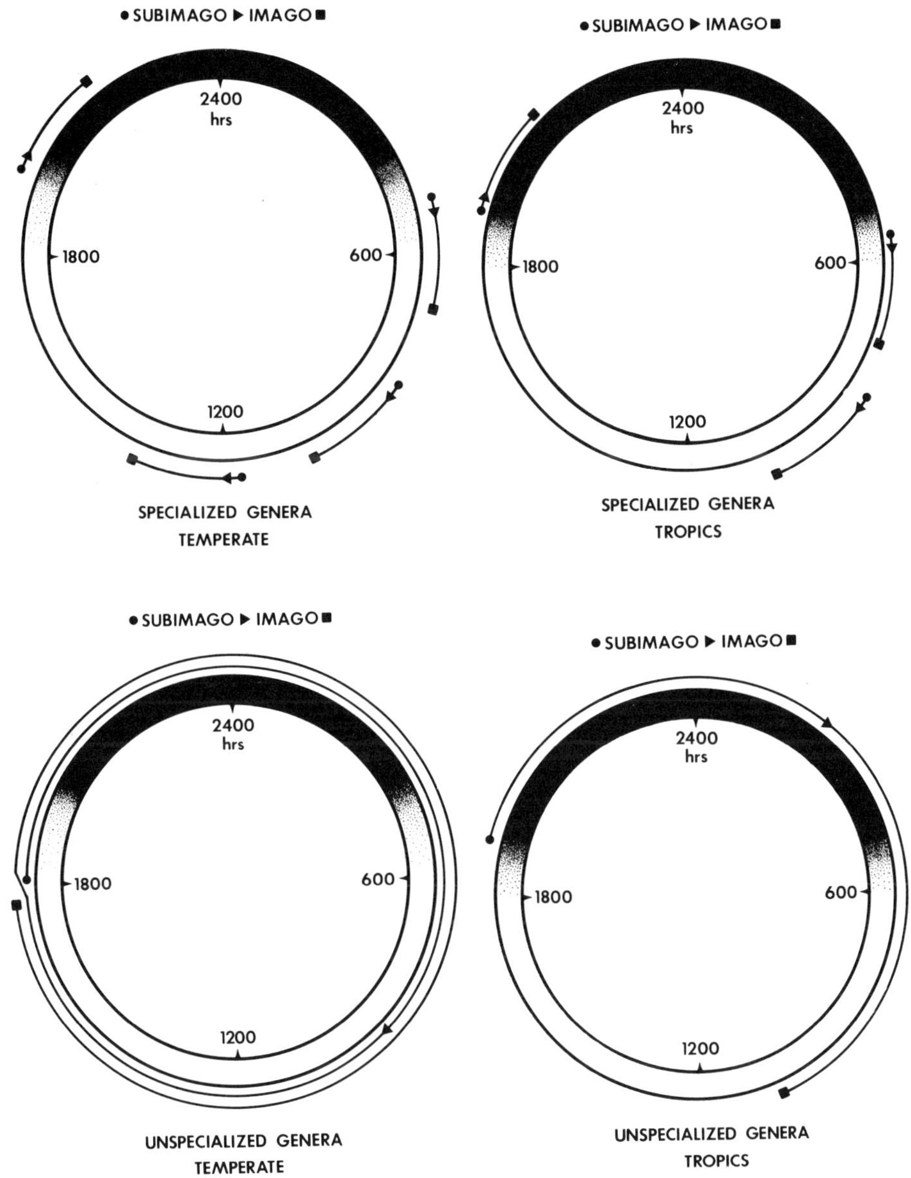
● SUBIMAGO ▶ IMAGO ■
2400 hrs
1800
600
1200
SPECIALIZED GENERA TEMPERATE
● SUBIMAGO ▶ IMAGO ■
2400 hrs
1800
600
1200
SPECIALIZED GENERA TROPICS
● SUBIMAGO ▶ IMAGO ■
2400 hrs
1800
600
1200
UNSPECIALIZED GENERA TEMPERATE
● SUBIMAGO ▶ IMAGO ■
2400 hrs
1800
600
1200
UNSPECIALIZED GENERA TROPICS

morning and evening swarming species of *Brachycercus* and *Caenis* or morning and midday swarming of *Lachlania*.

It is well known that many species of mayflies emerge seasonally. This occurs most frequently in temperate areas but appears to be true also of many tropical species. W.L. Peters (personal communication) noted a coordinated seasonal emergence of *Campylocia* in Peru. All known records of the adults of *Cheirogenesia decaryi* of Madagascar are February. Seasonal emergence results in greater concentration of adults at one time and increases the likelihood of mating success and tends to reduce the percentage of individuals taken by predation.

Mayflies also use water temperature, lunar cycles and possibly other environmental cues for coordinated mass emergence. Endogenous rhythms are also probable in some forms, e.g. *Dolania* (Peters and Peters 1977). Notable mass emergences have been reported in *Povilla*, *Campsurus*, *Ephoron*, *Eatonigenia*, *Hexagenia*, *Plethogenesia*, *Palingenia*, *Caenis*, *Oligoneuriella*, *Lachlania* and others. Mass emergence increases the probability of finding a mate and serves to satiate predators.

Remote nuptial flight is a common phenomenon in mayflies. We have frequently found nuptial swarms of *Baetis*, *Paraleptophlebia*, *Rhithrogena* and species of other genera at distances of 1.5 km from the nearest larval habitat. Probably many species whose nuptial flight has not been observed (e.g. some common species of *Ephemerella*, *Ameletus*, etc.) have nuptial swarming sites that are remote either laterally or vertically, high above the larval habitat. Remote nuptial swarming has energetic costs and increased danger of the female not finding her way back to the stream for oviposition. There is obviously some selection pressure against swarming over the stream, and we speculate that it is the higher concentration of predators near streams, rivers or lake margins.

We suspect that part of the explanation of the rarity of observations of swarming adults in Malaysia is because of remote nuptial flight. Corbet (1961) reported adults of *Adenophlebiodes burgeoni* swarming 6-10 feet above a 120 foot steel tower projecting above the forest canopy in Mpanga Forest, Uganda, Mass emergences that result in predator satiation obviate the streamside predation problem and it appears that all mass emergent species swarm over rivers (e.g. *Ephoron*, *Lachlania*) or river banks (e.g. *Traverella*).

Remote nuptial flight poses another problem; the mayflies must have some common orientation toward the remote swarm site. Many mayflies that swarm over or near rivers and streams orient to specific swarm markers: white water, slicks on the water, trees, shadows of trees or bushes, areas free of vegetation, and areas darker than the general background or lighter than the background. Each species is specific in its choice of swarm marker. Savolainen (1978) has given data on swarm markers for mayflies in Finland and W.L. Peters and the senior author gathered a substantial amount of data that remains unpublished. Remote nuptial flight seems to place a high premium on swarm markers and a reasonable prediction is that swarm marker orientation is an important aspect of remote swarming. Another facet of nuptial flight activity is timing; nuptial flight is most effective if both the place and time are restricted. In temperate mayflies, it is a combination of light and temperature that controls swarming time; in the lowland tropics, temperature would appear to be less restrictive and light intensity may be the prime environmental cue.

The importance of differences in emergence and swarming activities in different climates is obvious for those who are qualitatively sampling for Ephemeroptera in a region. The nuptial flight of adult *Prosopistoma* remained unknown until 1954 (Gillies 1954) primarily because the circadian rhythms of *Prosopistoma* and man did not coincide. In the Malay Peninsula and in Sabah, subimagos of Baetidae were first to emerge, just before full darkness, and were rapidly followed by species of a variety of other families. Emergence (as measured by arrival of subimagos at lights) declined markedly after one hour of darkness. In resuming collecting at lights at 0400 h large numbers of imagos of a variety of families appeared at lights, but very few subimagos. The specialized short-lived mayflies may emerge from larvae either in the first hour of darkness (e.g. *Polyplocia, Teloganella,* some Caenidae) or at dawn (*Prosopistoma,* some Caenidae) or during morning (*Isca*) but most long-lived species emerge in the first hour of darkness, transform to adults by dawn, and mate in the morning (Fig. 1). In Malaysia, most of the species that were subimagos in the first hour of darkness were collected as imagos in the morning and many species not seen the night before were collected. Between 0700 and 0715 h two females of *Prosopistoma* came to the mercury vapor light. Some short-lived species, however, were collected only by collecting just after it became dark. In Malaysia, it appears that subimagos of *Potamanthodes* and *Rhoenanthus* (Potamanthidae) do not transform the first night, but no other exceptions were noted.

The use of a light in Malaysia at dusk and the early hours of darkness resulted in imagos of specialized short-lived species and in many subimagos that had to be placed in subimago boxes until emergence of the imago. By using lights before dawn the specimens collected were imagos and much time was saved. It is now overwhelmingly clear that the adult mayflies of the tropical lowlands will become well known only by collecting after dusk and before dawn. This appears to be equally true of Trichoptera, Plecoptera and other insect groups.

RESUME

La vie brève des éphéméroptères adultes est une espèce d'adaptation propre à ces insectes pour réduire au minimum le temps où ils sont sujets à être victimes des prédateurs. La vie d'adulte de huit à dix familles à évolution indépendante et à morphologie complète ne dure que deux heures tout au plus. Les Odonates et les oiseaux prédateurs sont particulièrement voraces dans les basses-terres tropicales, et la plupart des formes non-différenciées à durée de vie plus longue émergent au stade de subimago dès les deux premières heures d'obscurité, passent au stade d'imago avant l'aube, s'accouplent et pondent leurs oeufs vers le milieu de l'avant-midi. Dans les régions tempérées, les nuits fraîches s'opposent souvent au processus de développement propre aux régions tropicales. En outre, le fait qu'il y ait moins de prédateurs rend possibles d'autres modes d'évolution. C'est ainsi que les éphéméroptères des climats tempérés ont des périodes d'émergence et d'essaimage variées. Les éphéméroptères à vie brève sont sujets à moins de contraintes aux périodes d'émergence et d'essaimage et leur morphologie tropicale ressemble à leur morphologie des régions tempérées. L'émergence saisonnière et leur émergence quasi-simultanée en masse sont des mécanismes destinés à rassassier les prédateurs. Les vols nuptiaux vers de lointaines destinations se sont probablement présentés comme une stratégie pour échapper aux prédateurs, et l'on suppose que pour s'envoler au loin les essaims ont dû nécessairement suivre des guides. Il est probable qu'il est plus commun de trouver des guides chez les éphéméroptères tropicales que chez celles des régions tempérées. Il semble enfin qu'en se dispersant sur de nouveaux territoires, les éphéméroptères conservent certaines habitudes ancestrales d'émergence et d'essaimage.

ZUSSAMENFASSUNG

Die Abkürzung des Erwachsenenlebens von Ephemeroptera stellt eine Adaption dar zum Zwecke der Verringerung der Zeit während der die Tiere Prädatoren ausgesetzt sind. Bei acht bis zehn spezialisierten Stämmen ist das Leben ausgewachsener Fliegen auf zwei Stunden oder weniger reduziert. Die Prädation durch Odonata und Vögel ist besonders intensiv in tropischen tiefgelegenen Gegenden. Die meisten länger lebenden Formen treten in den ersten beiden Stunden der Dunkelheit als Subimagos zutage, verwandeln sich vor der Morgendämmerung in Imagos, paaren sich und legen Eier im Laufe eines Vormittages. Dieses tropische Verhaltensmuster wird aber in gemäßigten Gegenden oft durch kühle Nachttemperaturen verhindert; auch bildet sich ein anderes Verhaltensmuster wenn verringerte Prädation besteht. Eintagsfliegen in gemäßigten Zonen haben daher verschiedene Emergenz- und Schwärmzeiten. Kurzlebige, spezialisierte Gattungen sind nicht so vielen Einschränkungen bezüglich Emergenz- und Schwärmzeiten unterworfen. Auch sind sich tropische Formen und solche von gemäßigten Zonen sehr ähnlich. Saison Emergenz und koordinierte Massenemergenz sind Mechanismen zur Sättigung der Prädatoren. Der entfernte Hochzeitsflug entstand wahrscheinlich als Mechanismus gegen die Prädatoren. Man nimmt an, daß Schwärmanzeiger wesentlich sind zur Ermöglichung des entfernten Fluges. Es ist denkbar, daß Schwärmanzeiger häufiger bei tropischen Eintagsfliegen auftreten als bei solchen der gemäßigten Zonen. Es gibt Beweise dafür, daß angestammte Emergenz und Schwärmgewohnheiten bestehen bleiden, wenn Eintagsfliegen sich auf neue Gebiete verbreiten.

REFERENCES

Bishop, J.E. 1973. Limnology of a small Malayan River Sungai Gombak. *Monog. Biol.* 22: 1-485.

Corbet, P.S. 1961. Entomological studies from a high tower in Mpanga Forest, Uganda XII. Observations on Ephemeroptera, Odonata, and some other orders. *Trans. R. Entomol. Soc. Lond.* 113: 356-362.

Edmunds, G.F. 1948. A new genus of mayflies from western North America (Leptophlebiinae). *Proc. Biol. Soc. Wash.* 61: 141-148.

Gillies, M.T. 1954. The adult stages of *Prosopistoma latreille* (Ephemeroptera), with descriptions of two new species from Africa. *Trans. R. Entomol. Soc. Lond.* 105: 355-372.

Hall, R.J., L. Berner, and E.F. Cook. 1975. Observations on the biology of *Tricorythodes atratus* McDunnough (Ephemeroptera: Tricorythidae). *Proc. Entomol. Soc. Wash.* 77: 34-49.

Peters, W.L., and J.G. Peters. 1977. Adult life and emergence of *Dolania americana* in northwestern Florida (Ephemeroptera: Behningiidae). *Int. Rev. gesamten Hydrobiol.* 62: 409-438.

Savolainen, E. 1978. Swarming in Ephemeroptera: the mechanism of swarming and the effects of illumination and weather. *Ann. Zool. Fenn.* 15: 17-52.

Tjonneland, A. 1970. A possible effect of obligatory parthenogenesis on the flight activity of some tropical larvo-aquatic insects. *Arbok. Univ. Bergen Mat-Naturvittensk.* Ser. 3: 1-7.

DIEL EMERGENCE PATTERNS OF SOME MAYFLIES (EPHEMEROPTERA) OF THE ROSEAU RIVER (MANITOBA, CANADA)

M.K. Friesen, John F. Flannagan and P.M. Laufersweiler

Department of Fisheries and Oceans
Freshwater Institute
501 University Crescent
Winnipeg, Manitoba, Canada
R3T 2N6

ABSTRACT

Emergence of mayflies was monitored at 2 h intervals throughout one 48 h period in each of June, July and August, 1976. In all three study periods most emergence occurred from mid-day until dark, with peak emergence at dusk. The main exceptions to these general trends were *Baetis intercalaris* whose peak emergence occurred at mid-day in June, and some females of *B. intercalaris* and *Tricorythodes* nr. *allectus* which emerged at dawn. The time of day of emergence was similar for most species from day to day and month to month. In species where considerable numbers of males and females emerged, peak emergence of both sexes tended to occur during the same 2 h periods.

INTRODUCTION

The Roseau River rises in the northern U.S.A. and flows northwest through southern Manitoba into the Red River system, which eventually drains into Hudson Bay (Fig. 1). The river, which runs through an area of glacial outwash, is of rather an unusual type on the Canadian prairies, since it is permanent, clear, fast and boulder strewn. Further descriptions of this river can be found in Flannagan (1978). Timing of emergence from the nymphal to the short-lived terrestrial stage of the mayfly is of utmost importance to ensure the simultaneous presence of adequate numbers of both sexes of the same species for successful reproduction. The present study investigated the diel emergence patterns of mayflies over three-48 h periods during the summer of 1976 - one each in June, July and August as part of a phenological study of mayfly emergence throughout the open water season.

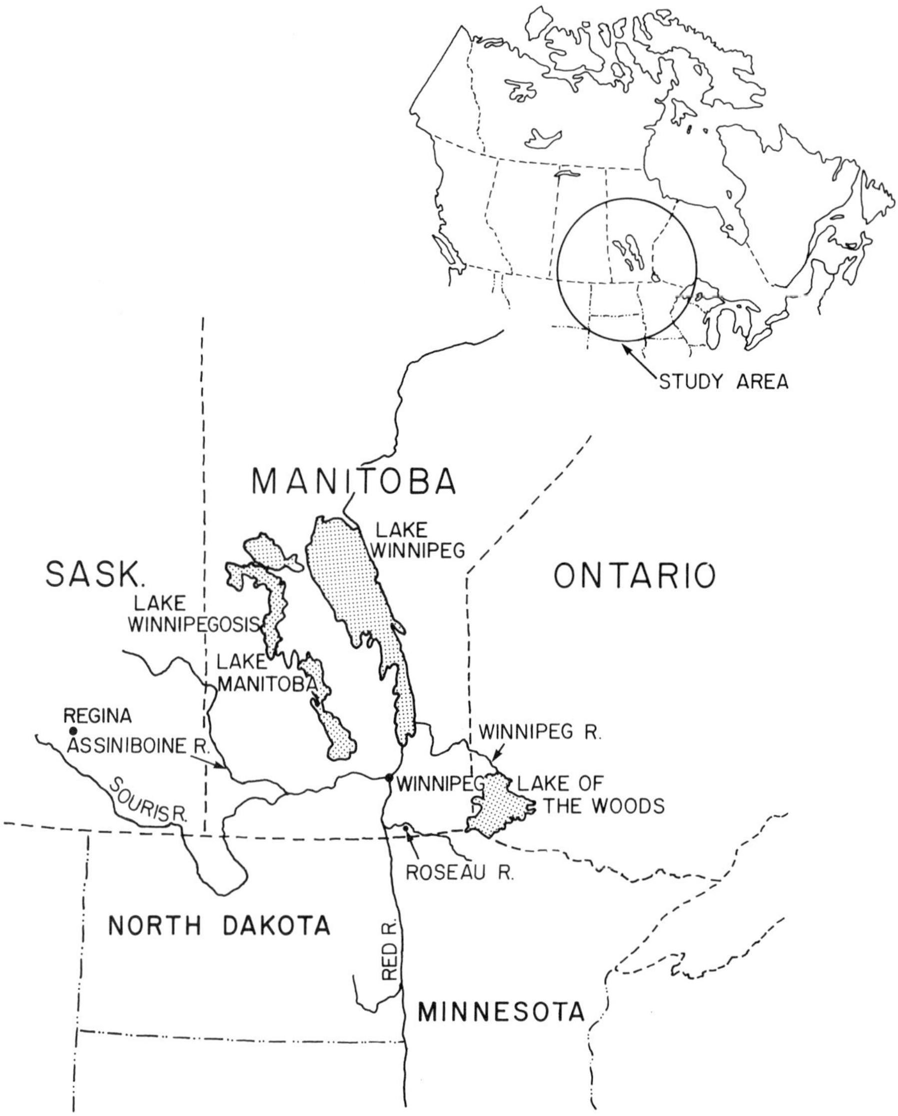

Figure 1. Location of the sampling site on the Roseau River, Manitoba.

MATERIALS AND METHODS

Four, 1 m^3 box emergence traps, located on different substrates (Flannagan 1978) were emptied every 2 h throughout 48 h periods on 13, 14, 15 June; 12, 13, 14 July and 10, 11, 12 August, 1976. Mayflies were preserved in 75% ethanol. Dissolved oxygen (except in June), water temperature, surface light and pH were recorded every 2 h.

RESULTS

Percentage emergence during each 2 h period of total emergence for each sample period, along with the recorded physical and chemical parameters, are given in Figures 2, 3 and 4. In all three months, most emergence occurred from mid-day to the dark period, with peak emergence at dusk. Emergence at dawn, when it occurred, consisted mostly of females of *Baetis intercalaris* and *Tricorythodes* nr. *allectus*. Emergence of most species was considerably reduced towards the end of the study period in each month probably as a result of disturbance of the substrate while sampling. In general, emergence of different families, of congeneric species, and of both sexes of the same species tended to occur in the same 2 h periods (Fig. 5). The time period of greatest emergence in July differed somewhat between *B. intercalaris* and *B. pygmaeus*, and between *Pseudocloeon dubium ?* and *P. punctiventris*. Emergence times of most species did not change substantially from day to day or month to month. An exception was *B. intercalaris* in which most emergence in June occurred at mid-day, in July peak emergence occurred at sunset and sunrise, and in August peak emergence occurred at sunset (Fig. 5). Moonlight and cloud cover in July caused changes in light intensity (Fig. 2) which did not occur during the other sample times.

DISCUSSION

The emergence at dusk of large numbers of mayflies of various species is a well known phenomenon. It is likely a "safety in numbers" strategy evolved to maximize survival from predation at one of the most vulnerable periods in the life cycle of these organisms. Emergence at other times of the 24 h period has been recorded for other mayfly species, and may be in response to presence of certain favourable environmental conditions e.g. temperature, or have some other adaptive advantage. For example, emergence at dawn consisted mainly of females of *Baetis intercalaris* and *Tricorythodes* nr. *allectus*. It was not noted whether mating swarms of these species species occurred at this time but emergence of females at dawn and subsequent mating have been recorded for other species (e.g. *T. atratus* McD. (Hall *et al.* 1975)). This type of behaviour would reduce the time females are exposed to predation before mating and

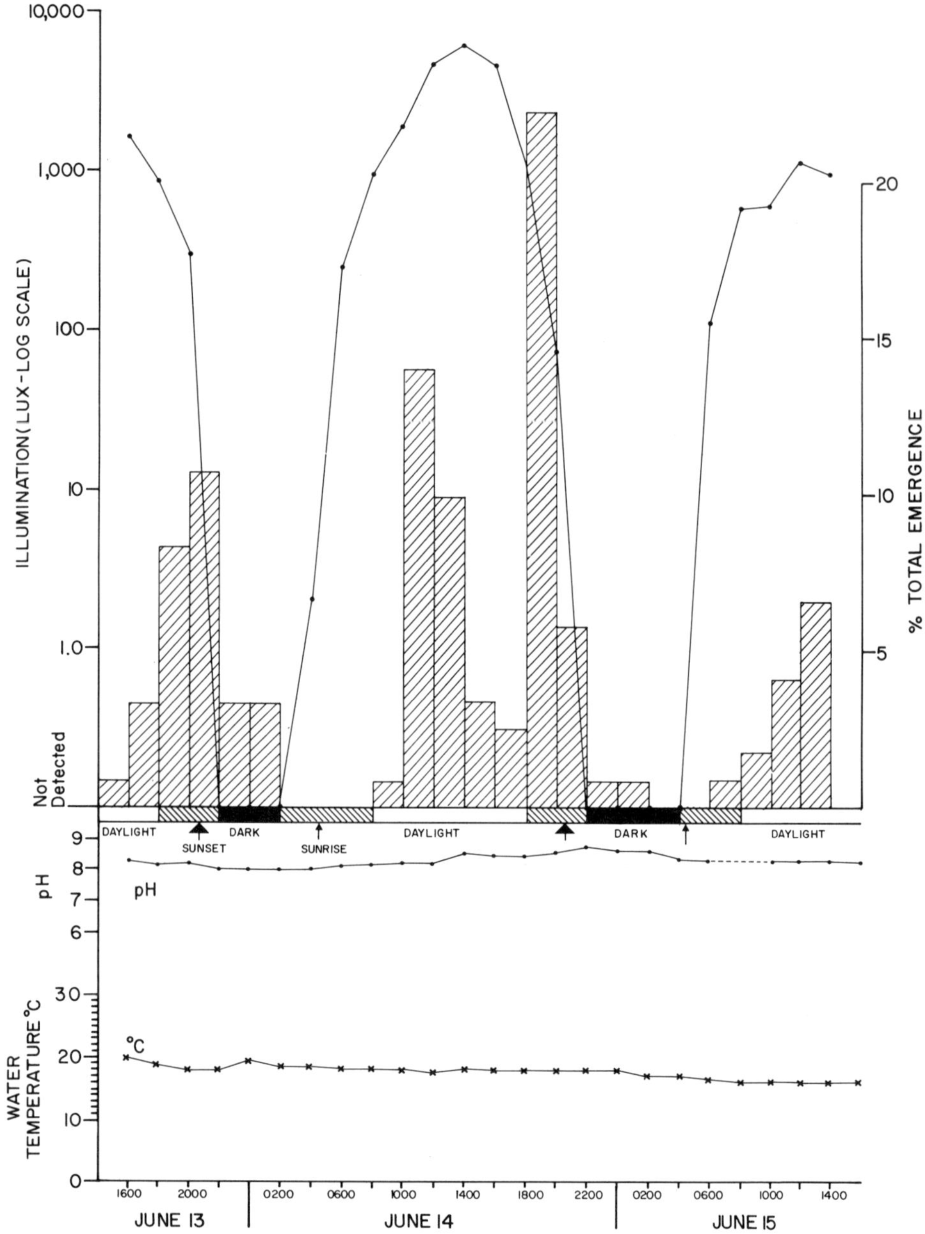

Figure 2. The relationship between emergence time of mayflies and various physical and chemical factors during the June, 48 h sample periods.

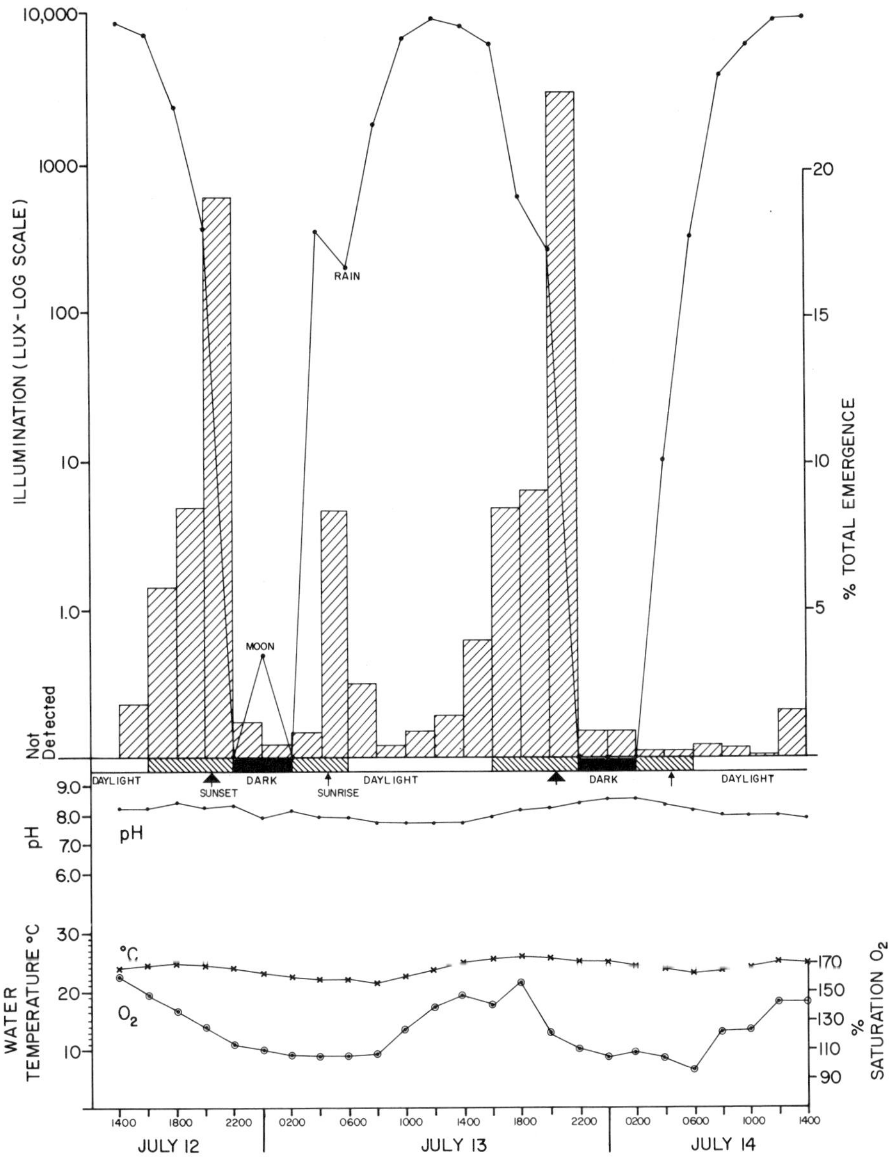

Figure 3. The relationship between emergence time of mayflies and various physical and chemical factors in the July 48 h sample periods.

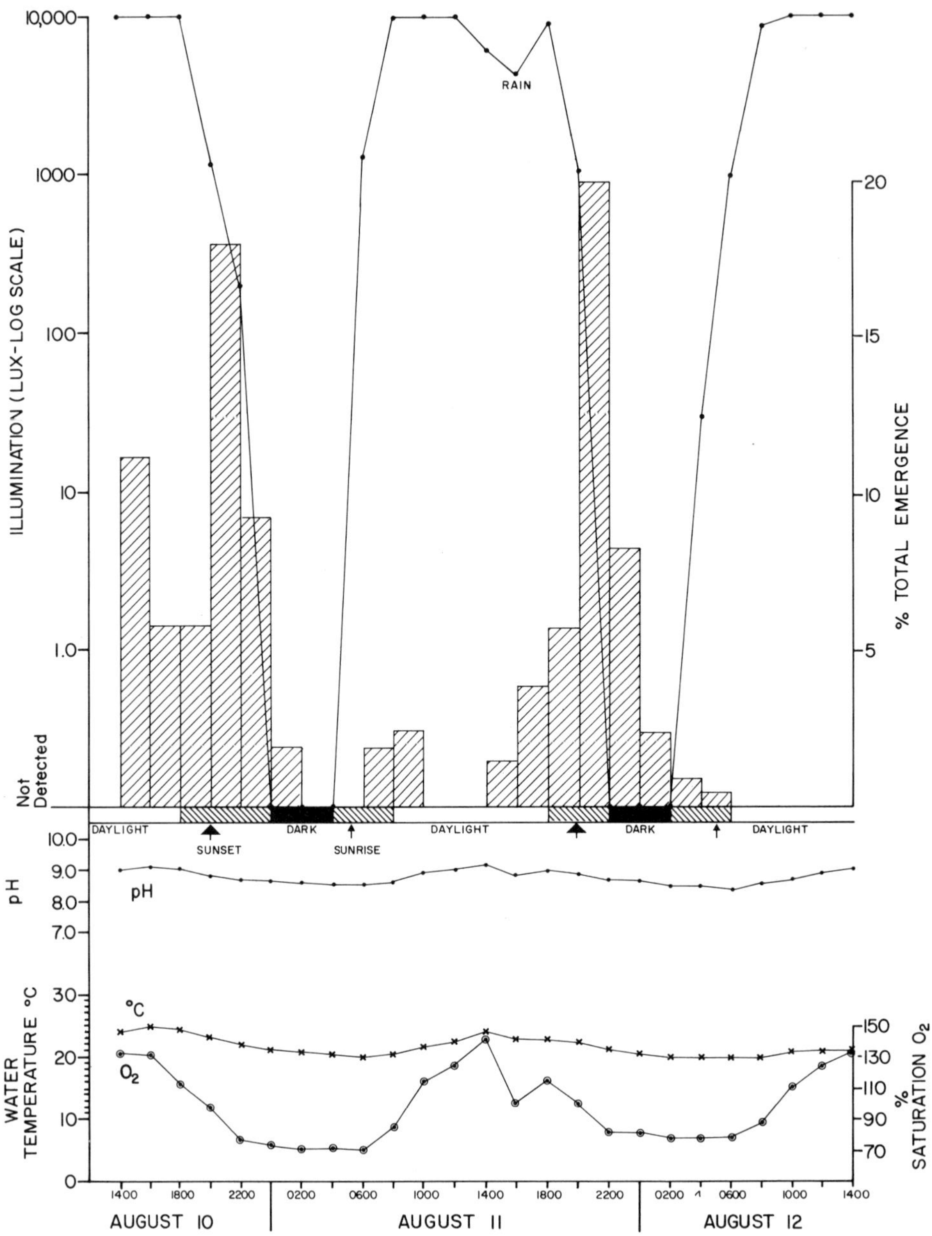

Figure 4. The relationship between emergence time of mayflies and various physical and chemical factors during the August 48 h sample periods.

egg-laying.

Temporal separation over the emergence season may serve to reproductively separate species. In the present study, there is no distinct temporal separation over a 24 h period between congeneric species or between families. Possible exceptions are males of *Baetis spinosus* and *Rhithrogena pellucida* which emerged at times quite different from most of the other species. However, the sample numbers are too low to draw any firm conclusions. There does seem to be a slight separation in time of emergence peaks between *B. intercalaris* and *B. pygmaeus*, and *Pseudocloeon dubium ?* and *P. punctiventris*, but whether this small difference is significant in terms of isolation of species cannot be determined from these data.

Boerger and Clifford (1975) found no difference in the daily mean time of emergence of male and female *Baetis tricaudatus* Dodds when sampled at hourly intervals over 24 h periods. Synchronous emergence of both sexes occurred in most species studied here and is an obvious selective advantage in ensuring the presence of both sexes and increasing the chances for intraspecific reproductive encounters.

Mayflies have evolved mechanisms which lead to the emergence of large numbers of the same species over brief time periods to ensure successful mating. The consistent emergence at dusk of the various species indicates light intensity as an important environmental cue, since it seems unlikely that other factors could act on the different species in a consistent, yet unique enough manner to lead to such a uniform response. The emergence of *B. intercalaris*, which coincides with sunset and sunrise in July and sunset in August also suggests a response related to light. Emergence at dawn in July may also have been influenced by presence of moonlight or cloud cover. Peak emergence of this species at mid-day in the June sample period indicates that if light is important, it is not the only influencing factor. The possibility exists that this shift in time of day of emergence is an adaptation to ensure that the adult emerges at a time of day at which the air temperatures are suitable for flight activity (see Flannagan 1978).

Emergence at a similar time of day of different mayfly species and synchronous emergence of sexes of the same species are probably important adaptations to ensure survival of adequate numbers of organisms for successful reproduction. The role of environmental factors, particularly,light, an initiating emergence needs further investigation and will probably require laboratory studies to fully elucidate this phenomenon.

Figure 5

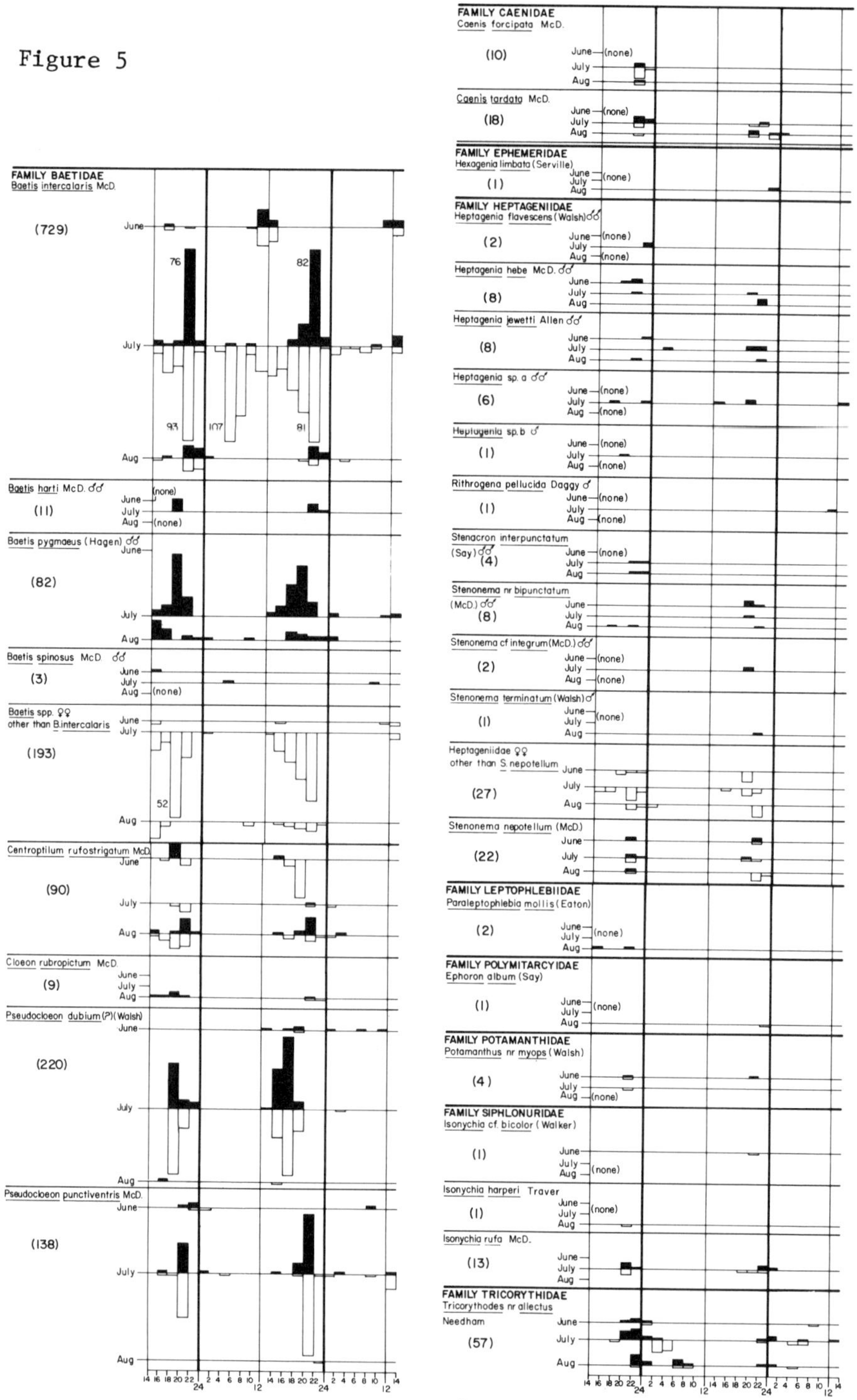
FAMILY BAETIDAE
Baetis intercalaris McD.
(729)
76
82
93
107
81
Baetis harti McD. ♂♂
(11)
Baetis pygmaeus (Hagen) ♂♂
(82)
Baetis spinosus McD. ♂♂
(3)
Baetis spp. ♀♀ other than B. intercalaris
(193)
52
Centroptilum rufostrigatum McD.
(90)
Cloeon rubropictum McD.
(9)
Pseudocloeon dubium (?) (Walsh)
(220)
Pseudocloeon punctiventris McD.
(138)
FAMILY CAENIDAE
Caenis forcipata McD.
(10)
Caenis tardata McD.
(18)
FAMILY EPHEMERIDAE
Hexagenia limbata (Serville)
(1)
FAMILY HEPTAGENIIDAE
Heptagenia flavescens (Walsh) ♂♂
(2)
Heptagenia hebe McD. ♂♂
(8)
Heptagenia jewetti Allen ♂♂
(8)
Heptagenia sp. a ♂♂
(6)
Heptagenia sp. b ♂
(1)
Rithrogena pellucida Daggy ♂
(1)
Stenacron interpunctatum (Say) ♂♂
(4)
Stenonema nr bipunctatum (McD.) ♂♂
(8)
Stenonema cf integrum (McD.) ♂♂
(2)
Stenonema terminatum (Walsh) ♂
(1)
Heptageniidae ♀♀ other than S. nepotellum
(27)
Stenonema nepotellum (McD.)
(22)
FAMILY LEPTOPHLEBIIDAE
Paraleptophlebia mollis (Eaton)
(2)
FAMILY POLYMITARCYIDAE
Ephoron album (Say)
(1)
FAMILY POTAMANTHIDAE
Potamanthus nr myops (Walsh)
(4)
FAMILY SIPHLONURIDAE
Isonychia cf. bicolor (Walker)
(1)
Isonychia harperi Traver
(1)
Isonychia rufa McD.
(13)
FAMILY TRICORYTHIDAE
Tricorythodes nr allectus Needham
(57)
June
July
Aug
(none)
14 16 18 20 22 24 2 4 6 8 10 12 14 16 18 20 22 24 2 4 6 8 10 12 14

Figure 5. Diel emergence patterns of mayfly species for the 48 h sample periods in June, July and August. Numbers in parentheses indicate total number of animals collected. Histograms (black) above the line represent males and those below (white) indicate females. Numbers are given where the scale has been adjusted.

RESUME

Un contrôle de l'émergence des éphéméroptères a été effectué à des intervalles de 2 h durant des périodes de 48 h aux mois de juin, juillet et août 1976. On a constaté au cours de ces trois périodes d'observation que l'emergence des éphéméroptères avait lieu davantage de midi jusqu'à ce qu'il fasse noir et surtout au crépuscule. En général, les moments de la journée où elle s'est produite étaient les mêmes pour chacune des espèces de mois en mois. On a toutefois signalé trois exceptions importantes à ces tendances générales: le mâle adulte de l'espèce *Baetis intercalaris* émergeait au milieu de la journée en juin, tandis que la femelle adulte de la même espèce ainsi que la femelle de l'espèce *B. intercalaris* et *Tricorythodes* nr. *allectus* émergeaient à l'aube au cours de la période d'étude de juillet. Dans le cas des espèces dont une quantité considérable de mâles et de femelles se transformaient en adultes, l'emergence des représentants des deux sexes avait tendance à se produire au cours des mêmes périodes.

ZUSSAMENFASSUNG

Zwischen Juni und August 1976 wurde das Auftreten von Eintagsfliegen über eine 48 Stunden Periode in zweistündigen Abständen jeden Monat überwacht. In allen drei Beobachtungsperioden kamen die Fliegen am häufigsten zwischen Mittag und Einbruch der Dunkelheit zum Vorschein. Dabei traten sie am zahlreichsten zur Zeit der Dämmerung auf. Im allegemeinen war die Tageszeit, zu der die Fliegen von Monat zu Monat auftraten, für alle Arten die gleiche. Eine Ausnahme zu diesem allgemeinen Trend machten zunächst *Baetis intercalaris*, die im Juni um die Mittagszeit zum Vorschein kamen, und dann die Weibchen von *Baetis intercalaris* und *Tricorythodes* nr. *allectus*, die im Juli zur Zeit der Morgendämmerung erschienen. Von Arten, bei denen eine beträchtliche Anzahl von männlichen und weiblichen Fliegen in Erscheinung traten, hielten sich beide Geschlechter an die gleichen Zeitabstände.

ACKNOWLEDGMENTS

We thank Mr. L. Taite for drafting the figures.

REFERENCES

Boerger, H. and H.F. Clifford. 1975. Emergence of mayflies (Ephemeroptera) from a northern brown-water stream of Alberta, Canada. *Int. Ver. theor. angew. Limnol. Verh.* 19: 3022-2028.

Flannagan, J.F. 1978. Emergence of caddisflies from the Roseau River, Manitoba. p. 183-197. *In:* Proceedings 2nd International Symposium on Trichoptera. Junk, the Hague.

Hall, R.J., L. Berner and E.F. Cook. 1975. Observations on the biology of *Tricorythodes atratus* McDunnough (Ephemeroptera: Ephemeridae). *Proc. Entomol. Soc. Wash.* 77: 34-49.

DIURNAL VARIATIONS IN THE ACTIVITY OF *MIRAWARA PURPUREA* RIEK (EPHEMEROPTERA, SIPHLONURIDAE) IN THE ABERFELDY RIVER, VICTORIA AUSTRALIA

Ian C. Campbell*

Zoology Department
Monash University

ABSTRACT

Although *Mirawara* is a widespread genus in eastern Australia, the nymphs are frequently under-represented in benthic samples. Evidence is presented which indicates that *Mirawara* nymphs descend into the substrate during the day, returning to the surface at night to feed. Relatively large numbers of nymphs can be collected at night. Light intensity is probably the factor which controls the movements of the nymphs. The significance of these results for ecological studies is discussed.

INTRODUCTION

Although species of the genus *Mirawara* Harker are widespread in eastern Australia (Fig. 1), the nymphs are not common in conventional stream macroinvertebrate samples. Smith *et al.* (1977) recorded *Mirawara* from the Mitta Mitta valley, but collected only a few specimens. Macmillan (1975) did not collect the genus from the Acheron River, although, from the river bed characteristics and the location of the river, it would be expected to occur there; and no specimens were reported in a study of the Thomson River (M.M.B.W. 1975), only 2 km from the site of the present study in a similar river with similar substrate. Dean (cited in M.M.B.W. 1975) collected nymphs from

---

* Present address: Centre for Water Studies,
Caulfield Institute of Technology
P.O. Box 197
East Caulfield, 3145
Victoria, Australia

Figure 1. Map of Australia with the known distribution of *Mirawara* indicated by shading. The top of the figure is north.

the Aberfeldy River in 1974, but Jackson (1978), in a study on the Aberfeldy at a site approximately 10 km upstream of the present study, also failed to record the presence of the genus.

The nymphs would not be readily overlooked in collections, since full grown specimens are about 30 mm long, and moderately large nymphs may be collected at all times of the year. Their characteristic head shape and unique gill structure make it unlikely that they would be misidentified. In the case of three of the studies cited, those of Smith *et al.*, Macmillan and Jackson, I have been able to examine the material collected and confirm the identifications.

The genus *Mirawara* was first erected by Harker (1954) for the single species *Mirawara aapta*. The material on which the description was based was collected by Tillyard in 1936 from the Murrumbidgee River at Adaminaby in southern New South Wales. In his revision of the Australian siphlonurids, Riek (1955) described two further species: *M. purpurea*, from the Australian Alps south of Canberra (the same general area as *M. aapta*) and *M. megaloprepia*, from Queensland. He records *M. megaloprepia* from as far north as Mackay, and I have collected *Mirawara* nymphs from the Atherton Tableland near Cairns. The genus does not occur in Tasmania.

Two species of the genus occur in Victoria: *Mirawara aapta* Harker and *M. purpurea* Riek. Of these, *M. purpurea* is the most common. In 1977 I commenced a study of the taxonomy, life histories and distributions of siphlonurids in southeastern Australia and found that nymphs of *M. purpurea*, although not abundant during the day, could be collected readily at night. Samples collected from the Aberfeldy River at 1730, 1930 and 2130 h (before, at and after dusk) on 24th February, 1978, showed a marked increase in the number of nymphs collected from a 4 m^2 area as the light faded, mean numbers being 1.8 at 1730, 2.0 at 1930 (dusk) and 14.3 after dark at 2130. The experiments described here were initiated to investigate this phenomenon.

THE STUDY AREA

The site selected for the study is a section of the Aberfeldy River, which is a tributary of the Thomson River, about 150 km east of Melbourne (Fig. 2). The Aberfeldy flows about 48 km in a southerly course draining a steep, forested catchment of 308 km^2. Although the river was the scene of considerable mining activity as a result of the discovery of gold in the area in 1861, there has been little human activity in the catchment since 1914, apart from some logging operations. The areas damaged by mining are now largely revegetated. The catchment is vegetated by open forest dominated by *Eucalyptus obliqua* and *E. radiata*. More detailed discussions of the vegetation and topography of the catchment can be found in L.C.C. (1973) and M.M.B.W. (1975).

The stretch of river studied is located at Beardmore's (37° 51.5'S, 146°26'E), about 150 m upstream from a Victorian State Rivers and Water Supply Commission gauging station. The mean annual flow at this point is 64.7 ML and the mean salinity is 45 ppm (Bibra and Riggs 1971). The substrate is predominately shingle: a mixture of small (< 30 cm diameter) stones and coarse sandy gravel. The river ranges from 3 m to 20 m wide and is mostly shallow, with a maximum depth of 60 cm.

Figure 2. Map of Victoria indicating the location of the Aberfeldy River. The top of the figure is north.

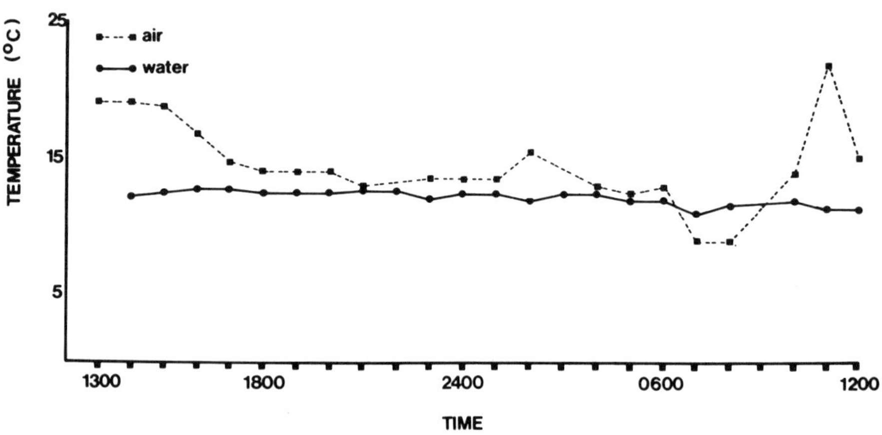

Figure 3. Air and water temperatures taken at hourly intervals from 1300 h on 22nd to 1200 h on 23rd April 1978.

METHODS

Samples were collected on two occasions: on the 22nd and 23rd of April, 1978, and again from 30th March to 1st April, 1979. On each occasion a grid was marked out, using string stretched across the river, defining metre square quadrats. The grid was constructed such that the water in each quadrat was no more than 35 cm and no less than 10 cm deep. The current on the grid varied from 0.08 to 0.50 m/sec in 1978, and from 0.05 to 0.29 m/sec in 1979. In 1978 one group of 18 quadrats were 50 m downstream of the rest, but in 1979 all of the quadrats were located in the one section of stream. The quadrats to be sampled at each time interval were chosen using random number tables. In 1978 six quadrats were sampled each hour for 24 hours, whilst in 1979 six were sampled every 2 hours for 48 hours. Each quadrat was sampled only once.

The sampling apparatus consisted of a wooden frame 1 m^2 and 40 cm high with the sides covered in 300 μm nylon mesh. The frame was pushed firmly against the substrate to obtain as good a seal as possible. The water inside the frame was then vigorously agitated and repeated swept with a 300 μ mesh FBA net for 2 minutes. The contents of the net were emptied into 1 L plastic jars and preserved with Kahle's solution. In the laboratory the samples were sorted by hand under a binocular microscope.

At the same time as each set of six samples were collected, air and water temperatures were recorded using a pair of mercury thermometers. Water velocity was recorded with a stopclock and an orange tied to a 1 m length of string.

From 2030 h on the second evening (31st March, 1979) until 0500 h the following morning the grid was illuminated with four 2.5 kva floodlights, to test the effect of illumination on the number of animals collected.

RESULTS

The water and air temperature data for the 1978 sampling are presented in Figure 3 and for 1979 in Figure 4. The mean numbers of *Mirawara* nymphs found in the six samples collected at each interval are presented in Figure 5 (1978) and Figure 6 (1979).

DISCUSSION

The number of animals collected in the 1978 samples was fairly small. Nineteen samples contained no *Mirawara*, and the largest number in any of the samples was 22. The low overall numbers of nymphs collected on this occasion probably accounts for the somewhat

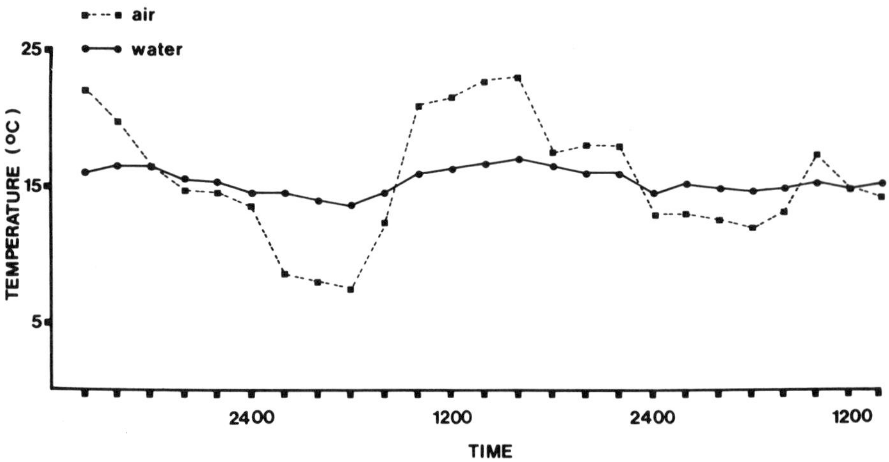

Figure 4. Air and water temperatures taken at two hourly intervals from 1400 h on 30th March to 1400 h on 1st April 1979.

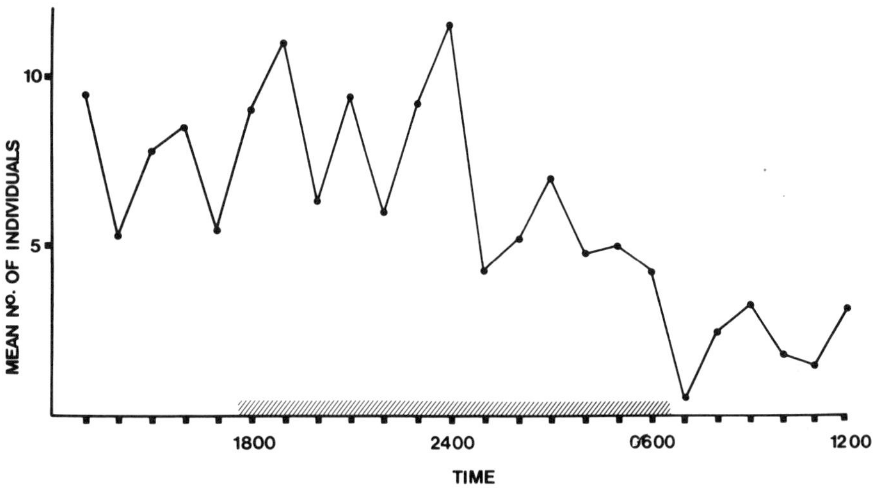

Figure 5. Mean numbers of *Mirawara* nymphs collected per sample in sets of six samples taken hourly from 1300 h on 22nd to 1200 h on 23rd of April 1978. The dark period is indicate by the shading.

erratic form of Figure 5; however, there is a fairly clear peak at 1900 h, just after dusk, and another at midnight with a drop between the two. A second very marked drop in numbers occurred after midnight, the numbers being relatively depressed until dawn, when there was a further marked decrease. The drop between midnight and 0100 h may have been due to the overcast sky clearing, allowing bright moonlight to illuminate the grid. Anderson (1966), Waters (1962) and Bishop and Hynes (1969) have all shown that bright moonlight is sufficient to depress drift activity of stream insects. The further drop in mean numbers of *Mirawara* collected per sample, which occurred between 0600 and 0700 h, was presumed to be a response to daylight.

The daylight results on the first day are markedly higher than those on the second day. This difference was tested using the median test for two independent samples (Siegel 1956) and found to be significant at the 0.005 level. The first day of the study was overcast and the second clear and sunny, and the difference in natural illumination is probably at least partially responsible for the difference in the number of nymphs collected.

Because of this difference between the two sets of light results, the dark results were tested against each of them separately, again using the median test. The difference between the dark results

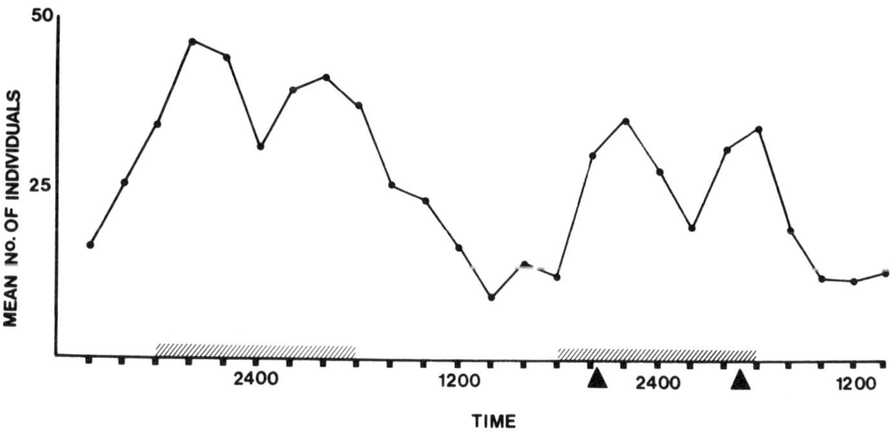

Figure 6. Mean numbers of *Mirawara* nymphs per sample in sets of six samples taken every two hours from 1400 h on 30th March to 1400 h on 1st April 1979. The dark period is indicated by shading, and the two arrows indicate the period during which the stream was illuminated.

and the light results on April 23rd was found to be significant at the 0.001 level, but no significant difference was found between the results of the light period on April 22nd and the dark period.

In 1979 the experiment was repeated over a longer time interval, with the intention of using floodlighting to test whether changes in illumination could influence the activity pattern noted in 1978. On this occasion, surprisingly large numbers of animals were collected. Only one sample contained no *Mirawara* and the highes number collected in any single sample was 74, an extremely high density for a carnivorous insect. The reasons for this large increas in numbers between 1978 and 1979 are not known; however, an extremely large flood, which occurred in September 1978, moved large amounts of substrate and, although the distribution and apparent composition of substrate was largely unchanged, subtle changes may have made the habitat more suitable for *Mirawara*.

The 1979 results show very clear dusk and dawn peaks with reduced numbers occurring at, or shortly after, midnight. The peaks were lower on the second night when the grid was illuminated, and this difference, when tested with the median test, was significant at the 0.005 level.

The sampling technique collects insects on top of stones in the quadrat and also those sheltering under any loose stones, since thes were turned over by the vigorous swirling of the water. I believe that the difference between the night and day samples can be attributed to the movement of nymphs into the substrate during light periods, and up to the surface again at night. *Mirawara* nymphs are always found in streams with a loosely packed shingle or cobble substrate which has abundant, large interstices into which the insects can move. The presence of large numbers of insects and other aquati invertebrates in, and their movement into and out of, the hyporheic environment is well established (e.g. Bishop 1973, Williams and Hynes 1974, Stanford and Gaufin 1974).

A number of the features of *Mirawara* nymphs could be interprete as adaptations to hyporheic life. Included amongst these are dorsoventral flattening, which is uncharacteristic of siphlonurids, the decided wedge shape of the head and the presence of both gill filaments and plates. The presence of gill filaments and plates is not shared by the related genus, *Ameletopsis* Phillips, which has only gill plates, although these are more highly ramified than those of *Mirawara* (Phillips 1930). The presence of filaments provides a large respiratory surface whilst the plates are used to create a water current over the respiratory surfaces.

The obvious difference in numbers of *Mirawara purpurea* collect during dark and light periods in both experiments, together with the significant depression of the numbers collected during the period in which the stream was illuminated in 1979, may be interpreted, therefore, as evidence of a negative phototaxis combined with nocturnal activity peaks probably associated with feeding behaviour. A number of studies have been carried out on the influence of light on the behavioural patterns of ephemeropterans. Moon (1940) showed that nymphs of *Ecdyonurus* and *Baetis* are more active during the night than the day. Harker (1953), in a series of laboratory experiments on *Ecdyonurus torrentis* Kimmins, *Heptagenia lateralis* (Curtis) and *Baetis rhodani* (Pictet) showed distinct diurnal rhythmicity of activity which, once established in the early instars, was not directly affected by the light regime. However, Harker's experiments were carried out using animals obtained from running water but observed in still water, and Elliott (1968) later showed that quite different results are obtained under flowing water conditions. He found that both activity and position on stones were directly contolled by light intensity for five species of Ephemeropterans: *Baetis rhodani* (Pictet), *Ephemerella ignita* (Poda), *Ecdyonurus venosus* (Fabricus), *Rhithrogena semicolorata* (Curtis) and *Heptagenia lateralis* (Curtis). Under laboratory conditions, significantly more individuals of all of these species were present on the top of a stone in an artificial stream during dark than light periods. Field counts of the numbers of *B. rhodani* present on the tops of stones in a stream in light and dark periods indicated that the same phenomenon also occurs under natural conditions. These results are in general agreement with those of the present study, although Elliott recorded a single nocturnal maximum in both activity and numbers of animals on top of stones, rather than a pair of maxima as was found here.

A considerable amount has been published on the relationships between light intensity, diurnal activity and drift of aquatic insects and other invertebrates. The literature on drift has been reviewed by Hynes (1970), Waters (1972) and Müller (1974). A common pattern is for maximal numbers of drifting animals to occur shortly after dusk and again shortly before dawn, with a slight decrease around midnight and minimal numbers during the day.

It has been widely suggested that these peaks correspond to activity peaks of the animals, which, in turn, are controlled by light intensity (e.g. Elliott 1967, Chaston 1968, Bishop 1969, Bishop and Hynes 1969), although this interpretation is disputed by Corkum (1978). The activity cycles found by the authors previously mentioned are consistent with the results of the present study, with the dawn and dusk drift and activity peaks corresponding to the periods in which *Mirawara* moves to the surface of the substrate to actively seek food.

Clifford (1972) compared two sets of Surber samples collected from the Bigoray River (Alberta, Canada) during the day and night and concluded that there were no significant differences between them with respect to total numbers, total number of taxa and total volume-biomass. The results presented here indicate that the time at which benthic samples are collected from eastern Australian streams could have a marked effect on the composition of the sample, and lead to a significant underestimate of the proportion and absolute numbers of insect carnivores.

CONCLUSIONS

Significantly higher numbers of *Mirawara purpurea* nymphs were found in samples collected at night than in samples taken during the day. This is probably due to a negative phototaxis, and the animals probably move down into the substrate during the day and ascend to the surface again to feed at night. Negative phototaxes and nocturnal activity peaks are common amongst ephemeropterans.

Results of this study indicate that samples of stream benthos collected during the day may significantly underestimate the numbers of *Mirawara* nymphs present in the stream or fail to collect them altogether. This may be the reason why a number of studies of Australian stream benthos found only small numbers of *Mirawara* nymphs or collected none at all.

ACKNOWLEDGMENTS

I would like to thank Dr. P.S. Lake for useful advice on the experimental design. Peter Breen, Pam Brewster, Moffy Doeg, Ian McKelvie, Luisa Macmillan, Yaneth Mayorga, Merrion Morrisey and Jenny Smith all assisted with the sampling, and to all of them I am extremely grateful. I am indebted to Helen Clausen, who assisted with the sampling, drawing of the figures and calculations, and remained cheerful throughout it all.

Work described in this paper is being carried out as part of a Ph.D. programme at Monash University.

RESUME

Bien que *Mirawara* soit un genre extrêmement répandu en Australie orientale, bien souvent les larves se trouvent en nombre inférieur dans les échantillons de faune benthique. L'auteur démontre, à l'aide de certains indices, que les larves *Mirawara* descendent sous terre pendant le jour et remontent le soir à la surface pour se nourrir. On peut alors en prendre d'assez grandes quantités. L'intensité de la lumière est problement le facteur qui règle le mouvement des larves. L'auteur montre enfin l'importance de ces découvertes dans le domaine des études écologiques.

ZUSSAMENFASSUNG

Obwohl *Mirawara* eine weitverbreitete Gattung im östlichen Australien ist, treten die Nymphen oft nur in geringen Mengen in benthischen Proben auf. Hier soll nun ein Nachweis erbracht werden, der zeigt, daß *Mirawara* Nymphen während des Tages tiefer in das Substrat hinabsteigen und nachts an die Oberfläche zurückkehren, um Nahrung zu erwerben. Verhältnismäßig große Zahlen von Nymphen können daher in der Nacht gesammelt werden. Wahrscheinlich ist die Lichtintensität der Faktor, der die Bewegungen der Nymphen kontrolliert. Die Bedeutung dieser Ergebnisse für ökologische Studien wird in der vorliegenden Arbeit erörtert.

REFERENCES

Anderson, N.H. 1966. Depressant effect of moonlight on activity of aquatic insects. *Nature (Lond.)* 209: 319-320.

Bibra, E.E. and H.C.W. Riggs. 1971. Victorian River gaugings to 1969. State Rivers and Water Supply Commission, Melbourne.

Bishop, J.E. 1969. Light control of aquatic insect activity and drift. *Ecology* 50: 371-380.

Bishop, J.E. 1973. Observations of the vertical distribution of the benthos in a Malaysian stream. *Freshw. Biol.* 3: 147-156.

Bishop, J.E. and H.B.N. Hynes. 1969. Downstream drift of the invertebrate fauna in a stream ecosystem. *Arch. Hydrobiol.* 66: 56-90.

Chaston, I. 1968. Endogenous activity as a factor in invertebrate drift. *Arch. Hydrobiol.* 64: 324-334.

Clifford, H.F. 1972. Comparison of samples of stream fauna collected during the day and at night. *Limnol. Oceanogr.* 17: 479-481.

Corkum, L.D. 1978. Is benthic activity of stream invertebrates related to behavioural drift? *Can. J. Zool.* 56: 2457-2459.

Elliott, J.M. 1967. Invertebrate drift in a Dartmoor stream. *Arch. Hydrobiol.* 63: 202-207.

Elliott, J.M. 1968. Daily activity patterns of mayfly nymphs (Ephemeroptera). *J. Zool. (Lond.)* 155: 202-221.

Harker, J.E. 1953. The diurnal rhythm of activity of mayfly nymphs. *J. Exp. Biol.* 30: 525-533.

Harker, J.E. 1954. The Ephemeroptera of eastern Australia. *Trans. R. Entomol. Soc. Lond.* 105: 241-268.

Hynes, H.B.N. 1970. The ecology of running waters. Liverpool University Press, Liverpool.

Jackson, P.D. 1978. Benthic invertebrate fauna and feeding relationships of brown trout, *Salmo trutta* Linnaeus, and river blackfish, *Gadopsis marmoratus* Richardson, in the Aberfeldy River, Victoria. *Aust. J. Mar. Freshw. Res.* 29: 725-742.

L.C.C. 1973. Report on the Melbourne study area. Land Conservation Council, Victoria, Melbourne.

M.M.B.W. 1975. Report on environment study into the Thomson Dam and associated works. Vol. I. Melbourne and Metropolitan Board of Works, Melbourne.

Macmillan, L.A. 1975. Longitudinal zonation of benthic invertebrates in the Acheron River, Victoria. B.Sc. (Hons.) Thesis, Zoology Department, Monash University.

Moon, H.P. 1940. An investigation of the movements of freshwater invertebrate faunas. *J. Anim. Ecol.* 9: 76-83.

Müller, K. 1974. Stream drift as a chronobiological phenomenon in running water ecosystems. *Annu. Rev. Ecol. Syst.* 5: 309-323.

Phillips, J.S. 1930. A revision of New Zealand Ephemeroptera. Part I. *Trans. N.Z. Inst.* 61: 271-333.

Riek, E.F. 1955. Revision of the Australian mayflies (Ephemeroptera) I. Subfamily Siphlonurinae. *Aust. J. Zool.* 3: 266-280.

Siegel, S. 1956. Nonparametric statistics for the behavioural sciences. McGraw-Hill, New York.

Smith, B.J., H.E. Malcolm and P.B. Morison. 1977. Aquatic invertebrate fauna of the Mitta Mitta Valley, Victoria. *Victorian Nat.* 94: 228-238.

Stanford, J.A. and A.R. Gaufin. 1974. Hyporheic communities of two montana rivers. *Science (Wash. D.C.)* 185: 700-702.

Waters, T.F. 1962. Diurnal periodicity in the drift of stream invertebrates. *Ecology* 43: 316-320.

Waters, T.F. 1972. The drift of stream insects. *Annu. Rev. Entomol.* 17: 253-272.

Williams, D.D. and H.B.N. Hynes. 1974. The occurrence of benthos deep in the substratum of a stream. *Freshw. Biol.* 4: 233-256.

THE MECHANISMS THAT MAINTAIN POPULATION STABILITY OF SELECTED SPECIES OF EPHEMEROPTERA IN A TEMPERATE STREAM

Eric C. Gyselman

Freshwater Institute
501 University Crescent
Winnipeg, Manitoba, Canada

ABSTRACT

The study was designed to determine whether either Muller's 'colonization cycle' or Waters' theory of intraspecific competition adequately explained why the population densities of eight species of Ephemeroptera in the Rat River, Manitoba, remained stable in spite of high drift rates. The flight patterns of the adults, the density of nymphs in the substrate and the drift rates were studied over two summer seasons. The results supported Waters' hypothesis that drifting was a mechanism of reducing population densities when the carrying capacity was reached. The proportion of the population in the drift was density dependant and at high drift rates size selective. There was no evidence of a 'colonization cycle' as Muller had proposed.

INTRODUCTION

All benthic invertebrates, which spend at least part of their life cycle in streams, show some degree of adaptation to maintaining their position in the substrate. However, a proportion of each population can always be found in the water column moving downstream or 'drifting'.

Muller (1954) demonstrated that drifting was not simply a passive process but for most species followed predictable cycles and patterns. Later studies (Bishop and Hynes 1969, Ulfstrand 1968, Elliott 1967, Waters 1965) showed that because of drifting, many species should have displayed successive downstream shifts in population density with each generation. However, no such shifts were observed and this led to speculation that there must be some mechanism which compensated for drifting.

Three theories have been proposed to explain why population numbers appear to remain stable in spite of high drift rates: (1) adult insects upon emerging, fly primarily upstream before laying their eggs and thus compensate for the distance that they drifted during the aquatic stages of their life cycle (Muller 1954), (2) drifting is density related and it is a mechanism of reducing intraspecific competition (Waters 1961), (3) individuals simply walk back upstream either on or in the substrate. However, it has been shown by Elliott (1971), Bishop and Hynes (1969), Hultin *et al.* (1969) and Minkley (1964) that walking back upstream can only partially compensate for drifting and it is probably only of importance on a localized scale. Therefore, this study was undertaken to determine whether either of the other two hypotheses adequately explained population stability in streams.

METHODS

As the Rat River in southeastern Manitoba, Canada, descends the beach ridge formations created by glacial Lake Agassiz, a short 7 km zone of riffle-pool type habitat is formed. Both upstream and downstream from this zone the river is slow and meandering. Species which are specifically adapted to a riffle-pool habitat are therefore confined to this one area of the Rat River.

The riffle-pool habitat zone formed the study area. Four permanent stations were established in riffles. Station 1 was at the beginning of the zone, Station 4 at the end and Stations 2 and 3 spaced approximately equidistant between 1 and 4.

The study was restricted to the eight most abundant species of Ephemeroptera that were found exclusively in the study area: *Baetis hageni* Eaton, *Baetis intercalaris* McDunnough, *Pseudocloeon myrsum* Borks, *Paraleptophlebia praepidita* (Eaton), *Ephoron album* (Say), *Stenacron interpunctatum canadense* (Walker), *Stenonema nepotellum* (McDunnough) and *Heptagenia maculipennis* Walsh.

To adequately evaluate Muller's and Waters' hypotheses, three aspects of mayfly ecology were studied: (1) the flight patterns of the adults, (2) the relative change in benthic density among the stations with time, (3) the proportion of the population that was drifting compared to the benthic standing crop.

Traps similar to those in Madsen *et al.* (1973) were used to determine whether there was a directional migration to the flight patterns of the adults (subimagos and imagos). Each trap consisted of a vertical wooden frame, 1 m^2, that was covered with clear polyethelene film and sprayed with Tanglefoot^R. One trap was set up at each station and the frames were changed at approximately 10 day intervals throughout the 1974 sampling season. The adult mayflies that had flown into the traps, and adhered to the plastic, were removed by dissolving the Tanglefoot^R with turpentine, then identified and counted. To determine whether a migration was taking place, a series of Chi-squared analyses were calculated comparing the number of individuals on either side of the traps. A separate analysis was done for males and females of each species caught in the traps.

Estimates of the benthic standing crop were obtained using a 0.11 m^2 surber sampler with a 202 μm net. Nine samples were taken at each station on July 9, August 13 and September 22, 1973 and July 2, July 22 and September 25, 1974. The nine samples were allocated equally among three substrate types: rocks, gravel and sand. Each sample was floated following the procedures recommended by Flannagan (1973) and subsampled using a volume type splitter (Gyselman 1976). Each sample was successively split three times so that the 'working subsample' was 1/8 the volume of the original sample. A regression analysis of the predicted number of individuals of each species in the subsample to the actual number showed that the technique was 95.3% efficient. The mayflies from the working subsamples were identified and the pronotum length of each individual was measured using an ocular micrometer. For each species, the range of pronotum lengths was arbitrarily divided into eight length groups. These groups were used as a measure of relative age with group 1 being the youngest and group 8 the oldest.

A four-way factorial analysis of variance of the benthic density data was calculated for each species in each year to determine the effects of sampling period, station, substrate type and age (pronotum length group). As with most age distribution data, the variance among the replicates within the period-station-substrate combinations for each species was found not to be homogeneous and a log (X+1) transformation of the data was used.

During the first two sampling periods of 1973, drift samples were taken over a 24 hour period at each station. The samplers were 33 cm wide with 202 μm nets and sampled the whole water column. There were two samplers at each station and each was emptied every-four hours. The drift samples were subsampled using the same technique as that used for the Surber samples. The mayflies from each working subsample were identified and the pronotum of each was measured. There was sufficient data on *Baetis hageni*, *Baetis intercalaris* and *Pseudocloeon myrsum* to carry out analyses on the relationship between the benthic density and the number of individuals in the drift. A curvilinear regression was done for each species to determine whether the number of individuals in the drift was linearly related to the benthic density. A second degree polynomial was proposed as an alternative model. In addition, a series of Student's *t*-tests were done to compare the mean pronotum length of the benthic populations with that of the individuals caught in the drift.

RESULTS

Aerial Migration

Only 5 species, *Ephoron album, Stenacron i. canadense, Stenonema nepoltellum, Heptagenia maculipennis* and *Pseudocloeon myrsum,* were caught in the aerial migration traps. In none of the Chi-squared analyses was there a significant different ($P < 0.05$) in the number of individuals caught on either side of the traps.

The Relationship of Density and Age Between the Drift and the Benthic Population

Four species were caught in significant numbers in the two sets of drift samples: *Baetis hageni, Baetis intercalaris, Pseudocloeon myrsum* and *Paraleptophlebia praepidita. Ephoron album, Heptagenia maculipennis,* and *Stenonema nepotellum* were not found in any of the drift samples and only five individuals of *Stenonema i. canadense* were caught.

The probability of Type I error in the curvilinear regressions was less than 0.01 for *P. myrsum* and *B. intercalaris* and less than 0.05 for *B. hageni*. Therefore, in each case the second degree polynomial is a better model of the data than is the linear equation.

The results of the Student's *t* analyses are summarized in Table 1.

Benthic Densities and Age Structures

A four-way factorial analysis of variance generates four main factor effects and 11 interaction effects. However, only the period/station/age interaction which measures the change in age structure differences among the stations throughout the sampling season is important in this study. The period/station/age interaction was significant for *Baetis hageni* in 1973, *Baetis intercalaris* in 1973, *Pseudocloeon myrsum* in 1974, *Paraleptophlebia praepidita* in 1974, and *Stenacron i. canadense* in 1974.

The drought in August, 1974 that caused the Rat River to dry up was the obvious cause of the significant period/station/age interaction for *P. praepidita* and *S. i. canadense*. Both of these species

Table 1. A comparison of the mean pronotum length of individuals caught in the drift and the benthic population.

	DATE	STATION	SURBER $\overline{X}$	SURBER N	DRIFT $\overline{X}$	DRIFT N	t
Baetis hageni	July 9 1973	1	0.65	5	0.82	5	2.50
		2	0.67	17	0.60	9	1.08
		3	0.67	37	0.66	12	0.60
		4	0.66	43	0.66	6	1.82
	August 13 1973	1	0.64	266	0.75	201	172.21**
		2	0.58	117	0.64	32	34.19**
		3	0.64	105	0.65	57	2.55*
		4	0.52	22	0.61	16	13.07**
Baetis intercalaris	July 9 1973	1	0.84	70	0.66	5	9.15**
		2	0.88	76	0.61	7	12.50**
		3	0.72	87	0.99	3	24.90**
		4	0.87	189	0.80	9	16.12**
	August 13 1973	1	0.94	199	0.75	43	106.38**
		2	0.88	36	0.78	2	3.56**
		3	0.96	59	0.80	7	28.93**
		4	0.91	41	0.87	7	6.99**
Pseudocloeon myrsum	July 9 1973	1	0.69	21	0.80	3	15.65**
		2	-	--	-	-	-
		3	-	-	-	-	-
		4	0.65	21	0.73	6	5.69**
	August 13 1973	1	0.79	59	0.85	117	16.34**
		2	-	-	-	-	-
		3	-	-	-	-	-
		4	0.62	6	0.63	6	0.16

$\overline{X}$ = Mean Pronotum Length
N = Total Number of Animals Caught by the Sampling Method
* = P < 0.05
** = P < 0.01

were common in all of the Surber samples taken in 1973 including those taken on September 22. They were also abundant in the samples taken on July 2 and July 22, 1974 but were absent from those taken after the drought on September 25, 1974.

The significant period/station/age interaction for *B. hageni* in 1973 appears to be the result of differential hatching and development of the nymphs (Fig. 1). There is an increase in the density of young individuals at station 1 on August 13 but the corresponding increase does not occur at Stations 2 and 3 until September 22. This was a feature of the 1973 sampling season only and it was not repeated in 1974.

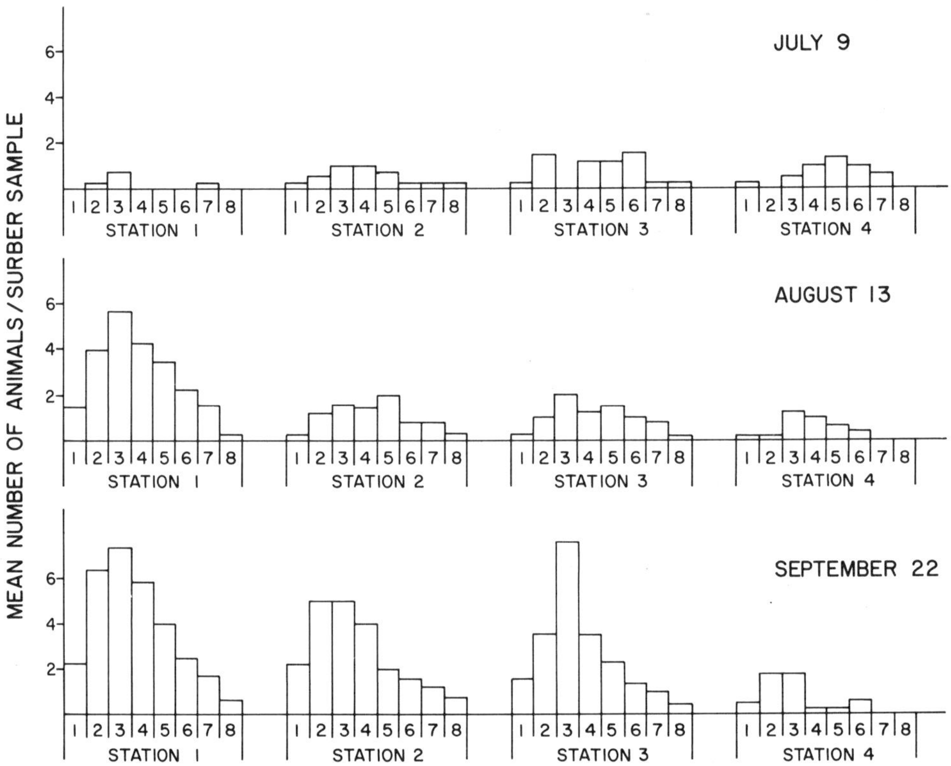

Figure 1. The mean density of *Baetis hageni* in each age cohort at each station during 1973.

There is a significant decline in density at station 4 over the sampling season of both *P. myrsum* in 1974 (Fig. 2) and *B. intercalaris* in 1973 (Fig. 3). While there is no explanation for these declines, they do not appear to be the result of an upstream shift in the population. In neither case was the decline repeated during the other sampling season.

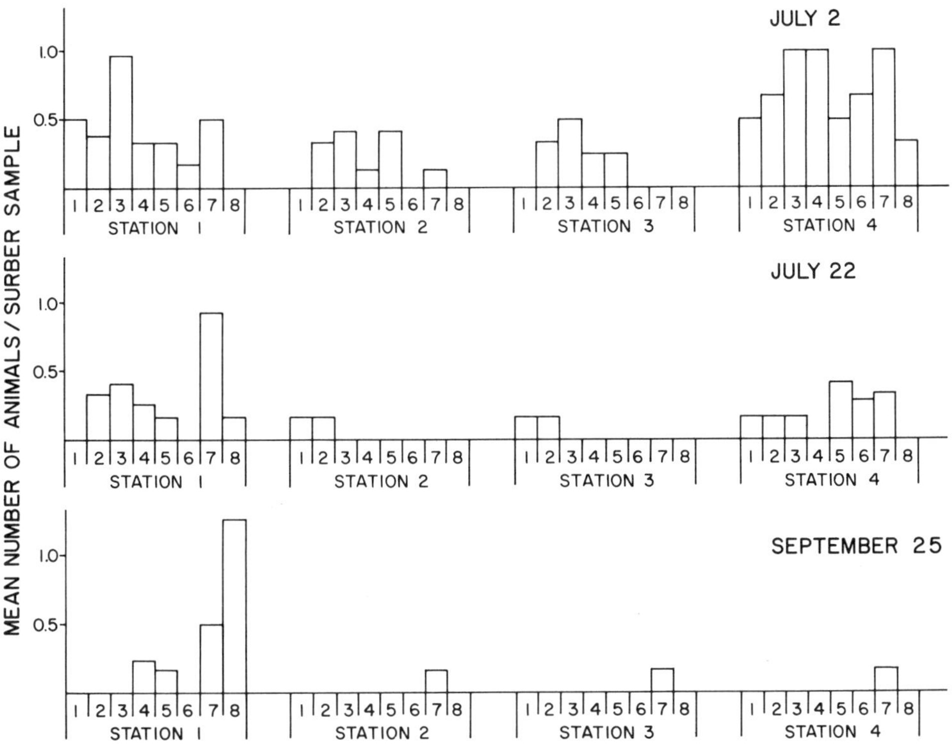

Figure 2. The mean density of *Pseudocloeon myrsum* in each age cohort at each station during 1974.

It is important to note that there is no evidence of either an upstream or downstream shift in the population over the sampling season in any of the cases where the period/station/age interaction was significant.

DISCUSSION

The results of this study do not support Müller's theory of a 'colonization cycle'. The analyses of the data from the aerial migration traps show that there was no upstream flight by the adults of the five species caught. Furthermore, the factorial analyses of the benthic density data indicate that the relative density among the stations remains stable throughout the summer for 11 of the 16 species/year groups. Of the five species/year groups that did have a significant period/station/age interaction, two were caused by the drought in August, 1974 and an examination of the age distribution histograms for the other three shows no indication of an upstream or a downstream shift in the population.

Waters' (1961) hypothesis is supported by some of the data from this study. The curvilinear regressions show that for *Baetis hageni*, *Baetis intercalaris* and *Pseudocloeon myrsum* the proportion of the population in the drift increases as the benthic density increases. This idea fits Waters' model well. He states (Waters

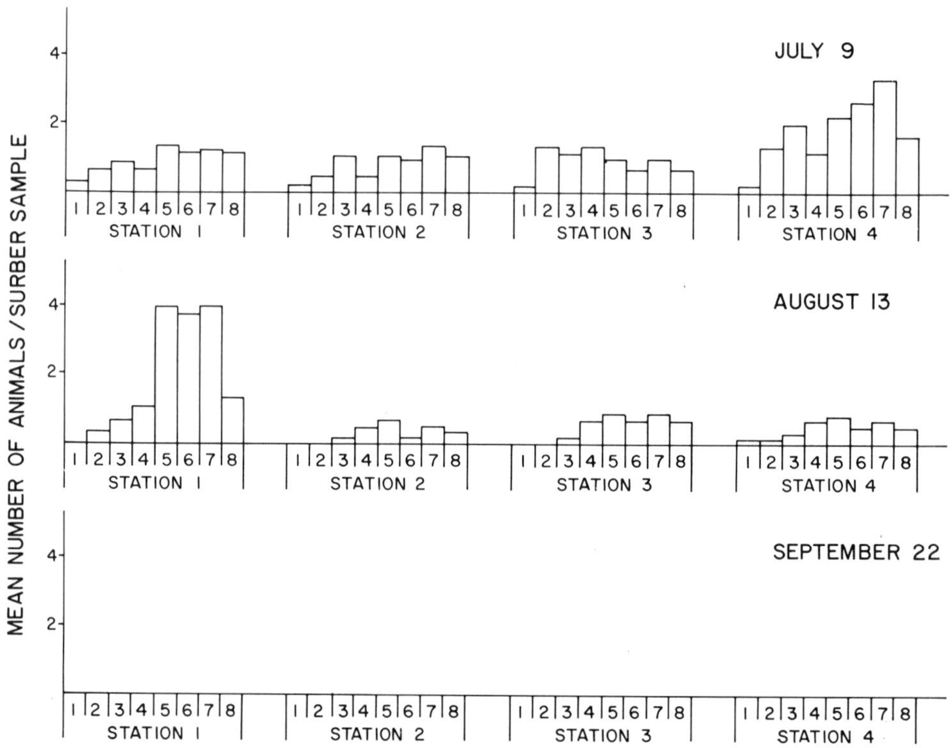

Figure 3. The mean density of *Baetis intercalaris* in each age cohort at each station during 1973.

1972) that the observed drift is a combination of 'constant drift' which is the result of involuntary dislodgement of individuals by the current and 'behavioural drift' which is made up of individuals who voluntarily enter the current because of population pressures. Apparently, the densities of *E. album, P. praepidita, S. i. canadense, S. nepotellum* and *H. maculipennis* were below their carrying capacities because they were rare in the drift although they were abundant in the sediments. Presumably there was very little behavioural drift. However, for *B. hageni* and *B. intercalaris*, the two most abundant species of mayflies in the Rat River, and for *P. myrsum*, the proportion of the population in the drift increased with the benthic desnity. This suggests that these species must have been near their carrying capacities. The high drift rate for these species was apparently due to behavioural drifting.

If the high drift rates for *B. hageni, B. intercalaris* and *P. myrsum* were the result of behavioural drift and if behavioural drift is caused by intraspecific competition, then there may have been a size selection for those individuals who were drifting. The Student's t analyses show that this is the case when the benthic density was relatively high (Table 1). The pronotum length was significantly greater for the *B. hageni* and *P. myrsum* in the drift and significantly less for the *B. intercalaris* with the exception of Station 3 on July 9, 1973. However, when the benthic density was relatively low there was no significant differences in size.

In summary, the benthic populations of the mayfly species used in this study seem to remain relatively stable over the summer months. As Waters (1961) had postulated, the proportion of the population in the drift is low as long as the benthic density is low. However, as the benthic density increases and presumably the carrying capacity is reached, the proportion of the population in the drift greatly increases. The fact that there was a size selection process taking place when the drift rate was high suggests some kind of intraspecific competition. There was no evidence of a 'colonization cycle' as Muller (1954) had proposed.

ACKNOWLEDGMENTS

I wish to thank Dr. R.H. Green for his assistance in analysing the data and Dr. G. Vascotto for his help in setting up the sampling program. I appreciate the critical reviews of the text by Ms. B. DeMarch, Dr. D. Rosenberg and Mr. A. Wiens. This study was funded by a grant from the National Research Council to Dr. R.H. Green.

RESUME

L'étude avait pour but de vérifier si le 'cycle de colonisation' proposé par Muller ou la théorie de la 'concurrence entre les espèces' soutenue par Waters expliquait adéquatement pourquoi la densité de la population de huit espèces d'éphéméroptères dans la Rat River (Manitoba) demeurait stable en dépit de taux élevés de déplacement. On observa deux étés durant les habitudes de vol des adultes, la densité des nymphes dans le substrat ainsi qui les taux de déplacement des éphéméroptères. Les résultats corroborent l'hypothèse de Waters selon laquelle le déplacement constitue un mécanisme de réduction de la densité des populations quand la capacité de charge est atteinte. Les mouvements de la population sont proportionnels à sa densité alors que les grands déplacements sont affaire de sélection liée à leur taille. Rien ne permet de conclure à un 'cycle de colonisation' tel quie le propose Muller.

ZUSSAMENFASSUNG

Mit der vorliegenden Arbeit sollte entschieden werden, ob Müllers 'Kolonisationszyklus' oder Waters 'Theorie der intraspezifischen Konkurrenz' hinreichen erklären, warum die Populationsdensitat acht verschiedener Ephemeropterenarten im Rat River (Manitoba) trotz hoher Driftraten stabil bleiben. Die flugpattern der voll ausgewachsenen Tiere, die Dënsitat der Nymphen im Substrat und die Driftraten wurden zwei Sommer hindurch beobachtet. Die Ergebnisse bestätigten Waters Hypothese, daß das Driften ein Mechanismus zur Reduzierung der Populationsdensität ist, sobald die Tragkapazität erreicht wurde. Die Proportion der Population beim Driften war densitätsabhängig und bei hohen Driftraten größenselektiv. Es gab keinen sicheren Beweis eines 'Kolonisationszyklus', wie ihn Müller vermutet hatte.

REFERENCES

Bishop, J.E. and H.B.N. Hynes. 1969. Upstream movements of the benthic invertebrates in the Speed River, Ontario. *J. Fish. Res. Board Can.* 26: 279-298.

Elliott, J.M. 1967. The life histories and drifting of the Plecoptera and Ephemeroptera in a Dartmoor stream. *J. Anim. Ecol.* 36: 343-362.

Elliott, J.M. 1971. Upstream movements of the benthic invertebrates in a Lake District stream. *J. Anim. Ecol.* 40: 235-252.

Flannagan, J.F. 1973. Sorting benthos using a floatation media. *Fish. Res. Board Can. Tech. Rep.* 354: 1-14.

Gyselman, E.C. 1976. The mechanisms which maintain population stability of selected species of Ephemeroptera in a temperate stream. MSc. Thesis, University of Manitoba, Winnipeg, Canada.

Hultin, L., B. Svensson and S. Ulfstrand. 1969. Upstream movements of insects in a South Swedish small stream. *Oikos* 20: 553-557.

Madsen, B.L., J. Bengtsen and I. Butz. 1973. Observations on upstream migration by imagines of some Plecoptera and Ephemeroptera. *Limnol. Oceanogr.* 18: 678-681.

Minckley, W.L. 1964. Upstream movements of *Gammarus* (Amphipoda) in Doe Run, Meade County, Kentucky. *Ecology* 45: 195-197.

Müller, K. 1954. Investigations on the organic drift in North Swedish streams. *Inst. Freshw. Res. Drottningholm Rep.* 35: 133-148.

Ulfstrand, S. 1968. Benthic animal communities in Lapland streams. *Oikos Suppl.* 10: 1-120.

Waters, T.F. 1961. Standing crop and drift of stream bottom organisms. *Ecology* 42: 532-537.

Waters, T.F. 1965. Interpretation of invertebrate drift in streams. *Ecology* 46: 327-334.

Waters, T.F. 1972. The drift of stream insects. *Annu. Rev. Entomol.* 17: 253-272.

IMPORTANCE OF BEHAVIOUR TO THE RE-ESTABLISHMENT OF DRIFTING EPHEMEROPTERA

Jan J.H. Ciborowski* and Lynda D. Corkum*

Department of Zoology
Erindale College
University of Toronto

ABSTRACT

Experiments with mayfly larvae in a laboratory stream revealed significant differences among settling capabilities of four species. Quantitative measures of overall settling capacity, contribution of morphology and that of behaviour were derived for comparative purposes. Importance of behaviour to re-establishment was greater in species which frequently show diel drift periodicities than those which do not. Although morphological differences between species were large, mean size of nymphs was more important in determining rates of return of dead animals. Intraspecific variations in the ability of drifting individuals to become re-established may partially account for changes in the relative distribution of populations over time.

INTRODUCTION

Several workers have investigated the pattern in which invertebrates released into running water return to the substrate (Madsen 1968, McLay 1970, Elliott 1971, McKone 1975). All have found that a fixed proportion of animals introduced into the water column settle to the bottom for every unit of time or distance which passes. The greater the current velocity, the longer

*Present address: Zoology Department, University of Alberta
Edmonton, Alberta, Canada. T6G 2E9

re-establishment takes or the farther the animals travel before all have settled. Elliott (1971) observed that live animals settle more quickly than dead individuals, and that settling capabilities differ between species. Elliott attributed this to behavioural dissimilarities since settling rates did not vary among dead individuals of different species.

Our study was conducted to evaluate settling rates of four mayfly species and to estimate the relative importance of behaviour to the settling ability of each.

METHODS

All experiments were conducted in an artificial stream which was housed within an environmental chamber maintained at 5°C. The elliptical stream was constructed of Plexiglass and powered by an Archimedes' screw (Ciborowski *et al.* 1977). Individuals released in the stream settled and remained only where a 15 x 30 cm area of sieved gravel (16-23 mm) was placed on the stream bottom. Water depth was 28.5 cm.

The settling ability of four widespread species of Ephemeroptera was tested. *Baetis vagans* McDunnough, *Ephemerella subvaria* McDunnough, *E. invaria* (Walker) and *Paraleptophlebia mollis* (Eaton) were all common or abundant at a collection site near the Forks of the Credit, Peel County, Ontario. Experiments were conducted with *E. subvaria* nymphs in November, 1976 and with all four species in March, 1977. Experiments consisted of introducing 50 live nymphs into the stream and recording the number suspended within the water column at one-minute intervals until 90 percent of the individuals had settled. This was determined by counting the number of animals which passed a fixed point in the stream during a time interval required for a hypothetical individual to complete one circuit of the stream at a particular current velocity.

Independent trials were run for each species at mean current velocities of 10, 15, 20, 25, and 26.8 cm/sec. Three replicates were conducted at each velocity. Individuals from each trial were killed in Kahle's fluid and experiments were repeated with the dead insects. Mean head widths of the four species at time of testing are given in Table 1.

RESULTS

Rates of settling for each trial were determined by using regression analysis to predict the relationship between the natural logarithm of the number of animals no longer in suspension and the estimated number of times individuals had passed over the substrate.

Table 1. Significance of regressions and analyses of covariance of ${}_rS$ against current velocity for 4 mayfly species. Equations taken the form: ${}_rS = A\cdot(\text{current velocity})^{-B}$.

SPECIES		A	B (95% C.I.)	$F_{reg.}$	$F_{cov.}$	MEAN HEAD WIDTH (mm) (95% C.I.)
E. subvaria (Nov.)	Live	23.96	1.9583±0.2302	733.02***	45.79***	1.62±0.04
E. subvaria (Nov.)	Dead	12.05	2.0638±0.3814	296.45***		
E. subvaria (Mar.)	Live	12.08	1.8678±1.2371	23.08*	9.50**	1.99±0.04
E. subvaria (Mar.)	Dead	0.22	0.6427±0.5688	12.93		
E. invaria	Live	4.03	1.5631±0.5380	85.46**	23.89***	1.12±0.03
E. invaria	Dead	0.22	0.7550±0.5886	16.66*		
B. vagans	Live	2.81	1.2583±0.7272	30.32	429.57***	0.80±0.03
B. vagans	Dead	0.27	1.0386±0.6250	27.96*		
P. mollis	Live	1.54	1.5232±0.2981	264.35***	38.46***	0.74±0.02
P. mollis	Dead	1.31	1.6519±0.6793	59.87**		
All species	Live				65.51***	
All species	Dead				27.54***	

* P <0.05　　** P <0.01　　*** P <0.001

The slope of each regression was the instantaneous settling rate ($_rS$) for that trial.

Replicated least-squares regression analysis was then used to test the relationship between $_rS$ and current velocity for each species. Logarithmic transformations of both variables gave the best regressions in all cases. Slopes of the regressions for live and dead animals of each species were significantly different (analysis of covariance, Table 1).

Visual examination of the curves (Fig. 1) suggests that some species were more capable that others of returning to the substrate at any given current velocity. Analysis of covariance indicated that there were significant differences among species in both live and dead insect trials (Table 1). *B. vagans* and *E. subvaria* (Nov.) had the highest curves (indicating most rapid settling capabilities), *E. subvaria* (March) and *E. invaria* were intermediate, and *P. mollis* was the least able to settle. These trends can be better elucidated by determining the area beneath the curve for each species (integrals of the regression equations). This provides a numerical index of the insects' capabilities of settling over the entire range of current velocities tested, which we call the settling capacity (SC).

Similarly, by calculating the area beneath the curve for dead insect trials of each species, one can determine that portion of the SC due to morphological characters (MC). The area between these two curves (SC - MC) must therefore be the contribution of behaviour to the SC. We term this the behavioural scope for settling (BC) after Fry (1947) who used a similar method to determine physiological tolerances of fishes.

These values permit calculation of the relative contribution of behaviour to the SC of an organism, or behavioural index (BI), i.e. BI = BC ÷ SC.

Table 2 shows the derived values for the four species tested, as well as values calculated from field $_rS$ data of Elliott (1971). Elliott pooled the regressions of his dead invertebrate trials and all indices from his data have been determined using this single value.

Figure 1. Relationships between $_rS$ and current velocity for live (circles: upper curves) and dead (squares: lower curves) individuals of 4 mayfly species. Each point represents geometric mean of three trials.

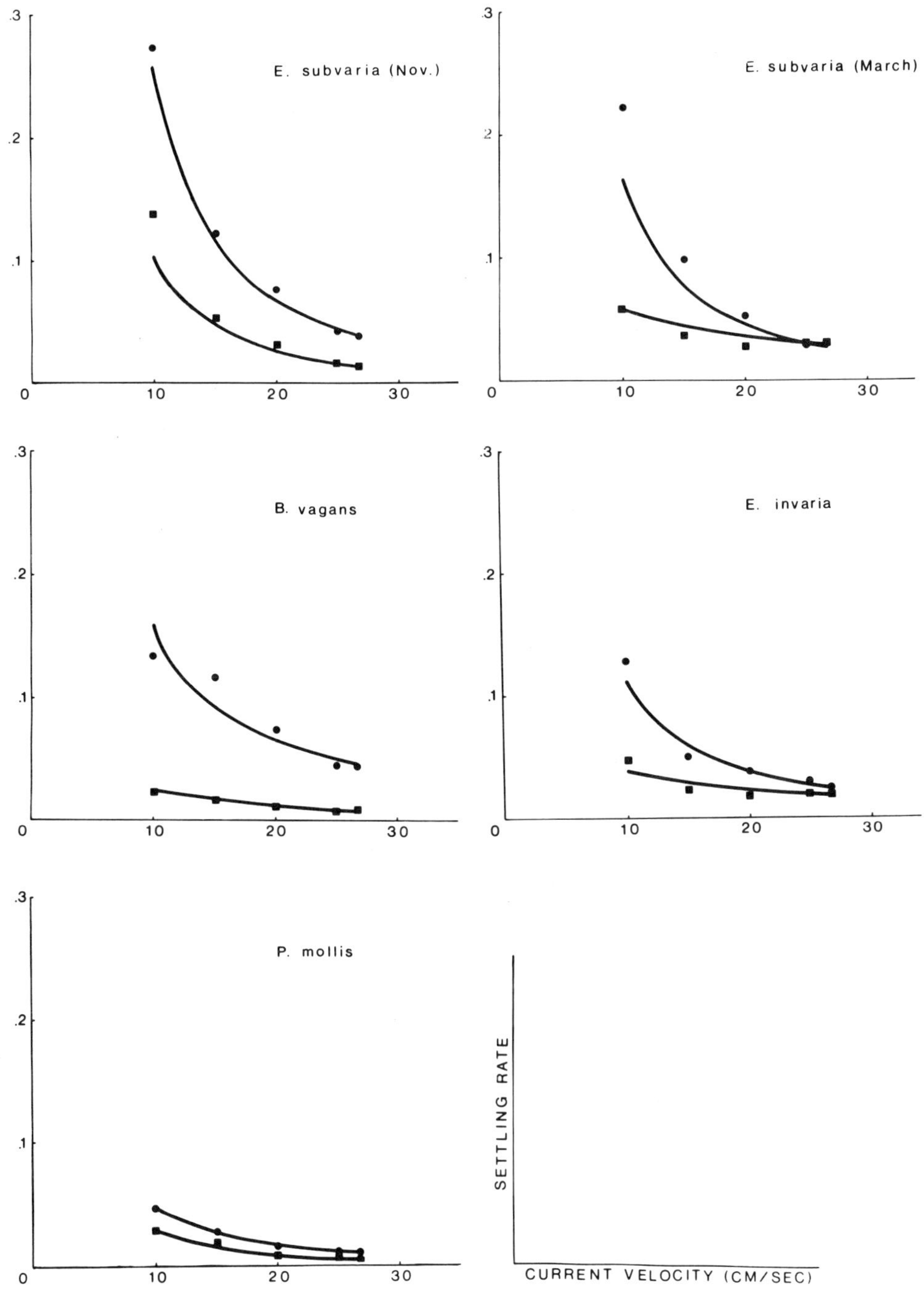
E. subvaria (Nov.)
E. subvaria (March)
B. vagans
E. invaria
P. mollis
.3
.2
.1
0
10
20
30
SETTLING RATE
CURRENT VELOCITY (CM/SEC)

Figure 1

Table 2. Measures of settling capabilities of various species determined from this study and from Elliott (1971); and frequency of diel drift periodicities for North American genera of aquatic invertebrates.

SPECIES	SC	MC	BC	BI	REL. FREQ. OF DIEL DRIFT %	N
This study:						
Baetis vagans	1.35	0.24	1.11	.82	53	40
Ephemerella subvaria (Nov)	1.68	0.64	1.05	.62	47	28
Ephemerella subvaria (Mar)	1.09	0.62	0.47	.43	47	28
Ephemerella invaria	0.83	0.41	0.42	.51	47	28
Paraleptophlebia mollis	0.36	0.21	0.15	.41	25	12
Elliott (1971):						
Amphipoda						
Gammarus pulex	19.12	5.49	13.63	.71	73	11
Ephemeroptera						
Baetis rhodani	18.84		13.34	.71	53	40
Ecdyonurus venosus	10.87		5.37	.49	33*	9
Rhithrogena semicolorata	6.27		0.78	.12	50	6
Plecoptera						
Protonemoura meyeri	6.32		0.83	.13	31**	61
Chloroperla spp.	6.28		0.79	.13	31**	61
Leuctra spp.	6.20		0.71	.11	31**	61
Coleoptera						
Elmidae	5.53		0.04	.01	54	13
Trichoptera						
Agapetus fuscipes	5.49		0.00	.00	20***	5
Diptera						
Simulium	6.07		0.57	.09	18	28
Chironomidae	5.48		0.00	.00	8	52

* for *Stenonema* spp. ** for all Plecoptera *** for *Glossosoma* spp.

To provide a comparative measure of the relationship between BI and behavioural drift tendencies of the taxa studied, we undertook a search of the North American drift literature. We tabulated by genus the total number of published accounts of taxa taken in 56 drift studies. The percentage of these citings for which diel drift periodicities were reported was used as an index of how prevalent behavioural drift is in that genus (Table 2). A Spearman's rank correlation test was performed and showed a significant association ($r = 0.55$, $P < 0.05$) between BI and the relative frequency of diel drift periodicities.

DISCUSSION

There appears to be a general correlation between BI and the reported tendency for the various organisms studied to show diel drift periodicities. *Baetis* spp. and *Gammarus* frequently exhibit strong behavioural drift patterns. Our data indicate that both have high BI and among the greatest overall SC (Table 2). Other animals (*Ephemerella, Ecdyonurus, Paraleptophlebia)* are intermediate both in BI and in their reported tendency to drift, whereas Plecoptera, Simuliidae, Chironomidae and Elmidae have conspicuously lower indices and, with the exception of the elmids, are less likely to show diel periodicities (Adamus and Gaufin 1976). Thus, species most prone to drift consistently may also be those best able to remove themselves from the drift. This suggests that behavioural drift may be more than the result of accidental displacement from vulnerable positions on the substrate (Stein 1972).

Our estimates for *E. subvaria* in November and March suggest that SC decreased through time, even though MC remained virtually unchanged. In March, the curve for live and for dead insects intersected at about 26 cm sec^{-1}, and at this point a live insect is no better at settling than its dead counterpart. This contrasts with a much greater estimated point of intersection for the November experiments. We would predict that, given an equal propensity to drift, older and large *E. subvaria* nymphs would tend to become more concentrated in slower water areas than young smaller larvae (perhaps reducing mortality during emergence). In fact, older *E. subvaria* in the laboratory are considerably more prone to drift than are young ones (Ciborowski 1979). Kovalak (1978) however found no differences in relative sizes of *E. subvaria* nymphs among fast and slow water areas in the Pigeon River, Wisconsin; and greater numbers were consistently located in the fast water area at all times of the year. Perhaps differences, if they exist, would be more apparent along a stream margin-to-centre gradient than along an upstream-downstream gradient (Waters 1972).

The four mayfly species which we examined were of widely different morphological forms. *B. vagans* is a streamlined animal,

E. subvaria and *E. invaria* are thick-bodied, and *P. mollis* is dorsoventrally flattened. We had initially hypothesized that such variety in body shape might differently affect the contribution of MC to the SC of each species. However, variations in MC may be almost completely the result of dissimilarities in the size rather than the shape of the animals tested. MC appeared to be closely correlated ($r = 0.96$, $P < 0.05$) to the mean head width of the animals tested. This relationship, if valid for all aquatic invertebrates, is of significance in interpreting field drift data. Several workers (Clifford 1972, Stein 1972, Waters 1972, Reisen 1977) have reported on variations in the abundance of animals of different sizes in the drift. In instances where behaviour makes a minimal contribution to SC (e.g. Chironomidae, Simuliidae; Table 2) smaller individuals may, rather than being more prone to entrainment in the drift, simply take longer to resettle and are therefore more likely to be captured in a net than larger animals.

Our analysis of settling was concerned only with the influence of a single variable (current velocity) on the settling abilities of mayflies, but this type of analysis need not be so restricted. Luedtke and Brusven (1976), Corkum *et al.* (1977), and Walton (1978) have all shown that different substrate types can also influence the drift and (or) settling of invertebrates. By performing the appropriate series of experiments, substrate type could be added to the analysis as a second independent variable to provide a similar measure of SC and the importance of behaviour. Indeed, any number of parameters could be incorporated and indices defined to give an overall estimate of the importance of behaviour and morphology of a species to its re-establishment on the substrate. Such estimates would be quantitatively comparable to values obtained for other species examined in the same way. Furthermore, regression curves used to estimate indices can be integrated over selected intervals of the independent variables to assess the probability of animals becoming established in particular habitats. Thus, one could predict possible changes in the distribution of animals within a stream as a result of drift and subsequent resettling.

ACKNOWLEDGMENTS

We wish to thank Drs. P.J. Pointing, W.G. Sprules, H.F. Clifford and Mr. R.L. Baker, all of whom provided valuable suggestions and criticisms of this work. The research was partially funded by grants from the National Research Council of Canada and Erindale College to Dr. P.J. Pointing.

RESUME

Grâce à des expériences réalisées sur des larves d'éphéméroptères dans un cours d'eau de laboratoire, on a découvert que quatre espèces accusent des différences importants dans leur aptitude à s'établir. Des mesures quantitatives de la capacité moyenne d' adaptation ont été prises et l'apport de la morphologie et du comportement évalué aux fins de comparaison.

Le comportement joue un rôle plus important dans le rétablissement des espèces qui accusent des périodicités plus fréquentes que les autres des mouvements de dérivation par vingt-quatre heures. Bien qu'il y ait des différences appréciables de morphologie entre les espèces, la grosseur moyenne des larves acquiert une importance plus grande lorsqu'il s'agit de déterminer le taux de mortalité des insectes. Des variations au sein d'une même espèce dans la capacité de rétablissement des insectes pris individuellement peuvent expliquer en partie les changements qui se produisent dans la répartition relative des populations au cours du temps.

ZUSSAMENFASSUNG

Experimente mit Larven von Eintagsfliegen in einem Laboratoriumsgewässer ergaben bemerkenswerte Unterschiede in Bezug auf das Ansiedlungsvermögen von vier verschiedenem Arten. Zur Vergleichung herangezogen wurden quantitative Meßwerte der Gesamtansiedlungskapazität, morphologische Eigenarten und spezifische Verhaltensweisen. Es zeigte sich, daß die Fähigkeit zur Re- Etablierung für diejenigen Arten von großerer Wichtigkeit ist, die im Verlauf eines 24 Stunden Zyklus gewissen Veränderungen durch Abtriften unterworfen sind, als für solche, die davon unbeeinflußt bleiben. Obwohl große morphologische Unterschiede zwischen den einzelnen Arten bestanden, war doch die mittlere Größe wichtiger bei der Bestimmung der Raten zurückkehrender, toter Tiere. Intraspezifische Variationen der Fähigkeit driftender Einzeltiere, sich wieder anzusiedeln, sind wahrscheinlich zum Teill verantwortlich für Veränderungen in der relativen Verteilung der Populationen über eine bestimmte Zeitspanne hinweg.

REFERENCES

Adamus, P.R. and A.R. Gaufin. 1976. A synopsis of nearctic taxa found in aquatic drift. *Am. Midl. Nat.* 95: 198-204.

Ciborowski, J.J.H. 1979. The effects of extended photoperiods on the drift of the mayfly *Ephemerella subvaria* McDunnough (Ephemeroptera: Ephemerellidae). *Hydrobiologia* 62: 209-214.

Ciborowski, J.J.H, P.J. Pointing, and L.D. Corkum. 1977. The effect of current velocity and sediment on the drift of the mayfly *Ephemerella subvaria* McDunnough. *Freshwater Biol.* 7: 567-572.

Clifford, H.F. 1972. A years' study of the drifting organisms in a brown-water stream of Alberta, Canada. *Can. J. Zool.* 50: 975-983.

Corkum, L.D., P.J. Pointing, and J.J.H. Ciborowski. 1977. The influence of current velocity and substrate on the distribution and drift of two species of mayflies (Ephemeroptera). *Can. J. Zool.* 55: 1970-1977.

Elliott, J.M. 1971. The distances travelled by drifting invertebrates in a Lake District stream. *Oecologia* 6: 350-379.

Fry, F.E.J. 1947. Effects of the environment on animal activity *Univ. Toronto Stud. Biol. Ser.* 55, *Publ. Ont. Fish. Res. Lab.* 68: 1-62.

Kovalak, W.P. 1978. Relationships between size of stream insects and current velocity. *Can. J. Zool.* 56: 178-186.

Leudtke, R.J. and M.A. Brusven. 1976. Effects of sand sedimentation on colonization on stream insects. *J. Fish. Res. Board Can.* 33: 1881-1886.

Madsen, B.L.A. 1968. A comparative ecological investigation of two related mayfly nymphs. *Hydrobiologia* 31: 337-349.

McKone, W.D. 1975. Quantitative studies of stream drift with particular reference to the McLay model. Ph.D. Thesis, Univ. British Columbia, 118 p.

McLay C. 1970. A theory concerning the distance travelled by animals entering the drift of a stream. *J. Fish. Res. Board Can.* 27: 359-370.

Reisen, W.K. 1977. The ecology of Honey Creek, Oklahoma: Population dynamics and drifting behaviour of three species of *Simulium* (Diptera: Simuliidae). *Can. J. Zool.* 55: 325-337.

Stein, I. 1972. The number and size of drifting nymphs of Ephemeroptera, Chironomidae and Simuliidae by day and night in the River Stranda, Western Norway. *Nor. Entomol. Tiddskr.* 19: 127-131.

Walton, O.E., Jr. 1978. Substrate attachment by drifting aquatic insect larvae. *Ecology* 59: 1023-1030.

Waters, T.F. 1972. The drift of stream insects. *Annu. Rev. Entomol.* 17: 253-272.

THE IMPORTANCE OF SPECIES ASSOCIATIONS AND SUBSTRATE TYPES TO BEHAVIOURAL DRIFT

Lynda D. Corkum and Hugh F. Clifford

Department of Zoology
University of Alberta
Edmonton, Alberta T6G 2E9

ABSTRACT

Laboratory experiments were conducted to study drift responses of two herbivorous mayflies and a carnivorous stonefly with different substrate types. Light was the only factor studied that influenced drift of immature *Baetis* sp.* nymphs. The predator, *Isogenoides elongatus* (Hagen), significantly influenced the drift of both mature *Baetis* sp. and *Leptophlebia cupida* (Say) nymphs. With the presence of artificial vegetation, *L. cupida* nymphs drifted in much lower numbers. The presence of *Leptophlebia* nymphs significantly influenced drift of *I. elongatus* specimens during dark periods. The two cohabiting herbivorous species showed no increased drift response to one another. It is suggested that during daylight hours there is a greater risk of predation for mature *Baetis* nymphs to enter the water column than to remain on the stream bed.

INTRODUCTION

Several workers have shown that some organisms have a greater propensity to drift than others (Waters 1969, Lehmkuhl and Anderson 1972, Corkum 1978). Most drifting animals reported in the literature are detritivores or herbivores, whereas invertebrate predators are only infrequently present in the drift (Waters 1969, Adamus and Gaufin 1976). Although changes in light intensity significantly influence the periodicity of drifting organisms (Waters 1972), perhaps some portion of the drift fauna is also avoiding benthic invertebrate predation.

---

* *Baetis* sp. is now known to be *Baetis tricaudatus* Dodds.

Corkum (1978) has shown a difference between swimming and crawling ephemeropteran nymphs in their tendency to leave the substrate and enter the water column. In this study, we suggest that these behavioural groups have different methods or strategies of avoiding predators. Will swimmers or crawlers more likely drift or seek cover to avoid predation? Since the behaviour of a nymph may also change with size, mature and immature nymphs of the same species may avoid predators in different ways.

Hildebrand (1974) has shown that reduced food levels tend to increase drift in some mayfly species. Although mayfly nymphs are frequently associated with macrophytes (Barber and Kevern 1973), plant cover has not been adequately examined as a factor influencing drift. By utilizing different substrate types (food and cover) and different organisms associations (herbivores and predators), we attempted to determine the factors that most influence the drift of the mayflies *Baetis* sp. and *Leptophlebia cupida* (Say) and the stonefly *Isogenoides elongatus* (Hagen).

METHODS

The Pembina River, near Entwistle, Alberta (53°36'N, 115°00'W), served as a source of *Baetis* sp. and *I. elongatus* nymphs used in the laboratory experiments. There are few *L. cupida* nymphs in the Pembina River; hence, we collected these nymphs from the Bigoray River (53° 31'N, 115°26'W). The Bigoray River flows into the Pembina upstream from the Entwistle site. Nymphs were collected no earlier than 2 days before any experiment and nymphs were never utilized in more than one experiment. Predators were starved for 24 h before each run.

Experiments were conducted in an artificial stream housed in an environmental chamber. A motor-driven paddlewheel within the stream provided a current of 10 cm/s; water depth was 15 cm. Two substrate trays (each 28 x 17.5 x 2.5 cm) positioned in the stream contained one of four substrates: (1) stones (16-23 mm in diameter), (2) stones + food, (3) stones + cover, and (4) stones + food + cover. Prepared instant oatmeal (40 g) was rubbed over wet substrate surfaces just before the trays were inserted into the stream. This procedure ensured that food sources were restricted to the substrate trays. Artificial turf was used for cover.

Nymphal sample sizes were 100 *Baetis* sp., 100 *L. cupida* and 10 *I. elongatus*. Depending on the trial, nymphs of one or more species were poured through a funnel into the raceway upstream from the substrate trays. The nymphs suspended in the water column were counted for 10 s each minute for 45 min as they passed a reference point in the stream. Afterwards, a net with attached jar was inserted immediately downstream from the substrate trays. For the duration of the experiment (3 h of light, followed by 3 h of dark), the net was examined at 30-min intervals, at which time any drifting organisms were removed and preserved. Later, the foregut of each predator was examined.

From 28 September to 22 October 1977 (autumn experiments), experiments were carried out with immature *Baetis* sp. (Clifford's 1970, stage II nymphs) using a 2 x 2 x 2 x 2 factorial design (presence/absence of predators, food, cover and light). Three trials were conducted. Recordings of settling, however, were only entered for two trials. Water temperature was 5°C.

From 21 April to 25 May 1978 (spring experiments), experiments were conducted with mature mayflies (Clifford's 1970, stage III nymphs). The set-up for both *Baetis* and *L. cupida* was a 2 x 2 x 2 x 2 x 2 factorial design (presence/absence of predators, food, cover, light, and the second mayfly species). For *I. elongatus*, the set-up (2 x 2 x 2 x 2 x 2) represented the presence/absence of *Baetis* sp., *L. cupida*, food, cover and light. These spring experiments were performed twice. Water temperatures ranged from 5 to 10°C depending on field conditions.

RESULTS

Nymphal Number Suspended in the Water Column Versus Time

For all mature nymphs of *Baetis*, *L. cupida* and *I. elongatus*, 95% of each sample population settled onto all substrate types within 5 min of the nymphs being added to the stream. Immature *I. elongatus* nymphs (autumn experiments) also quickly settled onto the substrate. In contrast, immature *Baetis* nymphs tended to remain longer within the water column.

Regression equations were calculated for the natural logarithm of numbers of immature *Baetis* nymphs suspended in the water column versus time. In experiments both with and without predators, immature *Baetis* nymphs significantly differed in their ability to settle on the stream bed with different substrate types (Table 1). The nymphs settled more quickly when either food alone or food and cover were present on the stream bottom. It is likely that the immature *Baetis* nymphs actually became stuck to the oatmeal on the stream bed. With other substrate types, nymphs settled briefly, but then re-entered the water column. Immature *Baetis* nymphs were slowest in settling onto bare stones in the absence of a predator. For the other substrates, there was no significant difference in settling times with or without predators on the stream bed (Table 2).

Table 1. Significance of regression and analysis of covariance of the number of immature *Baetis* nymphs suspended in the water column versus time for all substrate types, with and without predators.

Equation format: $Y = Ae^{-BX}$

Factors*	A	B (± 95% conf.)	Reg (p)	Covariance Test of Slopes F_s	*p*
IBOS	30.81	0.171 ± 0.666	< 0.001		
IBOSF	43.54	0.761 ± 0.317	< 0.01	10.584	< 0.001
IBOSC	35.96	0.236 ± 0.121	< 0.01		
IBOSFC	59.46	0.783 ± 0.302	< 0.05		
IBPS	52.18	0.344 ± 0.114	< 0.001		
IBPSF	56.32	0.960 ± 0.360	< 0.05	4.950	< 0.05
IBPSC	46.89	0.423 ± 0.179	< 0.01		
IBPSFC	48.48	0.594 ± 0.263	< 0.05		

*Code for factors:
IBO: Immature *Baetis* with no predator
IBP: Immature *Baetis* with predator
S: stones
SF: stones + food
SC: stones + cover
SFC: stones + food + cover

Table 2. Significance of analysis of covariance of immature *Baetis* with and without predators for each substrate type.

Factors*	Covariance Test of Slopes F_s	*p*	Test of Intercepts F_s	*p*
IBOS & IBPS	5.840	< 0.05	0.504	n.s.
IBOSF & IBPSF	0.648	n.s.	0.224	n.s.
IBOSC & IBPSC	2.693	n.s.	1.676	n.s.
IBOSFC & IBPSFC	0.844	n.s.	0.440	n.s.

*See footnote of Table 1 for explanation of factors.

Drift Experiments

A 4-way analysis of variance (ANOVA) was used to determine the influence of predators, cover, food and light on the drift of immature *Baetis* nymphs. There were significantly more nymphs caught in the drift net during daylight hours ($p < 0.001$). No other factors were significant. These results were consistent even when we eliminated the first half hour of the light period, which probably contained nymphs still suspended in the water column.

A 5-way ANOVA was used to determine the factors influencing the drift of mature *Baetis*, *L. cupida* and *I. elongatus* nymphs (Table 3). The presence of the stonefly predator, *I. elongatus*, significantly increased the drift of *Baetis* ($p < 0.001$) and, to a lesser extent, *L. cupida* ($p < 0.05$). When stoneflies were present, more than twice as many *Baetis* nymphs drifted than did nymphs of *L. cupida*. When both herbivorous species were present on the stream bed at the same time, neither species influenced the drift of the other. Even the active crawling behaviour of *L. cupida* did not significantly affect the numbers of drifting *Baetis* nymphs. In trials with both *Baetis* and *L. cupida*, 100 nymphs of each species were present in the stream. Even though the benthic density was doubled, no increase in drift for either species occurred.

The absence of vegetative cover significantly increased the numbers of drifting *L. cupida* nymphs ($p < 0.001$). Even though predators significantly influenced the drift of *L. cupida* from all substrate types, fewer *L. cupida* nymphs drifted from substrates with vegetation.

Both the presence of the crawling mayfly *L. cupida* and the absence of light significantly influenced the drift of the stonefly predator ($p < 0.001$). Various second, third and fourth order interactions also significantly affected the drift of the three species (Table 3). Food alone had no effect on the drift of any species.

Foregut Analysis

Foregut analysis of *I. elongatus* from laboratory studies showed that the predator ate nymphs of each mayfly species under both light and dark conditions. To determine which species was most often captured, only those predators still remaining on the substrates at the end of each experiment were analysed statistically. Since the total surface area of mature *L. cupida* nymphs is 2.8 times that of mature *Baetis* nymphs, the larger species would be three times more likely to be encountered than the smaller one (data on total body lengths and head capsule widths are given in Table 4). From chi-square analysis, we cannot reject this 3:1 ratio of predator encounter with the two mayfly species. There might be a tendency for this tactile predator to have more success in capturing crawling nymphs (*L. cupida*).

Table 3. Significant factors and their interactions influencing the drift of mature *Baetis* (3A), *Leptophlebia cupida* (3B), and *Isogenoides elongatus* (3C).

Table 3A - *Baetis* sp. Leach
(5-way ANOVA: *L. cupida*, *I. elongatus*, Cover, Food, Light)

Factor	df	SS	MS	F	*p*
I. elongatus	1	2.64736	2.64736	58.441	< 0.001
I. elongatus x Food	1	0.27262	0.27262	6.018	< 0.025
Cover x Light	1	0.22806	0.22806	5.034	< 0.05
L. cupida x Cover x Light	1	0.28259	0.28259	6.238	< 0.025
L. cupida x *I. elongatus* x Light	1	0.26402	0.26402	5.828	< 0.025
I. elongatus x Cover x Light	1	0.36116	0.36116	7.973	< 0.01
L. cupida x *I. elongatus* x Cover x Light	1	0.31043	0.31043	6.853	< 0.025
Error	31	1.40441	0.04530		
Total	63	6.92894	0		

Table 3B - *Leptophlebia cupida* (Say)
(5-way ANOVA: *Baetis* sp., *I. elongatus*, Cover, Food, Light)

Factor	df	SS	MS	F	*p*
I. elongatus	1	0.43992	0.43992	4.425	< 0.05
Cover	1	2.70563	2.70563	27.217	< 0.001
I. elongatus x Food	1	0.56063	0.56063	5.640	< 0.025
I. elongatus x Food x Cover	1	0.61363	0.61363	6.173	< 0.025
Error	31	3.08172	0.09941		
Total	63	6.92894	0		

Table 3C - *Isogenoides elongatus* (Hagen)
(5-way ANOVA: *Baetis* sp., *L. cupida*, Cover, Food, Light)

Factor	df	SS	MS	F	*p*
L. cupida	1	0.56318	0.56318	16.405	< 0.001
Light	1	2.32047	2.32047	67.593	< 0.001
Baetis x *L. cupida*	1	0.27615	0.27615	8.044	< 0.01
Baetis x Light	1	1.02115	1.02115	29.745	< 0.001
L. cupida x Light	1	0.22675	0.22675	6.605	< 0.025
Error	31	1.06420	0.03433		
Total	63	6.44787	0		

After being touched by the antenna of a stonefly, *Baetis* nymphs frequently swam away.

DISCUSSION

It has been established that live aquatic insect larvae readily return to the stream bed in daylight hours (Luedtke and Brusven 1976, Walton 1978, Ciborowski and Corkum *this volume*). Our investigation demonstrated that small *Baetis* nymphs settled more slowly onto the stream bed than did larger nymphs of the same species. Mature *Baetis* and *L. cupida* nymphs settled so quickly that it was not possible to detect any substrate influence with the data collecting procedures that we used. Small nymphs settled more quickly when food was present on the stream bed. With the exception of immature *Baetis* settling onto bare substrates, our study and that of Walton's (1978) indicate that a stonefly predator had no influence on the settling behaviour of mayflies.

Our study clearly shows that tactile invertebrate predators such as *I. elongatus* can influence the drift of two species of mature herbivorous mayflies. The swimming form, *Baetis*, drifted in significantly greater numbers than the crawling form, *L. cupida*. *Leptophlebia* evidently prefers substrates with vegetation for cover. Our experiments also showed that the stonefly predator had no effect on immature *Baetis*. It is not known what influence stonefly predators would have if larger numbers of small *Baetis* had occupied the stream bed. Larger prey, however, are more likely to be detected by foraging predators and also provide a greater return per search than smaller prey (Siegfried and Knight 1976). Siegfried and Knight (1976) showed that the stonefly predator *Acroneuria californica* Banks selected large size classes of both chironomid larvae and baetid nymphs from the Squirrel River, California.

Table 4. Nymphal measurements (head capsule width and total body length) for representative specimens used in the autumn, 1977 and spring, 1978 laboratory experiments. (Values are in mm).

Species	Date	n	H.C.W. $\bar{x} \pm$ s.d.	T.B.L. (cerci & antennae excluded) $\bar{x} \pm$ s.d.
Baetis sp.	Oct. 1977	50	0.545 ± 0.054	2.713 ± 0.430
I. elongatus	"	50	2.523 ± 0.342	12.550 ± 2.221
Baetis sp.	Apr. - May 1978	50	0.967 ± 0.092	5.655 ± 0.804
L. cupida	"	50	1.812 ± 0.287	8.535 ± 1.644
I. elongatus	"	50	3.157 ± 0.321	18.760 ± 2.389

Baetis and *L. cupida* nymphs may respond differently to the presence of each other than they do to predators. No increase in drift was observed for either herbivore when the second mayfly species was present even though the nymphal numbers on the stream bottom were doubled. It is likely that a large actively foraging carnivore having long antennae and cerci would encounter mayflies more often than either mayfly species would encounter each other. In drift-density studies of a single herbivorous species, it is often assumed, though not always stated, that less abundant cohabiting species do not influence the drift of the species being investigated. Our study shows that this assumption can be accepted even when the cohabiting herbivores are equal in density.

There have been relatively few studies investigating the influence of riverine insect predators on the behaviour of prey species. Walton (1978) showed that the stonefly *Acroneuria abnormis* (Newman) had no effect on the drift of the mayfly *Stenonema interpunctatum* (Say). This mayfly, like other heptageniids, typically clings to the undersides of rocks and therefore would not be expected to avoid benthic predators by entering the water column. In another study, Corkum (unpublished data) found that the stonefly predator *Paragnetina media* (Walker) significantly influenced the drift of mature *Baetis vagans* McDunnough at night. Hildrew and Townsend (1977) showed that differences in predation rates of the caddisfly *Plectrocnemia conspersa* (Curtis) corresponded to differences in behavioural responses of the prey once prey entered the predator's net.

Predation is probably the greatest factor accounting for mortality of benthic insect populations (Benke 1978). Rheophilic mayflies are vulnerable to both visual and tactile predators. Visual predators (i.e. fish) are probably more successful during the day and capture insects primarily within the water column, whereas tactile predators hunting on the substrate are effective both day and night. Immature baetids present within the water column are unlikely to be detected by visual predators because of the small size and transparency of these mayflies (Allan 1978). Thus, immature baetids may drift at any time to avoid benthic tactile predators and not risk capture by visual predators. Mature drifting mayflies, however, could be highly vulnerable during the day and thus may use drift avoidance behaviour to tactile predators only at night. Charnov *et al.* (1976) showed that 80% of baetid nymphs occupied dark corners of an aquarium whenever a fish predator, *Oncorhynchus nerka*, was introduced, compared to only 20% occupying dark corners when the predator was absent. The *Baetis* sp. of our experiments were similar to *B. vagans* in their thoracic and abdominal pigmentation. The lateral view of these mature nymphs appears to mimic the eyes of a much larger invertebrate. This pigmentation pattern may in part protect *Baetis* from visual benthic predators. Nevertheless, it is evidently a greater risk for mature *Baetis* to enter the water column during daylight hours than to remain on the stream bed.

ACKNOWLEDGMENTS

We thank Eric R. Whiting for his diligence in collecting nymphs for the spring experiments. Jan J.H. Ciborowski contributed helpful suggestions throughout the study. This work was supported by a grant from the Boreal Institute for Northern Studies through the Alberta Grants-in-Aid Program for northern research.

RESUME

Des expériences ont été effectuees au laboratoire avec différents types de substrat pour étudier les déplacements de deux éphéméroptères herbivores et d'une plécoptérère carnivore. La lumière, l'unique facteur étudié, provoqua le déplacement de jeunes larves de l'espèce *Baëtis*. Le prédateur appelé *Isogenoides elongatus* (Hagen) agit puissamment sur les larves de l'espèce *Baëtis* et de l'espèce *Leptophlebia cupida* (Say) qui ont atteint le stade de maturité. En présence de végétations artificielles, les larves de l'espèce *L. cupida* se déplacèrent en nombre beaucoup moins considérable. La présence de larves de *Leptophlebia* agit fortement sur le déplacement de représentants de l'espèce *I. elongatus* en pleine obscurité. Les deux espèces herbivores associées ne jouent aucun rôle sur le déplacement d l'une ou l'autre. Il semble qu'au cours de la journée, les larves de l'espèce *Baëtis* au stade de maturité courent un plus grand risque d'être la proie de prédateurs si elles s'aventurent dans la colonne d'eau que si elles restent sur le fond du cours d'eau.

ZUSSAMENFASSUNG

Laboratoriumsexperimente wurden durchgeführt, um die Drift Reaktionen zweier pflanzenfressender Eintagsfliegen und einer fleischfressenden Steinfliege mit verschiedenen Substrat-Typen zu studieren. Licht war der einzige untersuchte Faktor, der die Drift unreifer *Baetis* sp. Nymphen beeinflußte. Der Preadator, *Isogenoides elongatus* (Hagen), hatte entscheidenden Einfluß auf die Drift der beiden reifen *Baetis* sp. und *Leptophlebia cupida* (Say) Nymphen. Bei kunstlicher Vegetation drifteten die *L. cupida* Nymphen in viel geringerer Zahl. Die Gegenwart von *Leptophlebia* Nymphen hatte großen Einfluß auf die Drift von *I. elongatus* Exemplaren während der Dunkelperioden. Die beiden ko- habitierenden, pflanzenfressenden Arten zeigten keine Drift Reaktion zueinander. Man vermutet, daß bei Tageslicht für reife *Baetis* Nymphen eine größere Gefahr der Praedation besteht, wenn sie sich in die Wassersäule begeben, als wenn sie im Flußbett bleiben.

REFERENCES

Adamus, P.R. and A.R. Gaufin. 1976. A synopsis of nearctic taxa found in aquatic drift. *Am. Midl. Nat.* 95: 198-204.

Allan, J.D. 1978. Trout predation and the size composition of stream drift. *Limnol. Oceanogr.* 23: 1231-1237.

Barber, W.E. and N.R. Kevern. 1973. Ecological factors influencing macroinvertebrate standing crop distribution. *Hydrobiologia* 43: 53-75.

Benke, A.C. 1978. Interactions among co-existing predators — a field experiment with dragonfly larvae. *J. Anim. Ecol.* 47: 335-350.

Charnov, E.L., G.H. Orians and K. Hyatt. 1976. Ecological implications of resource depression. *Am. Nat.* 110: 247-259.

Ciborowski, J.J.H. and L.D. Corkum. 1979. Importance of behaviour to the re-establishment of drifting Ephemeroptera. *This volume.*

Clifford, H.F. 1970. Variability of linear measruements throughout the life cycle of the mayfly *Leptophlebia cupida* (Say) (Ephemeroptera: Leptophlebiidae). *Pan-Pac. Entomol.* 46: 98-106.

Corkum, L.D. 1978. The influence of density and behavioural type on the active entry of two mayfly species (Ephemeroptera) into the water column. *Can. J. Zool.* 56: 1201-1206.

Hildebrand, S.G. 1974. The relation of drift to benthos density and food level in an artificial stream. *Limnol. Oceanogr.* 19: 951-957.

Hildrew, A.G. and C.R. Townsend. 1977. The influence of substrate on the functional response of *Plectrocnemia conspersa* (Curtis) larvae (Trichoptera: Polycentropodidae). *Oecologia* 31: 21-26.

Lehmkuhl, D.M. and N.H. Anderson. 1972. Microdistribution and density as factors affecting the downstream drift of mayflies. *Ecology* 53: 661-667.

Luedtke, R.J. and M.A. Brusven. 1976. Effects of sand sedimentation on colonization of stream insects. *J. Fish. Res. Board Can.* 33: 1881-1886.

Siegfried, C.A. and A.W. Knight. 1976. Prey selection by a setipalpian stonefly nymph, *Acroneuria (Calineuria) californica* Banks (Plecoptera: Perlidae). *Ecology* 57: 603-608.

Walton, O.E., Jr. 1978. Substrate attachment by drifting aquatic insect larvae. *Ecology* 59: 1023-1030.

Waters, T.F. 1969. Invertebrate drift-ecology and significance to stream fishes, p. 121-134, *In*: "Symposium on salmon and trout in streams", H.R. MacMillan lectures in fisheries, Vancouver. T.G. Northcote, ed., Univ. British Columbia, Vancouver, B.C.

Waters, T.F. 1972. The drift of stream insects. *Annu. Rev. Entomol.* 17: 253-272.

OBSERVATIONS ON THE USE OF THE RADIO-ISOTOPE ^{32}P IN THE STUDY OF FOOD UPTAKE BY SOME MAYFLIES AND OTHER BENTHIC MACRO-INVERTEBRATES IN A LABORATORY STREAM ECOSYSTEM

H.J. Schoonbee and J.H. Swanepoel

Research Group for Freshwater Biology
Department of Zoology
Rand Afrikaans University
P.O. Box 524
Johannesburg, 2000
South Africa

ABSTRACT

A laboratory model stream ecosystem was used to study the transfer of the radio-isotope ^{32}P in a food chain, from the water environment through the benthic algae and submerged vegetation to the primary and secondary consumer levels. This type of investigation, in which the nymphs of the mayflies *Baetis harrisoni, Neurocaenis discolor, Caenis* sp and *Euthraulus bugandensis* were used, revealed the value of ^{32}P (and possibly other isotopes) in the study of not only the rates and quantities of food taken up by mayfly nymphs but also the diurnal activities in feeding by these organisms.

INTRODUCTION

The value of isotopes in the study of the productivity of marine and freshwater environments, including the rate of uptake and accumulation of nutrients such as phosphates by aquatic plants has been demonstrated by a number of research workers (Rigler 1964; Lean and Nalewajko 1976). McRoy and Barsdate (1970) showed, by using radio-active phosphorus, that at least 33% of this isotope, absorbed by the roots of the marine macrophyte *Zostera marina*, was again released through its leaves to the surrounding water. Because of severe eutrophication of some rivers and lakes in South Africa and the excessive growths of certain water weeds resulting from it, Vermaak *et al.* (1976) also used ^{32}P to compare the relative uptake and accumulation of phosphates in nine common aquatic macrophytes.

Algal blooms and luxuriant growths of water weeds unfortunately distract from the fact that the entire food chain is affected by eutrophication. In the same way the utility of isotopes in the study of transfer of nutrients from the producer to the consumer levels in aquatic ecosystems has been largely overlooked.

In the present preliminary investigation, observations were made on the uptake of $^{32}PO_4$ by some selected plants and animals in a freshwater laboratory stream ecosystem.

THE LABORATORY STREAM

Because of the possible dangers involved in releasing radioactive phosphorus into a natural stream, an open air recirculating laboratory stream was constructed. It consisted of a 24.6 m long riffle section, 0.3 m wide and deep, interrupted by two pools each of 1.8 m diameter and 1 m depth. The galvanized steel plate framework of the stream system was lined with replaceable P.V.C. plastic. The stream, which has a capacity of 5300 L was filled with a mixture of tap water and water from a nearby river known to be rich in nutrients from a sewage works effluent. The addition of river water was considered necessary since a growth of algae was required for the algal feeding of the benthic macro-invertebrates which were introduced into the stream. A pump system was fitted to circulate the water continuously through the system. The pool sections were provided with soft bottom sediments whilst different 3 m sections of the riffles were provided with sandy and stony bottom habitats. Material for this was collected from nearby, unpolluted rivers.

ORGANISMS USED

The sago pond weed *Potamogeton pectinatus* was introduced into a 3 m section of the stream as well as in the two pond sections. In order to obtain easily removable algae, a number of glass slides, each 100 cm^2 in surface area,were placed at selected localities in the stream.

During the collection of stones from a nearby river (needed for the stony bottom habitat of the laboratory stream), all associated macro-invertebrate organisms were carefully removed and returned to the stones after their transfer to the laboratory stream. Organisms collected included Hirudinea, Chironomidae larvae, larvae of the caddisfly *Macronema* as well as nymphs of the Ephemeroptera *Euthraulus bugandensis*, *Neurocaenis discolor*, *Baetis harrisoni* and *Caenis* sp. The freshwater clam *Corbicula* sp. was introduced to a 3 m sandy bottom habitat similar to that from which it was originally removed. The mosquito fish *Aplocheilichthys johnstonii* was also released into the stream.

The pump was switched on and the system allowed to circulate for a week prior to the introduction of the isotope in order to allow the organisms to acclimatize.

Observations made during the one week period of acclimatization showed that most organisms survived and showed little signs of discomfort. There was also a marked development of algal growth in the stream itself and on the glass slides placed in the stream during this period.

STREAM CONDITIONS

Regular physical and chemical analyses were made of the stream water during the study. The results of parameters analyzed according to standard procedures for water and waste water (APHA 1971), showed conditions of the stream to be similar to those of clean, slightly enriched streams (Table 1). The water was well

Table 1. Physical and chemical analysis of water in the laboratory stream prior to the addition of the $^{32}PO_4$ isotope.

Parameters*	Mean Values
Temperature °C	22.0
pH	8.6-9.4
Conductivity μS/cm	453.0
Dissolved oxygen	8.1
B.O.D. (5-day)	2.0
Orthophosphate	0.253
Total phosphate	0.470
Ammonia N	0.050
Nitrate N	0.200

* mg/L unless otherwise mentioned

oxygenated throughout the study with little indication of severe microbial activity or much organic material in suspension in the water. The relatively high pH of the water could partly be attributed to the effects of algal growths which developed. It is also an indication of the alkalinity of the water. The values for phosphates, nitrate and ammonia compared closely with those of the river from which the material was collected for the laboratory stream.

METHODS

4.5 mCi of the carrier free isotope $^{32}PO_4$* (specific activity of 86 Ci/mg and mass of 52.3 ng) mixed with the organic dye fluorescein, was released at the head of the stream and the time noted for the water containing the dye and isotope to complete one full cycle of the stream. This was found to take 31 minutes. A further period of two hours was then allowed, i.e. approximately another four cycles of the laboratory stream, before the sampling period commenced. For the next 12 hours all organisms introduced into the stream were sampled every three hours. Two stones together with the algae and macro-invertebrates associated with it, were selected, at random, during each collection period for analysis.

Except for the algae on the stones and the glass slides, all other material collected was thoroughly rinsed with clean water, dried with absorbent paper and the wet biomass of each specimen determined. Samples were then digested in test tubes containining 10 mL 12N HNO_3 in an 80°C water bath. The contents of each test tube were then transferred to a counting vial and its radioactivity determined by Cerenkov counting in a Philips liquid scintillation counter model PW 4510. The required corrections to allow for quenching and the rate of decay of ^{32}P were made for each sample. Counts obtained were then converted to disintegrations per minute (dpm) per g or μg biomass of each organism analysed at a given time.

Since it was difficult to remove all the algae from the stones in the current and from the glass slides without the inclusion of other material, the biomass of the algae was determined according to the quantities of chlorophyll present per unit of stone surface area (as determined using aluminum foil). Algae from the stones and slides were scraped off with a scalpel and the chlorophyll extracted according to Golterman and Clymo (1969).

In order to eliminate the possible effects of colour quenching on samples during the Cerenkov counting process, one mL of

---

* Obtained from the Radiochemical Centre, Amersham

bleaching agent was added to each sample after centrifuging.

RESULTS

From the results on the ^{32}P activity of the organisms studied (Table 2) it is clear that the aquatic weed *P. pectinatus* was very efficient in accumulating the isotope. This plant showed its highest ^{32}P activity for the entire period only two hours after the release of the isotope into the stream. Fluctuations recorded in values of activity for this plant suggest some release of the isotope during day time with definite indications of a further higher release of ^{32}P after dark. The possible translocation of the isotope to the roots of *P. pectinatus* might also be partly resonsible for this decline in activity.

Only if the values of the activity of the algae are converted to dpm/g can one realize how important the benthic algae are in the removal of nutrients such as phosphate from streams. The activity of the algae on the slides was much higher than that on the stones and increased to reach a peak at 2000 h with some indication of a decline at 2300 h. The initial uptake of the isotope by algae on the stones was not only lower than that on the slides but showed a considerably lower level of increase in activity over the study period. Even so, these values, when converted to activity/g biomass, far exceeded those obtained for *P. pectinatus*. Peak values of 32-36 dpm/μg occurred during 2000-2300 h.

One of the first problems encountered during the sampling of macroinvertebrate organisms on the stones in the riffle, was that the removal of two stones per sampling period was not enough to collect all organisms studied. Since there were not enough stones placed into the stream originally, the sampling had to be limited to two stones per collection period. The Hirudinea showed ^{32}P activity for the first time 14 h after the release of the isotope. *Corbicula* sp., which is a filter feeder, already obtained food containing the isotope prior to 1700 h but like *Macronema* sp. did not take in as much food as the actively feeding mayflies and chironomids. The fact that *Corbicula* is a filter feeder might also explain the relatively low value in ^{32}P activity obtained for this species.

The mosquito fish, which feeds on insect larvae, (mainly Chironomidae) only commenced feeding after dark. The isotope activity of 21428 dpm/g at 2300 h suggests that this fish species, which was also collected at 1700 h and 2000 h (Table 2) only started feeding actively after 2000 hr.

Euthraulus bugandensis and *Neurocaenis discolor* were the only two mayfly species sampled during the first three sampling periods. Both already showed isotope activity at 1100 h. Values obtained

Table 2. Time of day, duration of experiment in hours, weather conditions and ^{32}P activity (dpm) for organisms over a period of 14 hours.

Time of Day	1100 h	1400 h	1700 h	2000 h	2300 h
Duration of experiment (hours)	2	5	8	11	14
Weather conditions	cloudy	cloudy	sunshine	night	night
	^{32}P activity of samples during each period in dpm				
FLORA					
Potamogeton pectinatus/g	86800	64250	78550	53850	47050
Algae on slides/μg	36	128	172	343	301
	76	289	283	359	-
	65	212	245	302	-
Algae on stones in current/μg	18	23	28	32	32
	20	17	20	31	36
FAUNA ON STONES IN CURRENT (dpm/g)					
Hirudinea	-	-	0	0	21428
Euthraulus bugandensis	3555	17419	30278	-	-
Neurocaenis discolor	3555	9837	29752	62004	62276
Caenis sp.	-	-	-	-	28360
Baetis harrisoni	-	-	-	92460	136363
Macronema sp.	-	753	-	-	2269
Chironomidae larvae	8288	25806	33026	143316	185660
Corbicula sp.	0	0	928	-	1539
Aplocheilichthys johnstonii	0	0	0	691	1733

over the first eight hours after the release of the isotope suggest that feeding activity in both species increased towards 1700 h and that, in the case of *N. discolor*, feeding activity which further increased after 1700 h, virtually stopped at 2300 h.

Only one sampling of *Caenis* sp. took place at 2300 h. The activity of 23260 dpm/g suggests that this species might not be such an active feeder during the period 1100 - 2300 h.

Baetis harrisoni is one of the most commonly found mayfly species in South African rivers and thrives under organically enriched conditions. Results on its ^{32}P activity, and thus its rate of food uptake suggests that this species is the most active feeder amongst the four species of mayflies studied. At 2300 h *B. harrisoni* had a ^{32}P activity twice that of *N. discolor*.

The caddisfly larvae, *Macronema*, had already obtained food when sampled at 1400 h. Isotope activity recorded for specimens collected at 2300 h suggests that this species was either not receiving sufficient food or did not feed as actively as the mayflies over the same period.

The Chironomidae, which include a large proportion of algal feeders,clearly showed the most active feeding of all macro-invertebrates especially after 1700 h. Values obtained on the ^{32}P activity for this group of organisms indicate that they are feeding mostly during the night.

OBSERVATIONS ON THE ALGAL/MACRO-INVERTEBRATE BIOMASS RATIOS OF ORGANISMS IN THE CURRENT HABITAT

From the results available on the biomass of algae and macro-invertebrates from stones collected during the sampling period, it was possible to calculate the ratio of the algal/macro-invertebrate biomasses for each individual stone. From Table 3 it is clear that the algal biomass (expressed as µg chlorophyll) varies considerably from one stone to another (Column A). There are also similar variations in the macro-invertebrate biomass values from the various stones (Column B). The ratio of algal/macro-invertebrate biomasses showed, almost without exception similar values for each set of stones (Column C). A comparison of values in Columns B and C further shows that there was a decline in macro-invertebrate biomass on the stones between 1100 h and 1700 h after which it increased again to reach maximum values of 13-20 at 2300 h. This implies that some macro-invertebrates leave the stones to feed somewhere else during that part of the day but that they gradually return again to the stones after 1700 h. Values in Column C further suggest that the macro-invertebrates tend to disperse evenly between stones in relation to available (algal) food present.

Table 3. Ratios of algal and macro-invertebrate biomasses collected from the same stones during the sampling period 1100 h - 2300 h.

(A = algal biomass in µg chlorophyll/cm^2 stone surface area;
B = macro-invertebrate biomass in g/cm^2 stone surface area;
C = ratio of chlorophyll/macro-invertebrate biomass µg/g).

Sampling Time	Stones Sampled	A	B(x10^{-5})	C(x10^2)
1100 h	1	0.264	17.0	1.55
	2	0.0596	6.2	0.96
1400 h	1	0.201	8.3	2.40
	2	0.168	7.1	2.40
1700 h	1	0.270	1.5	18.00
	.2	0.582	2.4	24.00
2000 h	1	0.120	3.8	3.20
	2	0.174	3.6	4.80
2300 h	1	1.078	20.0	5.40
	2	0.705	13.0	5.40

DISCUSSION

The present study confirms the value of the use of isotopes such as ^{32}P in investigations on the rate of uptake, accumulation and transfer of nutrients in freshwater ecosystems.

The efficient and rapid absorption of phosphates by the sago pondweed *P. pectinatus* within the first two hours of release into the laboratory stream might partly explain why this macrophyte becomes a problem under eutrophic conditions in lakes and rivers. However, results on the uptake and accumulation of the isotope by algae show that these organisms might be far more important than the water weeds in removing nutrients from the abiotic environment and making them available to the rest of the food chain.

This study also showed that minute quantities of nutrients can be followed through the food chain and that this type of investigation might throw more light on the ecology of the aquatic macro-invertebrate fauna of lakes and rivers. This particularly applies to the nymphs of mayflies which are known to be sensitive to changes in water quality and as such are considered as indicators of prevailing conditions in water bodies.

Although isotopes such as ^{32}P are usually not allowed to be used in natural streams and lakes because of possible dangers to man and animals, their application in a laboratory stream, where work can be done under controlled conditions and where much smaller quantities of isotope are needed to obtain quantitative data, has considerable advantages.

ACKNOWLEDGMENTS

The authors wish to thank The Rand Afrikaans University and the C.S.I.R. for financial support. Our sincerest thanks to Mr. J.J.M. Van Graan for his assistance during the study.

RESUME

On a eu recours à un modèle d'écosystème de cours d'eau en laboratoire pour étudier le transfert du radio-isotope ^{32}P dans une chaîne d'alimentation, depuis le milieu aquatique jusqu'aux niveaux primaire et secondaire de consommation en passant par les algues benthiques et la végétation submergée. Ces recherches, pour lesquelles les nymphes des éphéméroptères *Baetis harrisoni, Neurocaenis discolor* et *Euthraulus bugandensis* furent utilisées, montrèrent la valeur du ^{32}P (et peut-être d'autres isotopes) dans l'étude non seulement des taux et des quantités de nourriture absorbée par les nymphes des éphéméroptères mais aussi des activités diurnes relatives à l'alimentation de ces organismes.

ZUSSAMENFASSUNG

Ein Laboratoriums Modell Strom-Ökosystem wurde verwendet, um die Übertragung von Radio-Isotopen ^{32}P in eine Nahrungskette von der Wasserumwelt durch benthische Algen und überflutete Vegetation bis zu den primären und sekundären Verbraucherstufen zu studieren. Diese Art der Untersuchung, bei welcher Nymphen der Eintagsfliegen *Baetis harrisoni, Neurocaenis discolor* und *Euthraulus bugandensis* benutzt wurden, offenbarte den Wert von ^{32}P (und möglicherweise anderer Isotopen). Beim Studium der Nahrungsraten und -quantitäten, die von Eintagsfliegen aufgenommen wurden, und auch bei der Beobachtung der täglichen Aktivitäten im Zusammenhang mit der Nahrungsaufnahme erwies sich ^{32}P als besonders nützlich.

REFERENCES

APHA. 1971. Standard methods for the examination of water and waste water. 13th ed. American Public Health Association. Washington, D.C.

Golterman, H.L. and S. Clymo. 1969. Methods for chemical analysis of freshwater. Blackwell, Oxford.

Lean, D.R.S. and C. Nalewajko. 1976. Phosphate exchange and organic phosphorus excretion by freshwater algae. *J. Fish. Res. Board Can.* 33: 1312-1323.

McRoy, C.P. and R.J. Barsdate. 1970. Phosphate adsorption in eel-grass. *Limnol. Oceanogr.* 15: 6-13.

Rigler, F.H. 1964. The phosphorus fractions and the turnover time of inorganic phosphorus in different types of lakes. *Limnol. Oceanogr.* 9: 511-518.

Vermaak, J.F., J.H. Swanepoel and H.J. Schoonbee. 1976. Absorption and accumulation of ^{32}P *Oedogonium* and some aquatic macrophytes. *Water S.A. (Pretoria)*, 2(1): 7-12.

ULTRASTRUCTURAL STUDIES ON OLIGONEURIIDAE - TAXONOMIC APPLICATIONS

John D. Agnew

Department of Genetics
University of Witwatersrand
1 Jan Smuts Avenue
Johannesburg 2000, South Africa

ABSTRACT

The surface ultrastructure (excluding trophi) of two species each of the Afrotropical genera *Elassoneuria* and *Oligoneuriopsis* (Oligoneuriidae) has been studied by scanning electron microscopy. New methods of specimen preparation are described; these give improved results compared to those obtained by methods currently advocated. Five main kinds, containing variants, of micro- and macrotrichia are described; some appear to be intermediate or connecting types. The taxonomic distribution of the structures within the study material is given; their utility in taxonomy is discussed.

INTRODUCTION

The family Oligoneuriidae contains, *inter alia*, two genera endemic to the Afrotropical Region (this term is used in place of the old Ethiopian Region, following the recommendations of Crosskey and White 1977). The genera are *Elassoneuria* Eaton 1881 and *Oligoneuriopsis* Crass 1947. The nymphs are among the larger mayfly species encountered in the area, and are confined to swiftly flowing waters. The general trophic adaptation characteristic of these nymphs is the development of elaborate filterbrushes on the fore femora and tibiae. Although no study has been made of the feeding mechanism in living specimens, it is probable that material (presumably particulate organic matter) is strained from the current by these brushes and is transferred by the fringed maxillae. It is also possible that the brushes are used as rakes on the substrate.

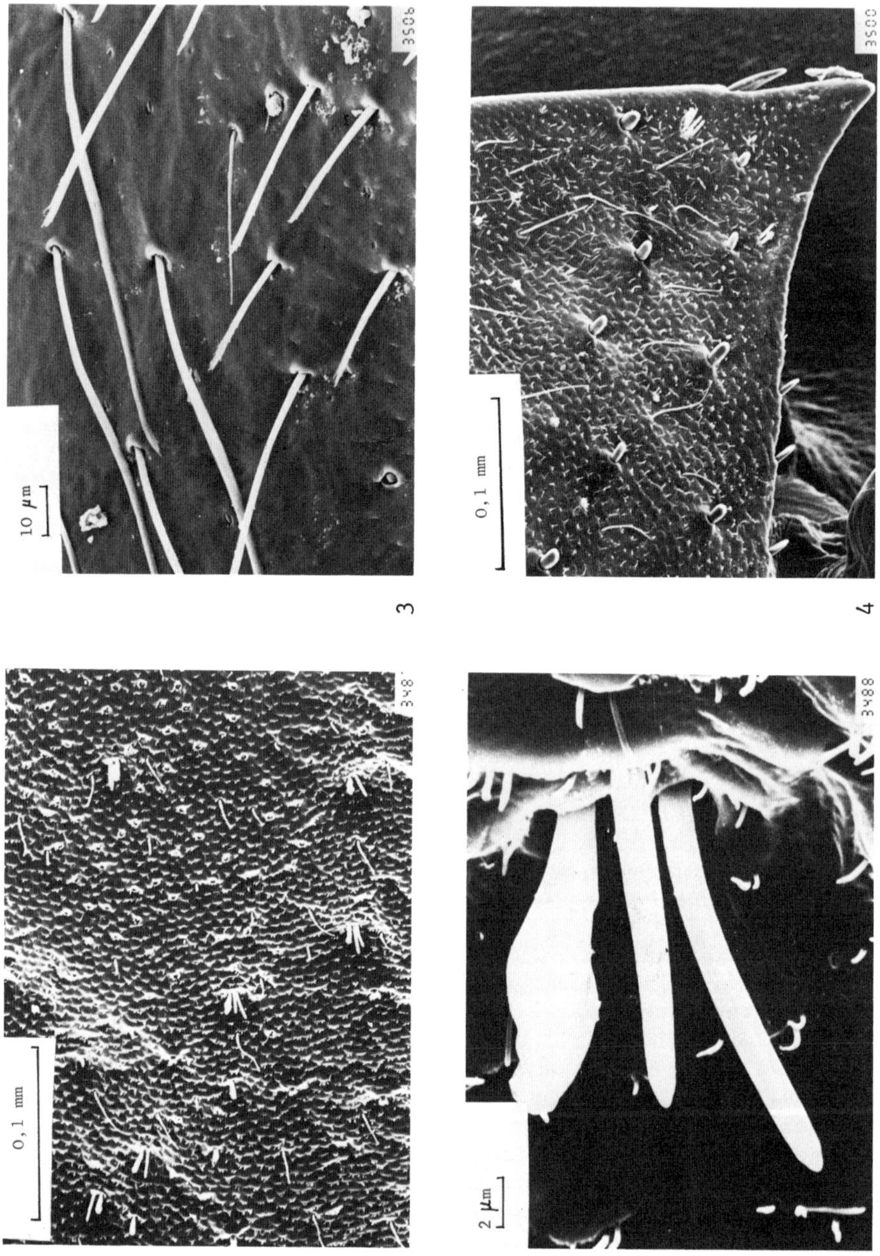

3 4 1 2

Plate 1

Other adaptations common to members of this family are the development of accessory gills of the fibrilliform or tuft type at the base of the maxillae, and protection of the lateral abdominal gill tufts by covering plates or lamellae. These plates articulate with, and seem to be derived from, the general body integument since they have the same basic surface ultrastructure as the latter, in both genera studied. This supports the old view that the gills are homologous with abdominal appendages which have been modified. Snodgrass (1935) states that there is no doubt that the gills arise from limblike rudiments of the embryo, and that their structure and musculature suggest that they are parts, at least, of true abdominal appendages.

Nymphs of the two genera may be distinguished by *Oligoneuriopsis* having a rounded frontoclypeus with no carina whereas *Elassoneuria* has a carina extending from between the antennal bases to the front of the head. The other differences are: in the adult, *Oligoneuriopsis* has 3-jointed ♂ forceps and 4 convex longitudinal veins in the forewing; *Elassoneuria* has 4-jointed forceps (sometimes with a minute extra joint) and 3 main veins. *Oligoneuriopsis* has been regarded as the more primitive of the two on adult features (Crass 1947) and is close to the European *Oligoneuriella* Ulmer, as noted by Edmunds (1961). In fact, the distinguishing feature between the last two genera in the nymphs, namely presence or absence of a lamella on the ventral gill of the first abdominal segment, seems no longer valid. I have found a lamella to be present on all mature nymphs from various parts of southern Africa, including those ascribable to the type species *O. lawrencei* Crass.

On the other hand, all *Oligoneuriopsis* nymphs have a well-developed tuft of hairs in the midline, toward the posterior border of abdominal sternites 2-4 or 2-5 inclusive. These are much better developed than those figured by Sowa (1973) for the European species *Oligoneuriella keffermuellerae* Sowa, *O. mikulskii* Sowa, *O. pallida* (Hagen) and *O. rhenana* (Imh.). The function of this tuft is undoubtedly to assist in attachment to the substrate under near-torrential conditions. In central and southern Africa, nymphs of

---

Plate 1. Figures 1-4. *Oligoneuriopsis lawrencei* Crass (GEN 596A. 24 Feb. 1960. Natal: Royal Natal National Park. 28°58'E, 28°42'S)

1. Fronto-clypeal area, dorsal view.
2. Detail of 1.
3. Lateral abdominal projection, under gill lamella, dorsal view.
4. Posterolateral projection of abdominal tergum, dorsal view.

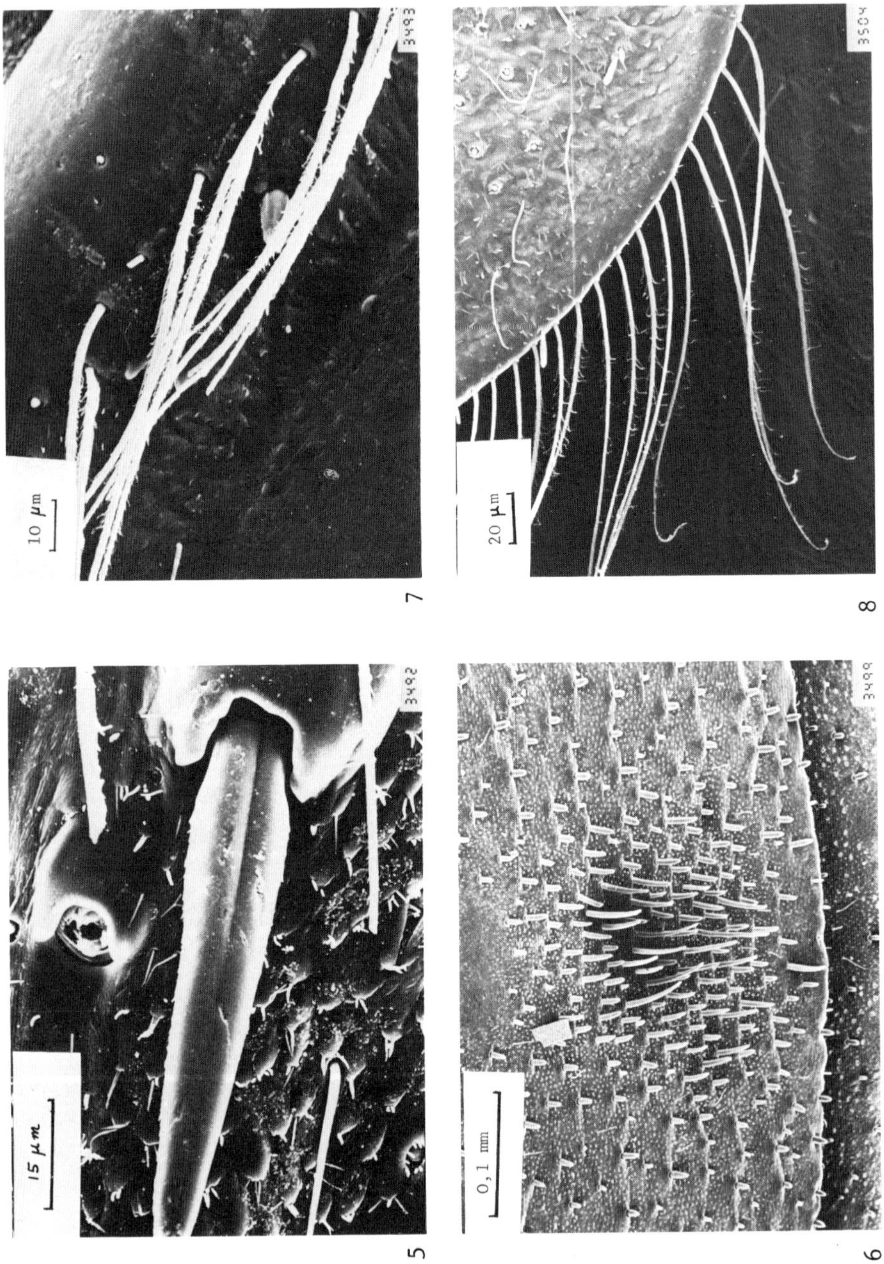

Plate 2

Oligoneuriopsis are cool-adapted, and found only at fairly high altitudes whereas those of *Elassoneuria* occur in the lower lying tropical and subtropical rivers. The requirements for cool conditions mean that the higher zones with their concomitant high current speeds must be chosen, hence the adaptive significance of the structures mentioned.

MATERIAL

The taxonomy of the nymphs, that is species discrimination, has been based largely on details of the gill lamellae. During routine examination of these structures using phase optics, it became clear that a wealth of further detail existed, which would require resolution higher than that offered by an optical system, for adequate study.

A fairly large sample of oligoneuriid nymphs belonging to both genera and collected at various localities throughout Southern Africa has been under study and these results will be reported elsewhere. Large numbers of nymphs of some species were collected and it was thus possible to prepare representative specimens for scanning electron microscopy (SEM). Where only a few individuals of a particular species were available, it was not feasible to do this because of the risk of damaging the specimen during preparation and viewing. The storage of coated specimens also presents difficulties due to their fragility, and the fact that they must be kept dry and dust-free if they are to retain their usefulness, means that one must usually discard preparations after study.

The catalogue numbers of specimens are those used by the hydrobiological survey teams of the National Institute for Water Research of the C.S.I.R. Two species of *Oligoneuriopsis*, namely *O. lawrencei* Crass and *O. elisabethae* Agnew and two as yet unnamed species of *Elassoneuria* (distinguished by their catalogue numbers) have been studied in this paper.

Plate 2. Figures 5-8. *O. lawrencei* Crass (GEN 596A. 24 Feb. 1960. Natal: Royal Natal National Park. 28°58'E, 28°42'S)

5. Detail of spination - posterior border of fore femur, dorsal view.
6. Abdominal sternite 5, posterior median area showing brush.
7. Fore femur, posterior border dorsal view.
8. Detail of inner border of gill lamella.

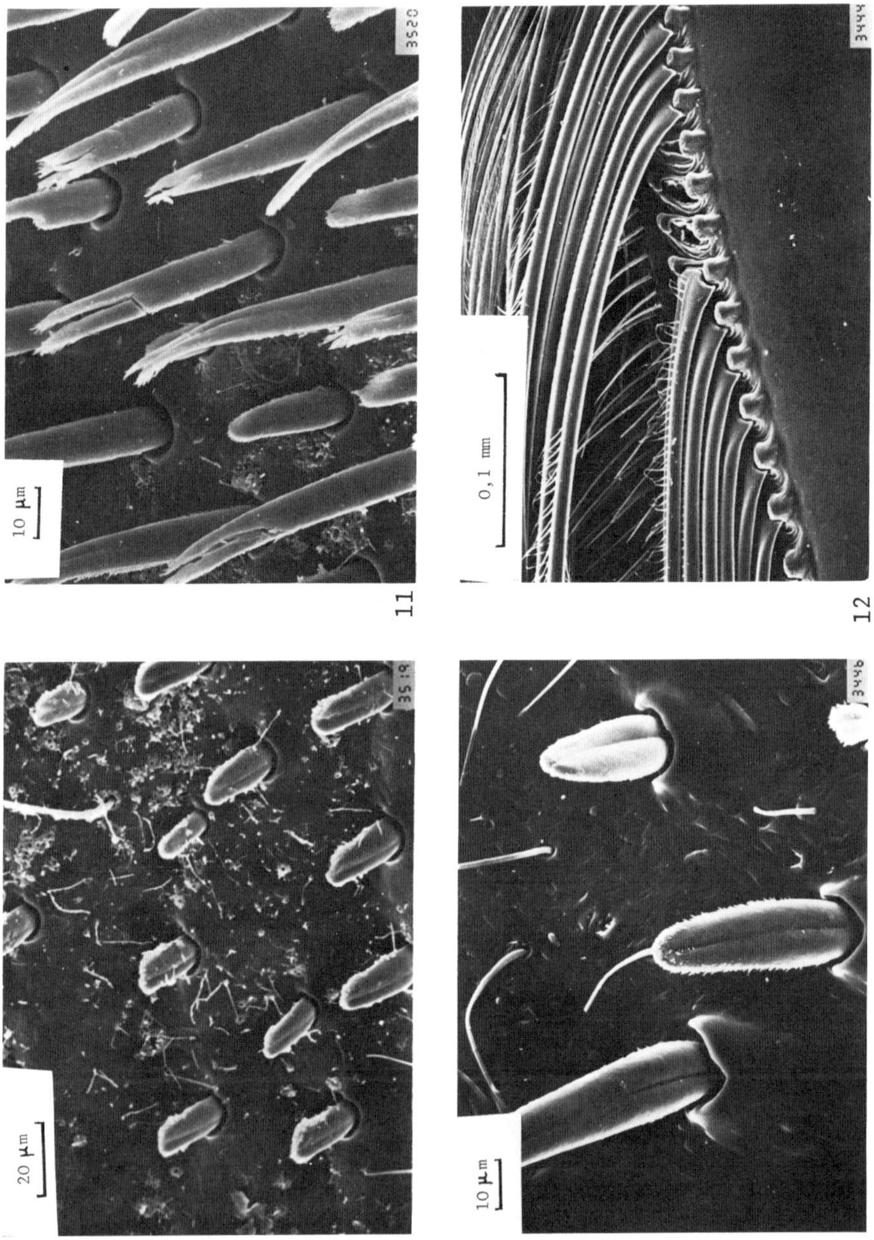

11 12 9 10

Plate 3

METHODS

Two severe problems encountered in specimen preparation were: (1) cleaning the specimens thoroughly since they are always covered with fine debris etc. and (2) dehydrating specimens without distortion, prior to coating. A search of the literature showed that many different methods have been advocated for overcoming these difficulties; in practice most of them are less than satisfactory. Regarding the cleaning of specimens, the best method was found to be the use of an ultrasonic tissue disintegrator or homogenizer (an MSE model was used) the probe of which is inserted into a beaker of 80% alcohol containing the specimen. With the amplitude of the probe set at a fairly low setting (4-6 μm), a few minutes gave adequate cleaning without damage to the specimen. Too high a setting caused parts of the specimen to detach.

With regard to specimen dehydration, air drying and critical-freeze-point drying after passage through a graded alcohol-amylacetate series were tried. Air drying tends to bring about collapse of the specimen which results in distortion, especially with large specimens such as Oligoneuriidae. Critical freeze-point drying is a lengthy process and yields a fragile specimen which disintegrates easily on handling and coating.

Good results were obtained using a method suggested by Professor H.E.H. Paterson. After cleaning, specimens are dehydrated through an alcohol series, then brought into xylene using one or two intermediate baths. Times used depend on specimen size. The xylene-hardened specimens retain their original shape after drying for a day in a desiccator. They may be affixed to the carrier stub using collidal graphite (I used DAG-580) suspended in absolute alcohol.

Specimens were coated with Au-Pd and viewed in a JEOL JSM-T-20 machine operating at 20 Kv.

Plate 3. Figures 9-12. *Oligoneuriopsis elisabethae* Agnew (HRG 30B. 16 Mar. 1965. Transvaal: Vaal River on farm Klipbank 104. 30°06'E, 26°42½'S)

9. Bristles next to tuft on sternite 2.
10. Detail of spines near base of fore femur.
11. Details of long bristles of brush; ventral abdominal sternite 2.
12. Details of brush on inner margin of fore tibia.

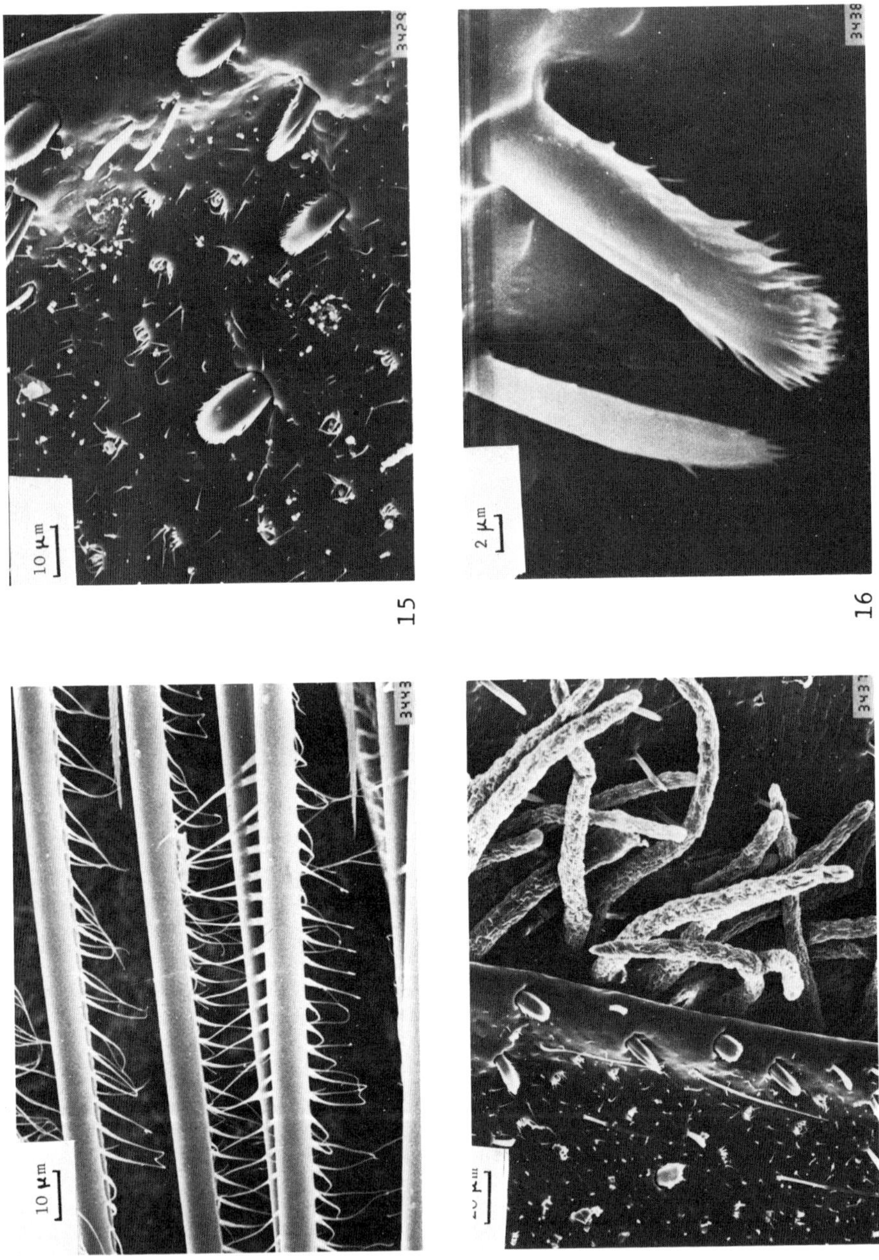

15

16

13

14

Plate 4

RESULTS

It is not possible to classify all structures found into clear cut types because of considerable overlapping in size. In fact it is likely that in evolution most of the types may have been derived one from the other or from one or more ancestral types. This matter can obviously only be resolved by comparative and/or developmental studies using a wide spectrum of representative material including fossil material.

Nevertheless, with the material at hand, it is possible to recognize the following main types:

(a) cuticular points: these are 1-5 μm in length and range in number from 2-4 per 'cell' in *Oligoneuriopsis* (Fig. 1) and 2-6, but sometimes up to 8 per 'cell', in *Elassoneuria* (background in Figs. 21 and 22). They are found in all parts of the surface with minor variations.

(b) hairs: these are scattered over most parts of the body surface. The simple kinds are 30-50 μm long in *Oligoneuriopsis* (Fig. 4) and up to 100 μm long in *Elassoneuria* (Fig. 20). Both genera have longer ones, up to 130 μm in length on the lateral parts of the abdominal terga, underlying the lamellae (Fig. 3); these are non-tapering with split ends and in addition are finely digitate along their length in both genera. Another type of hair may be termed the tinsel hair and this is virtually non-tapering and is found along the borders of the femora (rear edge) and gill lamellae (Figs. 7 and 8, *Oligoneuriopsis*; Fig. 24, *Elassoneuria*).

(c) clusters are distributed in varying degrees of density all over the general body surface. (Being found also on the antennal segments, gill lamella, appendages and caudal filaments. Clusters generally show up pure white in the e.m. photographs). They are fundamentally different in the two genera. In *Oligoneuriopsis* each cluster consists of three elements: an outer spine (the longest) 16 μm in length, which curves gently, a middle straight spine 12 μm in length and an outer, club-shaped spine, 12 μm in length (Fig. 2). In *Elassoneuria* the clusters consist of four elements, most pictures

---

Plate 4. Figures 13-16. *O. elisabethae* (HRG 30B. 16 Mar. 1965. Transvaal: Vaal River on farm Klipbank 104. 30°06'E, 26°42½'S)

13. Detail of filter brush on fore femur.
14. Inner margin of gill lamella, dorsal view.
15. Gill lamella, dorsal view of mine posterior area.
16. Detail of bristles on gill lamella.

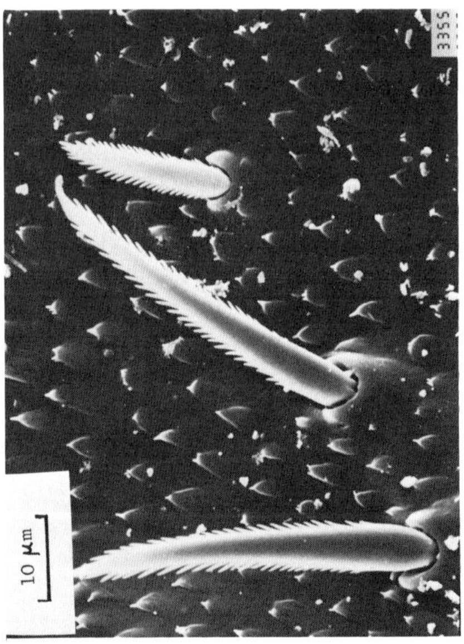

17

18

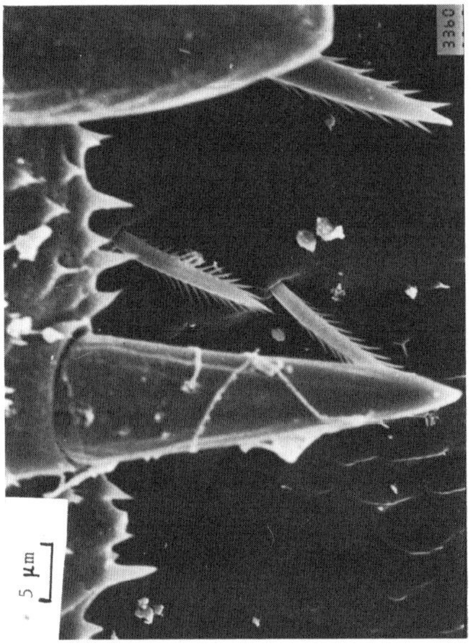

19

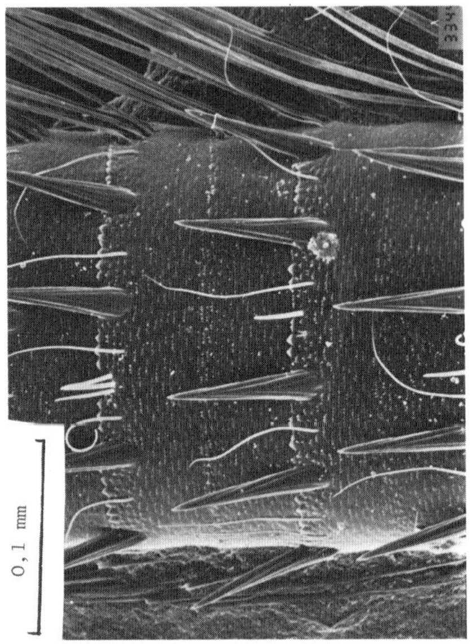

20

Plate 5

showing only three but it is possible that four are always present since the elements are packed in a circular arrangement, and sometimes one appears to be obscured. All elements are the same here, each consisting of a very gently tapered smooth spine about 40 μm in length, the ends coming together (Figs. 21 and 24). This does not occur in *Oligoneuriopsis*.

(d) feathers are single structures, found only on the proximal, mesial parts of the gill lamellae. In *Elassoneuria* (Figs. 17, 18, 19) they are ∿20 μm in length, tapering and digitate on either side while in *Oligoneuriopsis* (Figs. 14, 15, 16) they are more peg-like and split around the tips only.

(e) spines of various sizes and shapes are found all over the general body surface and gill lamellae. In *Elassoneuria* they vary in length from 35-90 μm and appear to be of multiple formation since they show a slightly spiral arrangement along their tapering length (Figs. 20, 21, 23). If indeed they are actually multiple in origin it is likely that the long bristles of the filtering brushes (Figs. 12 and 13) and the swimming hairs of the caudal filaments are modified from these spines. In *Oligoneuriopsis* the spines may be peg-like, about 10 μm long with swollen middles and most pictures show them to be double (Figs. 4, 9, 10); others are subtriangular and flattened, 20 μm long and occurring on the caudal filaments; others again, for example on the femora, are clearly double and up to 70-100 μm long (Fig. 5). In the ventral abdominal tufts of *Oligoneuriopsis* a gradation between spines from 20 μm long up to those 100-150 μm in length may be found, and again they all appear to be double in structure (Figs. 6, 9, 11). This appears to be a clear case of adaptive modification.

Morphologically, the types described above would fall into microtrichia i.e. without a basal socket and these would be the cuticular points, whereas all the others would be types of macrotrichia since they have sockets.

Plate 5. Figures 17-20. *Elassoneuria* sp. (HRBA34A. 19 Mar. 1965 Orange Free State: Wilge River at Frankfort. 28°29'E, 27°18'S)

17. Detail of bristles on lateral abdominal tergite.
18. Detail of bristle on posterior area of gill lamella (dorsal surface).
19. Gill lamella 4, inner posterior area of dorsal surface.
20. Detail of median filament.

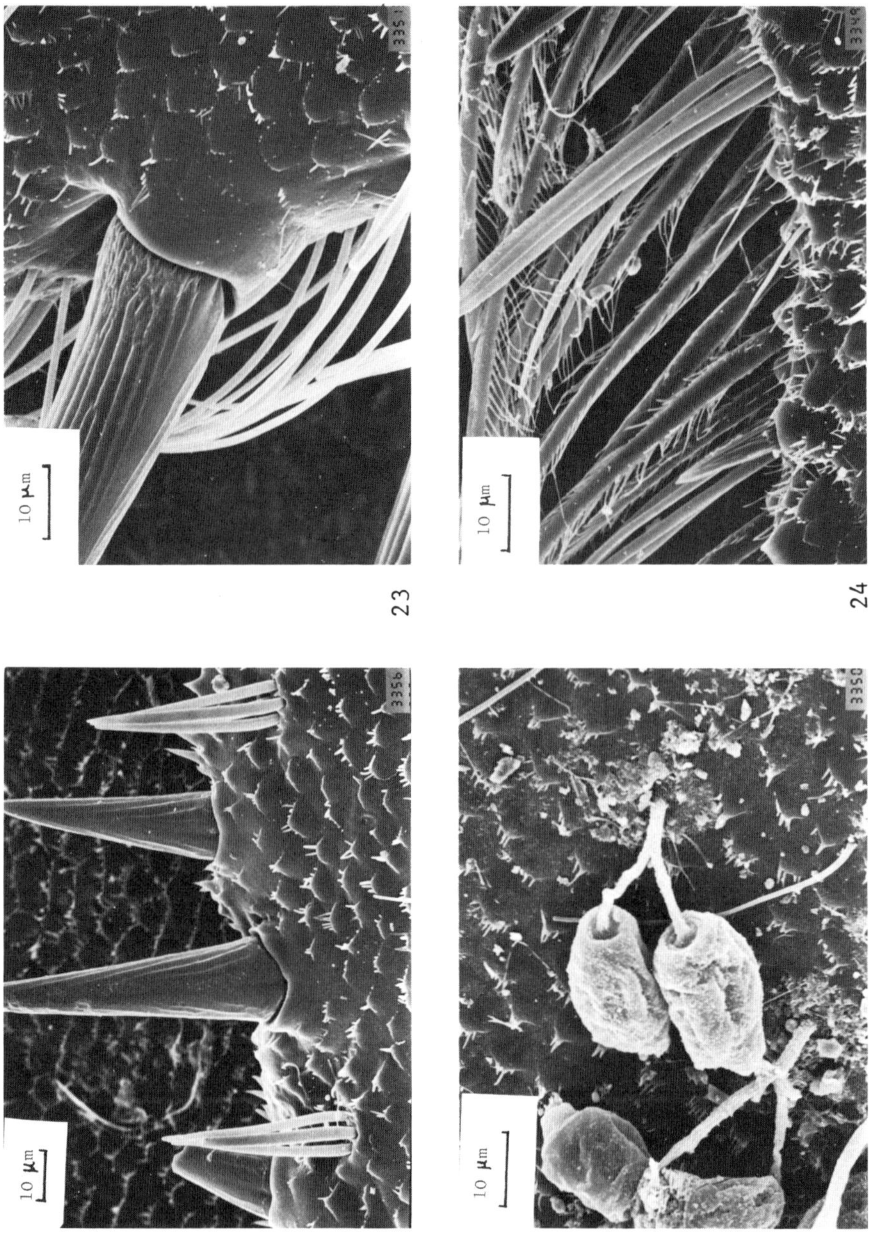

21 22 23 24

Plate 6

DISCUSSION

A discussion of these results is bedevilled by the fact that the functional significance of the types described is largely unknown. While the function of filtering and swimming paddle bristles is fairly clear, it is not certain what significance must be attached to the other types. What, for example, is the real function of clusters and feathers?

In the descriptive section it was suggested that some of the spine types may have given rise to others in evolution. Some counts of spines on the gill lamellae gave coefficients of variation of 10-15 percent, suggesting that these structures are under multifactorial control. This would form a genetic basis for natural selection to modify the structures adaptively for various purposes.

Whatever the physiological meaning and evolutionary relationships, it is seen that the two genera possess homologous yet recognizably different types of trichia and that their distribution confirms the formal taxonomic distinction. Studies of this kind would undoubtedly be important in the interpretation of fragmentary fossil material and in stimulating further studies on the adaptive significance of the structures which are at the interface of the animal and its external environment.

ACKNOWLEDGMENTS

I thank Professor H.E.H. Paterson for advice on specimen preparation for SEM; the University of the Witwatersrand, the Council for Scientific and Industrial Research and the Canada Department of Fisheries and Oceans for financial assistance enabling me to attend the Third International Conference on Ephemeroptera.

Plate 6. Figures 21-24. *Elassoneuria* sp. (HRGA34A. 19 Mar. 1965 Orange Free State: Wilge River at Frankfort. 28°29'E, 27°18'S)

21. Posterior border of tergite 9.
22. Gill lamella surface with attached stalked protozoa.
23. Gill lamella surface - anterior border.
24. Gill lamella, border just posterior to apex of lamella.

RESUME

La ultrastructure de surface (à l'exclusion des trophi) de deux espèces des deux genres afrotropicaux *Elassoneuria* et *Oligoneuriopsis* (Oligoneuriidae) a été etudiée au moyen de la microscopie par scintigramme électronique. De nouvelles méthodes de préparation d'échantillons y sont décrites qui donnent de meilleurs résultats que ceux que l'on obtient par les méthodes que l'on préconise actuellement. Cinq espèces prinicipales de micro- et de macrotrichies y sont examinées avec leurs variantes, certaines paraissant être de type intermédiaire ou faisant figure de trait d'union. La matière étudiée comprend la répartition taxonomique des structures et l'on y traite également de l'utilité de ces structures en taxonomie.

ZUSSAMENFASSUNG

Mit Hilfe von "scanning electron" Mikroskopie wurde die Oberflächen Ultrastruktur (Mundpartien ausgenommen) je zweier Arten der afrotropischen Gattungen *Elassoneuria* und *Oligoneuriopsis* untersucht. Neue Methoden der Präparierung von Specimina werden beschrieben, wobei sich im Vergleich zu bisher befürworteten Methoden weitaus verbesserte Resultate ergeben. Fünf Hauptarten, die Varianten von Mikro- und Makrotrichiasis enthalten, werden dargestellt. Einige davon scheinen Zwischen- oder Koppeltypen zu sein. Die Arbeit befaßt sich weiterhin mit der taxonomischen Verteilung von Strukturen innerhalb des Studienmaterials und erörtert ihren Nutzen in der Taxonomie.

REFERENCES

Crass, R.S. 1947. The Mayflies (Ephemeroptera) of Natal and the Eastern Cape. *Ann. Natal Mus.* 11: 37-110.

Crosskey, R.W. and G.B. White. 1977. The Afrotropical Region. A recommended term in zoogeography. *J. Nat. Hist.* 11: 541-544.

Edmunds, G.F. 1961. A key to the genera of known nymphs of the Oligoneuriidae (Ephemeroptera). *Proc. Entomol. Soc. Wash.* 63: 255-256.

Snodgrass, R.E. 1935. Principles of insect morphology. McGraw-Hill Book Co., New York.

Sowa, R. 1973. Contribution à l'étude des *Oligoneuriella* Ulm européennes (Ephemeroptera Oligoneuriidae). *Bull. Acad. Pol. Sci. Ser. Sci. Biol. Cl. II*, 21: 657-665.

A COMPARISON OF MAYFLY (EPHEMEROPTERA) INSTAR DETERMINATION METHODS

Thomas J. Fink

Laboratory of Aquatic Entomology
Univ. P.O. Box 111
Florida A&M University
Tallahassee, Florida
U.S.A. 32307

ABSTRACT

Mayfly instar determination methods have been examined to determine their accuracy and usefulness. The Janetschek and Cassie methods are shown not to be an improvement on the simple frequencies from which they are calculated. The Palmen body method suffers from great technical difficulties due to the body's structure and small size. The prospect of associating the number of molts with life history and environmental data is discussed.

INTRODUCTION

Mayflies generally have a large number of postembryonic molts. The number of nymphal instars* (the form of the insect between ecdyses) may range from a high of 40-45 as estimated for *Stenacron canadense* (Walker) (Ide 1935) to a low of 12 estimated for *Baetisca rogersi* Berner (Pescador and Peters 1974).

---

* The use of the term instar in this paper differs from that of Jones (1978) who defines an instar as the form of the insect between apolyses (i.e. the time the epidermis first separates from the old cuticle prior to the formation of the new cuticle). The term instar as used in this paper is consistent with general use of the term in mayfly literature.

The hardened body parts of mayflies and other arthropods may only increase in size and reveal morphological changes during the short period after a molt and before the new cuticle again becomes hardened. This yields an apparent stepwise growth pattern in which the size gaps or structural differences tend to isolate individuals of different instars. The investigator can determine the number of instars of field collected specimens (collected over the entire life cycle of a generation or cohort) by noting discrete size and morphological groups. This procedure is successful for species whose rate of development and number of instars are fairly homogenous, otherwise problems arise; the gaps become less distinctive as instars overlap in terms of size and morphology.

The rate of development and number of instars are probably controlled by genetic and environmental factors. Nutrition has been considered to be a factor affecting the number of instars in Lepidoptera and other arthropods (Gaines and Campbell 1935; Wigglesworth 1972; Gertsch 1979). Generally poorer nutrition seems to result in a greater number of instars. Cianciara (1979) found that food affected the number of molts of *Cloeon dipterum* (Linnaeus). Temperature, long implicated as a factor influencing size and growth rate in arthropods, may influence the number of molts (Wigglesworth 1972; Roberts *et al.* 1978), and during embryonic life may even affect adult size as discovered for the amphipod *Hyalella azteca* Saussure (de March 1978). Photoperiod has been thought to influence the number of instars in Plecoptera (Khoo 1968).

Genetic control may be implied in the following examples. Degrange (1959) reared, under apparently identical conditions, eggs from one female imago of *Cloeon simile* Eaton and obtained 43 imagoes which had passed through a range of 21 to 30 molts. Hunt (1953) reared *Hexagenia limbata* Serville in aquaria and although eggs were obtained around the same time and individuals in each aquarium were exposed to similar conditions, a large nymphal size range existed in each aquarium. Eason (1969) reared the spider *Pardosa lapidicina* Emerton in the laboratory at 24 ± 1.7^{o}C and found that maturity could be reached in the eighth, ninth or tenth instar.

Mayfly instar determination is particularly difficult due to the generally large number of instars and prolonged immature life cycle. The large number of instars leads to difficulty in resolution of the minute differences, if any, between instars. The prolonged immature life cycle of many mayflies may expose different generations or cohorts and delayed egg hatchlings to widely varying temperature, food and other environmental conditions. These factors may serve to affect the growth rates of the immature insects, the degree of development per instar and the number of instars. Sexual dimorphism is a further obscuring factor and cannot be resolved until late in the life cycle when the sexes can be identified.

The purpose of this paper is to briefly compare the accuracy and usefulness of those instar determination methods which have been used or may be used on Ephemeroptera. These methods include rearing, morphological characteristics, simple frequency distribution, Cassie, Janetschek, Dyar's law, allometry and Palmen body.

REARING

Rearing involves raising the insect throughout its life cycle and counting all ecdyses. The insects must be reared separately and closely monitored so as to procure all exuviae, the only conclusive proof of molting.

Pescador and Peters (1974) were able to rear *Baetisca rogersi* from eggs to the third nymphal instar. Further specimens were obtained by field-collecting nymphs and rearing them as far as possible. Similarly, Jones (1977), Rawlinson (1939) and Lubbock (1866) obtained most instars of *Neoephemera youngi* Berner, *Ecdyonurus venosus* (Fabricius) and *Cloeon cognatum* Stephens (as *dimidiatum* Curtis), respectively. Trost and Berner (1963) reared *Callibaetis floridanus* Banks throughout its life cycle but were unable to find the exuviae of tiny nymphs. Froehlich (1969) was not able to follow the growth of individual *Caenis cuniana* Froehlich nymphs past the fifth instar. Murphy (1922) successfully reared *Baetis posticatus* (Say) from egg to imago and determined 27 nymphal instars; however, results may be viewed with minor suspicion since nymphs were kept in small groups rather than singly. Degrange (1959) successfully reared and counted all ecdyses of 43 *Cloeon simile* individuals. Brittain (1976) determined 17-19 nymphal instars for reared specimens of *Leptophlebia vespertina* (Linnaeus). Illies and Masteller (1977) reared *Baetis vernus* Curtis to the dark wing padded stage which occurred in the 18th instar.

The future of rearing lies in the design and application of equipment expressly for the purpose of rearing such small aquatic insects. Only then will the conflicting needs of ease of observation and simulation of the natural environment be sufficiently balanced to ensure routine success.

The success of rearing means the accurate charting of mayfly development and its possible correlation with environmental conditions under controlled experiments. Even the partial rearing of a large number of nymphs from one clearly defined morphological stage to another equally well defined stage may yield illuminating information on the variability of the number of molts and relative change per molt.

MORPHOLOGICAL CHARACTERISTICS

If mayflies displayed a perfect stepwise growth pattern, instars would be readily separable by the discrete size and structural groupings of nymphs. This is the basis of the morphological characteristics method which entails the detailed description of periodically field-collected nymphs.

Pescador and Peters (1974) determined 12 nymphal instars for *Baetisca rogersi* and Jones (1977) determined 15 nymphal instars for *Neoephemera youngi* using this method in conjunction with rearing. In both studies it was realized that the calculated instars may actually be only developmental stages which are composed of several instars. Ide (1935) reared *Stenacron canadense* and *Ephemera simulans* Walker up to the eighth or ninth instar and obtained other nymphs from periodic field collections. Ide believed that minute changes in morphology were very constant and indicated true instars. The number of segments of the antennae and caudal filaments was believed correlated exactly with the number of molts but difficulty was encountered due to the fusion of segments, the cryptic nature of many sutures, and the ease with which these structures may be damaged. Using morphological criteria Ide estimated between 40-45 and 30 nymphal instars respectively for *S. canadense* and *E. simulans*. Rawlinson (1939) found with *Ecdyonurus venosus* that morphological characteristics were not instar specific, therefore to follow the general course of nymphal development, she designated broad stages based on gill and mesothoracic wing pad development. Similarly, other authors have abandoned instars in favor of broad developmental stages (e.g. Pleskot 1959, Bretschko 1965, Clifford 1970, Lehmkuhl 1970, Hall *et al.* 1975). These stages have been useful in analyzing complex life cycle patterns as demonstrated by McCafferty and Huff (1978) who separated out three overlapping broods found in the life cycle of *Stenacron interpunctatum* (Say).

SIMPLE FREQUENCY DISTRIBUTION

Simple frequency analysis for the determination of instars also relies on the idealized arthropod step-like growth pattern. The method involves frequent periodic field collection of specimens throughout the entire life cycle and the plotting of the number of individuals per size class (all collections combined) for a particular body part measurement. All specimens of a certain instar would be uniform in size and consequently be depicted in the histogram as an isolated peak if there was little or no variation in growth rate and number of instars in the population. Counting the number of peaks would yield the number of instars.

Certain Lepidoptera and other insects do show a homogeneous

rate of development and number of instars and therefore, instars may be adequately detected by the frequency distribution (e.g. Benke 1970, Elliott 1971, Cutler and Harris 1976, Poston *et al.* 1977, Bilapate *et al.* 1978, McNeil 1978, Ross and Craig 1979, Mizell and Nebeker 1979). This is not the case for other species (e.g. Metcalfe 1932, Gaines and Campbell 1935, Kishi 1971, Schmidt and Lauer 1977, Schmidt *et al.* 1977, Ross and Merritt 1978).

In mayflies, the simple frequency data base has been used in an altered form in almost every life history study to provide a meaningful growth trend throughout the life cycle for the purpose of detecting generations or cohorts, and estimating times of growth, egg hatching and emergence. The plot, which may be considered a life history diagram, does not combine collections as above but rather the size class data is plotted for each collection period (e.g. Macan 1957, Thorup 1963, Bretschko 1965, Thibault 1971, Brittain 1972, Pescador and Peters 1974, McClure and Steward 1976, Flowers and Hilsenhoff 1978). In those studies where the simple frequency method has been used in instar determination (McClure and Stewart 1976, Newell and Minshall 1978) only difficult-to-interpret histograms have resulted.

CASSIE METHOD

The Cassie method (Harding 1949, Cassie 1954) is an elaboration of the simple frequency method which attempts to better resolve overlapping distributions. The total distribution is assumed to be the result of a number of normally distributed components, such as the size of particular age classes or instars. The method attempts to uncover the limits of the component distributions.

The Cassie method uses the same data base as the simple frequency analysis except that cumulative size frequency percentages are calculated instead of frequency values. The cumulative frequency percentages are plotted on the probability scale of probability paper versus the size class on the arithmetic scale. A mode in the plot is due to a relatively rapid increase in cumulative frequency due to the accumulation of individuals in a restricted size range, and is assumed to be one component (e.g. age class or instar) distribution of the total distribution. Inflections in the plot are the gaps between the different age class or instar modes.

Harding (1949) and Cassie (1954) used the method with distributions in which the number of components was small and largely known, and both state that the method offers only approximate solutions. Bhattacharya (1967) who developed an additional graphical method for resolving a distribution into its Gaussian (normally distributed) components cautions that the Cassie, his and other methods adopted by fishery workers, "...are all approximate in the sense that they are applicable only when the components are adequately separated.", and that "The mathematical basis of the probability paper is not very clear.".

Recently the Cassie method has been applied to instar determination in mayflies and stoneflies. McClure and Stewart (1976) used it and the Janetschek method for *Choroterpes mexicanus* Allen, while Oberndorfer and Stewart (1977) used the same two methods for the stonefly *Hydroperla crosbyi* (Needham and Claassen).

The Cassie plots of the above two papers were traced on clean white paper (Fink in prep) but no modes or inflections could be determined. This method was also employed on *Baetisca rogersi* and simulated random number data and again no modes and inflections could be determined. Additionally, the Cassie plot of Cassie's (1954) example number 1 was retraced, but only 3 of the 4 inflections stated were present. Even Cassie (1954) admits difficulty in distinguishing the fourth inflection.

The major problem of the Cassie method is the detection of the modes and points of inflection. Although probability paper offers the advantage of estimating the ranges, means and standard deviations of the component distributions, it does so only when they are adequately separated. Probability paper offers a distinct disadvantage in detecting modes and points of inflection over that of the simple frequency method and for this reason the Cassie method cannot be recommended for mayfly instar determination.

JANETSCHEK METHOD

The Janetschek (1967) method is also an elaboration of the simple frequency method which attempts to more clearly resolve significant peaks. Gliding mean values per size class are calculated from the same simple frequency data base resulting in a histogram which supposedly displays a trendline view of growth through the life cycle. The gliding mean values are then subtracted from the simple frequency values to yield positive and negative values. This yields a histogram which is thought to show the periodic maxima of the body part measurement unmasked by the general growth pattern of the population (Janetschek 1967). Significant positive peaks may indicate instars.

The Janetschek method has been used with Collembola (Janetschek 1967), Ephemeroptera (McClure and Stewart 1976, Newell and Minshall 1978) and Plecoptera (Oberndorfer and Stewart 1977, Harper 1973). In these uses of the Janetschek method there has been a somewhat capricious choice of which positive peaks correspond to instar modes. The Janetschek method offers no objective means of determining significant peaks.

By superimposing the periodic maxima plot on the original simple frequency histogram for each of these studies and for *Baetisca rogersi* and simulated random number data, exact agreement between most peaks of both plots was found (Fink, in prep.).

Thus the Janetschek method is a mathematical exercise which is nothing more than a recapitulation of the simple frequency method, and as such offers no additional information.

DYAR'S LAW

Dyar's (1890) purpose in proposing his empirical generalization was to provide a means to discover overlooked instars of field-collected specimens. His rule states that the discontinuous or step-like increase of sclerotized parts of Lepidoptera larvae usually follow a regular geometric progression. A departure from this progression may then indicate a missing instar or instars. His rule has since been extended to other insects.

Dyar's law can be valid only for those insects whose rate of development and number of instars are homogeneous. Since growth through time of many animals is not constant (Simpson *et al.* 1960), departures from Dyar's law should be commonplace, especially for relatively long lived insects whose life cycles extend over different seasons.

Despite the shortcomings of Dyar's law it remains a useful tool in instar determination. The method is most easily applied by plotting the logarithm of the body measurement on the ordinate versus the suspected instar number on the abcissa. The data points are then connected and a relatively straight line will result if there is some semblance of a regular geometric progression (Forbes 1934). Any large gap between successive data points may indicate a missing instar or instars.

Jones (1977) used this rule for *Neoephemera youngi* and postulated the existence of two additional instars, numbers 3 and 4, between the first two reared instars and the remaining field-collected specimens. Specimens of such a small size ($\leq$ 1 mm in body length) are very difficult to collect in the field. McClure and Stewart (1976) discovered a possible missing second instar for *Choroterpes mexicanus*, which again would be difficult to obtain in field collections. Vaught and Stewart (1974) showed, using Dyar's law semi-logarithmic plot, that nymphal growth of the stonefly *Neoperla clymene* Newman appeared to exhibit 3 distinct phases. Froehlich (1969) appears to have reared the first 5 instars of *Caenis cuniana* and then estimated an additional 10 or 11 instars using Dyar's law.

ALLOMETRY

The allometric constant α is the ratio of the geometric growth rate of body part y to the geometric growth rate of body part x ($K_{gy}/K_{gx}=\alpha$) (Simpson *et al.* 1960). An allometric plot (logarithm of body measurement y versus logarithm of body measurement x) could be used in the same way as a Dyar's law plot to check for missing instars. The usefulness of this method would depend on the constancy of α over the period of time considered. α is not likely to be constant over the growth period of many organisms as has been shown for *Notonecta undulata* Say (Simpson *et al.* 1960) and *Porcellio laevis* (Latreille) (Nair 1978).

PALMEN BODY

The Palmen body is a ringed structure located in the head of Ephemeroptera at the junction of 4 tracheae. Counting the rings would yield the number of instars if one ring is secreted during the course of each instar. Discovered in 1877 by Palmén, and correlated with the number of molts by Wodsedalek in 1912, it wasn't until 1959 that Degrange tested the Palmen body. Forty-three *Cloeon simile* individuals were separately reared and the Palmen bodies of 1/3 of the specimens successfully counted. The number of instars determined by the Palmen body method and rearing were stated to be identical.

Despite the Palmen body's potential in instar determination only limited use has been made of it. Estimates of the number of molts through at least partial use of the Palmen body have been made by Rawlinson (1939), Taylor and Richards (1963), McLean (1970) and Benech (1972).

The Palmen body method possesses technical difficulties due to the structure's small size and unique morphology. The lower limit of imaginal Palmen body size is probably around 60 μm), as reported by Degrange (1959) for *Cloeon simile*. Consequently, the smallest inner rings are difficult to see, and may be near the limit of resolution in the light microscope.

Thin sectioning is the preferred technique for counting Palmen body rings; however, the problems of orientation of the Palmen to the knife and section thickness make it unlikely that any one section will pass through all rings. Scanning electron microscopy has also been attempted but has failed primarily due to the presence of a persistent coating which does not permit the penetration of electrons to the underlying rings.

The time of Palmen body formation adds an additional problem that should be investigated for each species. Rawlinson (1939) and Degrange (1959) have reported that the tracheal system and Palmen body do not appear until the third or second postembryonic instar respectively for *Ecdyonurus venosus* and *Cloeon simile*.

Presently the Palmen body method offers only an estimate of instar number and in many instances the technical difficulties preclude its routine use. It retains, however, the exciting potential of accurate instar determination for every individual and the generation of statistically treatable instar number data.

CONCLUSION

The process of molting is integrally coordinated with mayfly growth and differentiation, and by compartmentalizing development into small units known as instars, may enable the charting of development and the correlation with environmental factors on a fine scale not possible if development were more gradual.

The number of molts or relative change per molt could provide some insight into several important phenomena known or thought to occur in mayflies. One widespread phenomenon is the often large size differences of imagos throughout the emergence period, or of size differences between individuals of different generations or cohorts (e.g. Rawlinson 1939, Macan 1957, Gledhill 1959, Elliott 1967, Minshall 1967, Coleman and Hynes 1970, Bengtsson 1973, Fahy 1973, Sweeney 1978). A related phenomenon has been discussed by Clifford (1970) and Svensson (1977) who have postulated the existence of transitory periods in which cohorts are differentiated or fused. Nymphs having reached a certain physiological stage and minimum size synchronize and increase their development, whereas nymphs not reaching this stage are left behind to complete their development at a later date.

The future of mayfly instar determination should look to the application of new specially designed rearing techniques and the continued testing of the Palmen body. Broad developmental stages and life history diagrams will remain practical means by which mayfly development may be followed, but will result in a more generalized and perhaps schematic history of mayfly development.

ACKNOWLEDGMENTS

This research was supported by a research program (Flax 79009) of SEA/CR U.S.D.A., at Florida A&M University, William L. Peters research leader. I am also grateful for the ideas and assistance of W.L. Peters, M.D. Hubbard and L. Yasui in preparation of this paper.

RESUME

Des études ont été effectuées sur le sujet pour déterminer l'exactitude et l'utilité des différentes méthodes. Les méthodes de Janetschek et Cassie n'ajoutent rien aux fréquences simples à partir desquelles elles ont été mises au point. Celle de Palmen, dite "du corps" se heurte à de sérieuses difficultés techniques en raison de la structure du corps et de sa petite taille. L'ouvrage fait le point sur cette tentative d'association du nombre des mues à la genèse de leur existence et aux données de l'environnement.

ZUSSAMENFASSUNG

Methoden zur Bestimmung von Eintagsfliegen in verschiedenen Stadien zwischen den Häutungen wurden einer Untersuchung unterzogen, um ihre Genauigkeit und Nützlichkeit zu bestimmen. Die Janetschek und Cassie Methoden erweisen sich nicht als Verbesserung gegenüber den einfachen Frequenzen von denen sie errechnet wurden. Die Palmen Körpermethode leidet an großen technischen Schwierigkeiten auf Grund der Struktur des Körpers und der geringen Größe. Vorliegende Arbeit befaßt sich mit der Möglichkeit, die Zahl der Häutungen zu Lebensgeschichte und Umgebungsdaten in Beziehung zu setzen.

REFERENCES

Benech, V. 1972. Le polyvoltinisme chez *Baetis rhodani* Pictet (Insecta, Ephemeroptera) dans un ruisseau a truites des Pyrénées-Atlantiques, le Lissuraga. *Ann. Hydrobiol.* 3: 141-171.

Bengtsson, J. 1973. Vaekst og livscyklus hos *Baëtis rhodani* (Pict.) (Ephemeroptera). *Flora Fauna* 79: 32-34.

Benke, A.C. 1970. A method for comparing individual growth rates of aquatic insects with special reference to the Odonata. *Ecology* 51: 328-331.

Bhattacharya, C.G. 1967. A simple method of resolution of a distribution into its Gaussian components. *Biometrics* 23: 115-135.

Bilapate, G.G., V.M. Pawar, and A.K. Raodeo. 1978. Biometrical analysis of growth in the larvae of *Heliothis armigera* Hubner. *Proc. Indian Acad. Sci., Sect.* 87: 349-355.

Bretschko, G. 1965. Zur larvalentwicklung von *Cloeon dipterum, Centroptilum luteolum* und *Baetis rhodani*. *Z. wiss. Zool.* 172: 17-36.

Brittain, J.E. 1972. The life cycles of *Leptophlebia vespertina* (L.) and *L. marginata* (L.) (Ephemeroptera) in Llyn Dinas, North Wales. *Freshw. Biol.* 2: 271-277.

Brittain, J.E. 1976. Experimental studies on nymphal growth in *Leptophlebia vespertina* (L.) (Ephemeroptera). *Freshw. Biol.* 6: 445-449.

Cassie, R.M. 1954. Some uses of probability paper in the analysis of size frequency distributions. *Aust. J. Mar. Freshw. Res.* 5: 513-522.

Cianciara, S. 1979. Some study on the biology and bioenergetics of *Cloeon dipterum* (L.) Ephemeroptera (preliminary data). p. 175-192 *In*: K. Pasternak and R. Sowa, ed. Proc. Second Int. Conf. Ephem.; August 23-26, 1975: Polish Academy of Sciences Laboratory of Water Biology, Kracow. p. 175-192.

Clifford, H.F. 1970. Analysis of a northern mayfly (Ephemeroptera) population, with special reference to allometry of size. *Can. J. Zool.* 48: 305-316.

Coleman, M.J. and H.B.N. Hynes. 1970. The life histories of some Plecoptera and Ephemeroptera in a southern Ontario stream. *Can. J. Zool.* 48: 1333-1339.

Cutler, B.L., and M.K. Harris. 1976. Head capsule widths of the walnut caterpillar. *J.N.Y. Entomol. Soc.* 84: 156-158.

Degrange, C. 1959. Nombre de mues et organe de Palmén de *Cloeon simile* Etn. (Ephéméroptères). *C.R. Hebd. Seances Acad. Sci.* 249: 2118-2119.

de March, B.G.E. 1978. The effects of constant and variable temperatures on the size, growth, and reproduction of the freshwater amphipod *Hyalella azteca* (Saussure). *Can. J. Zool.* 56: 1801-1806.

Dyar, H.G. 1890. The number of molts in lepidopterous larvae. *Psyche* 5: 420-422.

Eason, R.R. 1969. Life history and behaviour of *Pardosa lapidicina* Emerton. *J. Kans. Entomol. Soc.* 42: 339-360.

Elliott, J.M. 1967. The life histories and drifting of the Plecoptera and Ephemeroptera in a Dartmoor stream. *J. Anim. Ecol.* 36: 343-362.

Elliott, J.M. 1971. The life history and biology of *Apatania muliebris* McLachlan (Trichoptera). *Entomol. Gaz.* 22: 245-251.

Fahy, E. 1973. Observations on the growth of Ephemeroptera in fluctuating and constant temperature conditions. *Proc. R. Ir. Acad. Sect. B* 73: 133-149.

Flowers, R.W., and W.L. Hilsenhoff. 1978. Life cycles and habitats of Wisconsin Heptageniidae (Ephemeroptera). *Hydrobiologia* 60: 159-171.

Forbes, W.T.M. 1934. A note on Dyar's law (Lepidoptera: larvae). *Bull. Brooklyn Entomol. Soc.* 29: 146-149.

Froehlich, C.G. 1969. *Caenis cuniana* sp. n., a parthenogenetic mayfly. *Beitr. neotrop. Fauna.* 6: 103-108.

Gaines, J.C., and F.L. Campbell. 1935. Dyar's rule as related to the number of instars of the corn ear worm, *Heliothis obsoleta* (Fab.), collected in the field. *Ann. Entomol. Soc. Am.* 28: 445-461.

Gertsch, W.J. 1979. American spiders, second ed. Van Nostrand Reinhold, New York. 274 p.

Gledhill, T. 1959. The life-history of *Ameletus inopinatus* (Siphlonuridae, Ephemeroptera). *Hydrobiologia* 14: 85-90.

Hall, R.J., L. Berner, and E.F. Cook. 1975. Observations on the biology of *Tricorythodes atratus* McDunnough (Ephemeroptera: Tricorythidae). *Proc. Entomol. Soc. Wash.* 77: 34-49.

Harding, J.P. 1949. The use of probability paper for graphical analysis of polymodal frequency distributions. *J. Mar. Biol. Ass. U.K.* 28: 141-153.

Harper, P.P. 1973. Life histories of Nemouridae and Leuctridae in southern Ontario (Plecoptera). *Hydrobiologia* 41: 309-356.

Hunt, B.P. 1953. The life history and economic importance of a burrowing mayfly, *Hexagenia limbata*, in southern Michigan lakes. *Bull. Inst. Fish. Res.* 4: 1-151.

Ide. F.P. 1935. Post embryological development of Ephemeroptera (mayflies). External characters only. *Can. J. Res.* 12: 433-478.

Illies, J., and E.C. Masteller. 1977. A possible explanation of emergence patterns of *Baetis vernus* Curtis (Ins.: Ephemeroptera) on the Breitenbach-Schlitz studies on productivity, nr. 22. *Int. Rev. gesamten Hydrobiol.* 62: 315-321.

Janetschek, H. 1967. Growth and maturity of the springtail, *Gomphiocephalus hodgsoni* Carpenter, from South Victoria Land and Ross Island. p. 295-305 *in* Entomology of Antarctica. Antarct. Res. Ser. 10.

Jones, J. 1977. The ecology, life cycle, and seasonal distribution of *Neoephemera* (s.s.) *youngi* Berner (Ephemeroptera: Neoephemeridae). M.S. thesis. Florida State University. 152 pp.

Jones, J.C. 1978. A note on the use of the terms instar and stage. *Ann. Entomol. Soc. Am.* 71: 491-492.

Khoo, S.G. 1968. Experimental studies on diapause in stoneflies. I. Nymphs of *Capnia bifrons* (Newman). *Proc. R. Entomol. Soc. Lond., Ser. Gen. Entomol.* 43: 40-48.

Kishi, Y. 1971. Reconsideration of the method to measure the larval instars by use of the frequency distribution of the head capsule widths and lengths. *Can. Entomol.* 103: 1011-1015.

Lehmkuhl, D.M. 1970. The life cycle of *Rhithrogena morrisoni* (Banks) in western Oregon (Ephemeroptera: Heptageniidae). *Pan-Pac. Entomol.* 46: 124-127.

Lubbock, J. 1866. On the development of *Chloeon (Ephemera) dimidiatum*. *Trans. Linn. Soc. Lond.* 24: 61-78, 25: 477-492.

Macan, T.T. 1957. The life histories and migrations of the Ephemeroptera in a stony stream. *Trans. Soc. Br. Entomol.* 12: 129-156.

McCafferty, W.P., and B.L. Huff, Jr. 1978. The life cycle of the mayfly *Stenacron interpunctatum* (Ephemeroptera: Heptageniidae). *Great Lakes Entomol.* 11: 209-216.

McClure, R.G., and K.W. Stewart. 1976. Life cycle and production of the mayfly *Choroterpes (Neochoroterpes) mexicanus* Allen (Ephemeroptera: Leptophlebiidae). *Ann. Entomol. Soc. Am.* 69: 134-144.

McLean, J.A. 1970. Studies on the larva of *Oniscigaster wakefieldi* (Ephemeroptera: Siphlonuridae) in Waitakere stream, Auckland. *N.Z.J. Mar. Freshw. Res.* 4: 36-45.

McNeil, J.N. 1978. The number of larval stages of *Thymelicus lineola* (Lepidoptera: Hesperiidae) in eastern Canada. *Can. Entomol.* 110: 1293-1295.

Metcalfe, M.E. 1932. On a suggested method for determining the number of larval instars in *Sitodrepa panicea* L. *Ann. Appl. Biol.* 19: 413-419.

Minshall, J.N. 1967. Life history and ecology of *Epeorus pleuralis* (Banks) (Ephemeroptera: Heptageniidae). *Am. Midl. Nat.* 78: 369-388.

Mizell, R.F. III, and T.E. Nebeker. 1979. Number of instars of the southern pine beetle (Coleoptera: Scolytidae) and some comparisons of head capsule widths. *Ann. Entomol. Soc. Am.* 72: 313-316.

Murphy, H.E. 1922. Notes on the biology of some of our North American species of mayflies. *Bull. Lloyd Libr.* 22, *Entomol. Ser.* 2: 1-46.

Nair, G.A. 1978. Growth of different instars of *Porcellio laevis* (Latreille) (Isopoda, Porcellionidae). *Comp. Physiol. Ecol.* 3: 120-122.

Newell, R.L., and G.W. Minshall. 1978. Life history of a multivoltine mayfly, *Tricorythodes minutus:* an example of the effect of temperature on the life cycle. *Ann. Entomol. Soc. Am.* 71: 876-881.

Oberndorfer, R.Y., and K.W. Stewart. 1977. The life cycle of *Hydroperla crosbyi* (Plecoptera: Perlodidae). *Great Basin Nat.* 37: 260-273.

Palmén, J.A. 1877. Zur Morphologie des Tracheensystems. Wilhelm Engelmann, Leipzig. 149 p.

Pescador, M.L. and W.L. Peters. 1974. The life history and ecology of *Baetisca rogersi* Berner (Ephemeroptera: Baetiscidae). *Bull. Fla State Mus.* 17: 151-209.

Pleskot, G. 1959. Die Periodizität einiger Ephemeropteren der Schwechat. *Wasser Abwasser* 1958: 1-32.

Poston, F.L., R.B. Hammond, and L.P. Pedigo. 1977. Growth and development of the painted lady on soybeans (Lepidoptera: Nymphalidae). *J. Kans. Entomol. Soc.* 50: 31-36.

Rawlinson, R. 1939. Studies on the life-history and breeding of *Ecdyonurus venosus* (Ephemeroptera). *Proc. Zool. Soc. Lond. Ser. B* 109: 377-450.

Roberts, W.P., J.R. Proctor, and J.H.H. Phillips. 1978. Effect of constant temperatures on the number of larval instars of the oriental fruit moth, *Grapholitha molesta* (Lepidoptera: Tortricidae). *Can. Entomol.* 110: 623-626.

Ross, D.H., and D.A. Craig. 1979. The seven larval instars of *Prosimulium mixtum* Syme and Davies and *P. fuscum* Syme and Davies (Diptera: Simuliidae). *Can. J. Zool.* 57: 290-300.

Ross, D.H., and R.W. Merritt. 1978. The larval instars and population dynamics of five species of black flies (Diptera: Simuliidae) and their responses to selected environmental factors. *Can. J. Zool.* 56: 1633-1642.

Schmidt, F.H., and W.L. Lauer. 1977. Developmental polymorphism in *Choristoneura* spp. (Lepidoptera: Tortricidae). *Ann. Entomol. Soc. Am.* 70: 112-118.

Schmidt, F.H., R.K. Campbell, and S.J. Trotter, Jr. 1977. Errors in determining instar numbers through head capsule measurements of a Lepidopteran - a laboratory study and critique. *Ann. Entomol. Soc. Am.* 70: 750-756.

Simpson, G.G., A. Roe, and R.C. Lewontin. 1960. Quantitative zoology (revised ed.). Harcourt, Brace and Co., Inc., New York. 440 p.

Svensson, B. 1977. Life cycle, energy fluctuations and sexual differentiation in *Ephemera danica* (Ephemeroptera), a stream-living mayfly. *Oikos* 29: 78-86.

Sweeney, B.W. 1978. Bioenergetic and developmental response of a mayfly to thermal variation. *Limnol. Oceanogr.* 23: 461-477.

Taylor, R.L., and A.G. Richards. 1963. The subimaginal cuticle of the mayfly *Callibaetis* sp. (Ephemeroptera). *Ann. Entomol. Soc. Am.* 56: 418-426.

Thibault, M. 1971. Le développement des Ephéméroptères d'un ruisseau a truites des Pyrénées-Atlantiques, le Lissuraga. *Ann. Limnol.* 7: 53-120.

Thorup, J. 1963. Growth and life-cycle of invertebrates from Danish springs. *Hydrobiologia* 22: 55-84.

Trost, L.M.W., and L. Berner. 1963. The biology of *Callibaetis floridanus* Banks (Ephemeroptera: Baetidae). *Fla Entomol.* 46: 285-299.

Vaught, G.L., and K.W. Stewart. 1974. The life history and ecology of the stonefly *Neoperla clymene* (Newman) (Plecoptera: Perlidae). *Ann. Entomol. Soc. Am.* 67: 167-178.

Wigglesworth, V.B. 1972. The principles of insect physiology (seventh ed.). Chapman and Hall, London. 827 pp.

Wodsedalek, J.E. 1912. Palmen's organ and its function in nymphs of the Ephemeridae *Heptagenia interpunctata* (Say) and *Ecdyurus maculipennis* (Walsh). *Biol. Bull. (Woods Hole)* 22: 253-273.

THE EFFECT OF MESH SIZE ON THE INTERPRETATION OF THE LIFE HISTORY OF TWO MAYFLIES FROM SOUTH AUSTRALIA

P.J. Suter* and J.E. Bishop**

Department of Zoology
University of Adelaide

ABSTRACT

Using a Surber-type benthic sampler with a double net collecting system, the efficiency of nets having aperture sizes of 480 µm and 110 µm was assessed over a twelve month period in two streams in South Australia. Efficiency was defined as the percentage increase in yield (numbers) of animals obtained by use of 110 µm mesh instead of 480 µm mesh. The average annual increase in sampling efficiency of the fine net as compared with the coarse for mayfly nymphs was 412% and 235% for Spring Creek and Deep Creek respectively. The life cycles of two species of mayfly *Tasmanocoenis tillyardi* (Lestage) and *Baetis soror* Ulmer from Deep Creek are presented and the effect of the two mesh sizes on the interpretation of the life histories is discussed. Interpreted from the coarse mesh only, the life cycle of both species is bivoltine, each having one winter generation and one summer generation. Combining both fine and coarse net collections, the life cycle interpretation is distinctly different, illustrating the difficulties in drawing conclusions from coarse mesh samples. Not only is the number of generations misinterpreted in *B. soror*, but conclusions on the duration of the egg stage and length of each generation are also inaccurate for both species.

* Present address: National Museum of Victoria, Survey Department, 71 Victoria Crescent, Abbotsford, Victoria, 3067.

** Present Address: 23 Backhouse Avenue, Crafers, South Australia, 5152.

INTRODUCTION

The effect of net mesh size on sampling efficiency in lotic habitats has been discussed by various authors, including Jonasson (1955, 1958), Macan (1958), Hynes (1961), Maitland (1964), Tanaka (1967), Mundie (1971), Frost *et al.* (1971), Zelt and Clifford (1972), Clifford *et al.* (1973) and Barber and Kevern (1974). All of these

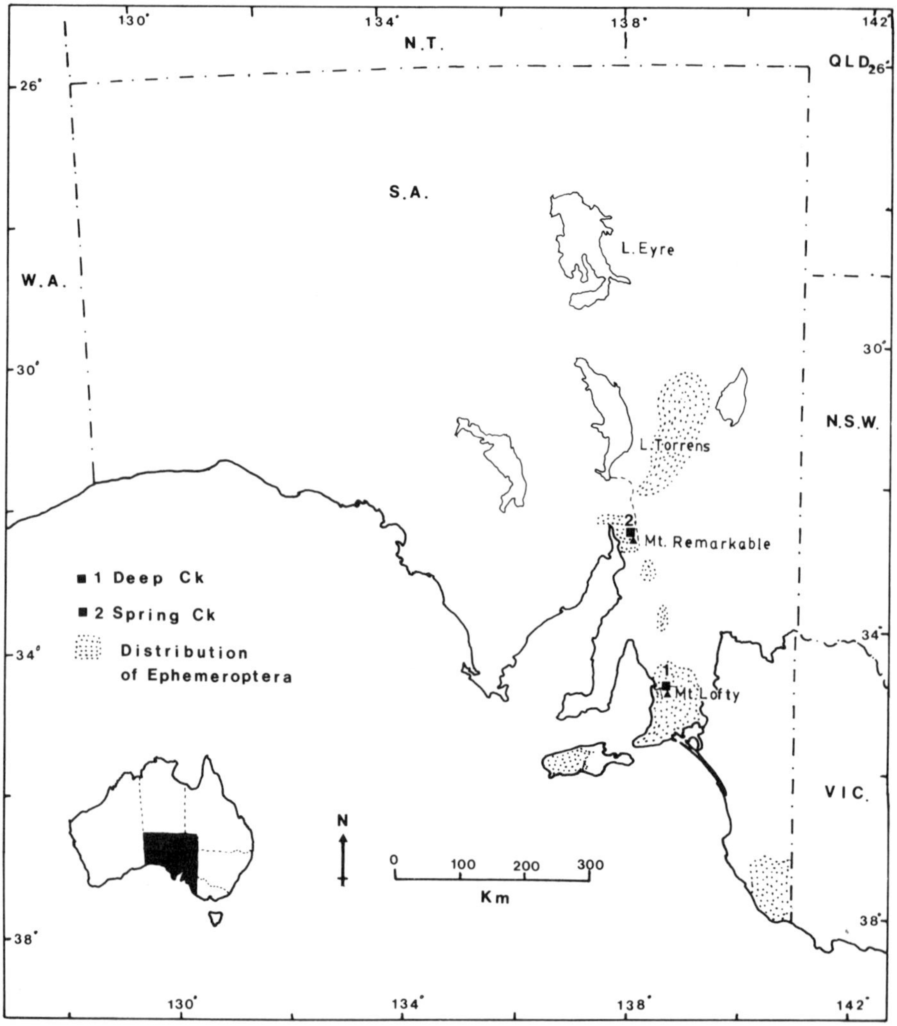

Figure 1. The location of the study sites and the distribution of Ephemeroptera in South Australia.

studies showed the deficiencies of coarse mesh sampling, but Zelt and Clifford (1972) and Clifford *et al.* (1973) showed specifically that inaccurate life cycle conclusions could be drawn using only coarse net samples.

In our study of the life cycles of some South Australian mayflies, a dual net Surber-type sampler was used. This allowed an assessment of the relative efficiency of retaining all nymphal size groups of both the inner 480 μm pore size net, and the outer 110 μm pore size net, in terms of numbers and life cycle interpretation.

STUDY SITES

The study was conducted from October 1976, to October 1977, in two creeks in South Australia; Deep Creek, draining the Mt. Lofty Ranges just north of Mt. Lofty; and Spring Creek, draining the slopes of Mt. Remarkable in the Southern Flinders Ranges (Fig. 1).

Deep Creek is a permanent flowing tributary of the R. Torrens draining a catchment of approximately 40 km^2. The creek is slightly alkaline (pH range 7.75-8.82) with ionic dominances $Mg^{2+}>Na^{+}>Ca^{2+}>K^{+}$ and $HCO_3^{-}>Cl^{-}>SO_4^{2-}$. The water temperature range observed during the study period was 5°C (August, 1977) to 23°C (March, 1977). Discharge ranged from 14 L/s in the summer to 696 L/s after a winter downpour. The study site was 24 km from Adelaide just above the confluence with the R. Torrens at an elevation of 140 m a.s.l. At the study section the stream is fifth order and drains the entire catchment. The stream is 10-15 m wide and the depth varied from 0.12-0.42 m in the riffles and pools respectively. Substrate was predominantly coarse to very coarse gravel (Φ = -4 to -6) and the erosional zones almost attained small cobble size (Φ = 6) as defined by Doeglas (1968).

Spring Creek is an intermittent or temporary stream which is a tributary of the Willochra Creek, a large intermittent stream flowing northwards towards the dry salt pan of Lake Torrens. Its catchment area is approximately 50 km^2. The creek is slightly alkaline (pH range of 7.4-8.3) with ionic dominances $Ca^{2+}>Na^{+}>Mg^{2+}>K^{+}$ and $HCO_3^{-}>Cl^{-}>SO_4^{2-}$. The Ca^{2+} and Na^{+} dominance was reversed, with Ca^{2+} dominant only when there was no surface flow. The stream water temperature was warmer than Deep Creek: minimum of 9°C in October 1976, and maximum of 29°C in March 1977. Discharge varied from zero in January to May 1977, to an estimated 30 cumec during flash flood periods. The study section was 13.5 km from the source at an elevation of 350 m a.s.l., some 8 km south of Wilmington. At the study site the stream is fifth order and drains the entire catchment. The stream bed is 15-30 m wide and depth varies from zero in summer months to approximately 2 m deep during flash floods, but is usually less than 0.5 m. Substrate was fine to coarse gravel (Φ = -3 to -5)

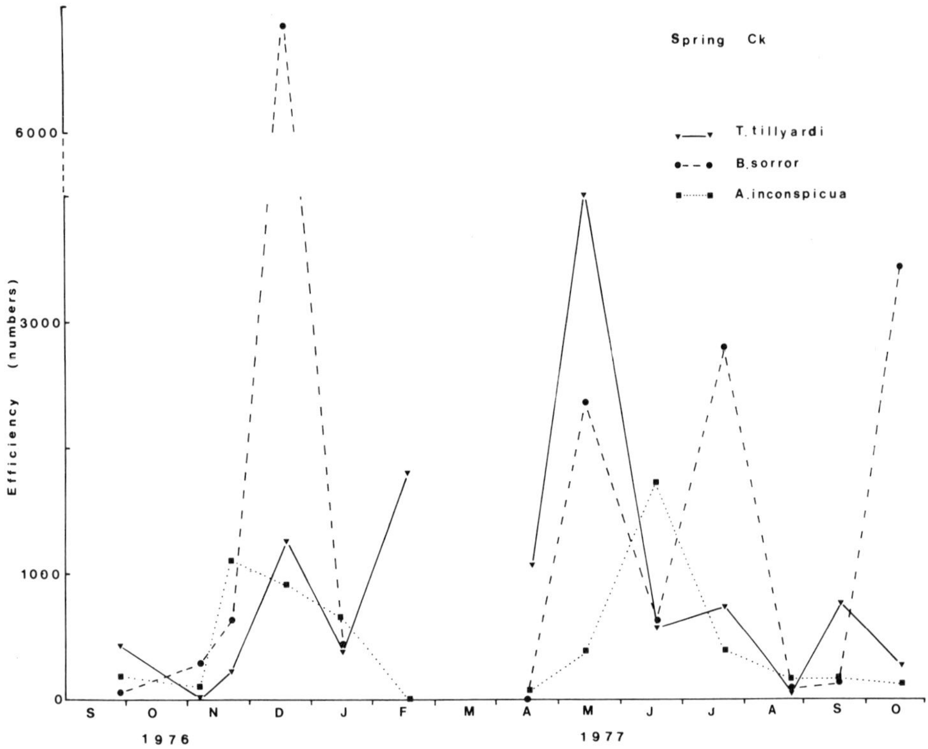

Figure 2. Seasonal variation in sieving efficiency of Ephemeroptera from Spring Creek.

as defined by Doeglas (1968).

Both localities were used in the assessment of collecting efficiency of the two nets, but life cycle interpretation is based only on material from Deep Creek.

MATERIALS AND METHODS

The general characteristics of the two streams were such that a Surber-type sampler could be used (Hellawell 1978). Although the primary aim of the study was to determine the life histories of the Ephemeropteran species occupying each stream, the selection of a dual net box sampler, based on the sampler described by Mundie (1971) and the sampler used by Bishop (1973), allowed a comparison

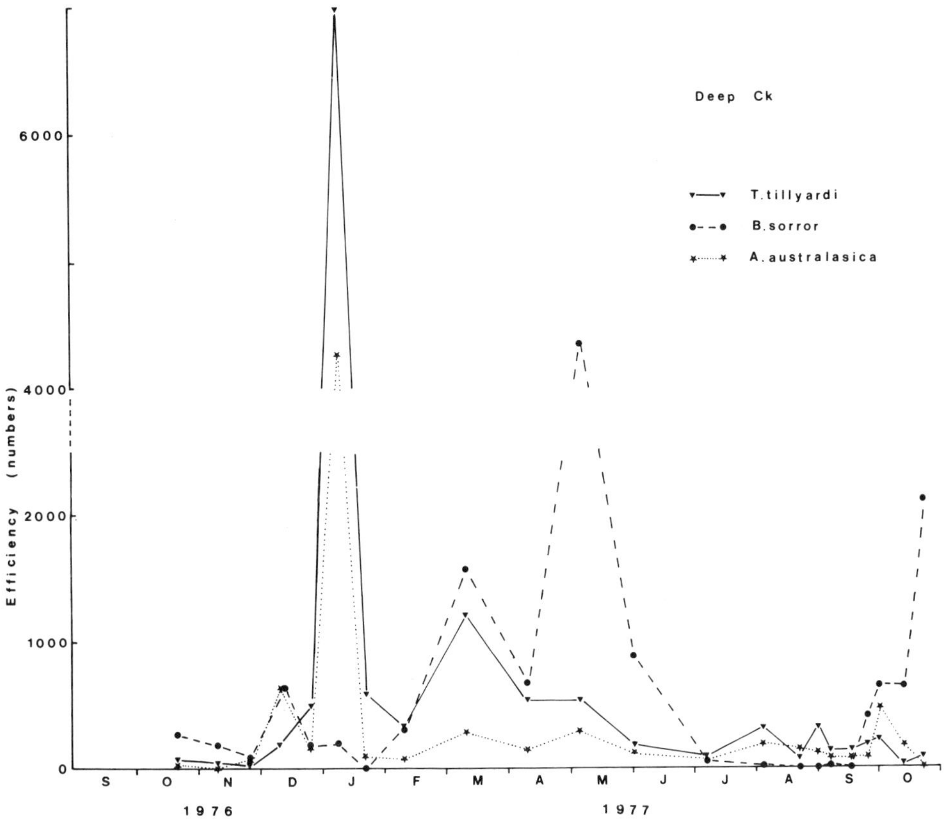

Figure 3. Seasonal variation in sieving efficiency of Ephemeroptera from Deep Creek.

of the efficiencies of different mesh apertures in the collection of the mayfly fauna.

The sampler was a galvanised steel box enclosing 0.1 m^2 of stream bed. Into a rear steel groove were inserted two nets, an inner coarse mesh net ("Nytal" 500 μm mesh aperture; actual measured aperture of 460-480 μm; diagonal aperture of 678 μm) and an outer fine mesh net ("Nytal" 106 μm mesh aperture; actual measured aperture of 98-120 μm; diagonal aperture of 155 μm). The coarse net inside the fine retained larger animals and detritus carried in by the current after physical disturbance of the substrate, and protected the fine net, which retained material down to 80 μm diameter, from damage.

Both nets were a tapering design with the inner coarse net 60 cm long and tapering to a 10 x 10 cm trailing end, and the outer net 2 m long also tapering to a 10 x 10 cm trailing end. This tapering design ensured that the inner coarse net did not come in contact with the outer net, thus maintaining the sieving effectiveness of both nets. The long length of the 110 µm net overcame the major problems experienced with fine net samplers, those of blockage by fine organic and inorganic material and creation of backwash.

At each site on each sampling date four samples were taken from visually selected habitats, broadly categorised as pool, run, flat and riffle (after Allen 1951). The sampler was forced into the substrate until an effective seal was formed, and the substrate was disturbed and brushed by hand. Once the sampling was completed, the nets were removed from the rear groove and the contents of each washed into separate polyethylene bags and preserved with 10% formalin. During the periods of zero flow at Spring Creek a 2 m kick sample using both nets (c.f. Hynes 1961), was used to collect the benthos in the pools and slack reaches.

In the laboratory each coarse net sample was poured into a 250 µm aperture copper sieve and washed with a large volume of water. The sample was then hand sorted for Ephemeropteran nymphs in a white sorting tray. When there were more than 300 specimens of any one species in a sample, a subsample was taken such that a minimum of 150 animals were measured in any subsample.

The fine mesh net samples were washed in a fine sieve and the organic material was separated from the inorganic fraction by repeated (5 x) flotation in saturated calcium chloride solution, and elutriation through the 110 µm mesh. The organic fraction was washed in water, and when the volume of material was too great to sort in its entirety subsampling was carried out using the technique of Bishop (1973). The sample or subsample was sorted in a plankton sorting tray, based on the Fenwick pattern as illustrated by Hellawell (1978), using 25 x magnification.

Each nymph, from both of the above sets of samples, was identified, the head with measured at 50 x magnification, and stored in ethyl alcohol-glycerol solution.

The sampling efficiency of each net was calculated as the percentage increase in yield (numbers) of nymphs obtained by use of 110 µm mesh instead of 480 µm mesh (c.f. Jonasson 1955,1958; Barber and Kevern 1974).

The effect of the two meshes on interpretation of the life cycle of *Tasmanocoenis tillyardi* (Lestage) (Caenidae) and *Baetis soror* Ulmer (Baetidae) was qualitatively assessed only from Deep Creek by comparing the life cycles as interpreted from collections made with the coarse net with the life cycle obtained with the combined fine and coarse nets. Interpretation of the life cycle necessitated the recognition of distinct polymodal distributions. A computer program NORMSEP, developed by Hasselblad (1966), for fitting a series of normal curves to frequency data, was used to aid the determination of modes. However, this technique depends on the decision on the number of modes present (Cohen 1966). Similar results to those obtained using NORMSEP were obtained using visual selection of distinct distributions from which a mean size class was calculated. NORMSEP also required at least 100 specimens (preferably 400) and when this requirement was not met the mean size classes were calculated from visually selected distributions. The calculated mean values were plotted to represent the seasonal growth pattern of each species.

RESULTS

The seasonal variations of sieving efficiency for the two study localities, Spring Creek and Deep Creek, are given in Figures 2 and 3. The average annual sampling efficiencies of the fine compared with coarse net for mayfly nymphs were 421% and 235% for Spring Creek and Deep Creek respectively. At both localities the average annual efficiencies, as measured for each species, were *T. tillyardi* 838% (19-4050%) and 575 (8-6997%) respectively; *B. soror* 1263% (0-6890%) and 595% (0-4350%) respectively; *Atalophlebia australasica* Pictet (Deep Creek only) 339% (0-4267%); and *Atalonella inconspicua* (Eaton) (Spring Creek only) 430% (0-1738%). These values represent large increases in the number of specimens collected throughout the annual cycle, but the larger efficiency increases were invariably associated with periods of recruitment. The Leptophlebiid species, *Atalophlebia australasica* and *Atalonella inconspicua* , with large preemergent size showed consistently lower efficiency results than the smaller *T. tillyardi* and *B. soror*.

Tasmanocoenis tillyardi

The size frequency histograms of the 'coarse net' population of *T. tillyardi* from Deep Creek, are given in Figure 4. From these histograms interpretation is difficult, but the seasonal growth curves (Fig. 7) show that *T. tillyardi* has a bivoltine life cycle with one generation (G1) emerging in November-December 1976, and a second generation (G2) emerging in April-May 1977. This suggests that G2 originated with eggs laid by generation G1, and developed completely over the summer months. At the end of the May emergence, another generation was found (G1'). This generation apparently originated from eggs oviposited during the latter half of the emergence period G1. Growth of G1' was rapid during the summer and autumn months but very slow during winter. With the onset of warmer

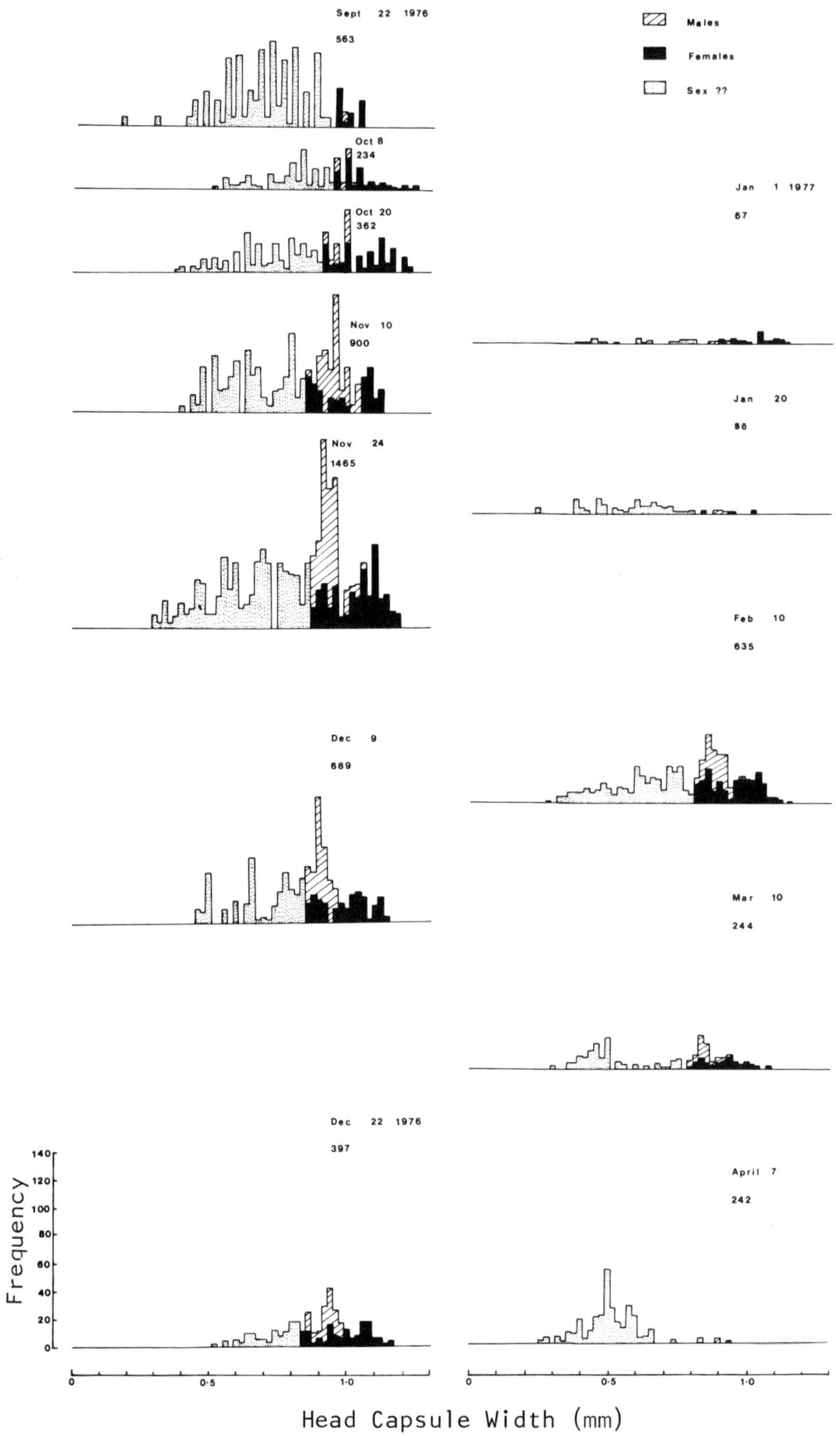
Sept 22 1976
563
Oct 8
234
Oct 20
362
Nov 10
900
Nov 24
1465
Dec 9
689
Dec 22 1976
397
Males
Females
Sex ??
Jan 1 1977
67
Jan 20
86
Feb 10
635
Mar 10
244
April 7
242
Frequency
140
120
100
80
60
40
20
0
0
0·5
1·0
Head Capsule Width (mm)

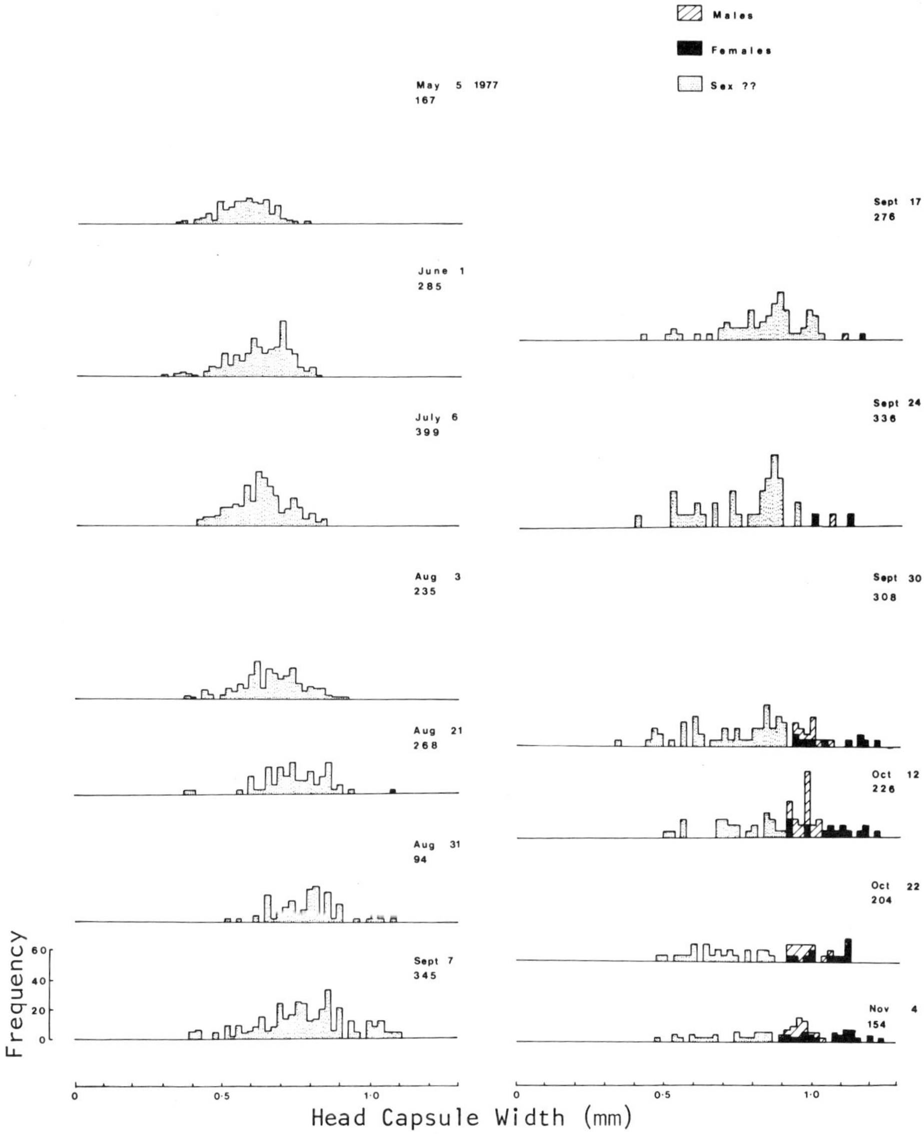

Figure 4a & 4b. Life cycle size frequency histograms of *Tasmanocoenis tillyardi* from 500 μm net samples only.

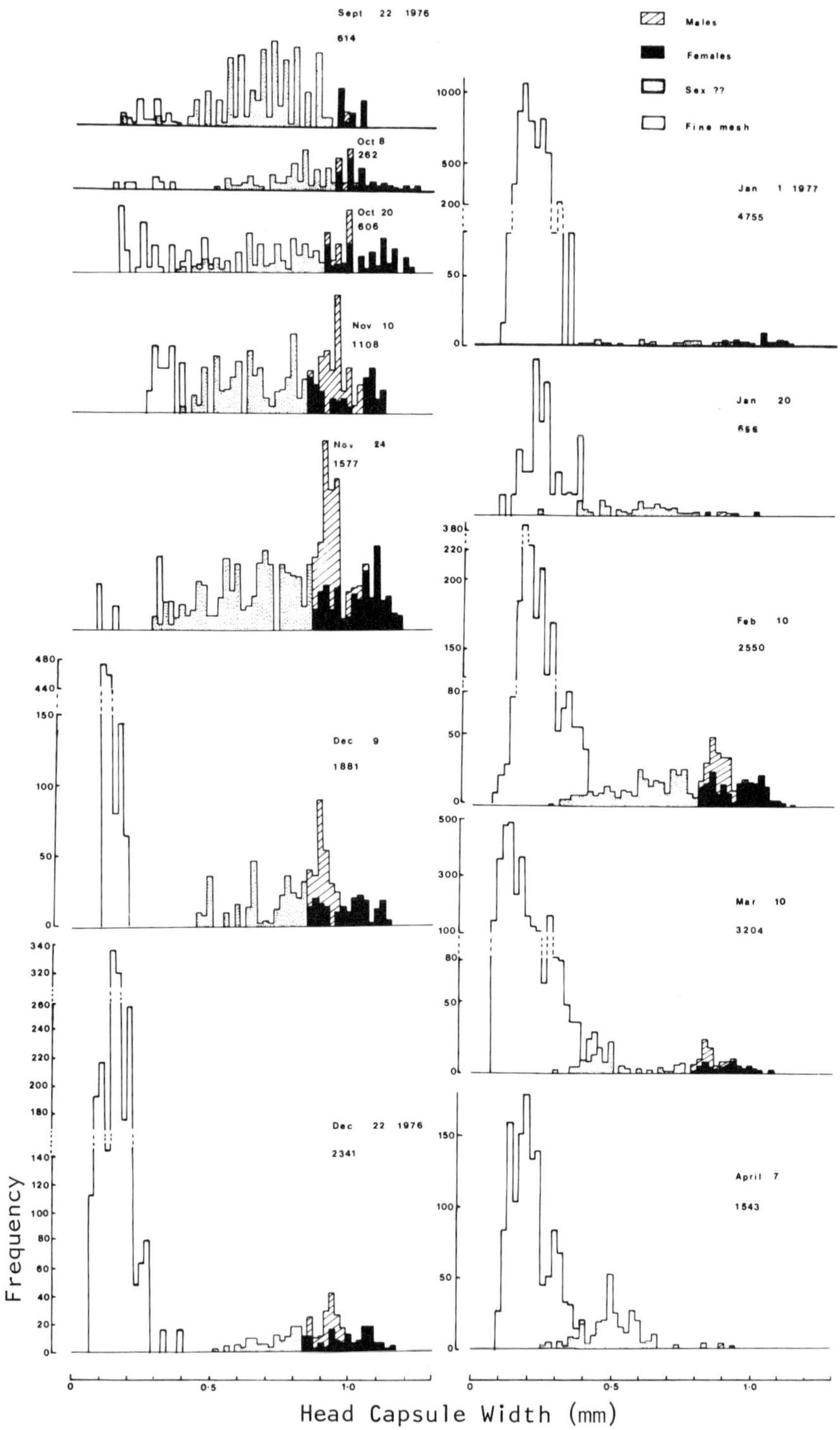

Males
Females
Sex ??
Fine mesh
Sept 22 1976
614
Oct 8
262
Oct 20
606
Nov 10
1108
Nov 24
1577
Dec 9
1881
Dec 22 1976
2341
Jan 1 1977
4755
Jan 20
656
Feb 10
2550
Mar 10
3204
April 7
1543
Frequency
Head Capsule Width (mm)

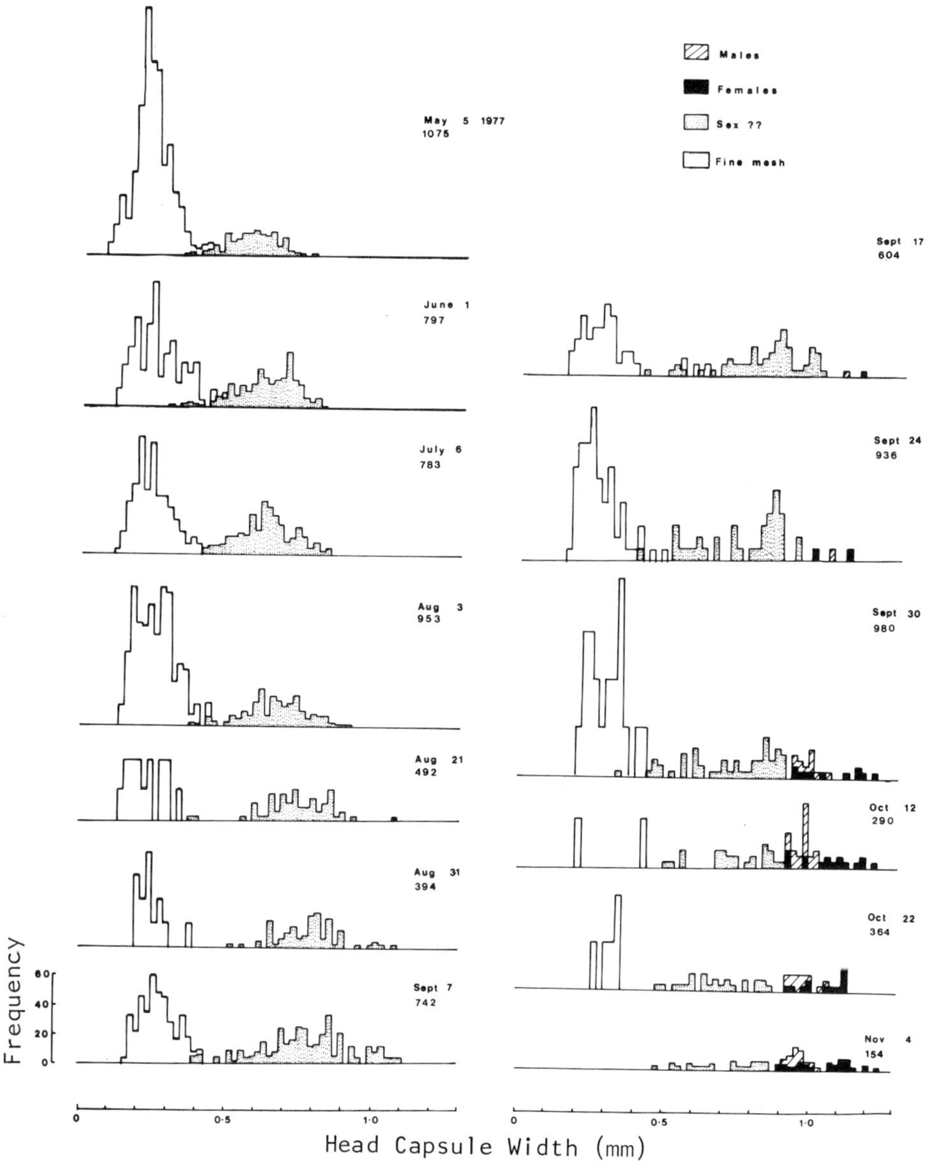

Figure 5a & 5b. Life cycle size frequency histograms of *T. tillyardi* from combined 110 μm and 500 μm net samples.

spring temperatures in August-September the growth rate increased and emergence commenced in September and continued to October 1977. Emergence continued after the completion of the sampling program, and with the exception of the precocious emergence in September-October, this generation can be equated with generation G1 of the previous year. The life cycle as described, is equivalent to the bivoltine life cycle classification B1 in Landa's (1968) scheme, with G1 giving rise to G2 and part of G1', and G2 giving rise to part of G1'.

In comparison, histograms from both coarse and fine nets combined (Fig. 5) and the seasonal growth curves (Fig. 8) show that *T. tillyardi* has, in fact, a bivoltine life cycle, but with both generations present throughout the annual cycle. The emergence times are similar, but the hatching times and duration of nymphal development are totally different from those derived from the coarse net collections.

The life cycle of *T. tillyardi* at Deep Creek is bivoltine with two generations present in September 1976, one nearing emergence (G1) and one of early instar nymphs (G2). Adults from G1 were first recorded in November 1976 and emergence continued until December, and possibly into January, 1977. Hatching of eggs laid by generation G1 was first recorded in November 1976 and hatching was continuous until May 1977. Laboratory egg development experiments suggest that the length of the recruitment period was exaggerated by the hatching of eggs laid by generation G2 in April-May 1977.

Generation G2 developed relatively rapidly in November 1976-January 1977, and although last instar animals were recorded in early February 1977 no adults were recorded until late February, early March. The emergence of G2 completed the two generations present at the beginning of the sampling program. Both were replaced, however, by their respective offspring forming distinct cohorts, G1' and G2'.

Generation G1' commenced in November 1976 and recruitment was continuous throughout the long emergence period, depressing the average size of this generation. Growth rate therefore appeared slower during the summer months that it was in March-April 1977, when recruitment ceased. Generation G2' was distinguishable in early March 1977, and recruitment continued until early May 1977.

Development of G1' and G2' was slow during the cooler months of May to August 1977, but as the water temperature began to increase in August and September (Fig. 6), growth rate increased, and a short period of recruitment was observed in September. These early instar nymphs were indistinguishable from G2' by early November 1977. This recruitment occurred at least one week before any last instar nymphs were collected. Laboratory egg development experiments suggested

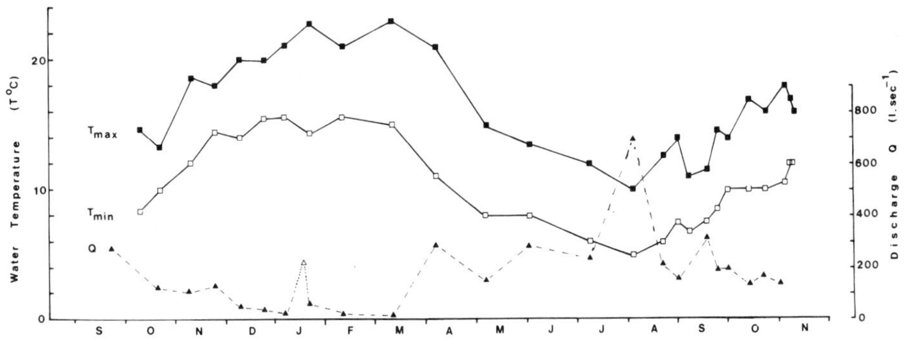

Figure 6. Seasonal changes of water temperature (°C) and discharge (L/sec) in Deep Creek.

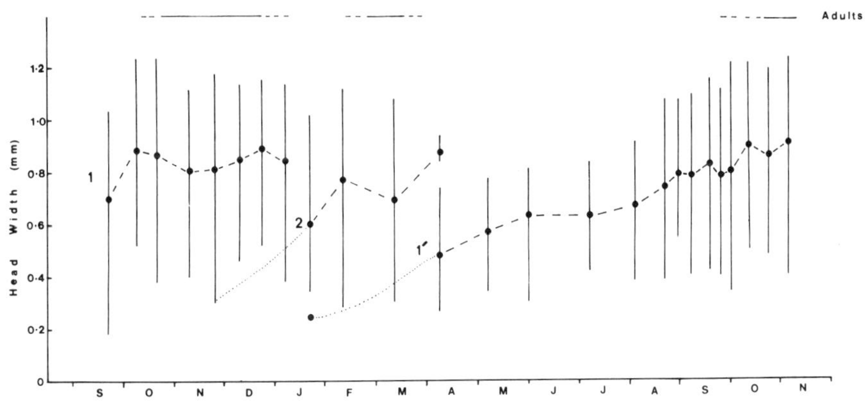

Figure 7. Seasonal growth curves of *T. tillyardi* from coarse mesh samples only.

that *T. tillyardi* eggs may become dormant at low temperature. Therefore this hatch was probably from eggs laid by G2 adults in late autumn, with the eggs over-wintering in a dormant state.

By October when the sampling program ended both G1' and G2' had attained similar sizes to generations G1 and G2 of October 1976, suggesting the above is an annual cycle. Consequently both generations G1 and G2 require 12-13 months for complete development.

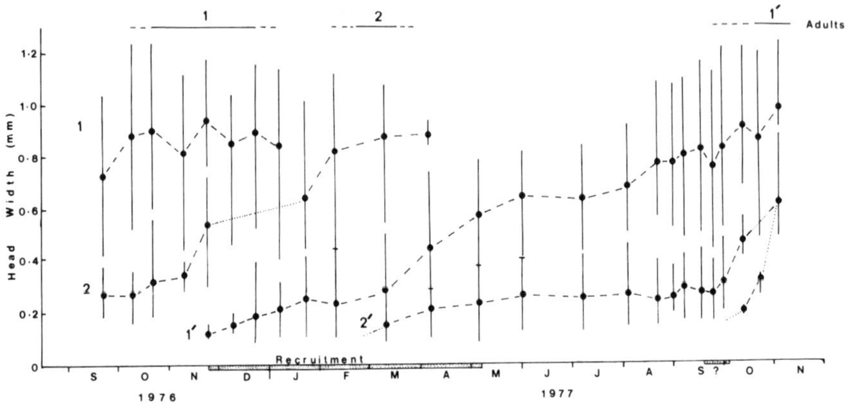

Figure 8. Seasonal growth curves of *T. tillyardi* from combined coarse and fine mesh collections.

Baetis soror

Field data for *B. soror* from Deep Creek, as collected by the coarse net only, are presented as size frequency histograms in Figure 9. Figure 11 illustrates the seasonal growth curves as interpreted from these histograms. The life cycle of *B. soror* appeared to be bivoltine with one generation (G1) having a long emergence period from October 1976 to January 1977, and a second generation (G2) emerging in April to May 1977. From the seasonal growth curves, generation G1 apparently emerged and gave rise to generation G1'. There was no evidence of the presence of this sibling generation during the summer months, but it appears in May to June 1977, suggesting that eggs laid by G1 did not hatch during the summer, but remained dormant until temperatures began to fall.

Generation G2 developed rapidly during the summer and emerged, and oviposited in April-May 1977, but no recruitment became evident until September 1977 when generation G2' appeared. Throughout the winter period only generation G1' developed slowly until emergence commenced in August 1977. Therefore it appears that the eggs laid by G2 remained dormant throughout the winter.

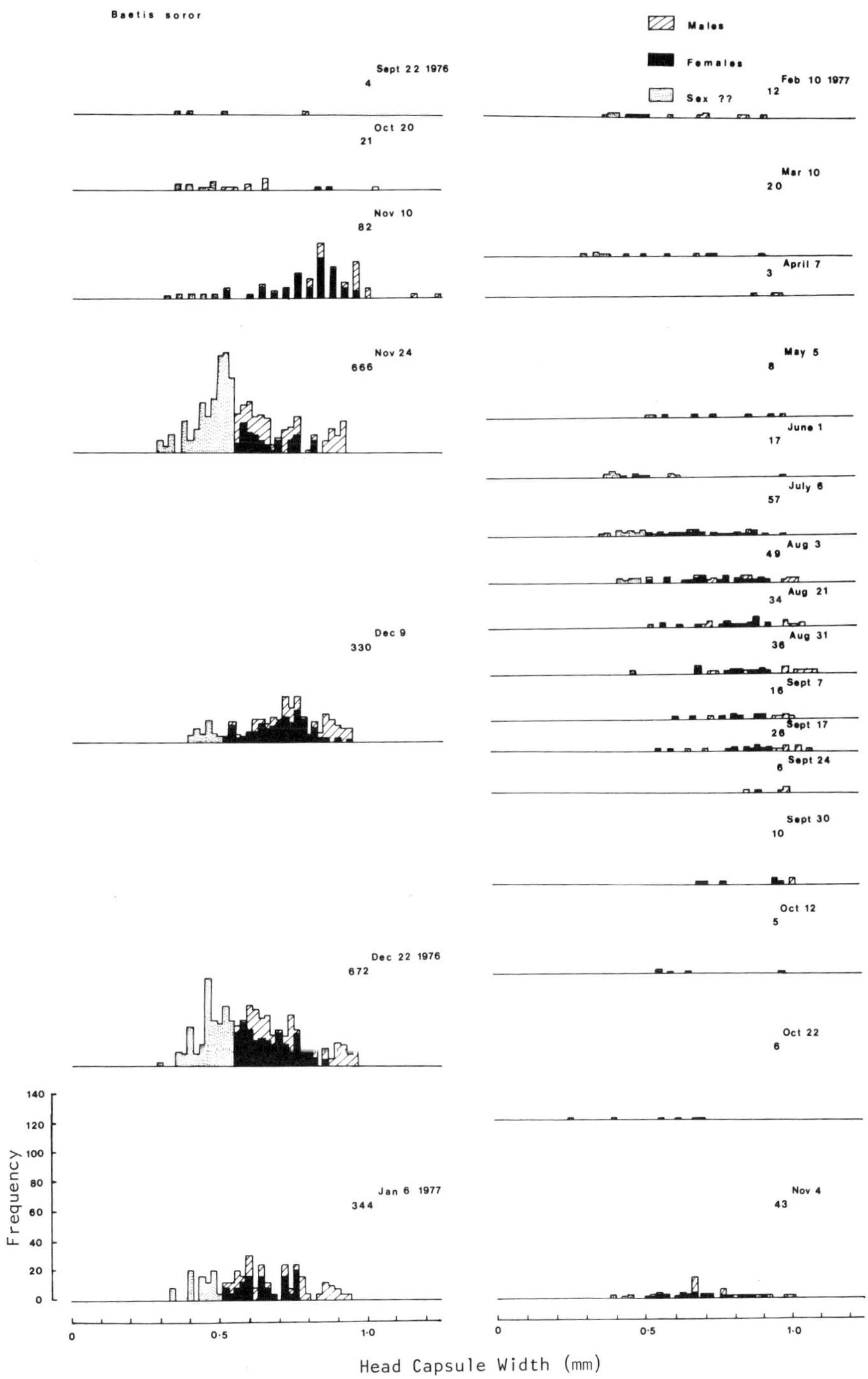

Figure 9. Life cycle size frequency histograms of *Baetis soror* from the coarse net samples only.

By October 1977 both generations G1' and G2' had attained the size of their counterpart generations of the previous year, 1976. Thus an annual cycle is indicated, with G1 giving rise to G1' and G2 giving rise to G2'.

The interpretation of the life cycle of *B. soror* differs dramatically once the fine mesh portion of the sample is included. Size frequency histograms, (Fig. 10) and seasonal growth curves (Fig. 12) both suggest a polyvoltine life cycle with three generations per year.

From Figure 10 a drastic decrease in the number of nymphs collected after January 6, 1977 is evident. During early January (2-4) 1977, a spate scoured the stream bed removing much of the *Cladophora*, clearly affecting the population numbers of *B. soror* (Fig. 6). The absence of subsequent recovery during the study period suggests that the effect of the spate was at least semi-permanent.

The field data for October 1976 suggest the presence of three generations, one emerging and with nymphs approaching last instar (G1); a second generation with head widths between 0.30 mm and 0.66 mm (G2) and a third generation of early instar animals (G3).

Generation G1 commenced emergence in November 1976 but continued over an extended period to January 1977. Generation G2 developed rapidly over this period and commenced emergence in late April to early May 1977. The growth curve discontinuity of the third generation, G3, during December and January was caused by recruitment from early hatchings from eggs laid by G1 females. Not until March could offspring of G1 be distinguished from G3, commencing generation G1'. Generation G3 developed rapidly during March and April 1977 and began to emerge in early June.

As mentioned, G1' had its origins from eggs laid by G1 females emerging in the second half of the emergence period. Hatchings from eggs of earlier emergers were indistinguishable from generation G3, and therefore were included in the calculation of the G3 mean size classes. Generation G1', therefore began in February-March 1977. Recruitment from eggs of G2, oviposited in April and May, caused the depression of the mean size of G1', and from June to August 1977 the two hatches, appeared as one generation G1'. Therefore, G1' was the sole generation developing throughout the winter period. Last instars were collected and emergence commenced in late August continuing through until the end of the study in November 1977.

Figure 10. Life cycle size frequency histograms of *B. soror* from combined coarse and fine net samples.

Figure 10

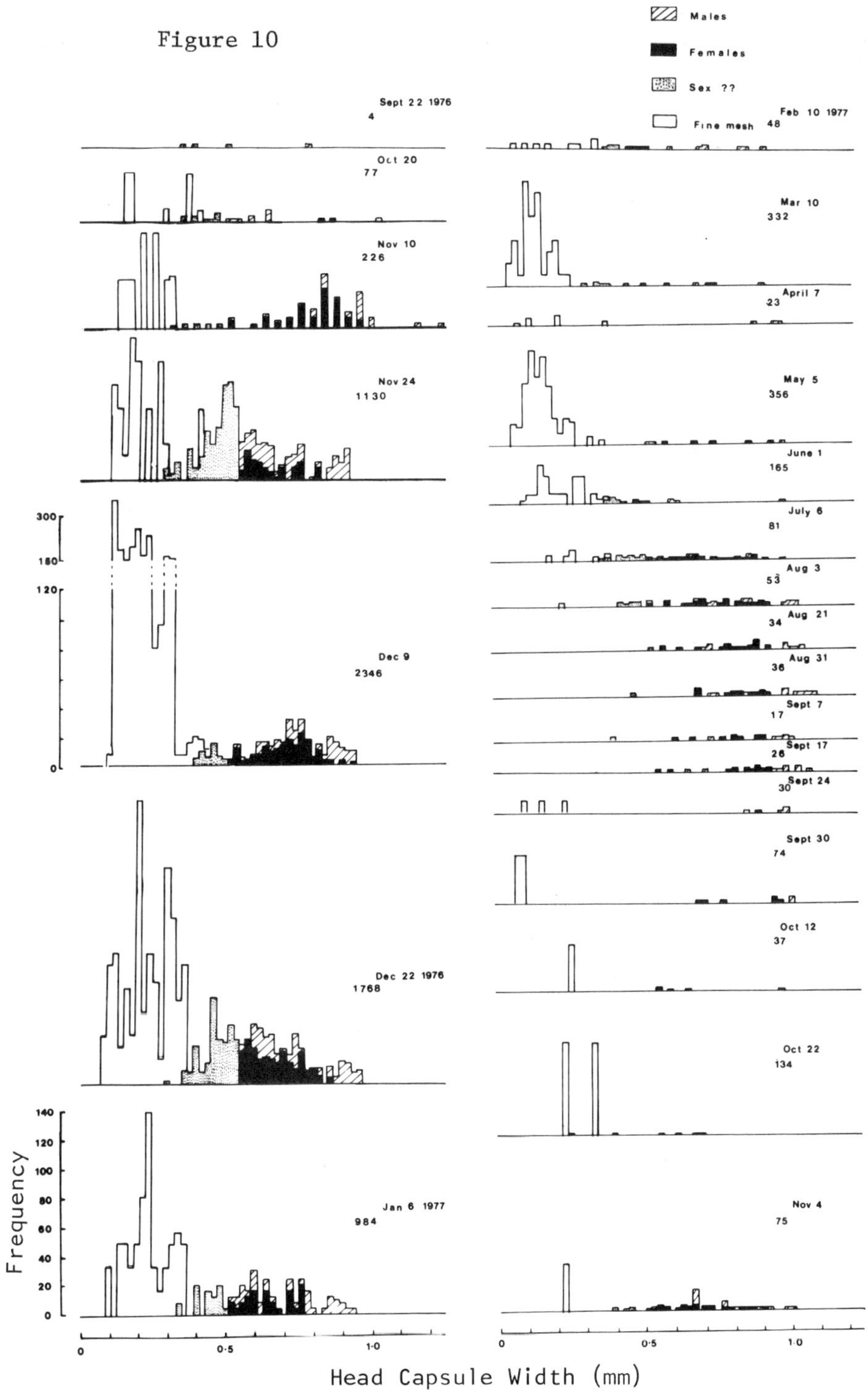
Males
Females
Sex ??
Fine mesh
Sept 22 1976
4
Oct 20
77
Nov 10
226
Nov 24
1130
Dec 9
2346
Dec 22 1976
1768
Jan 6 1977
984
Feb 10 1977
48
Mar 10
332
April 7
23
May 5
356
June 1
165
July 6
81
Aug 3
53
Aug 21
34
Aug 31
36
Sept 7
17
Sept 17
26
Sept 24
30
Sept 30
74
Oct 12
37
Oct 22
134
Nov 4
75
300
180
120
0
140
120
100
80
60
40
20
0
Frequency
0
0·5
1·0
Head Capsule Width (mm)

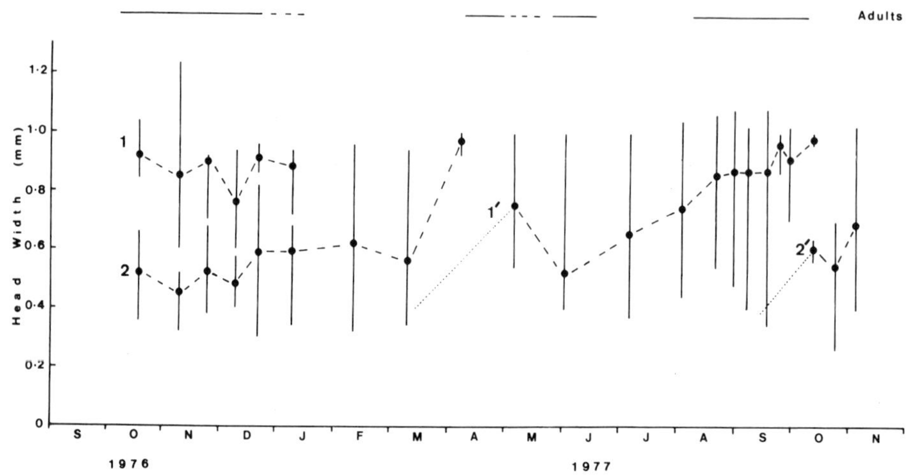

Figure 11. Seasonal growth curves of *B. soror*, interpreted from coarse net samples only.

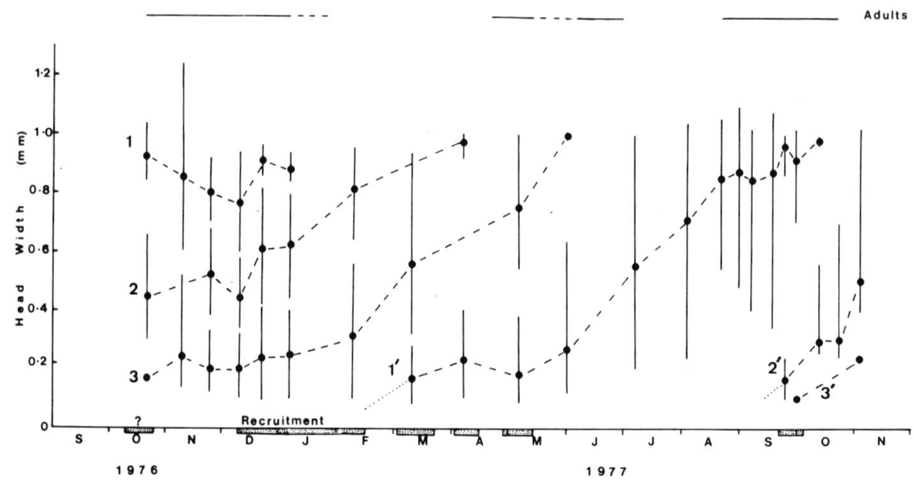

Figure 12. Seasonal growth curves of *B. soror*, interpreted from combined coarse and fine net samples.

Although first instar nymphs were missed in the August-early September 1977 period, evidence of a new generation (G2') in the form of 3rd$^+$ instar nymphs was apparent. These perhaps, recruited from the eggs laid by G3 females in May to June, developed slowly in the low temperatures of winter. Laboratory experiments showed that no quiescence or diapause occurred in eggs of *B. soror*, and at the winter temperatures experienced, (greater than 5°C) development was continuous although very slow. Recruitment was observed in late September-early October 1977, from eggs laid by the early emerging adults of G1' beginning generation G3' and completing the annual cycle of three generations.

The duration of each generation from earliest hatching to last emergence, as extrapolated from the annual cycle, is as follows, G1 (G1') eleven months (February-January); G2 (G2') eight months (September-May) and G3 (G3') eight months (November-June).

DISCUSSION

The effect of inadequate sampling with nets of large pore size has been discussed by many authors. Those studies by Macan (1958), Maitland (1964), Tanaka (1967), Mundie (1971) and Frost *et al.* (1971) were based on samples taken on a single occasion, and although in all cases the fine mesh collection showed the deficiencies of the coarse mesh samples, the influence on the interpretation of life cycle within the lotic habitat was not investigated.

Zelt and Clifford (1972) assessed the efficiency of Surber and dip nets with 720 μm and 320 μm pore size. Their rationale was that "Recent studies have shown that numerically much of the stream fauna can pass through nets having pore size as small as 300 μm. However in the past most of the inferences about stream faunas have been based on samples collected with coarser-mesh nets". Zelt and Clifford used a double net system attached either to a dip net, or to a Surber sampler and found that 45% of the total collections of insects passed through the coarse dip net, and 54% passed through the coarse mesh of the Surber sampler. Of the total animals passing through the 720 μm mesh in both samplers, the Ephemeroptera accounted for 58%. In terms of biomass, the lost animals were almost insignificant, only 3% and 6% for the dip net and Surber sampler respectively. The percentage increase in numbers of animals collected with the finer mesh (320 μm) was obviously significant. However, as they themselves discussed, the fine mesh net was itself relatively coarse, allowing many smaller animals to escape.

Barber and Kevern (1974), using preserved animals (therefore eliminating the active escape behaviour of nymphs and larvae), found that a sieve of 0.25 mm pore size instead of 0.50 mm increased the sampling efficiency by 300-600% for mayflies, chironomids, simuliids, "other dipterans", water mites and plecopterans throughout an annual cycle. "For the fauna as a whole, efficiency varied from 95% to 325% over the study period." This order of magnitude increase in sampling efficiency is obviously important to any study of benthic fauna. In terms of biomass, Barber and Kevern calculated the increase in efficiency to be 10% but reached 174%, 60% and 80% for chironomids, simuliids and water mites respectively. Consequently they concluded, "a smaller-mesh screen (about 0.25 mm or 0.20 mm mesh openings) should be employed if the purpose is to quantitatively estimate numbers or to obtain first instars of all invertebrates. Also the mesh size will somewhat influence the estimate of production if one used the method proposed by Hynes and Coleman (1968: modified by Hamilton 1969), a method which depends on the changes in numbers of organisms in different length categories, as numbers of the smaller lengths will not be estimated quantitatively."

This conclusion is substantially correct, but the present study has shown that a finer mesh than Barber and Kevern proposed (i.e. finer than 200 μm) would be required to obtain first instars of many living aquatic invertebrates. In the present study the mesh aperture sizes of 480 μm and 110 μm produced similar results to those obtained by Barber and Kevern (1974), but on many occasions the efficiency gained by the fine mesh was considerably greater than that recorded by those authors.

At both localities the increase in efficiency obtained using both nets throughout the annual cycle illustrated the effect of coarse net sampling in terms of numbers and life cycle interpretation. That the presence of one generation of *T. tillyardi* throughout the autumn, winter and spring of 1977 was not evident in the coarse net sample, leads to obvious incorrect conclusions based on those samples. Similarly, the incorrect assessment of the number of generations of *B. soror* in Deep Creek from the coarse net samples is also the result of the inadequate sampling. Conclusions drawn from the coarse net samples may not reflect the true nature of life histories. Clifford *et al.* (1973) and Zelt and Clifford (1972) recorded that the 720 μm mesh led to inaccurate interpretation of life cycles of some mayflies and stoneflies from the foothill streams of Alberta. Zelt and Clifford (1972) illustrated the erroneous conclusions drawn from coarse net samples with the stonefly *Nemoura decepta*. The life cycle of this species appeared to be a summer type with dormant eggs during the winter, however, fine net samples revealed small nymphs in February. The similarity between those observations and those made with *T. tillyardi* emphasize the need to use fine mesh nets in life history studies, especially when classification of life cycle (i.e. Landa 1968) is considered.

Using Landa's (1968) classification system the evidence obtained using the coarse net suggests a B1 cycle for *T. tillyardi* with two generations per year. One generation develops over the winter months, emerges and oviposits in spring, and eggs from this generation hatch giving rise to a second generation which completes its development during the summer months. However, the life cycle, as interpreted using the combined nets (which did successfully collect all instars of *T. tillyardi*), cannot be classified using Landa's scheme. There are in fact two winter generations present at any one time, and therefore a modification of Landa's classification is required by the inclusion of bivoltine species with two winter generations, one hatching in spring and emerging the following spring, and a second hatching in autumn and emerging in the following autumn (perhaps a B5?).

The life cycle of *B. soror*, as interpreted from coarse samples only, cannot easily be classified in Landa's system. Although bivoltine, the two generations differ. One commences in autumn and develops over winter to emerge during spring and summer, but its eggs appear to remain dormant during the summer months. The second generation commences in the spring and develops over summer to emerge in autumn; its eggs remaining dormant over the winter period. Data from both coarse and fine samples suggest not two, but three generations per year; one commencing in summer and emerging in the following spring; one commencing in early spring and emerging in early autumn; and one starting in late spring and emerging in late autumn. This system is similar to a B3 cycle as proposed by Landa. Although there are differences, especially in the origins of generation 2 (i.e. generation 2 in Landa's scheme has its origins from the overwintering generation 1, whereas for *B. soror* generation 2 has its origins from eggs laid by the second summer generation (3) which has slow development over winter) the similarities are obvious, with two "summer" generations and one "winter" generation.

In summary, evidence from this study therefore strongly supports the conclusions of Zelt and Clifford (1972) who recommended that nets of pore size of 100 μm even 76 μm be used for life cycle studies. The inaccurate interpretation of life cycles derived from coarse net samples alone are obvious from these studies on the two populations of the South Australian mayflies *T. tillyardi* and *B. soror*. Use of a fine mesh net (110 μm) allowed direct collection of hatching and early instars of these species and is preferred to the reliance on extrapolated conclusions, which may not be justified, from coarse net samples.

ACKNOWLEDGMENTS

We would like to thank Miss U. Gawronski for typing the manuscript, and the National Museum of Victoria for granting permission for one of us (P.J.S.) to attend the Third International Conference on Ephemeroptera. This study was supported in part by a Commonwealth of Australia Postgraduate Research Award held by Phillip J. Suter and by research funds of the Zoology Department, University of Adelaide.

RESUME

Utilisant un échantillonneur benthique de type Surber muni d'un système collecteur à double filet, on a évalué l'efficacité des filets d'une ouverture de 480 μm et de 110 μm durant une période de douze mois dans deux cours d'eau d'Australie méridionale. L'augmentation moyenne annuelle de l'efficacité de la prise d'échantillons des nymphes d'éphéméroptères dans le filet fin par rapport au gros filet fut de 412% pour le Spring Creek et de 235% pour le Deep Creek. Les cycles de vie de la *Tasmanocoenis tillyardi* (Lestage) et la *Baetis soror* Ulmer recueillis dans le Deep Creek sont décrites de même que le rôle de la grosseur des mailles dans l'interprétation de la vie des insectes.

Le cycle de vie des deux espèces, interprété uniquement à partir des grosses mailles, est celui de bivoltins. Si l'on combine les échantillons recueillis par les filets aux mailles fines et grosses, l'interprétation du cycle de vie est nettement différente, ce qui montre combien il est difficile de tirer des conclusions à partir de spécimens obtenus au moyen d'un filet à grosses mailles.

ZUSSAMENFASSUNG

Beim Gebrauch eines benthischen Samplers vom Type "Surber" mit Doppelnetz Sammelsystem wurde die Wirksamkeit von Netzen mit einer Öffnungsgröße von 480 μm und 110 μm während einer Periode von zwölf Monaten überprüft. Der durchschnittliche, jährliche Probe Effektivitätszuwachs des feinen Netzes betrug, verglichen mit dem groben Netz, beim Fang der Eintagsfliegen jeweils 412% und 235% beim Spring Creek und Deep Creek. Die Lebensyklen von *Tasmanocoenis tillyardi* (Lestage) und *Baetis soror* Ulmer aus dem Deep Creek werden hiermit in diesem Aufsatz dargestellt, und die Wirkung verschiedener Maschengrößen in Bezug auf die Interpretation der Lebensgeschichten wird erörtert. Der Lebenszyklus beider Arten, nur vom grobmaschigen Netz her interpretiert, ist bivoltin. Wenn Sammlungen von feinen und groben Netzen kombiniert verwendent werden, ergibt sich eine ganz unterschiedliche Lebenszyklus Interpretation. Daran erkennt man, welche Schwierigkeiten auftreten, wenn man Schlüsse zieht von Proben, die mit einem grobmaschigen Netz genommen wurden.

REFERENCES

Allen, K.R. 1951. The Horokiwi stream. A study of a trout population. *Fish. Bull. N.Z.* 10: 1-231.

Barber, W.E. and N.R. Kevern. 1974. Seasonal variation of sieving efficiency in a lotic habitat. *Freshw. Biol.* 4: 293-300.

Bishop, J.E. 1973. Limnology of a Small Malayan River, Sungai Gombak. Dr. W. Junk, The Hague. *Monogr. Biol.* 22.

Cohen, A.C. 1966. Discussion of "Estimation of mixture of normal distributions" by Victor Hasselblad. *Technometrics* 8: 445-446.

Clifford, H.F., M.R. Robertson and K.A. Zelt. 1973. Life cycle patterns of mayflies (Ephemeroptera) from some streams of Alberta, Canada. p. 122-131. *in*: W.L. Peters and J.G. Peters, ed. Proceedings of the First International Conference on Ephemeroptera. E.J. Brill, Leiden.

Doeglas, D.J. 1968. Grian-size indices, classification and environment. *Sedimentology* 10: 83-100.

Frost, S., A. Huni and W.E. Kershaw. 1971. Evaluation of a kicking technique for sampling stream bottom fauna. *Can. J. Zool.* 49: 167-173.

Hamilton, A.L. 1969. On estimating annual production. *Limnol. Oceanogr.* 14: 771-782.

Hasselblad, V. 1966. Estimation of parameters for a mixture of normal distributions. *Technometrics* 8: 431-444.

Hellawell, J.M. 1978. Biological surveillance of rivers. A biological monitoring handbook. Water Research Centre, England.

Hynes, H.B.N. 1961. The invertebrate fauna of a Welsh mountain stream. *Arch. Hydrobiol.* 57: 344-388.

Hynes, H.B.N. and M.J. Coleman. 1968. A simple method of assessing the annual production of stream benthos. *Limnol. Oceanogr.* 13: 569-573.

Jonasson, P.M. 1955. The efficiency of sieving techniques for sampling freshwater bottom fauna. *Oikos* 6: 183-207.

Jonasson, P.M. 1958. The mesh factor in sieving techniques. *Int. Ver. theor. angew. Limnol. Verh.* 13: 860-866.

Landa, V. 1968. Developmental cycles of Central European Ephemeroptera and their interrelations. *Acta Entomol. Bohemoslov.* 65: 276-284.

Macan, T.T. 1958. Methods of sampling the bottom fauna in stony streams. *Int. Ver. theor. angew. Limnol. Mitt.* 8: 1-21.

Maitland, P.S. 1964. Quantitative studies on the invertebrate fauna of sandy and stony substrates in the River Endrick, Scotland. *Proc. R. Soc. Edinb.* 68: 277-301.

Mundie, J.H. 1971. Sampling benthos and substrate materials, down to 50 microns in size, in shallow streams. *J. Fish. Res. Board Can.* 28: 849-860.

Tanaka, H. 1967. On the change of composition of aquatic insects resulting from difference in mesh size of stream bottom-samplers. *Bull. Freshw. Fish. Res. Lab. (Tokyo)* 17: 1-6.

Zelt, K.A. and H.F. Clifford. 1972. Assessment of two mesh sizes for interpreting life cycles, standing crop, and percentage composition of stream insects. *Freshw. Biol.* 2: 259-269.

TAXOCENES OF EPHEMEROPTERA IN UNPOLLUTED AND POLLUTED STREAMS OF THE TATRA MOUNTAINS

Andrzej Kownacki

Laboratory of Water Biology
Polish Academy of Sciences
31-016 Kraków, Poland

ABSTRACT

In the streams of the Tatra Mountains over 1550 m a.s.l. mayflies are generally lacking. At 900 to 1550 m a.s.l. *Baetis alpinus* and *Rhithrogena loyolaea* dominate in the current, and *Ameletus inopinatus* in the pools. In the submontane zone of the streams (below 900 m a.s.l.) many other species appear. In a polluted stream (1380 to 1390 m a.s.l.) below the outlet of sewage mayflies are absent. Thirty m below the sewage outlet *B. vernus* dominates from July to September (tourist season). One hundred to 500 m below the sewage outlet *B. rhodani* and *B. muticus* dominate. Single specimens of *B. alpinus* are also encountered but representatives of *Rhithrogena* are missing.

INTRODUCTION

The aim of this paper is to present the zonation, and distribution in time, of the taxocenes of Ephemeroptera in the Tatra mountain streams and to show the effect of sewage on these taxocenes.

STUDY AREAS AND METHODS

Investigations were carried out in several streams and rivers flowing down the northern slopes of the Polish High Tatra Mountains. Twenty-six sampling stations were chosen in these streams and a short description of them is given in Table 1. Year round investigations were carried out in the stream Sucha Woda and in the stream Olczyski, both of which have preserved their natural character to a great

Table 1. Physical and chemical aspects of the streams and rivers in High Tatra Mountains and Podhale Region. 2B - sewage outlet (chemical data after Bombówna 1968, 1971, 1977; Pasternak 1971).

Stream	No.	Altitude m	Gradient ‰	Temperature of water °C	pH	O_2%	BOD_5 O_2 mg/L	Ammonia $N\text{-}NH_4$ mg/L	Phosphate PO_4 mg/L	Calcium Ca mg/L
Olczyski Potok	1	1070	80	3.8- 4.8	7.0-7.6	91.7- 98.5	1.76-3.36	0-0.146	0.043-0.222	16.4-20.0
	2	1000	80	3.3- 5.6	7.0-7.6	88.3- 99.5	0-2.24	0-0.111	0.018-0.192	17.1-23.5
	3	920	80	1.2- 8.8	7.2-7.6	92.3- 97.7	0.48-2.40	0-0.116	0.012-0.185	18.5-25.7
Sucha Woda[1]	1	±1700	334	1.5- 6.4	6.2-6.4	-	-	0.08	0.038	2.5
	2	±1560	210	0.5-10.7	6.3-6.8	85.4	-	0.016-0.08	0.015	2.9
	2A	1540	80	4.0- 7.5	6.4-6.8	-	-	0.04	0.026	-
	3	1460	100	1.5- 8.0	6.5-7.0	85.1	-	0.016	0.010	7.2- 7.9
	4	1330	66	0.4- 7.0	6.7-7.5	88.3	-	0.010-0.096	0.005-0.012	7.5
	5	1180	66	0.4- 9.0	6.7-7.5	87.8	-	0.010	0.021	7.5
	6	880	20	1.5-10.0	7.4-8.3	91.5	-	0	0.010	14.3
	7	775	20	0.3-15.3	7.0-8.5	96.2	-	0.024	0.010	22.2-27.2
Potok Roztoka	1	1450	200	5.5	6.5	83.5	-	0	0.004	3.6
	2	1300	80	5.0	6.5	84.9	-	0	0.009	3.9
	3	1040	66	1.7- 8.8	6.4-6.9	85.8-108.3	1.04-2.36	0-0.068	0-0.012	3.6- 4.1
Rybi Potok[2]	1	±1450	321	0.6- 7.5	6.6-6.7	81.7- 85.9	0.26-1.60	0-0.056	0-0.020	4.1- 5.1
	2A	1393	6	1.2-12.4	6.4-7.0	67.2-114.5	0.48-3.52	0-0.106	0-0.034	4.3- 5.4
	(2B)	1393	6	1.2-12.6	6.4-6.9	69.2- 90.1	2.40-9.76	0.016-1.192	0.010-0.268	4.6- 6.4
	2C	1393	6	1.2-12.4	6.5-7.1	66.2- 95.6	1.66-3.42	0.008-0.356	0.005-0.111	4.6- 6.4
	2D	1393	6	1.1-11.9	6.5-7.1	64.3- 94.0	0.48-2.88	0.008-0.176	0.003-0.082	4.6- 6.1
	2E	1380	10	0.2-12.7	6.4-7.8	69.2- 95.1	0.90-5.82	0.008-0.136	0-0.090	4.6- 6.1
	3	1150	113	1.2-10.9	6.4-7.0	67.5- 91.8	0.16-3.04	0-0.106	0-0.100	4.6- 6.4
Białka Tatrzańska	1	960	20	9.4	7.1-7.8	86.1- 94.0	1.12-1.28	0-0.072	0-0.012	17.2-22.9
	2	840	8	-	-	-	-	-	-	-
	3	780	15	13.7	7.3-7.8	88.0- 99.7	0.64-1.44	0-0.068	0-0.034	22.5-32.2
	4	620	9	15.8	7.2-7.8	90.8- 93.2	0.64-1.92	0.002-0.260	0-0.038	25.4-32.2
	5	530	8	16.5	7.2-7.8	91.1- 97.4	0.96-1.98	0.005-0.184	0-0.104	26.6-32.9

[1] Station 1 - stream joining Zmarzty Staw Lake (1787 m) with Czarny Staw Gasienicowy (1619 m) (without name); Stations 2, 3 and 4 - Czarny Potok stream; Stations 5, 6 and 7 - Sucha Woda stream; Station 8 - Cicha Woda stream.

[2] Station 1 - stream joining Czarny Staw lake (1580.5 m) with Morski Oko lake (1393 m) (without name); Stations 2A - 2E and 3 - Rybi Potok streams.

extent, and in the stream Rybi Potok into which sewage is discharged. In the Roztoka, a stream joining the lakes Czarny Staw and Morskie Oko, and in the river Białka, qualitative samples were taken several times during the spring and summer season. At each station, and on all dates, several sample were collected with a hand net covered with 0.3 mm mesh bolting cloth. In the laboratory all specimens of mayflies were sorted from the samples, identified and counted from a 2 dm^3 volume of the substratum.

VERTICAL DISTRIBUTION OF TAXOCENES OF EPHEMEROPTERA IN UNPOLLUTED AND POLLUTED STREAMS

The zonal distribution of Ephemeroptera (Fig. 1) was examined in the system of the stream Sucha Woda which is the outflow from the lake Zmarzły Staw (1787 m above sea level (a.s.l.)) and flows into the stream Poroniec (775 m a.s.l.). Above 1550 m a.s.l. mayflies do not occur, as a rule. Single individuals of *Baetis alpinus* and *Rhithrogena loyolaea* were found; they do not, however, undergo their full life cycle there. At an altitude of 1000 to 1550 m a.s.l. in the subalpine forest zone, *B. alpinus*, constituting over 50 percent of the total number of Ephemeroptera, dominate in the taxocene. *R. loyolaea* is a fairly numerous subdominant. Single specimens of *Rhithrogena hybrida* occur below 1350 m a.s.l. and of *Rhithrogena iridina* below 1200 m a.s.l., while single specimens of *Ameletus inopinatus* are found in the pools. Below 900 m a.s.l. the taxocene of Ephemeroptera changed markedly. Although *B. alpinus* still dominates along with *Baetis melanonyx* (species difficult to separate in their juvenile stages), *R. loyolaea*, *R. hybida*, and *R. iridina* occur only rarely. In addition, a whole range of new species of the genus *Baetis* (*B. rhodani*, *B. muticus*, *B. scambus*, *B. sinaicus*), *Rhithrogena* (*R. diaphana*, *R. semicolorata*), *Ecdyonurus* (*E. venosus*, *E. torrentis*), *Epeorus*, *Ephemerella* and *Habroleptoides* are found here. In contrast, the densities of Ephemeroptera are almost equal along the whole stream length, varying from 50 to 80 individuals/2 dm^3 of stones. Only at an altitude 1550 m a.s.l., in a place where springs enrich the stream is their number higher (137 individuals/2 dm^3 of stones).

In the stream Olczyski (1070 to 920 m a.s.l.) which arises from a very rich spring,the taxocene of Ephemeroptera is basically similar to that in the Sucha Woda at the same altitude. The dominant species is *B. alpinus* occurring with subdominants *R. loyolaea*, in the upper course, and *R. hybrida* and *R. iridina* in the lower. Although the density of *B. alpinus* is higher here, especially in the spring (1100 individuals/2 dm^3 of stones), the nymphs of *R. loyolaea* occur only as individual specimens (< 1 percent of the total number of Ephemeroptera). By 400 m downstream from the spring,the density of *B. alpinus* decreases markedly (253 individuals/2 dm^3 of stones) and the importance of the nymphs of *R. loyolaea* and *R. hybrida*

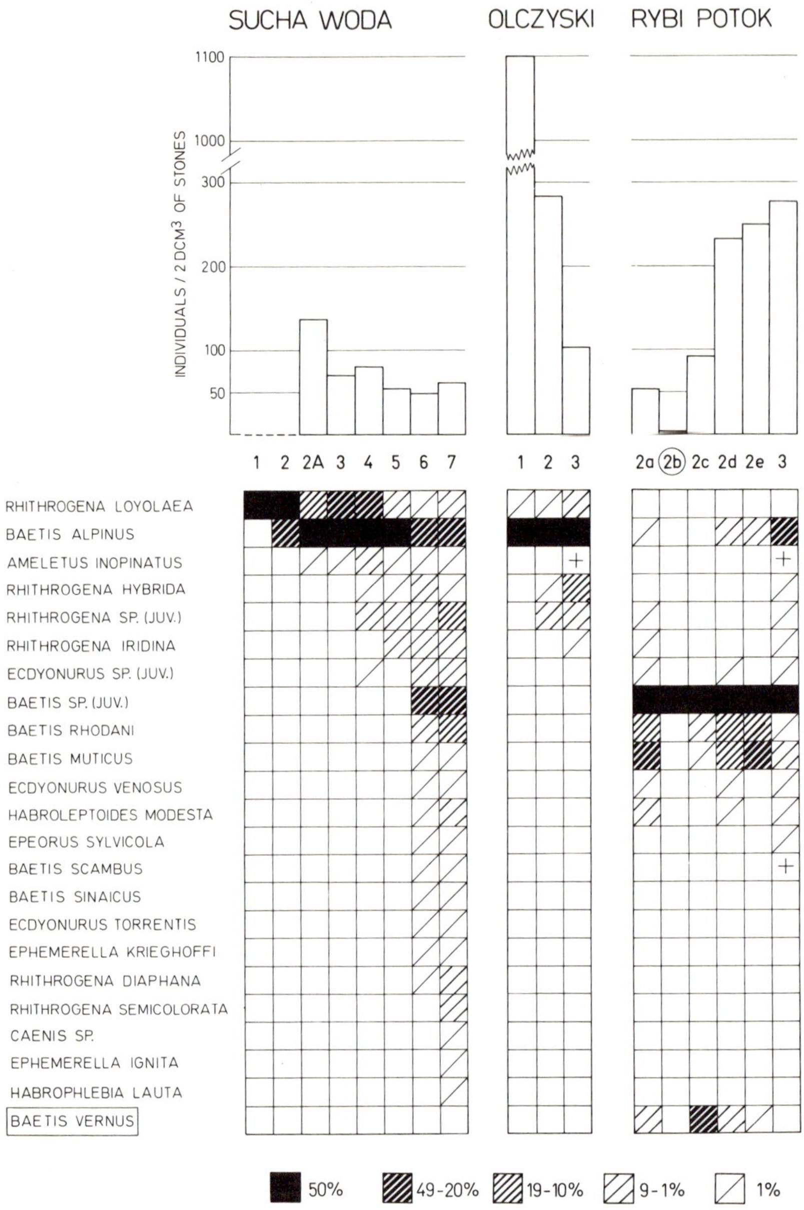

Figure 1. The altitudinal distribution of the density of Ephemeroptera and the percentage composition of the taxocenes of Ephemeroptera in unpolluted and polluted streams in High Tatra Mountains; based on the collections taken around the year. + data from the other publications, 2b is the sewage outlet.

increases in the taxocene (about 10 per cent of the total number of Ephemeroptera). By the third station, at the mouth of the valley, the density of *B. alpinus* (73 individuals/2 dm^3 of stones) is similar to the stream Sucha Woda and the subdominant species, *R. loyolaea*, *R. hybrida* and *R. iridina*, constitute 30 per cent of the total number of Ephemeroptera. No nymphs of *Ameletus* were present in the samples. They were, however, found by Sowa (1975) in the lower sector of this stream.

In other Tatra streams, the taxocenes of Ephemeroptera are similar to those discussed above (Table 2). In the stream Roztoka (1669-1030 m a.s.l.) *B. alpinus* co-dominates with *R. loyolaea* in the current and *A. inopinatus* in stagnant pools, in the upper and middle course; in the lower course, in addition to those species, *R. hybrida* and *R. iridina* dominate. *Ecdyonurus lateralis* imagines were also caught.

In the stream connecting the lakes Czarny Staw and Morskie Oko (1584-1393 m a.s.l.) *B. alpinus* and *R. loyolaea* dominate with subdominating species *R. hybrida* and *A. inopinatus*.

In the stream Rybi Potok (1393-1077 m a.s.l.), into which the sewage from the shelterhouse is discharged. A quite different taxocene of Ephemeroptera is found. Above the sewage, apart from juvenile stages of *Baetis* which dominate along the whole stream, *Baetis muticus* occurs commonly with the subdominant species *B. rhodani* and less commonly *Baetis vernus* and *Habroleptoides modesta*. *B. alpinus* and species of the genus *Rhithrogena*, so characteristic of the Tatra streams at this altitude, are rare in this stream. The total number of Ephemeroptera, however, is similar to that in other streams of the Tatra Mountains. Below the sewage discharge mayflies are almost absent. Individual juvenile specimens of the genus *Baetis* which found their way here in the drift, do not undergo their full life cycle here. By 30 m below the discharge, the total number of mayflies is the same as at the station above the discharge. *B. vernus* (missing in other streams of the Tatra) is the dominant species. Mayflies *B. alpinus*, and those of the genus *Rhithrogena*, are lacking. At a distance of 100-500 m downstream from the sewage discharge the Ephemeroptera taxocene gets more and more similar to that above the sewage discharge point. *B. rhodani* and *B. muticus* dominate again in the mayfly community and the importance of *B. vernus* gradually decreases. Individual nymphs of *B. alpinus* start to appear, although species of the genus *Rhithrogena* are still absent. However, the total number of Ephemeroptera increases markedly to about 250 individuals/2 dm^3 of stones). Only at 3200 m downstream from the sewage discharge, at an altitude of 1180 m a.s.l., does the Ephemeroptera taxocene become similar to that found in other Tetra streams. *B. alpinus* dominates, and individual specimens of *R. hybrida* and *R. iridina* are found. At the same time a number of species (*B. muticus*, *B. rhodani*, *E. venosus*, *H. modesta*) lacking in

Table 2. Distribution of the mayflies at stations in the streams and rivers in High Tatra Mountains and Podhale Region. + present, · occurs sporadically, o data according to other authors.

Stream	Olczyski			Sucha Woda								Roztoka			Rybi Potok							Białka				
Sation No.	1	2	3	1	2	2A	3	4	5	6	7	1	2	3	1	2A	2F	2C	2D	2E	3	1	2	3	4	5
Ephemera danica Müll.																									+	+
Rhithrogena germanica Etn.																										o
- *hercynia* Landa																								+	+	+
- *hybrida* Etn.		+	+					+	+	+	+			+							+			o		
- *loyolaea* Nav.	+	+	+	·	·	+	+	+	+			+	+	+	+											
- *semicolorata* Curt.											+													+	+	+
- *iridina* Kolenati			+					+	+	+	+			+		+					+	+	+	+	+	
- *ferruginea* Nav.																								+	+	+
- *diaphana* Nav.										+	+											+		+	+	+
Epeorus sylvicola Pict.										+	+										+			+	+	+
Ecdyonurus torrentis Kimm.										+	+													+	+	+
- *venosus* F.			o							+	o					+			+		+		+	+	+	+
- *submontanus* Landa											o															
- *lateralis* Curt.														+											+	o
- *dispar* Curt.																										+
Habroleptoides modesta Hag.										+	+					+				+	+				+	+
Habrophlebia lauta Etn.											+															
Ameletus inopinatus Etn.			o			+	+	+	+	+	+	+	+	+	+						+		+	o		
Siphlonurus lacustris Etn.																										o
Baetis alpinus Pict.	+	+	+		·	+	+	+	+	+	+	+	+	+	+	+			+	+	+	+	+	+	+	+
- *melanonyx* Pict.									+	+	+															
- *lutheri* M.-L.																								+	+	+
- *sinaicus* Bog.										+	+												+	+	+	+
- *rhodani* Pict.										+	+					+		+	+	+	+			+	+	+
- *vernus* Curt.																+		+	+	+	+	+			+	+
- *scambus* Etn.										+	+									+				+	+	+
- *muticus* L.										+	+					+		+	+	+	+		+	+	+	+
Centroptilum luteolum Müll.											o													o	o	+
Oligoneuriella rhenana Imm.																									+	+
Ephemerella ignita Poda.											+														+	+
- *krieghoffi* Ulm.										+	+												+	+	+	+
- *major* Klap.																									+	
Caenis sp.											+														+	+

other streams at that altitude, occur at that station. The total density of Ephemeroptera is still high.

SEASONAL CYCLES OF EPHEMEROPTERA

In the streams Olczyski, Sucha Woda, and Rybi Potok, seasonal changes in the number and structure of Ephemeroptera groups were recorded and the life cycle of the more important species investigated.

In the stream Olczyski, at all three stations, the maximum numbers of Ephemeroptera occur in the winter months, while in the summer the number is much lower. This regularity is very distinct at Station 1 (Fig. 2). The maxima observed here in November were followed by a gradual decrease, reaching the lowest value in August, whereas, maximal densities at Stations 2 and 3 take place in December.

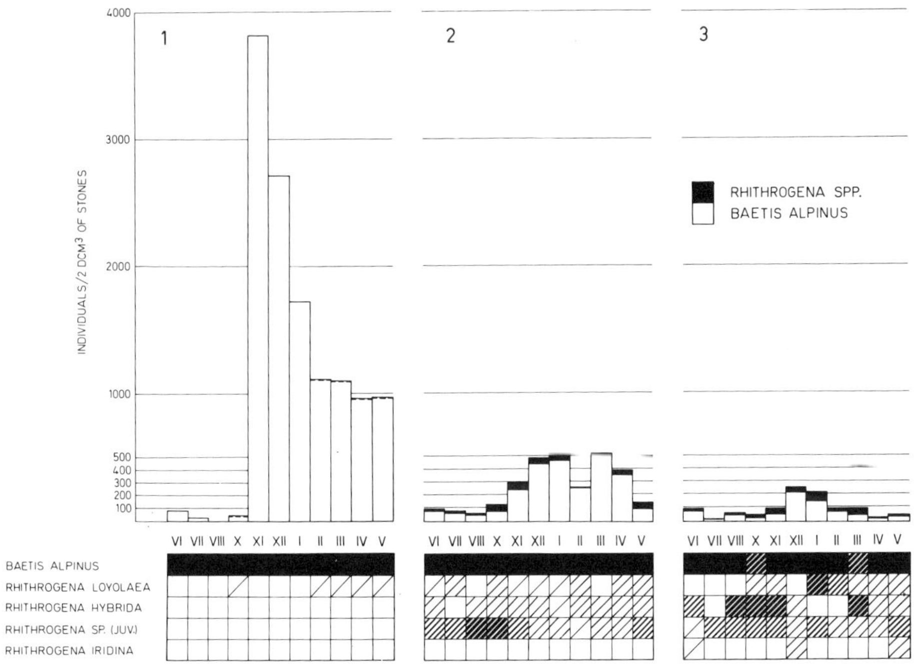

Figure 2. Annual fluctuations in the density of Ephemeroptera and the percentage composition of the taxocenes of Ephemeroptera in stream Olczyski (symbols as in Fig. 1).

At Station 1 their density is the highest of these recorded in the Tatra streams. Later the number of the individuals decreases so rapidly that it could hardly be caused by the mortality of nymphs. At Station 2 the density of Ephemeroptera remains approximately on the same level from December till March. It seems probable that the females in their compensating upstream flight, lay eggs in masses in the headwaters spring of the stream Olczyski. The juvenile forms which hatch there then drift downstream making the number of mayflies at Station 2 (400 m downstream) remain at a constant level all the winter long. At all stations in the stream Olczyski the nymphs of *B. alpinus* are the main component of the fauna of Ephemeroptera all year round. Species of the genus *Rhithrogena* are a little more numerous at Stations 2 and 3 only towards the end of summer and in autumn. The life cycle of *B. alpinus*, which dominates the total number of Ephemeroptera at these stations, was followed in this stream (Fig. 3). Individual specimens in juvenile stages (below 1 mm) appear in August, but their mass appearance takes place as late as November. During the winter and spring a gradual growth of nymphs takes place. From May on, acceleration of nymphal growth occurs, and from June till the end of August very large and pre-emergent nymphs are found. At Stations 1 and 2, this species has only one generation in a year. At Station 3, in contrast, two cohorts of *B. alpinus* or perhaps parallel occurrence of *B. melanonyx*, occurs.

In the stream Sucha Woda (Fig. 4), above 1500 m a.s.l. juvenile stages of *R. loyolaea* or *B. alpinus* are collected only in autumn and perish with the freezing of the stream. After the thaw in spring,

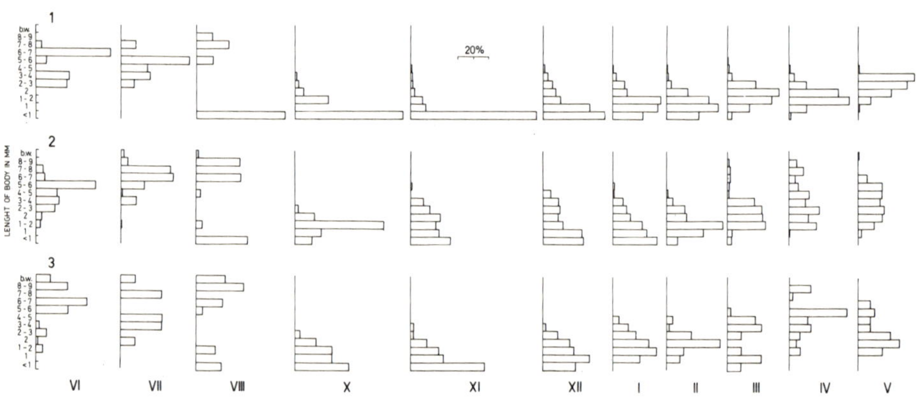

Figure 3. The percentage size-distributions of *Baetis alpinus* in approximately monthly collections from the stream Olczyski; b.w. = nymphs before emergence.

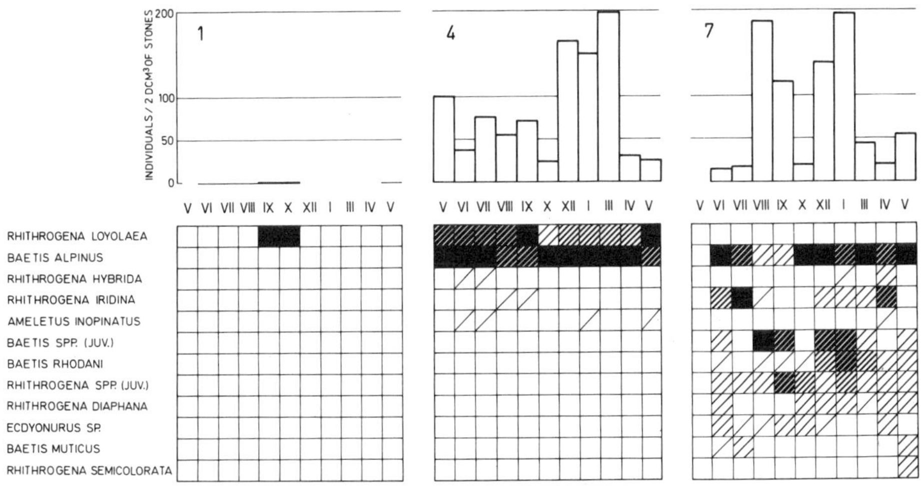

Figure 4. Annual fluctuations in the density of Ephemeroptera and the percentage composition of the taxocenes of Ephemeroptera in the stream Sucha Woda at the stations 1, 4 and 7. (Symbols as in Fig. 1).

and in summer, no representatives of the order Ephemeroptera are found. In the zone of subalpine forests, as in the stream Olczyski, *B. alpinus* dominates all year round. The importance of *Rhithrogena loyolaea* in the Ephemeroptera fauna is much greater here, especially in summer. Thus the change in the density of Ephemeroptera shows a small peak in summer and a greater one in winter. In this zone *B. alpinus* has one generation per year (Fig. 5). Juvenile stages are common in December and occur during the whole winter. Large and pre-emergent nymphs were caught from June till September; while small-and medium-sized nymphs of *R. loyolaea* occurred during the whole year in approximately the same numbers; large nymphs were found from May till December (Fig. 5). In the stream, at the foot of the Tatra Mountains, seasonal changes in Ephemeroptera are complicated. For a greater part of the year *B. alpinus* dominates along with *B. melanonyx* but at certain periods other species become dominants too.

In the stream Rybi Potok (Fig. 6) below the sewage outlet (Station 2B) individual juvenile specimens of the genus *Baetis* were found, these, however, do not undergo their full life cycles at this station. 30 m below the sewage, maximum density of Ephemeroptera occurs in spring (May, June) when a reduction in the number of

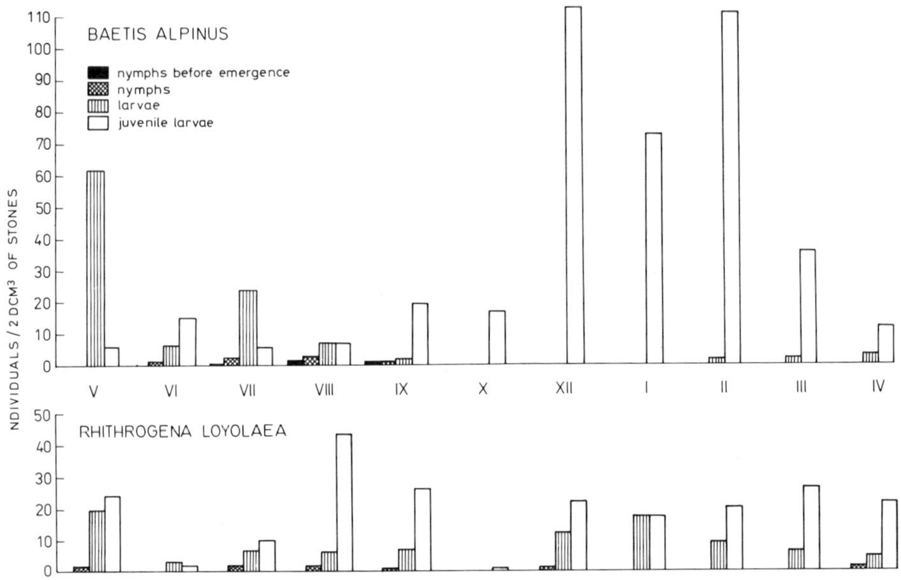

Figure 5. Annual changes of four size groups of *Baetis alpinus* and *Rhithrogena loyolaea* in the stream Sucha Woda at station 4 (data from Kownacka 1971).

tourists causes a reduction in sewage outflow. This peak is caused by juvenile stages of the genus *Baetis*. From July to December a decrease in the total number of Ephemeroptera occurs, probably related to the greater inflow of sewage. During the tourist peak (July to September) and later (October to December) *B. vernus* and *B. rhodani* dominate respectively. From January till March no mayflies were found at this station. With increasing distance from the sewage effluent (100 to 500 m) the curve representing the changes in the total number becomes similar to that for other stations. The maximum density occurs in autumn - winter months. It is interesting that *B. vernus* appeared at these stations during the tourist peak when the number of the species (*B. rhodani*, *B. muticus*) which dominated at all other periods decreased. The nymphs of *B. alpinus* too, are absent during intensive tourist traffic (July - September) at these stations. In the lower sector of the stream, about 3000 m below the sewage inflow, nymphs of *B. alpinus* are dominant all year round. Only at the beginning of September, at the end of the tourist season, does their importance in the fauna of Ephemeroptera decrease. This is caused by the biology of the species (a considerable part of juvenile stages of *Baetis* at that station are *B. alpinus*) rather than by the sewage.

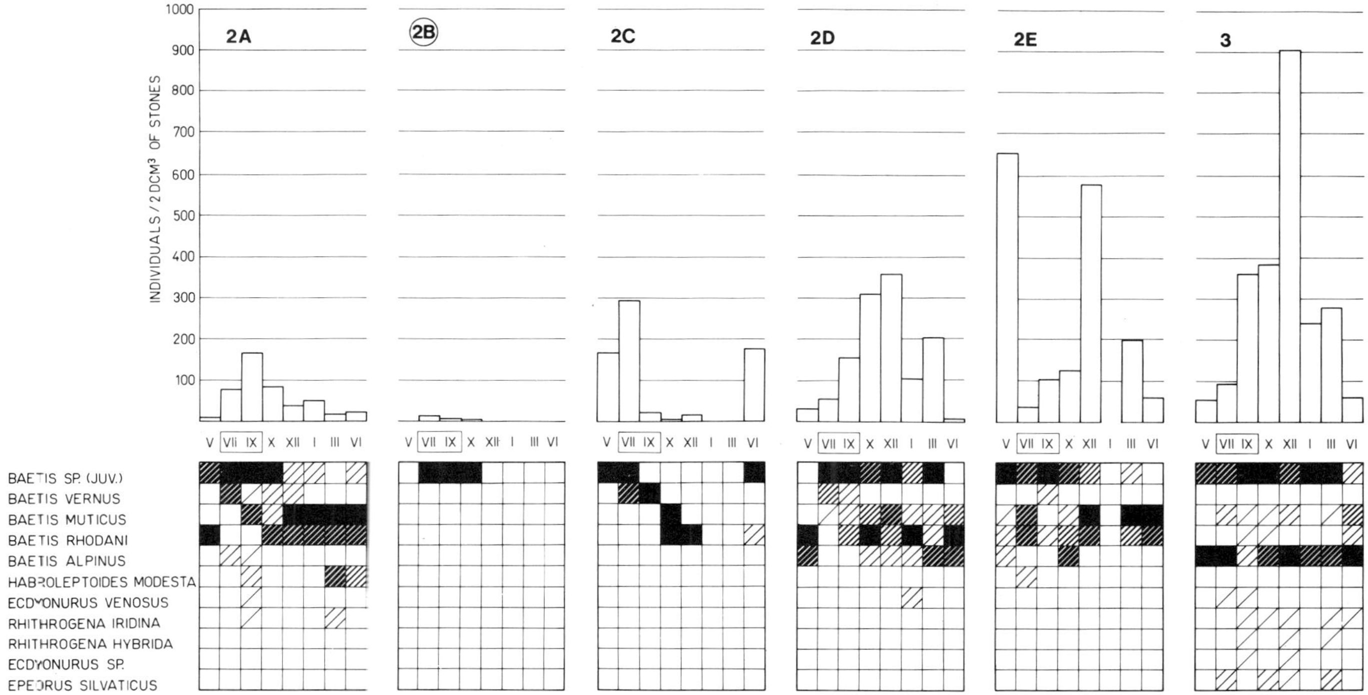

Figure 6. Annual fluctuations in the density of Ephemeroptera and the percentage composition of the taxocenes of Ephemeroptera in stream Rybi Potok (2b = sewage outlet), VII-IX is the period of increased tourist traffic (symbols as in Fig. 1).

CONCLUSIONS

Taking into consideration the species composition, the dominance structure expressed as a percentage, and the density of Ephemeroptera in the High Tatra streams and the rivers flowing at their foot, two taxocenes of Ephemeroptera, allowing defination of three zones, were distinguished.

Above 1550 m a.s.l. mayflies appear only periodically and do not form a typical taxocene. The majority of species of Ephemeroptera found in the Tatra Mountains have a one year (or longer) life cycle and require a continuous flow of water. Freezing of the streams in winter eliminates them. Sowa (1975) defines this zone as 0. But other differences, apart from lack of mayflies, between this zone of the Tatra Mountains and the rest of the Carpathians are so large that this zone should be included into the kryal (Steffan 1971, Kownacka and Kownacki 1972).

At an altitude of 900 to 1550 m a.s.l. a typical taxocene develops in which *B. alpinus* dominates with *R. loyolaea* in the upper course, and *R. hybrida* and *R. iridina*, in the lower one, as subdominants. *A. inopinatus* dominates in pools.

At an altitude of 750 to 900 m a.s.l. nymphs of *B. alpinus* and *B. melanonyx* dominate the taxocene. Numerous other species (19 species) are found simultaneously; these, however,apart from *B. rhodani*, do not reach great densities.

This distribution of the taxocenes, with small modifications, is found in all of the investigated streams of the High Tatra Mountains.

The inflow of sewage into the streams in the zone of subalpine forests causes disturbances, both in the structure and density of mayflies in the natural taxocene of Ephemeroptera. The species typical of the Tatra Mountain streams: *B. alpinus*, *R. hybrida* and *R. loyolaea* are absent in the polluted stream sector or occur as individual specimens, whereas many species characteristic of streams flowing at the foot of the Tatra Mountains (*B. rhodani*, *B. muticus*, *H. modesta*) occur. *B. vernus* deserves special attention. Its most intensive development took place 30 m below the sewage discharge point during the tourist peak. It was not found in other streams of the subalpine forest zone, or in streams flowing at the foot of the Tatra Mountains. Individual specimens were caught only in the lower reaches of the river Białka below 650 m a.s.l. On one occasion some individuals were caught in the river Białka at an altitude 960 m a.s.l., below the buildings at the frontier crossing at Łysa Polana, from which sewage is released into the river. In other Carpathian streams and rivers they were found only as individual specimens. They occurred more numerously only in the rivers Wilga near Cracow

(300 m a.s.l.) and Drwinka (200 m a.s.l.), (Sowa 1975). It seems probable that a more common occurrence of this species could indicate the pollution of mountain rivers and streams with sewage.

Apart from changes in the structure of the taxocene the inflow of sewage also causes some changes in the total number of Ephemeroptera. Just below the sewage discharge Ephemeroptera are eliminated from the biocenosis. By some tens of meters downstream however, a rapid increase in the density of Ephemeroptera, not reported from any other Tatra streams, takes place. An increase in density of mayflies was also noticed at a great distance (> 3000 m) from the outlet of sewage, though the structure of the taxocene was already similar to that found in unpolluted streams.

From the discussion above, taxocenes of mayflies provide a very good index in distinguishing zones in streams and rivers. They can also be useful in detecting slight pollution. On the other hand, heavy organic pollution causes a total elimination of mayflies from the biocenosis (e.g. the stream Kryniczanka - Szczęsny 1974).

ACKNOWLEDGMENTS

I am very grateful to Docent dr hab. Ryszard Sowa for his kind verification and assistance in identification of mayflies and Professor dr hab. Kazimierz Pasternak for providing the unpublished chemical data from stream Olczyski.

RESUME

Dans les torrents des Tatras au-dessous de 1550 m d'altitude les Ephemeroptera n'apparaissent que sporadiquement. Entre 900-1550 m d'altitude *Baetis alpinus* et *Rhithrogena loyolaea* dominent dans le courant et *Ameletus inopinatus* dans l'eau stagnante. Dans la zone du torrent en aval (au-dessous de 900 m) beaucoup de nouvelles especes apparaissent. Dans le torrent pollué (entre 1380-1390 m) au-dessous des égouts Ephemeroptera manquent. 30 m au-dessous, de juillet jusqu' a septembre (saison touristique) *B. vernus* domine. 100-500 m au-dessous des égouts *B. rhodani* et *B. muticus* dominent. On rencontre aussi quelques rares exemplaires *B. alpinus*, mais il n'y a guere de représentants de *Rhithrogena*.

ZUSSAMENFASSUNG

In Tatrabächen unter 1550 m begegnet man beinahe keine Ephemeroptera. Zwischen 900-1550 m dominieren im Strom *Baetis alpinus* und *Rhithrogena loyolaea* und im stagnierenden Wasser *Ameletus inopinatus*. In der Gebirgszone des Baches (unter 900 m) treten neue Arten auf.

In einem verunreinigten Bach (1380-1390 m) gleich unter dem Abwasserkanal befinden sich keine Ephemeroptera. 30 m unterhalb, von Juli bis September (Touristsaison) dominiert *B. vernus*. 100-500 m unterhalb des Abwasserabflusses dominieren *B. rhodani* und *B. muticus*. Einzelne Exemplare *B. alpinus* begegnet man auch, aber die Vertreter der *Rhithrogena* fehlen ganz.

REFERENCES

Bombówna, M. 1968. Hydrochemical characteristics of the Białka Tatrzańska. *Acta Hydrobiol.* 10: 27-37.

Bombówna, M. 1971. The chemical composition of the water of the Polish High Tatra Mountains, particularly with regard to the stream Sucha Woda. *Acta Hydrobiol.* 13: 379-391.

Bombówna, M. 1977. Biocenosis of high mountain stream under the influence of tourism. 1. Chemistry of the Rybi Potok waters and the chlorophyll content in attached algae and seston in relation to the pollution. *Acta Hydrobiol.* 19: 243-255.

Kownacka, M. 1971. The bottom fauna of the stream Sucha Woda (High Tatra Mountains) in the annual cycle. *Acta Hydrobiol.* 13: 415-438.

Kownacka, M. and A. Kownacki. 1972. Vertical distribution of zoocenoses in the streams of the Tatra, Caucasus and Balkans Mountains. *Int. Ver. theor. angew. Limnol. Verh.* 18: 742-750.

Pasternak, K. 1971. The physiography and character of the substratum of the drainage areas of streams of the Polish High Tatra Mountains. *Acta Hydrobiol.* 13: 363-378.

Sowa, R. 1975. Ecology and biogeography of mayflies (Ephemeroptera) of running waters in the Polish part of the Carpathians. 1. Distribution and quantitative analysis. *Acta Hydrobiol.* 17: 223-297.

Steffan, A.W. 1971. Chironomid (Diptera) biocenoses in Scandinavian glacier brooks. *Can. Entomol.* 103:477-486.

Szczęsny, B. 1974. The effect of sewage from the town of Krynica on the benthic invertebrate communities of the Kryniczanka stream. *Acta Hydrobiol.* 16: 1-29.

A PRELIMINARY REPORT FROM A STUDY ON THE ENVIRONMENTAL IMPACT OF PYRITE MINING AND DRESSING IN A MOUNTAIN STREAM IN NORWAY

Karl Jan Aanes

Norwegian Institute for Water Research
Box 333
Blindern, Oslo 3, Norway

ABSTRACT

The mining, milling and processing of complex zinc and copper pyrite ores causes water pollution in some districts of Norway To determine the effects of the wastewater from the mining and flotation process, and the acid mine drainage from an old, now exhausted pyrite mine, physical, chemical and biological investigations were performed in the Folla river during 1975-1976. The character of the stream is changed primarily by the inflow of sulphate, calcium and heavy metals. In addition to higher concentrations of a number of substances, increased sediment deposition occurs, which is most distinct in the upper regions, downstream from the tailings impoundment. The bottom fauna in the riffles, sampled by a Surber square foot sampler are qualitatively and quantitatively changed. Ephemeroptera, which seem to give the best picture of the pollution, are investigated at the species level. In this part of the stream the mayfly fauna is represented by 5 families and 14 species. The impact of mine drainage on the bottom fauna, particularly on the mayflies is discussed in relation to the physico-chemical parameters.

INTRODUCTION

Streams with drainage from metal-mining districts frequently contain trace elements in concentrations high enough to render the water unfit for fishery, irrigation, domestic and many industrial uses. These are some of the disadvantages accompanying the constantly accelerated mining of ore bodies to satisfy man's ever increasing utilization of the metallic elements in a growing industrial society.

The mining and processing of pyrite (FeS_2) and copper and zinc ores have been important factors in the development of the Folldal valley for nearly 250 years. This valley and its river was selected for an investigation to acquire a deeper knowledge of the biological magnitude and mechanisms of water quality deterioration associated with mines and mine spoils. Here, in the same catchment area, it was possible to study the impacts associated with a new mine and processing plant, and an old, now exhausted mine. An intense, one-year study started in July 1975 and was completed in September 1976. During this time there were 15 surveys in the water-course. Some background data were available, mainly on physico-chemical observations and fisheries. These data were based on an annual quality monitoring program conducted by the Norwegian Institute for Water Research (NIVA) since 1966.

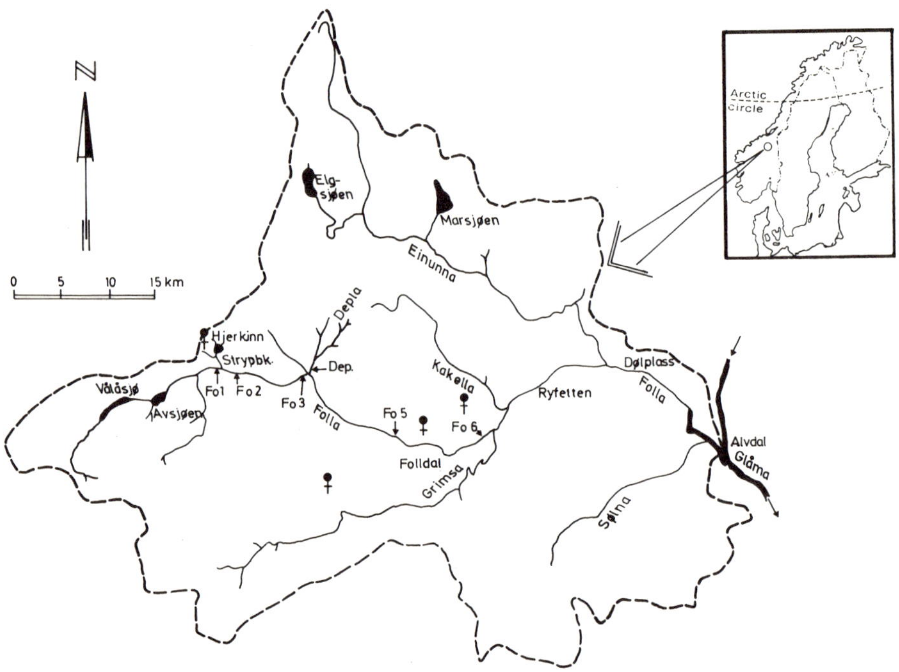

Figure 1. Folla river drainage system.

DESCRIPTION OF SAMPLING STATIONS

The Folldal valley is located in a mountain area about 410 km from Oslo. In Fig. 1, the stretch (36 km) of the river Folla, which was the object of investigation, can be seen. The Folla is the third largest tributary to the river Glåma which is the largest river in Norway.

The headwaters of the Folla are located 1470 m above sea level, between mountain peaks that reach a height of 1650-1700 m. The Folla meets the river Glåma at Alvdal 108 km from its headwaters. Here it has a mean yearly flow of 28 m^3/sec. The total catchment area is 2110 km^2, the investigated area is 600 km^2 of which 350 km^2 are higher than 1000 metres above sea level.

The area is one of the most arid parts of Norway and, at Hjerkinn, the annual mean precipitation is 222 mm. This region has a typical inland climate, with cold winters and relatively hot summers. The heaviest rainfall comes during the summer.

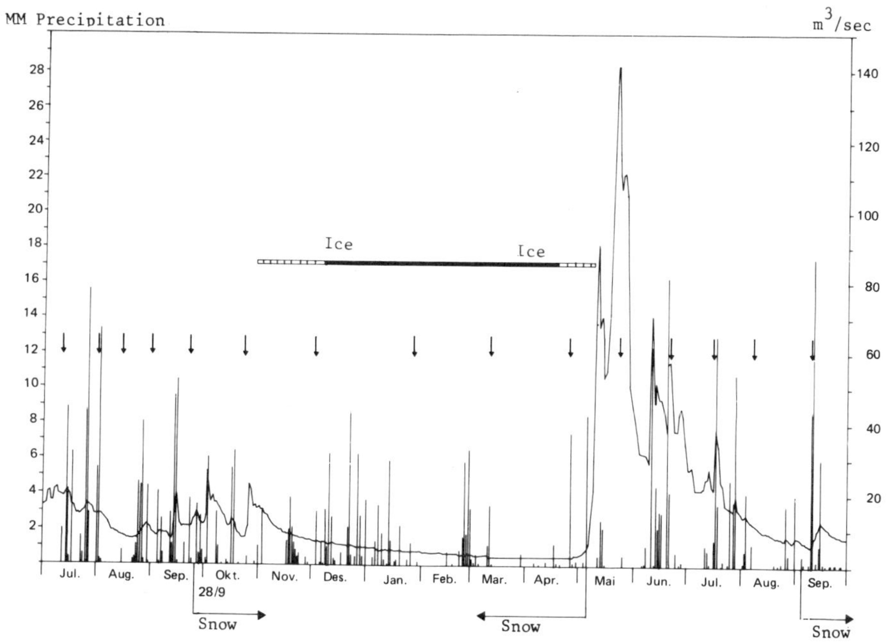

Figure 2. Daily precipitation at Folldal, and daily water flow at Ryfetten water gauge in m^3/sec. ↓Sampling time.

In Fig. 2 the daily precipitation at the Folldal village and the water flow at Ryfetten, five stream-km below station 6, is shown. Characteristic of the water flow is a large spring flood, a relatively high summer flow, some smaller floods during the autumn and a low winter flow, fed mainly from groundwater.

In this part of the country there are very large deposits of fluvio-glacial materials left from the last glacial period. During deglaciation, ice dammed lakes deposited considerable amounts of fine materials in the valley. The geology consists mainly of the so-called Trondheim stone, which consists of cambro-silurian sediment minerals of clay origin. These appear as fylitts and glimmer slates. Limestone occurs only in thin layers and south of Hjerkinn there are stones built up of gabbro minerals. Between the gabbro and glimmer slates ore-minerals, particularly pyrites, can be found. Cuprous ores and zinc blend also exist in this area.

The main industrial activity in the valley is mining, with forestry and farming being carried out on a lesser scale. Mining started in Folldal in 1748 where the village later grew up. The mine was operated with varying success until 1940. Activity was then moved to some small mines in the neighbourhood. In 1968 all the old mines were closed down and mining was continued only in the new mine at Hjerkinn which is very easy to operate and where the ore reserve is quite large. The ore has an average content of 30% S, 0.9% Cu and 1.1% Zn. Today, 770,000 tons of raw ore is moved out of the mine annually. This yields 17,000 tons of Cu concentrates (Cu content 25.5%), 300,000 tons of iron pyrites (50% S), 12,000 tons of Zn concentrates (52% Zn) and large amounts of very finely ground material ($\sim$75 µm) with traces of ore minerals. Residue from the flotation process is carried to an artificial impoundment (700,000 m^2, maximum depth 5 m) which is instrumental in reducing settleable solids from the mining effluents prior to discharge into the river. To prevent oxidation of iron pyrites it is important to deposit the tailings under water.

There are two main kinds of pollution from ore mines, and it is important to know the difference between them. The first type is drainage water from mines (waste deposits and tailings). This has a low pH and high Fe, Cu and Zn content. As the pH increases, ochre deposits, which are characteristic of this kind of pollution, are formed. The ochre deposits in the Folla river can be seen for several miles downstream from the old mines, and the greatest concentrations of heavy metals were measured from this part of the river. The second type is the return-water from the flotation process with the tailings - a compound of finely ground stone particles, some leftovers from iron pyrites and organic and inorganic chemicals from the plant. The characteristics for the water from the overflow of the dam are high conductivity, Ca and sulphate values, a characteristic odour, pH above 7 and from time to time some suspended matte

In addition to the pollution from the mines, from which the first type of pollution usually increases with time, there is an organic load coming from the dairy in Folldal village and, more or less untreated, sewage emanating from the population along the river.

The effect of this organic load is clearly seen downstream from the village (1550 person equivalents) where the river has a heterotrophic growth and blue-green algae (*Oscillatoria* sp.).

Five sampling stations were located along the main stream of the Folla plus one station in the feeder stream coming from the tailings impoundment and another in the unpolluted tributary Depla (Fig. 3). The uppermost mainstream station was situated above the polluted feeder stream from the tailings dam and was used

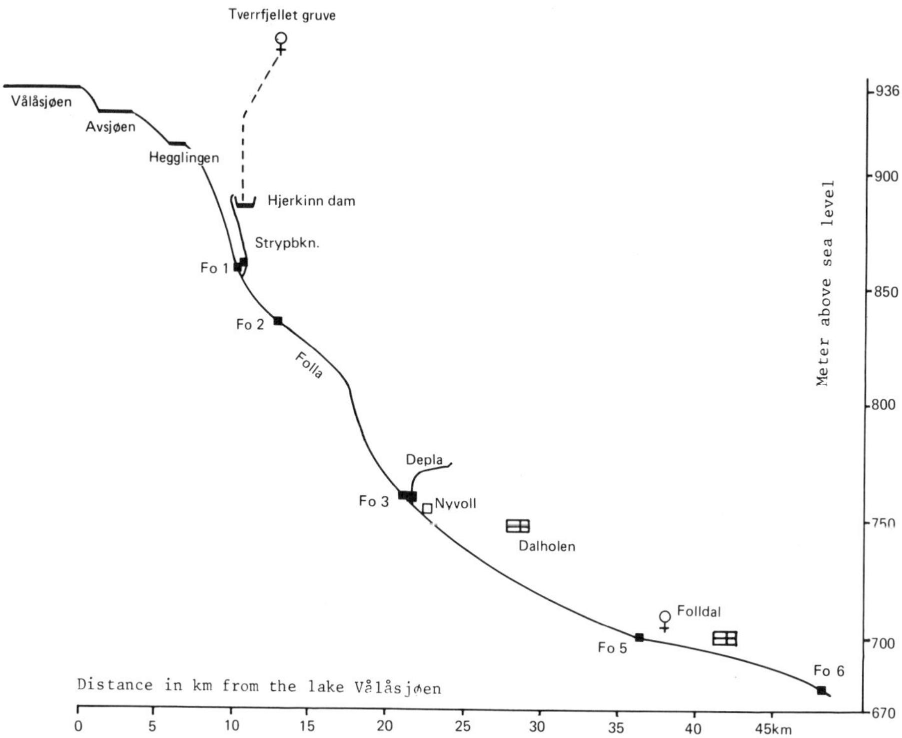

Figure 3. Slope curve for the Folla River with the sampling stations in the rapids marked.

along with the Depla to describe the natural background levels of the water quality in the drainage area. With this distribution of stations it was possible to monitor the main problems in the recipient river. When the effluents reach the Folla as point sources, the work is simplified considerably. Stations 1 and 2, and 5 and 6 bracketed these sources (problems). Station 3 was established 8 stream-km below Station 2 in order to cover the area in which the stream seems to recover from the effects of the effluents upstream from Station 2.

The river is a typical mountain torrent. (Harrison and Elsworth, 1958) and Illies (1961) would have classified it as Epirhithron, while Huet (1949, 1954) would have divided the investigated area into two zones somewhere between station 3 and station 5. The upper part would have been the trout zone with the grayling zone below.

All the sampling sites were located in places with rather fast riffles. The flow, which was measured 10 cm below the surface of the water, was between 0.8 and 1.6 m/sec (measured with an Ott flow meter). At the selected stations the Folla is a wide (20-30 m) but not very deep river (maximum depth when sampled, 75 cm). The substrate was of a multilayered stone/gravel composition (10-15 cm) on a hard packed sand and stone bottom. Stability of the substrate is an essential factor for the development of the bottom fauna. By looking at the development and the distribution of the Bryophyta, and comparing this to the stone sizes on which the aquatic mosses grow, one can get information about the stability in the substrate at the sampling station. In the Folla, Station 3 is the only station where the development of aquatic mosses is restricted because of instability.

MATERIALS AND METHODS

Physico-chemical

From July 1975 through September 1976, fifteen water samples were collected at each station and transported to the laboratory (NIVA) for analysis. Determinations of oxygen (*in situ)*, temperature, pH and conductivity (μS/cm - 20 °C) were made in the field. Other parameters measured were: total alkalinity,turbidity, colour (filtered), organic and inorganic seston, total organic carbon (filtered), COD, total Fe, Mn, Zn, Cu, sulphate, Ca, Mg, K, Na, Cl, total N and total P. The sampling frequency for some of the parameters is rather low (see Table 1) because of their late incorporation into the program. The methods used are described in the 14th edition of Standard Methods (American Public Health Association 1975). The heavy metals were collected in vials (25 mL) specially cleaned, and at the laboratory 0.2 mL of 3.5 N HNO_3 was added to

Table 1. Physico-chemical characteristics of the river Folla and tributaries 1975-1976.

		Fo 1		Strypbekken		Fo 2		Fo 3		Depla		Fo 5		Fo 6	
		$\bar{x}$ (N)	range	$\bar{x}$ (N)	range	$\bar{x}$ (N)	range	$\bar{x}$ (N)	range	$\bar{x}$ (N)	range	$\bar{x}$ (N)	range	$\bar{x}$ (N)	range
pH		7,3(15)	7,6-6,7	7,3(14)	7,6-7,0	7,3(15)	7,7-7,0	7,2(11)	7,5-6,7	7,5(11)	7,7-7,0	7,5(15)	7,8-7,0	7,4(15)	7,8-6,9
$Cond_{20}$	µS/cm	34(15)	48-21	935(14)	1485-259	156(15)	460-38	154(11)	328-44	62(11)	85-26	108(15)	238-44	137(15)	258-56
Alkalinity	meq/l	0,3(11)	0,4-0,2	0,8(11)	1,0-0,5	0,3(11)	0,5-0,2	0,3(11)	0,4-0,2	0,6(10)	0,8-0,2	0,5(11)	0,7-0,2	0,6(11)	0,8-0,2
Color filt	mg Pt/l	15(7)	38-2	18(10)	46-3	17(7)	38-5	16(7)	36-3	14(7)	31-0	11(7)	31-0	16(7)	38-5
Turbidity	FTU	0,4(12)	0,8-0,2	4,4 (12)	24,0 -1,6	0,8(12)	2,2-0,2	0,7(11)	2,1-0,2	0,6(11)	3,0-0,2	0,9(12)	3,6-0,2	2,1(12)	5,9-0,5
Ca	mg/l	5,4(12)	8,2-3,0	205(12)	400-40	22(12)	45-5	21(11)	39-7	11(11)	14-5	18(12)	36-8	19(12)	34-8
Mg	mg/l	0,7(10)	1,0-0,4	5,3(10)	7,4-2,2	1,2(10)	2,6-0,5	1,2(10)	2,0-0,6	1,0(10)	1,5-0,5	1,3(10)	2,1-0,6	1,7(10)	2,5-0,7
SO_4	mg/l	4,0(12)	5,5-2,4	566(12)	1100-89	72(12)	210-6	58(11)	150-5	5,6(11)	11-1,9	31(12)	83-6	34(12)	78-10
Fe	µg/l	68(12)	195-15	350(12)	890-150	80(12)	270-25	76(11)	350-30	47(11)	270-5	67(12)	315-20	380(12)	700-110
Mn	µg/l	3,2(3)	3,5-3,0	320(3)	475-55	24(3)	50-5	11(3)	15-5	1,7(3)	4,0-0,5	5,3(3)	10-2,0	30(3)	49-15
Cu	µg/l	2,4(12)	4,5-0,5	32(12)	155-7	4,3(12)	13,0-2,0	4,8(11)	14,0-1,9	3,1(11)	9,0-1,5	3,9(11)	8,7-1,5	15,3(12)	50-6,0
Zn	µg/l	5,1(12)	10,0-2,5	84(12)	250-10,0	7,4(12)	30-5,0	6,7(11)	20-4,0	4,6(11)	5,0-2,5	5,8(11)	15,0-3,5	68(12)	110-35
COD	mg O/l	6,3(12)	11,6-2,5	27(12)	62-9,0	9,8(11)	19,4-4,4	8,0(10)	15,1-2,5	6,7(11)	12,8-2,5	5,6(12)	14,5-2,2	11,4(12)	36-2,5
Org.seston	mg/l	0,28(12)	0,50-0,05	0,76(12)	1,90-0,28	0,35(12)	0,67-0,10	0,35(11)	0,80-0,05	0,19(11)	0,40-0,07	0,34(12)	0,90-0,10	0,61(12)	1,60-0,10
Inorg. "	mg/l	0,3(12)	1,1-0,03	2,1(12)	11-0,2	0,5(12)	1,9-0,08	1,2(11)	8,5-0,03	4,3(11)	43,4-0,00	1,5(12)	8,4-0,03	3,6(12)	30-0,1
TOC-filt.	mg/l	2,4(10)	5,2-1,1	3,4(10)	5,5-1,1	2,5(10)	4,7-0,8	2,5(10)	5,1-1,0	2,2(10)	5,5-0,4	2,1(10)	5,6-0,7	2,1(10)	5,0-0,6
K	mg/l	0,65(2)	± 0,10	8,9(2)	± 0,4	1,3(2)	± 0,5	1,4(2)	± 0,4	1,4(2)	± 0,25	1,5(12)	± 0,15	1,6(2)	± 0,2
Na	mg/l	0,87(2)	± 0,03	8,8(2)	± 1,0	1,4(2)	± 0,3	1,4(2)	± 0,3	1,0(2)	± 0,0	1,2(2)	± 0,1	1,4(2)	± 0,15
Cl	mg/l	0,55(2)	± 0,05	7,3(2)	± 1,0	0,65(2)	± 0,05	0,75(2)	± 0,15	0,45(2)	± 0,05	0,80(2)	± 0,00	0,95(2)	± 0,05
TOT-N	µg/l	175(11)	280-90	780(11)	1490-340	243(11)	510-110	200(11)	400-65	179(11)	490-50	52(11)	110-10	234(11)	540-110
TOT-P	µg/l	4,8(10)	9,0-2,0	13,4(11)	33-4,0	6,4(11)	11-3,0	9,1(11)	22-2,0	5,0^{x}(11)	195-2,5	15(11)	100-2,0	19(11)	98-4,0

$\bar{x}$ aritmetic mean

(N) no of analyses

x - median ($\bar{x}$ = 24)

the samples as a preservative (Henriksen & Balmér 1977). Concentrations of Cu, Mn and Zn were determined using a Perkin-Elmer 300 atomic absorption spectrophotometer equipped with a HGA 72 graphite furnace. Daily values for stream flow were available from a national water gauge situated at Ryfetten, 5 stream-km downstream from Station 6 (Fig. 2).

Biological

A square foot Surber sampler (mesh size 250 μm) was used to collect benthic macroinvertebrate samples at each station from July 1975 to July 1976. During the cold season, from November through May, it was impossible to sample because of the ice conditions. Similar stream substrates were sampled and no two samples were taken from the same spot before at least two months had elapsed. Five samples were collected at each station along a cross section of the river. In the field the samples were preserved (70% ethanol). Back at the laboratory in Oslo, the smaller numbers of the larger individuals were removed from the samples by hand under a magnifying lamp (2 x) before subsampling.

With the aid of a device described by Södergren (1974a) the material was subsampled and as a rule, two (one) subsamples out of ten were counted using a stereomicroscope. The subsampler functioned very well and a chi-square test for random distribution was significant at the 1% level.

RESULTS AND DISCUSSION

Physico-chemical data for the five stations in the river Folla, the tributary Depla and the feeder stream from the tailings dam, are listed in Table 1.

Effluents

The overflow from the tailings impoundment reflects a composition which is usual for return-waters after basic flotation. The high conductivity is characteristic mainly because of the great amount of $CaSO_4$ (Fig. 5) used in the flotation process. It also contains appreciable quantities of the trace elements which mostly seem to be "bound" to suspended particles. The total alkalinity is high and the pH is always greater than 7.0. The content of nitrogen compounds is high, from the explosives used in the mine and although the phosphorus content is also relatively high, there are no noticeable signs of eutrophication below the outlet. Further, although the chemical oxygen demand is rather high, the oxygen saturation was never below 93% in the daytime due to the great turbulence.

The drainage from the old mine and large refuse piles in the Folldal village are of a typical acid mine drainage (AMD). This is formed when sulphur in the ore is exposed to the oxidizing action of air, water and chemosynthetic bacteria which utilize the energy obtained from the conversion of inorganic sulphur to sulphate and sulphuric acid. This results in a very low pH and a high content of sulphate and heavy metals, particularly iron.

The main stream

The temperature of the water was registered at Nyvoll which is approximately in the middle of the investigated part of the river. Figure 4 is a plot of the daily average of the max. and min. water temperatures. It is important to notice that for 6 months of the hydrological year, the water temperature in the Folla is around zero degrees. The number of degree-days for the year (1975-1976) was 1350, and the maximum temperatures in 1975 and 1976 were 18.5°C and 17.4°C respectively. The river Folla at Station 1 and the Depla, has in comparison to typical Norwegian conditions, a relatively high pH value, conductivity and total hardness. The

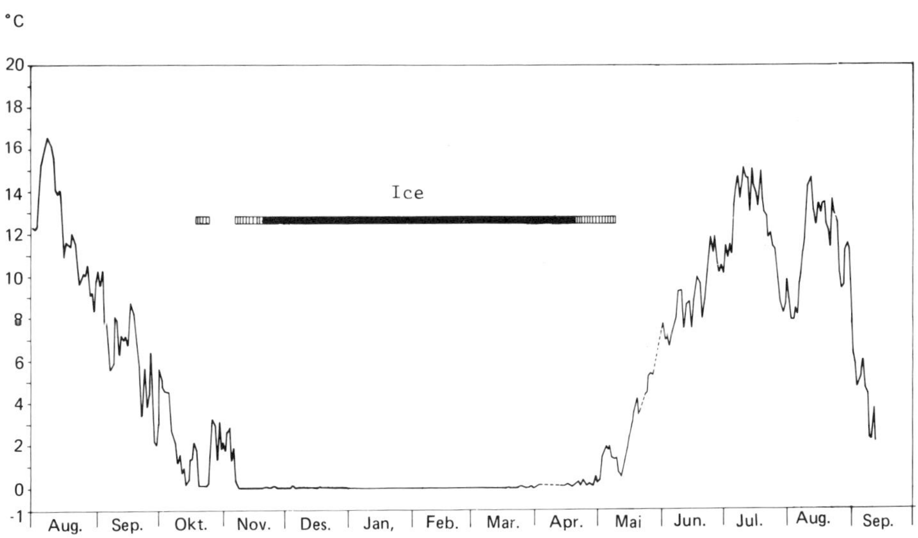

Figure 4. Mean daily water temperature at Nyvoll, Folla River, 1975-1976.

effluents do not seem to affect the pH significantly although there is a small decrease at Station 3 and 6. Between Station 3 and the tributary Depla there is a statistically significant difference ($P < 0.05$, Mann-Whitney U test) in the pH and alkalinity which is mainly caused by the greater groundwater influence in the Depla.

With the aid of a modified Maucha diagram (Broch and Yake 1969) the major elements are presented in Fig. 5. The diagrams show the ionic composition and the total ion concentration in the river where the circle's area corresponds to the specific conductivity. Each polygon represents the equivalent per cent contribution of that particular ion to the total ion concentration. Although the ionic composition at Station 1 is dominated by the calcium bicarbonate - carbonate system, the river changed totally below the entrance of the effluent from the tailings dam where the sulphate dominates the anions. In spite of tributaries with a high total alkalinity, the river has a characteristic type of sulphate water, and this influences the water quality all the way down to the confluence with the river Glåma.

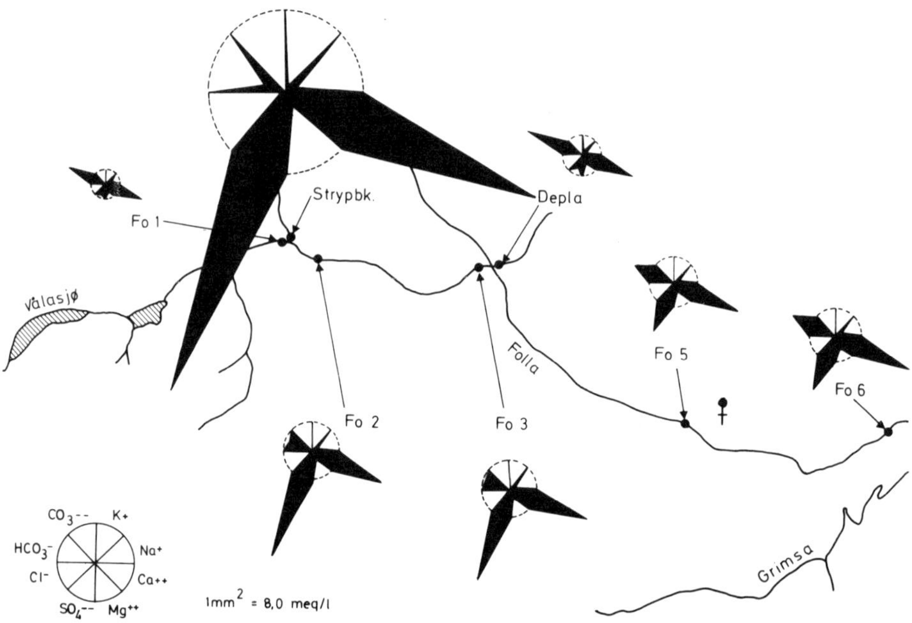

Figure 5. Modified Maucha diagram of mean ionic composition and total ionic concentration for Folla river.

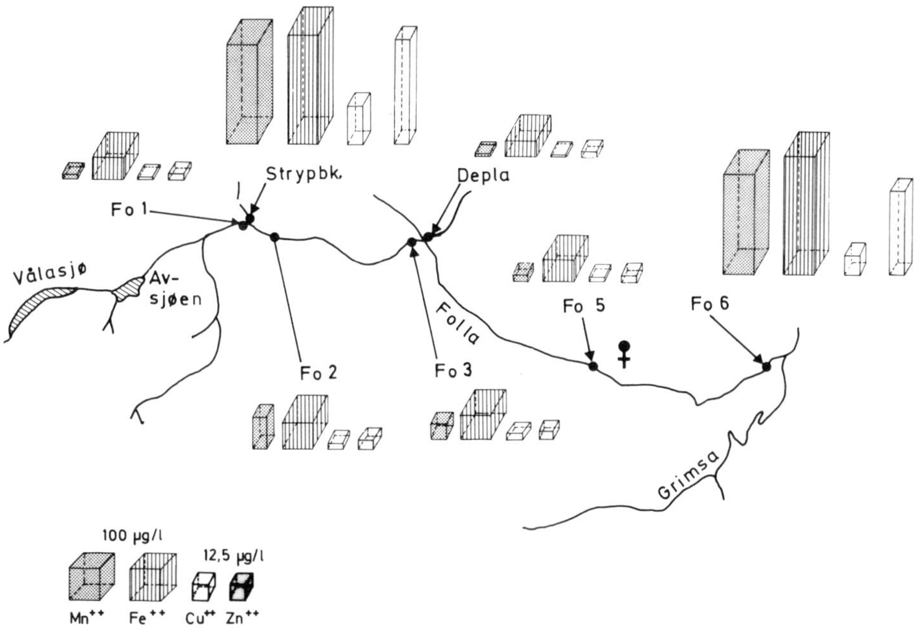

Figure 6. The heavy metal content in the river Folla presented as mean values for the period August 1975 - September 1976.

The heavy metal content in the Folla (Fig. 6) shows a small, insignificant increase at Station 2, mostly marked by manganese. Trace metals from the tailings dam are bound to suspended particles and settle out rather quickly. At Station 6 the effect of the mine drainage is clearly seen, and there is a significant difference ($P < 0.05$) in the heavy metals Fe, Zn and Cu between Station 5 and 6. The sampling frequency is too low to use the test on manganese, but the difference is considerable.

The transport of seston, shown by the parameters organic and inorganic seston and the turbidity in the river, is graphically presented in Fig. 7. For all three parameters there is a small, but marked increase at Station 2. For turbidity this increase is statistically significant at the 5% level. The substrate at station 2 is clearly marked by this - something which is evident particularly in April before the spring flood. The Depla which is meandering in a glacio-fluvial deposit about 300 m upstream from the sampling site, shows an increased inorganic seston transport. This is particularly seen during the spring flood when the transport value is 43.4 mg/L.

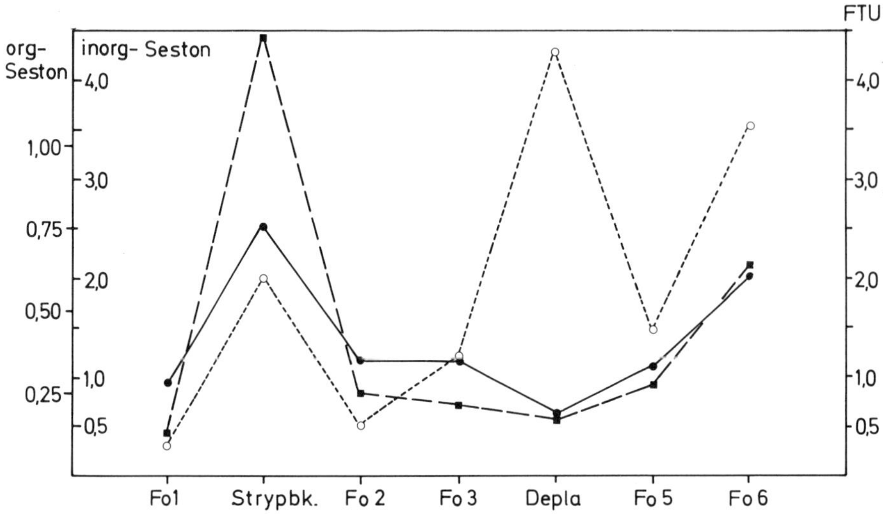

Figure 7. ●——● Organic seston, o----o Inorganic seston, ■— —■ Turbidity. Mean values during the period Aug. 1975 - Sept. 1976.

The most marked difference in the seston transport in the main stream is between Station 5 and 6. Here there is a statistically significant difference ($P < 0.05$) in both the organic seston load and the turbidity. Although the Folla passes some extensive glacio-fluvial deposits above Station 6, this seems to influence the seston load only during exceptionally high water flow. The conclusion is, therefore, that the differences in the water are due to the acid mine drainage and sewage from the village below Station 5.

Biological

It is well known that bottom-dwelling animals are effective monitors of the environmental conditions prevailing during the period of their residence. By examining changes in total numbers of the major groups and species composition in the Ephemeroptera, the effects of the outflow from the tailings dam, drainage from the old mine and sewage from the village of Folldal, the extent of pollution can be evaluated. Table 2 gives the abundance of the major groups of benthic macroinvertebrates, and the total fauna is

Table 2. Means of individuals at the different sampling sites given as individuals per m^2 for the period July 1975 - July 1976.

	Fo1	Fo2	Fo3	Depla	Fo5	Fo6
Nematoda	527	1125	186	185	845	124
Oligochaeta	605	715	841	798	1088	85
Gastropoda	6	-	-	-	19	-
Bivalvia	11	6	-	-	1	-
Plecoptera	566	384	602	601	710	1107
Ephemeroptera	2205	1205	3190	4348	1883	1865
Trichoptera	306	288	224	254	563	280
Coleoptera	44	2	20	23	585	36
Simuliidae	532	1658	1045	105	243	107
Chironomidae	14798	8261	6705	6569	9008	7153
Tipulidae	123	99	112	205	229	137
Arachnida	588	357	236	673	671	508
Σ	20311	14100	13161	13761	15845	11402

presented as yearly means at the different sampling sites, given as individuals per square meter. In Fig. 8, the total fauna and the mayfly population are graphically presented.

From Station 1 with a high number of individuals, mainly insects, to Station 2, the density dropped by 31%. There was a further decrease at Station 3, but the number of individuals was almost the same as in the river Depla, an unpolluted tributary. At Station 5 there was an increase but still the numbers are not as great as they were at Station 1, mainly because of the high numbers of chironomids at that station. As pointed out earlier (in the physico-chemical part) Station 6, located 11 stream-km below the village in Folldal, is the most polluted stretch of the river. This is confirmed by the low density of macroinvertebrates. In addition, Station 6 had the lowest number of individuals per square meter as a mean value for each sampling trip and the figure is 28% lower than at the next upstream station.

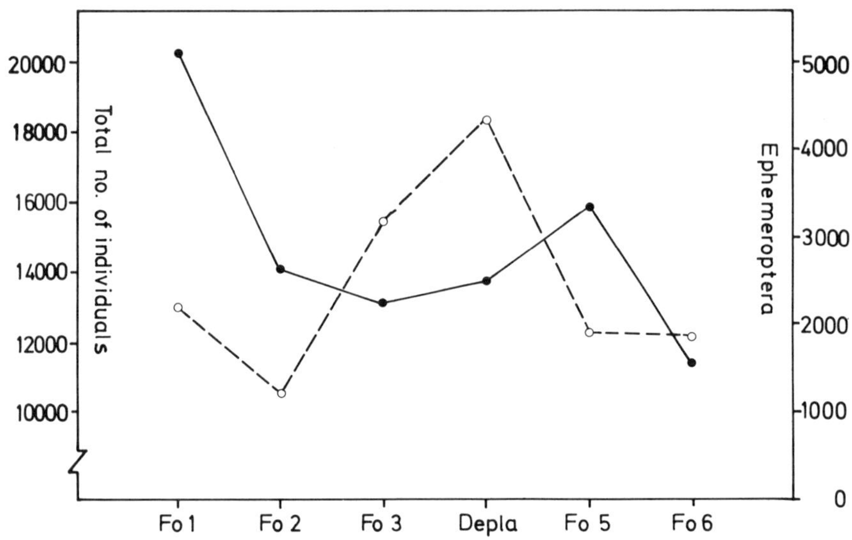

Figure 8. ●————● Means of individuals and o----o means of mayflies at the different sampling sites given as number per m^2 for the period July 1975 - July 1976.

In ecological evaluations of macrobenthic responses it is also important to know the composition of the communities in question (Hynes 1966). Table 2 shows some qualitative, and quantitative changes in the community structure below the outlets and there are distinct differences in the recovery of each group.

The nematodes and oligochaetes prefer a finer substrate (Brinkhurst and Cook 1974) but are markedly affected by heavy metal pollution at Station 6 (Aston 1973). As frequently mentioned in the literature, the molluscan fauna (see Fuller 1974 and Harman 1974) is often severely affected by mining activities (Simmons 1972). Stretches where silt and sand constantly scour the river bottom are, as a rule, inimical to both bivalves and gastropods. They are also particularly sensitive to the heavy metals, and Wurtz (1962) considered zinc followed by copper to be the most toxic to them, of the heavy metals.

At Station 2 the larvae of the important groups Plecoptera, Ephemeroptera, Trichoptera and Coleoptera along with the Arachnida are markedly affected by the effluents from the tailings dam. The Coleoptera, represented by the Elmidae, are practically wiped out, and the mayfly population is nearly halved. Great reductions are also found in the populations of stone- and caddisflies. Among the dipterans, the chironomids whose larvae are found in almost all imaginable habitats and are undoubtedly the most ubiquitous of all aquatic insects, show a significant drop from nearly 15,000 per m^2 at Station 1 to 8,000 per m^2 at Station 2. The opposite is however the case for the Simuliidae that show a remarkable increase to more than 3 times the population at Station 1 (also distinctly felt during the field work at Station 2!). Suspended organic materials from the tailings impoundment are able to sustain this great population of blackflies.

At Station 3 there is a recovery for some of the groups, though there are still some differences, particularly among the Ephemeroptera and Simuliidae, compared to the reference station, Depla.

At Station 5, just upstream of the Folldal village, recovery seems to be complete, and the station has a naturally developed macrobenthic fauna.

The activity upstream of Station 6 severely affects the environmental conditions at this station. The "yellow boy" deposit found on the riverbed is characteristic and large brown flocs of ferric-detritus which have accumulated periodically in areas of slow moving water among stones and other debris along the riverbank. With the exception of the mayflies and the stoneflies, all the other major groups show a more or less drastic reduction in abundance. The molluscs are completely wiped out and in other groups the reduction varies from about 94% (Coleoptera) to nearly

25% (Arachnida) compared to the population at the upstream station. Water-mites showing such a tolerance to heavy metal pollution is interesting. In their investigation of mine effluents, Thorp and Lake (1973) also found the water-mites to have great tolerance to heavy metals. along with the drainage from the mine, untreated sewage is discharged from nearly 1600 person equivalents. This reduces the effect of the heavy metal pollution but it does not seem to produce favourable conditions for the growth of a saprophelic community.

The stoneflies, as shown in Table 2, are the only group which increased its density at Station 6. Although the abundance is larger, the diversity is reduced. Letterman and Mitsch (1978) found in their investigation of Ben's Creek, a small mountain stream receiving coal mine drainage in western Pennsylvania, that the Plecopteran fauna showed a rapid initial recovery. Koryak *et al.* (1972) and Tomkiewicz and Dunson (1977) describe a similar relationship, and it seems that the family Nemouridae in particular, flourished better in moderately polluted water from acid mine drainage than in the non-polluted upstream area.

The Ephemeroptera populations are nearly identical at station 5 and 6, but as the species composition will show later, the mayflies at station 6 consist almost entirely of *Baetis rhodani*, a very eurytopic species. In contrast to the oligochaetes, the chironomids, as a group, seem to be very little affected and are usually recognised for their general tolerance to acid mine drainage (Koryak *et al.* 1972, Nichols and Bulow 1973, McPhilliamy and Green 1973, Dills and Rogers 1974) and Roback (1974) shows that chironomid larvae have representatives tolerant of many chemical extremes.

Although Table 2 and Fig. 8 give a rough impression of the situation in the river, there can be great variations in species composition of the different groups which will severely affect the system. We are unable to derive that kind of information unless we are working at the species level. It would be desirable to work with more of the most important groups, but resources are often limited and it is my feeling that the mayflies are able to integrate the environmental conditions in the biotope very well. They are also, for all or part of their life cycles, undoubtedly the most truly aquatic of all the insects that have adapted to the aquatic environment (Roback 1974).

It is perhaps a possibility that factors other than the effluents along the river are responsible for the differences in the results, for example, the configuration of the substrate (the structur of the microhabitat) and the kind and amount of nutrition available. The sampling sites were selected in places which ecologically should be as much alike as possible unless secondary effects of pollution had changed the conditions of the area. Another factor which we should have in mind is that in a mountain stream there can be differences in periods of emergence and hatching of eggs at different altitudes. With a sampling period of 3-4 weeks this can affect the results. As is well known, running waters, as an ecosystem, have a strong potential for colonization and drift. But even though the drift can affect the results, it is obvious that even the downstream drift of nymphs cannot compensate for nymphal deaths. The nymphs are unable to swim long distances to escape toxic elements, so their presence or absence in otherwise suitable habitats reflects the quality of the water which flows over them.

The Ephemeroptera are generally considered to be very sensitive to pollutants, though usually meaning organic loading. Roback (1974) has the impression from years of work with mayflies, that this group is not as sensitive as commonly believed. At the same time he points out that consistently reliable nymphal determinations are not yet possible for central groups such as e.g. Baetidae. Further, it is also unfortunate that the data available on heavy metal tolerance are not more reliable. Fremling (1970) found that the mayflies are excellent indicators of general water quality in the Mississippi River. Investigations on lotic macroinvertebrate fauna in connection with mine drainage and heavy metal pollution show that this impact seems to strike the mayflies most severely (Sprague *et al.* 1965, Roback and Richardson 1969, Warnick and Bell 1969, Bell 1971, Weed and Rutschky 1972, Nichols and Bulow 1973, Carrithers and Bulow 1973, Södergren 1974b, Winner *et al.* 1975 and Brånin 1979).

In Norway, the ephemeropteran fauna consists of 43 species (Dahlby 1973) and among them there are nine (ten) species of the genus *Baetis*. During the investigation in the river Folla, 14 mayfly species were collected of which seven were *Baetis*. Among the total population of Ephemeroptera the dominant genus was *Baetis* with 88.3%. *Baetis rhodani* was the most abundant species comprising nearly 60% of the total mayfly population. Varying degrees of mine pollution were reflected by changes in the species composition as seen in Table 3. Here the total number of the different mayfly species at each station as mean values for the sampling trips are presented as species per m^2. In Fig. 9 the composition of the different species at each station is shown in percent. Although the dominance changed among the species downstream from the effluents, *B. rhodani* is always found in much greater numbers than any other species. During the season, the overall picture is as in Fig. 9. But although seasonal changes in species diversity correspond with the life history patterns, the study gave the impression that summer species such as *Baetis vernus*, *B. subalpinus* and *B. scambus* are less affected than mayflies with a winter generation. This is in agreement with the fact that the discharge of heavy metal gives higher winter than summer values, and that the spring flood washes

Table 3. Means of the Ephemeroptera species at the different sampling sites given as individuals per m^2 for the period July 1975 - July 1976.

			Fo1	Fo2	Fo3	Depla	Fo5	Fo6	Σ	%	
Ameletus inopinatus	Eaton	1885	135	140	175	3	12	3	468	3,19	
Baëtis lapponicus	(Bengtsson)	1912	17	4	6	122	19	13	181	1,23	88,32 (B. lapponicus – B. muticus)
B. rhodani	(Pictet)	1843-45	1401	448	1991	2805	827	1120	8592	58,50	
B. vernus	Curtis	1834	29	25	210	70	170	146	650	4,43	
B. subalpinus	Bengtsson	1917	47	102	147	193	186	125	800	5,45	
B. scambus	Eaton	1870	121	212	400	934	350	276	2293	15,61	
B. niger	(Linneaus)	1761	120	41	35	51	6	-	253	1,72	
B. muticus	(Linneaus)	1758	89	39	14	50	9	1	202	1,38	
Centroptilum luteolum	(Müller)	1776	105	109	120	-	-	-	334	2,27	
Heptagenia sulphurea	(Müller)	1776	102	22	5	10	63	12	214	1,46	
Ephemerella aurivillii	(Bengtsson)	1908	37	63	87	110	237	166	700	4,77	
=) Σ			2203	1205	3190	4348	1879	1862	14687	100,01	

Others: *Siphlonurus lacustris* Eaten 1870, *Cloëon dipterum* (Linneaus) 1761, *Paraleptophlebia* sp. } ≈ 0,4 %

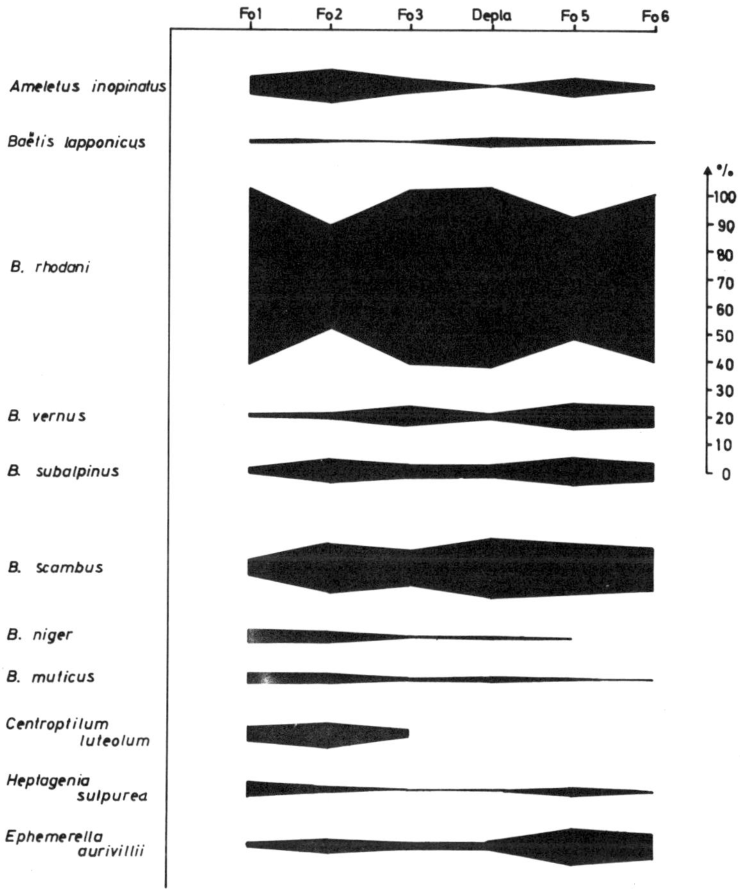

Figure 9. The % composition among the Ephemeroptera species at the different sampling sites in the river Folla, given for the total amount of material collected during July 1975 - July 1976.

away from the substrate the silt and other fine materials deposited during late autumn and winter.

That heavy metal pollution seems to strike the Ephemeroptera with a winter generation most severely, seems to be a general picture. Södergren (1974b) found that in connection with Cd and Ni pollution, *Ephemerella mucronata* was eliminated downstream from the effluent while *E. ignita*, a summer species, did not seem to be affected at all.

Ameletus inopinatus was missing in all samples collected below Station 5 with exception of the August sampling where some very small larvae were found. The obvious explanation to this phenomenon is that the nymphs drift down the river, away from the unpolluted areas. This is very probable, especially in spring and early summer when the water flow is large. At Station 6 this is seen particularly in species such as *B. muticus* which was found only in samples collected after the spring flood in 1976, and in *Ephemerella aurivillii* where only 8 small individuals (0.7%) were found in months other than June and July. *Heptagenia sulphurea* also seems to have a distribution at Station 2 and 6, mainly built up around small drifting larvae without the possibility to establish a sustaining population at these stations.

Centroptilum luteolum was only registered in the upper part of the investigated area. The species are usually found on stony shores of lakes and in streams with a slower flow of water than the river Folla (Kimmins 1972). Thus it is likely that the nymphs have drifted from the lentic biotopes upstream from Station 1.

SUMMARY

A small mountain stream named Folla in the southern part of Norway has been receiving mine effluents for nearly 250 years. Today there are two major sources of mine drainage. First is seepage from the old, now exhausted mine and from the dumps of tailings and slag in the Folldal village. Secondly there is the return water from a new pyrite mine and a flotation plant situated about 28 stream-km upstream from the old mine. The study in the Folla thus made it possible to investigate the environmental impact of a new, modern mine industry with a tailings impoundment for which there has been great concern about the pollution problems. The well-known effect from old mines with acid mine drainage was here associated with an organic load from a dairy and sewage from the village running into an otherwise unpolluted stream.

The feeder stream from the tailings impoundment had no significant effect on the pH, but changed the water quality considerably. Below the outlet the water is characterized as a sulphate-water. At Station 2 there is a statisticall significant difference ($P < 0.05$) between conductivity, turbidity, SO_4 and Ca compared to the unpolluted, upstream station. Along with the dissolved salts there is at times a heavy outflow of suspended particles in which the organic part is able to sustain a large population of Simuliidae. The inorganic part causes a silting of the substrate, especially marked in the low-water period before the spring flood. The water temperature is low, and for about 6 months of the year the river is ice covered. The effluents from the old mine are typical acid mine drainings with a heavy content of trace elements, mainly iron.

Deposits of ferric hydroxide on the substrate and organic - ferric flocs are seen along the river bank. Compared to the upstream station there is a significant difference ($P < 0.05$) in the parameters:turbidity, Cu, Zn, Fe and organic seston, at the station below the village. The oxygen conditions are good, and the lowest O_2 saturation was measured in the effluent from the tailings impoundment (93%).

The altered chemistry of the stream and the associated sedimentation in the upper regions and ferric hydroxide precipitation which blanketed several kilometres of stream bottom, had a significant impact on the stream ecosystem. Below the effluents there is a drastic decrease in the total fauna. How the different major groups are affected by the environmental impact is discussed. The Ephemeroptera particularly have been the object for a more thorough study. Among the total population of mayflies the dominant genus was *Baetis* with 88%. *Baetis rhodani* was the most abundant species with nearly 60%. A total of 14 species were found and of these the summer species seem to be less affected by the mine drainage. The possibility that factors other than the effluents along the river are responsible for the differences is discussed along with some general comments on how the community responds to this kind of environmental impact.

ACKNOWLEDGMENTS

The author gratefully acknowledges the good working conditions at the Norwegian Institute for Water Research during this study. The financial support given to the investigation in the period 1975 to 1976 from the mining company Folldal Verk A/S at Hjerkinn, is greatly appreciated. My sincere thanks also to Ellen Hand who is responsible for improving the English and to Ingjerd Borch who typed the manuscript.

RESUME

Les effets des eaux usées résultant de l'extraction et du traitement du zinc et du cuivre sur la faune invertébrée de la rivière Folla ont été étudiés en 1975-1976. Les éphéméroptères ont été reconnus, tant du point de vue de la qualité que du point de vue de la quantité, comme les meilleurs indicateurs de cette pollution. L'impact des effluents de la mine sur les invertébrés (particulièrement les éphéméroptères) de la rivière est étudié en fonction des transformations physico-chimiques causées par ces effluents.

ZUSSAMENFASSUNG

Zwischen 1975 und 1976 wurde untersucht, welche Wirkung Abwasser, die von Abbau und Verarbeitung von Zink- und Kupfererzen herrühren, auf die Invertebraten Fauna des Flusses "Folla" haben. Ephemeropteren erwiesen sich qualitativ und quantitativ als die besten Indikatoren der daraus resultierenden Verunreinigung. In der vorliegenden Arbeit wird die Einwirkung der Bergbauabwasser auf die Invertebraten, besonders die Eintagsfliegen, des Flusses erörtert mit Bezug auf physio- chemikalische Veränderungen, die durch diese Abwasser hervorgerufen wurden.

REFERENCES

American Public Health Association. 1975. Standard methods for the examination of water and waste water. 14th Ed., New York, American Public Health Association.

Aston, R.J. 1973. Field and experimental studies on the effects of a power station effluent on Tubificidae (Oligochaeta, Annelida). *Hydrobiologia* 42: 225-242.

Bell, H.L. 1971. Effect of low pH on the survival and emergence of aquatic insects. *Water Res.* 5: 313-319.

Brinkhurst, R.O. and D.G. Cook. 1974. Aquatic Earthworms (Annelida: Oligochaeta). p. 143-156. *In:* Pollution ecology of freshwater invertebrates. C.W. Hart, Jr. and S.L.H. Fuller. (ed.) Academic Press, New York.

Broch, E.S. and W. Yake. 1969. A modification of Maucha's ionic diagram to include ionic concentrations. *Limnol. Oceanogr.* 14: 933-935.

Brånin, B. 1979. Physiocochemical and bottom fauna conditions in a Northern Swedish stream influenced by discharge from pyrite mining and milling (Swedish text). Thesis for a doctor's degree, Unit. Umeå, Sweden, 99 pp.

Carrithers, R.B. and F.J. Bulow. 1973. An ecological survey of the west fork of the Obey River, Tennessee, with emphasis on the effects of acid mine drainage. *J. Tenn. Acad. Sci.* 48(2): 65-72.

Dahlby, R. 1973. A check-list and synonyms of the Norwegian species of Ephemeroptera. *Nor. entomol. Tidsskr.* 20: 249-252.

Dills, G. and D.T. Rogers, Jr. 1974. Macroinvertebrate community structure as an indicator of acid mine pollution. *Environ. Pollut.* 6: 239-62.

Fremling, C.R. 1970. Mayfly distribution as a water quality index. *U.S. EPA Water Pollution Control Research Series* 16030 DQH 11/70: 1-39.

Fuller, S.L.H. 1974. Clams and mussels (Mollusca: Bivalvia). p. 215-276. *In:* Pollution ecology of freshwater invertebrates. C.W. Hart, Jr. and S.L.H. Fuller (ed.) Academic Press, New York.

Harman, W.N. 1974. Snails (Mollusca:Gastropoda) p. 275-312. *In:*

Pollution ecology of freshwater invertebrates. C.W. Hart, Jr. and S.L.H. Fuller, (ed.) Academic Press, New York.

Harrison, A.D. and J.F. Elsworth. 1958. Hydrobiological studies on the Great Berg River, Western Cape Province. Part I. General descriptions, chemical studies and main features of the flora and fauna. *Trans. R. Soc. S. Afr.* 35: 125-226.

Henriksen, A. and K. Balmer. 1977. Sampling, preservation and storage of water samples for analysis of metals. *Vatten* 33: 33-38.

Huet, M. 1949. La pollution des aux l'analyse biologique des aux polluées. *Bull. Cent. Belg. Etude Doc. Eaux* 5 and 6: 1-31.

Huet, M. 1954. Biologie, profils en long et en travers des eaux courantes. *Bull. franç. Pisic.* 175: 41-53.

Hynes, H.B.N. 1966. The biology of polluted waters. Liverpool Univ. Press, Liverpool.

Illies, J. 1961. Versuch einer allgemein biozönotischen Gleiderung der Fliessgewässer. *Int. rev. gesamten. Hydrobiol.* 46: 205-213.

Kimmins, D.E. 1972. A revised key to the adults of the British species of Ephemeroptera, with notes on their ecology. *Freshw. Biol. Assoc. Sci. Publ.* 15: 1-75.

Koryak, M., M.A. Shapiro and J.L. Sykora. 1972. Riffle zoobenthos in streams receiving acid mine drainage. *Water Res.* 6: 1239-1247.

Letterman, R.D. and W.J. Mitsch. 1978. Impact of mine drainage on a mountain stream in Pennsylvania. *Environ. Pollut.* 17: 53-73.

McPhilliamy, S.C. and J. Green. 1973. A chemical and biological evaluation of three mine drainage treatment plants. *U.S. Environmental Protection Agency Region III, Wheeling Field Office Work. Doc.* No. 47.

Nichols, L.E., Jr. and F.J. Bulow. 1973. Effects of acid mine drainage on the stream ecosystems of the east fork of the Obey River, Tennessee. *J. Tenn. Acad. Sci.* 48(1): 30-39.

Roback, S.S. 1974. Insects (Arthropoda:Insecta). p. 313-376 *In:* Pollution ecology of freshwater invertebrates. C.W. Hart, Jr. and S.L.H. Fuller, (ed) Academic Press, New York.

Roback, S.S. and J.W. Richardson. 1969. The effects of acid mine drainage on aquatic insects. *Proc. Acad. Nat. Sci. Phila.* 121: 81-107.

Simmons, G.M., Jr. 1972. A preliminary report on the use of the sequential comparison index to evaluate acid mine drainage on the macrobenthos in a pre-impoundment basin. *Trans. Am. Fish. Soc.* 101: 701-713.

Sprague, J.B., P.F. Elson and R.L. Saunders. 1965. Sublethal copper-zinc pollution in a salmon river - a field and laboratory study. *Int. J. Air. Water Pollut.* 9: 531-543.

Södergren, S. 1974a. A subsampler for stream-bottom-fauna samples. *Arch. Hydrobiol.* 73: 549-551.

Södergren, S. 1974b. Ecological effects of heavy metal discharge in a salmon river. Thesis for a doctor's degree at the Univ. of Umeå, Sweden. 125 pp.

Thorp, V.J. and P.S. Lake. 1973. Pollution of a Tasmanian river by mine effluents. Distribution of macrovertebrates. *Int. Rev. gesamten. Hydrobiol.* 58(6): 885-892.

Tomkiewicz, S.M., Jr. and W.A. Dunson. 1977. Aquatic insect diversity and biomass in a stream marginally polluted by acid strip mine. *Water Res.* 11: 397-402.

Warnick, S.L. and J.L. Bell. 1969. The acute toxicity of some heavy metals to different species of aquatic insects. *J. Water Pollut. Control Fed.* 41: 280-283.

Weed, C.E. and C.W. Rutschky, III. 1972. Benthic macroinvertebrate community structure in a stream receiving acid mine drainage. *Proc. Pa Acad. Sci.* 46: 41-47.

Winner, R.W., J.S. van Dyke, N. Caris and M.P. Farrel. 1975. Response of the macroinvertebrate fauna to a copper gradient in an experimentally polluted stream. *Int. Ver. theor. angew. Limnol. Verh.* 19: 2121-2127.

Wurtz, C.B. 1962. Zinc effects on fresh-water mollusks. *Nautilus* 76: 53-61.

EFFECTS OF CRUDE OIL CONTAMINATION ON EPHEMEROPTERA IN THE TRAIL RIVER, NORTHWEST TERRITORIES, CANADA

David M. Rosenberg, Allen P. Wiens and John F. Flannagan

Freshwater Institute
Department of Fisheries and Oceans
501 University Crescent
Winnipeg, Canada R3T 2N6

ABSTRACT

Artificial substrates were soaked in crude oil and placed in the Trail River, N.W.T. for periods up to almost 1 year. Ephemeroptera nymphs usually occurred in lower numbers, fewer taxa and lower diversity on oiled substrates than on unoiled substrates. Communities of nymphs were significantly different except for the substrates left in the river for the longest period (335 days). Numbers of *Heptagenia (flavescens* Walsh?), *Stenonema vicarium* (Walker), *Ameletus* sp. 1, and *Baetis* spp. decreased in response to the oil while numbers of *Pseudocloeon* sp. 1 and *Ephemerella aurivillii* Bengtsson increased. Numbers of *E. (bicolor* Clemens?) and *E. simplex* McDunnough were unaffected. *H. (flavescens?)* should be the best indicator (reduced numbers) of low level contamination of aquatic systems by oil and petroleum products although *S. vicarium* (reduced numbers) and *E. aurivillii* (increased numbers) may also be useful. Responses of these three species to oil contamination are related to the presence of food and to quality of the substrate.

INTRODUCTION

Development of petroleum reserves in Alaska and northern Canada will increase the contamination of northern freshwater habitats by oil and petroleum products. A large-diameter oil pipeline has already been built across Alaska; a large-diameter natural gas pipeline is planned for Alaska and the Yukon Territories; another is planned for the Canadian Arctic Islands; and mining of the vast oil sands reserves in northern Alberta has begun. Little is known about the responses of freshwater invertebrates to such contamination, especially in northern areas where only a few studies have been done (Rosenberg and Wiens 1976; Mozley and Butler 1978; Barton and Wallace 1979). Our objectives in this paper are to describe the responses of Ephemeroptera experimentally exposed to crude oil and to identify species that might be used as indicators of low-level contamination by oil and petroleum products.

METHODS

Experiments were done in the Trail River, N.W.T., ≃1 km above its confluence with the Mackenzie River (62°06'N, 122°11'W). The river is in the boreal forest of the Upper Mackenzie Valley lowlands (150-300 m a.s.l.) and drains a watershed (mainly muskeg) of 660 km^2.

Artificial substrate samplers (Anderson and Mason 1968) were filled with approximately 30-35 dry stones (5-8 cm diameter) taken from the river bank. Some of the samplers were dipped in a pan of Norman Wells crude oil, allowed to drain, and installed on the river bottom (≃1 m depth). An equal number of untreated (control) samplers were placed upstream to avoid contamination. Sufficient samplers were installed to provide three replicates of oiled and control substrates (= 1 set) per sampling. Samplers usually were removed after ≃1 month colonization and/or for varying periods thereafter. Experiments were begun on three dates (Table 1). After removal of the samplers, invertebrates were brushed from the surface of the stones, preserved in 70% ethyl alcohol and returned to the laboratory for identification and enumeration.

Diversity of Ephemeroptera was characterized using Brillouin's H (Pielou 1966; Lloyd *et al.* 1968), an index based on information theory. Kendall's τ (Siegel 1956; Ghent 1963), a non-parametric measure of rank correlation, was used to determine similarity of communities on oiled and control substrates. Categories of responses of species of Ephemeroptera to the crude oil and evaluation of indicator potential are adapted from Rosenberg and Wiens (1976).

Further details of the study area and methods used can be found in Brunskill *et al.* (1973) and Rosenberg and Wiens (1976).

RESULTS

Numbers of individuals, numbers of taxa, and diversity (Brillouin's H) of Ephemeroptera on oil-treated artificial substrates were generally lower than on the untreated controls (Table 1). Communities of Ephemeroptera on the two types of

Table 1. Comparison of communities of Ephemeroptera on oil-treated (O) and untreated (U) artificial substrates.

Date Artificial Substrates Installed	No. of Days of Colonization	% Oil Remaining[a]	No. of Individuals[b]		No. of Taxa[b]		Diversity (Brillouin's H)		Kendall's τ		Are Communities Similar?[c]
			U	O	U	O	U	O	τ Corrected	P-Value	
July 14, 1972	28	69.5	190	138	9	7	2.33	1.80	.354	.260	No
	56	33.2	179	88	10	6	2.43	2.24	.167	.600	No
	132	4.7	144	23	10	4	2.19	1.43	.196	.612	No
	335	28.3	680	392	7	7	1.38	1.38	.823	.009	Yes
Sept. 8, 1972 (Set A)	279	28.2	619	318	7	7	1.82	1.59	.704	.015	No
(Set B)	279	4.4	616	407	9	6	1.78	1.58	.530	.060	No
July 12, 1973	28	28.3	193	361	6	4	1.61	1.02	.500	.272	No
	57	15.8	146	250	7	7	1.66	1.87	.333	.472	No

[a] See Table 1 in Rosenberg and Wiens (1976)

[b] Total number on three artificial substrates

[c] Decision based on $P \leq .01$

substrate were significantly different (Kendall's τ) on seven of the eight sets. Discrepancies in these trends occurred in samplers with the longest colonization times (279 and 335 days) and in the sets from 1973. Residues of oil remaining on these substrates were <30% of the amount originally applied (Table 1; Rosenberg and Wiens 1976) probably due to longer exposure to river water and/or the higher annual discharge in 1973 ($\approx$69.5 x 10^6 m^3) than in 1972 ($\approx$26.6 x 10^6 m^3). A threshold quantity of oil may therefore be necessary to cause changes in characteristics of communities of Ephemeroptera.

Table 2. Taxa of Ephemeroptera occurring on artificial substrates

Taxa
Siphlonuridae
Ameletus sp. 1
Siphlonurus sp. 1
Baetidae
Baetis spp.
Pseudocloeon sp. 1
Heptageniidae
Epeorus (Iron) sp. 1
Heptagenia (flavescens Walsh?)
Rhithrogena sp. 1
Stenacron (pallidum (Traver)?)
Stenonema (fuscum (Clemens)?)
Stenonema (ithaca (Clemens & Leonard)?)
Stenonema pulchellum (Walsh)
Stenonema vicarium (Walker)
Leptophlebiidae
Leptophlebia cupida (Say) or *L. nebulosa* (Walker)
Paraleptophlebia sp. 1
Ephemerellidae
Ephemerella (Dannella) simplex McDunnough
Ephemerella (Ephemerella) aurivillii Bengtsson
Ephemerella (Eurylophella) (bicolor Clemens?)

Seventeen taxa of Ephemeroptera, representing five families, colonized the artificial substrates (Table 2). Out of a possible of 16 occurrences (eight sets of oiled substrates, eight sets of controls), nine taxa occurred on four sets or less (0-<25% occurrence) and were not used in the following consideration of responses because of their erratic occurrence. Five taxa occurred in 12-16 sets (≥75-100% occurrence, Group I), one species occurred on 8-12 sets (≥50-<75% occurrence, Group II) and two species occurred on 4-8 sets (≥25-<50% occurrence, Group III). Two species showed a positive response to the oil (i.e. numbers higher on oiled substrates than on controls in more than half the occurrences), four species were negatively affected (i.e. numbers lower on oiled substrates than on controls in more than half the occurrences), and two species were unaffected (i.e. approximately equal numbers on oiled and control substrates) (Table 3). No species showed a strong positive response to the oil but three species showed a strong negative response. These results were verified, in a general way, when the total numbers of each species on oiled and control substrates were compared (Fig. 1).

DISCUSSION

Oil contamination can affect invertebrates in a number of ways (Moore and Dwyer 1974). Lighter fractions of the oil may be directly toxic. Substrate surfaces may be coated by oil and made unuseable. The oil may enhance microbial and algal growth (Rosenberg and Wiens 1976, Barton and Wallace 1978) thus increasing the food supply for invertebrates but also altering the microhabitat (e.g. by lowering dissolved oxygen concentrations, covering substrate surfaces and filling interstices).

Ephemeroptera appear to be sensitive to oil contamination, their numbers usually being reduced (Hynes 1965, U.S.E.P.A. 1973, Barton and Wallace 1979). Certainly, in this study, more species were reduced in number than increased by the oil treatment (Table 3, Fig. 1). This aversion to oil by Ephemeroptera is even more evident when the distribution of the species not considered in the above assessment (i.e. those occurring on 1-4 sets, the 0-<25% occurrence group) is examined (Table 4).

It would be useful to predict those species which may be potentially indicative of low-level, chronic oil contamination in lentic fresh waters. Species listed in the positive and negative response categories of Table 3 were assessed using three criteria: taxonomic soundness, zoogeographic distribution and abundance (for rationale, see Rosenberg and Wiens 1976). *Baetis* spp., *Pseudocloeon* sp. 1, and *Ameletus* sp. 1 were excluded because of their uncertain taxonomic status. No additional species were obtained for assessment from the literature because species identifications were rarely

Table 3. Responses of Ephemeroptera to oil.

Group[a]	Response Index[b]								
	Positive				Unaffected	Negative			
	Strong 1.0	Strong .875	Moderate .750	Weak .625	.500	Weak .375	Moderate .250	Strong .125	Strong 0
I			*Ephemerella aurivillii*	*Pseudocloeon* sp. 1	*Ephemerella (bicolor?)*		*Baetis* spp.		*Heptagenia (flavescens?)*
II								*Stenonema vicarium*	
III					*Ephemerella simplex*				*Ameletus* sp. 1

[a] Based on frequency of occurrence on all sets of artificial substrates (See text for explanation).

[b] Based on number of occurrences of higher numbers on oiled artificial substrates than on controls out of a maximum of 8 possible dates (e.g. *E. aurivillii* occurred in higher numbers on oiled artificial substrates than on controls 6 out of 8 times for an index value = .750 and a "positive, moderate" response designation).

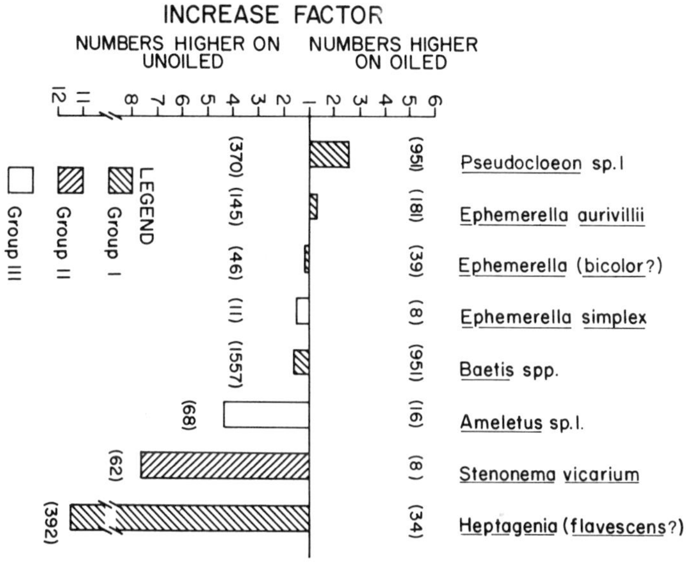

Figure 1. Responses of Ephemeroptera to crude oil. Numbers in parentheses are totals occurring on oiled (top) and control (bottom) artificial substrates. See text for explanation of Groups I, II, III.

given in studies on the effects of oil on Ephemeroptera or the data for species were inconclusive (e.g. U.S.E.P.A. 1973).

Of the three species assessed (Table 5), *Heptagenia (flavescens?)* offers the most potential, provided its identification is correct. Numbers of *H. (flavescens?)* should be reduced by oil contamination. *Stenonema vicarium* should be less useful because, although it was obviously reduced in number by the oil treatment (Fig. 1), it occurred in lower numbers and less frequently than *H. (flavescens?)*. *Ephemerella aurivillii* should increase in numbers as a result of oil contamination although the minimal response shown in this study detracts from its usefulness as an indicator.

Toxic lighter fractions of oil are short-lived (Moore and Dwyer 1974). Thus is is unlikely that they are involved in the responses described here because of the relatively long colonization periods involved in this study. Responses by the three species

Table 4. Distribution of Ephemeroptera in the 0-<25% occurrence group on oil-treated and control artificial substrates.[a]

Species	None found on: Oiled	Control
	Artificial Substrates	
Stenonema pulchellum	x	
Stenonema ithaca	x[b]	
Rhithrogena sp. 1	x[c]	
Epeorus sp. 1		x[c]
Siphlonurus sp. 1	x[b]	
Leptophlebia cupida or *nebulosa*	x[c]	
Paraleptophlebia sp. 1	x	

[a] *Stenonema (fuscum?)* - one specimen on each type of substrate
Stenacron (pallidum?) - approximately twice as many on control substrates (12) as oiled (7) but only present on two sampling dates.

[b] One specimen only.

[c] Present on one artificial substrate only.

assessed as indicators are probably linked to increased food or to changes in the quality of the substrate. *E. aurivillii* usually occurred in higher numbers on oiled than control artificial substrates, the differences being greatest during the first two months of colonization in 1972 and 1973 (Table 6). Extensive algal growth occurred on the oiled artificial substrates during these periods (Rosenberg and Wiens 1976). *E. aurivillii* is known to eat diatoms and filamentous algae (Shapas and Hilsenhoff 1976) and could be exploiting this additional food resource on the oiled artificial substrates. Numbers of *S. vicarium and H. (flavescens?)* were almost always dramatically higher on control artificial substrates than on oiled (Table 6). Heptageniidae need clean water and clean substrates with direct access to interstices (Dodds and Hisaw 1924). *S. vicarium* can withstand some organic enrichment providing the dissolved oxygen concentration is >5 mg l^{-1} (Lewis 1978). Persistence of residues in the oiled artificial substrates (Table 1) and the large amount of organic matter resulting from algal growth probably created an unfavorable habitat for these mayflies on oiled artificial substrates. More information on the autecology of species is required to explain fully the responses of Ephemeroptera observed in this study.

Table 5. Evaluation of some species of Ephemeroptera as potential indicators of oil contamination.

Species	Response Category	Agreement with Criteria: Taxonomically Sound	Widely Distributed	Numerous	Predicted Usefulness as Indicator
Ephemerella aurivillii	Positive (moderate)	Yes. (Burks 1953, Allen & Edmunds 1965, Edmunds *et al.* 1976).	Yes. Holarctic, generally distributed in Nearctic; found north into Alaska (Needham *et al.* 1935, Burks 1953).	Yes. (this study). Infrequent in Barton and Wallace (1978).	Moderate-high
Stenonema vicarium	Negative (strong)	Yes. (Burks 1953, Lewis 1978).	Yes. Mainly Nearctic; mostly in east and central U.S.A. and Canada (Edmunds *et al.* 1976).	Moderately (this study, Barton and Wallace 1978).	Moderate-high
Heptagenia (flavescens?)	Negative (strong)	Probably. Species complexes exist in this genus and some nymphs have not been described (Edmunds *et al.* 1976).	Yes. Holarctic and widely distributed in Nearctic (Needham *et al.* 1935).	Yes. (this study)	High

Table 6. Numbers of *Ephemerella aurivillii*, *Stenonema vicarium*, and *Heptagenia (flavescens?)* on oil-treated (O) and untreated (U) artificial substrates.[a]

Date Artificial Substrates Installed	No. of Days of Colonization	Numbers of Individuals					
		Ephemerella aurivillii		*Stenonema vicarium*		*Heptagenia (flavescens?)*	
		U	O	U	O	U	O
July 14, 1972	28	5	25	14	0	46	3
	56	4	7	13	0	10	5
	132	4	6	32	0	63	1
	335	34	33	6	0	67	4
Sept. 8, 1972 (Set A)	229	65	18	7	1	17	1
(Set B)	229	25	26	0	3	80	4
July 12, 1973	28	4	27	14	0	36	2
	57	4	27	7	4	74	14

[a] Total number on three artificial substrates.

RESUME

Des substrats artificiels furent trempés dans de l'huile brute et placés dans la Trail River (T.N.-O.) pour des périodes allant jusqu'à un an. On a généralement trouvé moins de larves d'éphéméroptères, moins de taxons et moins de variétés sur les substrats huilés que sur les substrats non huilés. Les populations de larves accusaient des différences notables sauf en ce qui concerne les substrats qui restèrent submergés le plus longtemps (335 jours). Les populations d'*Heptagenia (flavescens* Walsh?), de *Stenonema vicarium* (Walker), d'*Ameletus* sp. 1, ainsi que des *Baetis* spp. accusèrent une diminution en raison d l'huile alors que les populations de *Pseudocloeon* sp. 1 et d'*Ephémérella aurivillii* Bengtsson augmentèrent. Les populations d'*E.* (*bicolor Clemens?)* et d'*E. simplex* McDunnough ne subirent par de changements. L'*Heptagenia (flavescens?)* se révéla le meilleur indice du faible niveau de pollution des eaux dû à l'huile et aux produits dérivés du pétrole. Ajoutons que les *S. vicarium* et les *E. aurivillii* peuvent aussi s'avérer utiles. Les réactions de ces trois espèces à la pollution par l'huile dépendent de la présence de nourriture et de la qualité du substrat.

ZUSSAMENFASSUNG

Künstliche Substrate wurden mit Rohöl durchtränkt and im Trail River N.W.T. bis zu einem Jahr ausgesetzt. Ephemeropterennymphen kamen gewöhnlich an geölten Substraten in Kleineren Mengen, niedrigeren Taxa und in geringerer Vielfalt zum Vorschein, als an ungeölten Substraten. Die Nymphengemeinschaften wiesen bedeutende Underschiede auf. Nur die Substrate, die die längste Zeit im Fluß waren, (335 Tage), zeigten keinerlei Unterschied. Die Zahlen von *Heptagenia (flavescens* Walsh?), *Stenonema vicarium* (Walker), *Ameletus* sp. 1 und *Baetis* spp. verringerten sich als Reaktion auf das Öl, während *Pseudocloeon* sp. 1 und *Ephemerella aurivillii* (Bengtsson) zunahmen. *E. (bicolor* Clemens?) und *E. simplex* (McDunnough) blieben vom Öl unbeeinflußt. *H. (flavescens?)* erwies sich als der beste Indikator einer leichten Verschmutzung der aquatischen Systeme durch Öl und Petroleumprodukte, obwohl *S. vicarium* und *E. aurivillii* ebenfalls brauchbar sein mögen. In der vorliegenden Arbeit soll die Reaktion der genannten drei Arten auf Ölverschmutzung in Beziehung gesetzt werden zu dem Vorhandensein von Nahrung und zur Qualität des Substrats.

REFERENCES

Allen, R.K., and G.F. Edmunds, Jr. 1965. A revision of the genus *Ephemerella* (Ephemeroptera, Ephemerellidae). VIII. The subgenus *Ephemerella* in North America. *Misc. Publ. Entomol. Soc. Am.* 4: 243-282.

Anderson, J.B., and W.T. Mason, Jr. 1968. A comparison of benthic macroinvertebrates collected by dredge and basket sampler. *J. Water Pollut. Control Fed.* 40: 252-259.

Barton, D.R., and R.R. Wallace. 1978. Ecological studies of the aquatic invertebrates of the Alberta Oil Sands Environmental Research Program study area of northeast Alberta. *Alberta Oil Sands Environmental Research Program, Edmonton, Alberta.* (Unpublished draft report): 1-245 p.

Barton, D.R., and R.R. Wallace. 1979. The effects of an experimental spillage of oil sands tailings sludge on benthic invertebrates. *Environ. Pollut.* 18: 305-312.

Brunskill, G.J., D.M. Rosenberg, N.B. Snow, G.L. Vascotto, and R. Wagemann. 1973. Ecological studies of aquatic systems in the Mackenzie-Porcupine drainages in relation to proposed pipeline and highway developments. *Can. Task Force North. Oil Dev. Environ. Soc. Comm.* Vol. I Rep. 73-40: 1-131. Vol. II Append. Rep. 73-41: 1-345.

Burks, B.D. 1953. The mayflies or Ephemeroptera, of Illinois. *Bull. Ill. Nat. Hist. Surv.* 26: 1-216.

Dodds, G.S. and F.L. Hisaw. 1924. Ecological studies of aquatic insects. 1. Adaptations of mayfly nymphs to swift streams *Ecology* 5: 137-150.

Edmunds, G.F., Jr., S.L. Jensen and L. Berner. 1976. The mayflies of North and Central America. University of Minnesota Press, Minneapolis.

Ghent, A.W. 1963. Kendall's "Tau" coefficient as an index of similarity in comparison of plant or animal communities. *Can. Entomol.* 95: 568-575.

Hynes, H.B.N. 1965. The significance of macroinvertebrates in the study of mild river pollution. *Publ. U.S. Pub. Hlth. Serv.* 999-WP-25: 235-240.

Lewis, P.A. 1978. Taxonomy and ecology of *Stenonema* mayflies. *US EPA*-670/4-74-006: 1-81.

Lloyd, M., J.H. Zar and J.R. Karr. 1968. On the calculation of information-theoretical measures of diversity. *Am. Midl. Nat.* 79: 257-272.

Moore, S.F. and R.L. Dwyer. 1974. Effects of oil on marine organisms: A critical assessment of published data. *Water Res.* 8: 819-827.

Mozley, S.C. and M.G. Butler. 1978. Effects of crude oil on aquatic insects of tundra ponds. *Arctic* 31: 229-241.

Needham, J.G., J.R. Traver and Y.-C. Hsu. 1935. The biology of mayflies with a systematic account of North American species. Comstock Publishing Co., Ithaca, N.Y. 759 p.

Pielou, E.C. 1966. The measurement of diversity in different types of biological collections. *J. Theoret. Biol.* 13: 131-144.
Rosenberg, D.M. and A.P. Wiens. 1976. Community and species responses of Chironomidae (Diptera) to contamination of fresh waters by crude oil and petroleum products, with special reference to the Trail River, Northwest Territories. *J. Fish. Res. Board Can. 33: 1955-1963.*
Shapas. T.J. and W.L. Hilsenhoff. 1976. Feeding habits of Wisconsin's predominant lotic Plecoptera, Ephemeroptera, and Trichoptera. *Great Lakes Entomol.* 9: 175-188.
Siegel, S. 1956. Nonparametric statistics for the behavioural sciences. McGraw-Hill, Inc. New York, N.Y. 312 p.
U.S. Environmental Protection Agency. 1973. Boone Creek oil spill. *U.S. Environ. Prot. Agency, Surveillance Anal. Div., Athens, GA*: 1-37.

EVALUATION OF THE ACUTE TOXICITY OF THE HEAVY METAL CADMIUM TO NYMPHS OF THE BURROWING MAYFLY, *HEXAGENIA RIGIDA*

Sharon L. Leonhard, Sharon G. Lawrence, M.K. Friesen
and John F. Flannagan

Department of Fisheries and Oceans
Freshwater Institute
501 University Crescent
Winnipeg, Manitoba R3T 2N6

ABSTRACT

Laboratory-cultured, yearling nymphs of the mayfly *Hexagenia rigida* were tested in static, no-replacement 96 h bioassay. The 96 h LC_{50} was estimated to be 1.0 mg/L by a graphical method and 6.2 mg/L by the O-Trimmed Spearman-Karber calculation. The latter was easier to apply and gave a more precise estimate.

INTRODUCTION

The mayfly, *Hexagenia rigida*, is being investigated as one of the organisms in a toxicological screening system to evaluate the effect of contaminants using a range of sensitive to tolerant organisms and/or stages within their life cycle. In this study, yearling nymphs, from a stock culture of animals, were tested with cadmium, a toxic heavy metal widely distributed in the environment due to increased industrial use. The object of the study was two-fold: to determine the acute toxicity of cadmium (96 h LC_{50}) to yearling nymphs in order to assess their efficacy as a test stage and to compare two popular estimates of LC_{50} used by bioassay researchers; the "routine bioassay method" given in "Standard Methods" (APHA *et al.* 1965) and further described by Sprague (1969), and the trimmed Spearman-Karber calculation (Hamilton *et al.* 1977).

MATERIALS AND METHODS

Nymphs were sieved from stock cultures. These stock cultures had been initiated one year earlier by extruding eggs from fertilized adult females that had been gathered along the Red River at Winnipeg, Manitoba, Canada. The cultures were held at 20°C in approximately 3 cm of silt, from the Red River, covered with 3L of water. The pH of the culture system was 7.0. The animals were fed undefined amounts of a water suspension of Tetramin, fish food flakes, weekly.

Prior to testing, the experimental animals (nymphs 24 mm in length from the tip of the frontal process to the end of the abdomen) were acclimated for 24 h in 1 L pyrex glass beakers (5 animals/beaker) containing 500 mL of water at 18°C. No food substrate or aeration was provided. Light (12 $\mu E/m^2$sec) was provided by overhead fluorescent light banks for 14 h daily. Winnipeg city water dechlorinated by sodium thiosulphate additions (Table 1) was used in the stock culture maintenance, the acclimation and preparation of dilutions and controls in the toxicity tests.

Exposure system:

The cadmium stock solution was prepared by dissolving 1.0 g/L of the pure metal cadmium in acidified deionized water. The test dilutions of cadmium (mg/L) were as follows: 0.01, 0.1, 1.0, 5.0, 10, 20, and 40. These were made by diluting the stock solution

Table 1. Water chemistry data for thiosulphate-dechlorinated Winnipeg city water (18°C).

Analysis	Units	
chloride	4.2	mg/L
sulphate	12.4	mg/L
sodium	3.68	mg/L
potassium	1.30	mg/L
calcium	22.7	mg/L
magnesium	5.44	mg/L
pH	7.96	
conductivity	170	μS/cm
dissolved oxygen	8.0	mg/L
hardness (by calculation)	79.1	mg/L

* This sulphate value is an overestimate caused by thiosulphate interference.

one hour prior to the bioassay. Two 100 mL samples were removed from each test container for water chemistry and heavy metal (Table 2) analysis respectively.

The test containers were uncovered, 600 mL pyrex glass beakers containing 500 mL of test solution. There were seven concentrations of cadmium, plus a control and two replicates per concentration. Ten animals were tested at each concentration. The number of mortalities was recorded at eight hour intervals over a 96 h period.

The criteria for determining death were: cessation of respiratory movements of the gills, and no response to abdominal prodding with a Pasteur pipette tip for a period of 60 sec. If no response was visible, the animals were removed with 5 mL of test solution to a 50 mL disposable plastic beaker for further observation at the subsequent mortality check. Confirmed mortalities were discarded.

Statistical analysis:

Mortality data were analysed by two methods. One, using the "routine bioassay method" given in "Standard Methods" (APHA *et al.* 1965). This method yields median tolerance limits (TLM's or LC_{50}'s). For each fixed time, percentage mortalities are plotted against the test concentrations. The concentration lethal to half of the test organisms is estimated by interpolation. Confidence limits have customarily been omitted for LC_{50}'s estimated by this method (Sprague 1969).

Table 2. The concentration of cadmium (mg/L) in the test containers.

Nominal Value	Actual Initial Value*
0 (control)	0.0010
.01	0.0072
0.1	0.036
1.0	0.94
5.0	4.6
10.0	8.5
20.0	18.0
40.0	34.0

* no change after 96 h exposure.

The second method, more recently adopted, has four simple steps:

1. Adjust the mortality proportions p_i, ..., pk to form $\tilde{P}_i$'s in a monotone nondecreasing sequence.
2. Plot the (x_i, $\tilde{p}_i$) points and connect them with straight lines. The polygonal figure formed is an estimate of P(x), the cumulative relative frequency curve for the tolerance distribution
3. Choose α or the % trim desired. Trim off the upper α percent and lower α percent of the polygon. Adjust the ordinate scale appropriately. Where α=0, as herein, no adjustments are required.
4. Calculate the mean associated with the cumulative relative frequency polygon to yield the α% - trimmed Spearman-Karber estimate of μ. The antilog of μ will yield the estimate of LC_{50}.

RESULTS AND DISCUSSION

No mortality was observed at cadmium concentrations of 0 - 0.1 mg/L for the entire 96 h test period. One hundred per cent mortalit occurred in the 40.0 mg/L test vessels after 12 h. exposure and in the 20.0 mg/L containers after 30 h. Figure 1 illustrates mean survival time in hours (the calculated time to 50% death in each concentration) as a function of the concentration of cadmium (mg/L) in the test series. The 96 h LC_{50} by interpolation is read as 1.0 mg/L cadmium.

This method is the simplest routine procedure. Increased in formation may be obtained by fitting a straight line to the data plotted on logarithmic - probability graph paper (Litchfield and Wilcoxon 1949). This refinement is useful when survival percentages are found to change little or erratically with progressive increases of tested concentrations. It has become one of the most widely accepted procedures in calculating and reporting bioassay results (APHA *et al.* 1965). A second popular calculation method is the trimmed Spearman - Karber (Hamilton *et al.* 1977).

Table 3 shows an example of the data set used in the trimmed Spearman - Karber calculation of the LC_{50}. Table 4 demonstrates the application of a 0-trim in the trimmed Spearman-Karber method to estimate μ (the mean of the population tolerance distribution). Using this procedure, the estimate of the LC_{50} is, then, the antilog of the estimate of μ, or, 6.2 mg/L cadmium. Hamilton *et al.* (1977) discuss this procedure in depth and show that this method is a better overall procedure which eliminates the target criticisms of Probit and Logit models; namely, that the maximum likelihood iterative procedure can be very unstable if the data do not conform to the assumed model and that one may not always accurately compare

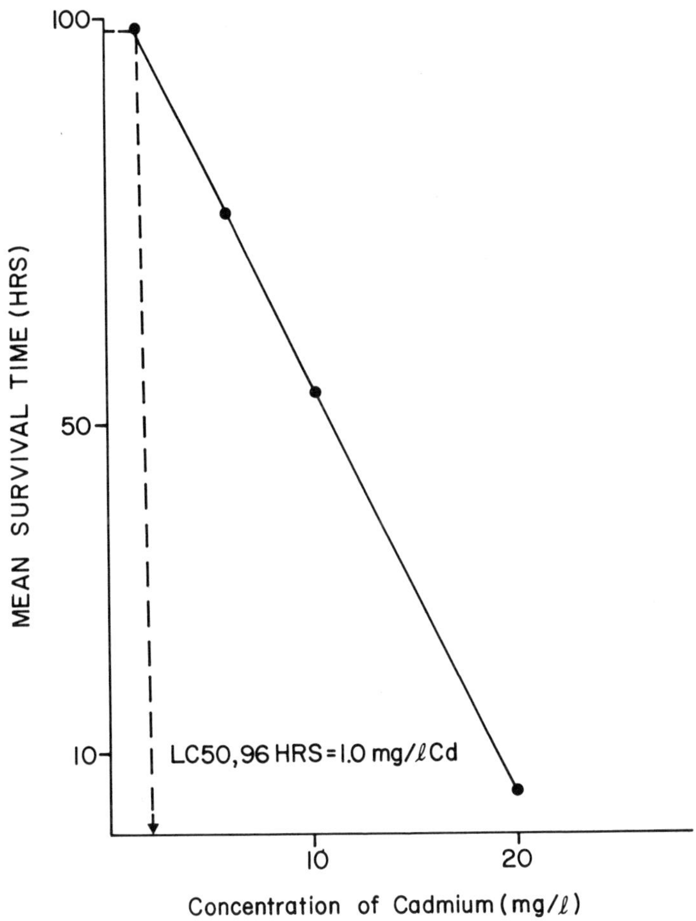

Figure 1. Graphical estimation of the 96 h LC_{50} for cadmium to nymphs of the burrowing mayfly, *Hexagenia rigida*.

results from duplicate experiments.

Thus, the 96 h LC_{50} for the nymphs of the burrowing mayfly, *Hexagenia rigida* lies between 1.0 and 6.2 mg/L cadmium when tested in thiosulphate dechlorinated water and using two estimators. These estimates may be high since tests conducted using thiosulphate water as diluent may yield higher estimates of the LC_{50} than those conducted using ultraviolet dechlorinated water or a reconstituted medium as diluent. This elevation of LC_{50} in thiosulphate dilutions was observed by Friesen (personal communication) in defining the

Table 3. Example of Trimmed Spearman-Karber calculations. Data set at 96 h exposure.

Concentration	mg/L	0	.01	0.1	1.0	5	10	20	40
Log_e conc.	x_i	0	-4.605	-2.302	0	1.609	2.302	2.996	3.689
No. of nymphs	n_i	10	10	10	10	10	10	10	10
No. of mortalities	r_i	0	0	0	5	7	7	10	10
Mortality proportion	p_i	0.0	0.0	0.0	0.5	0.7	0.7	1.00	1.00
Adjusted mortality proportion	$\tilde{p}_i$	not necessary; $\tilde{p}_i$'s are in a monotone nondecreasing sequence							

Table 4. Calculation of 0-Trimmed Spearman-Karber estimates of μ (the mean (and median) of the population tolerance distribution).

(1) Log_e of conc. interval: (x_{j-1}, x_j)	0 to -4.605	-4.605 to -2.302	-2.302 to 0	0 to 1.609	1.609 to 2.302	2.302 to 2.996	2.996 to 3.689
(2) Relative frequency: $P_j - P_{j-1}$	0	0	.5	.2	0	.3	0
(3) Midpoint of interval: $(x_{j-1} + x_j)/2$	-2.302	-3.454	-1.151	.804	1.955	2.649	3.342
(4) (2) x (3)	0	0	-0.575	1.608	0	.7947	0

The estimate of μ is 1.8277 and the estimate of LC_{50} is thus $e^{1.8277} = 6.2$ mg/L incipient LC_{50} for early (1 mg/L) and mid-stage eggs (3.5 mg/L) of *Hexagenia rigida* exposed to cadmium. Apparently, the yearling nymphs of *Hexagenia rigida* are not any more sensitive to cadmium exposure than the developing eggs.

In this study, care was taken not to select late instar nymphs (identified by dark wing pads), since Fremling (1975) reported that physiological stresses involved in the transformation to the adult stage may produce atypical results in tests done with such animals. Further studies should be designed to assess the sensitivity of newly hatched (hours to days old) nymphs with the continuing aim of identifying the most vulnerable stage in the life cycle of this mayfly.

CONCLUSIONS

The 96 h LC_{50} for cadmium exposure with *Hexagenia rigida* yearling nymphs lies between 1.0 and 6.2 mg/L depending on the estimator used.

The yearling nymphs of this species do not appear to be any more sensitive than the developing eggs in determining the acute toxicity of cadmium. Further assessments should be made on newly hatched nymphs.

ACKNOWLEDGMENTS

The authors wish to express their sincere thanks to the analytical chemistry units at the Freshwater Institute for the routine water chemistry and selected heavy metal analyses. Thanks also to Mrs. Grace Decterow, Toxicology Section, for typing this manuscript and to Laurie Taite, Graphics Section, for preparing the illustration.

RESUME

Développées en milieu de culture, des nymphes d'éphéméroptères d'un an appartenant à l'espèce *Hexagenia rigida* ont été soumises à des tests biologiques statiques et sans modification d'éléments d'une durée de 96 heures. La LC_{50}, après 96 h, a été évaluée à 1.0 mg/L selon une méthode graphique et à 6.2 mg/L selon la méthode de calcul dite "0-Trimmed" de Spearman-Karber. Cette dernière s'est révélée d'application plus simple et a permis une estimation plus précise.

ZUSSAMENFASSUNG

Im Labor gezuchtete einjährige Nymphen der Eintagsfliege *Hexagenia rigida* wurden in statischen 96 Stunden Bioassayas ohne Neuzufuhr getestet. Die 96 Stunden LC_{50} wurde an Hand einer graphischer Methode auf 1.0 mg/L geschatzt. Die 0-Trimmed Spearman-Karber Kalkulation ergab den Schätzwert von 6 2 mg/L. Letztere war leichter anzuwenden und ergab präzisere Schätzwerte.

REFERENCES

APHA *et al.* 1965. Standard Methods for the Examination of Water and Wastewater including Bottom Sediments and Sludges. Am. Pub. Health Assoc., New York, 12th edition.

Fremling, C.R. 1975. Acute toxicity of the lampricide 3-trifluoromethyl-4-nitrophenol (TFM) to nymphs of mayflies *(Hexagenia* sp.). *U.S. Fish. Wildl. Serv. Invest. Fish Control* 58: 1-8.

Hamilton, M.A., R.C. Russo, and R.V. Thurston. 1977. Trimmed Spearman-Karber method for estimating median lethal concentrations in toxicity bioassays. *Environ. Sci. Technol.* 11(7): 714-719.

Litchfield, J.T., Jr. and F. Wilcoxon. 1949. A simplified method of evaluating dose-effect experiments. *J. Pharmacol. Exp. Ther.* 96: 99-113.

Sprague, J.B. 1969. Measurement of pollutant toxicity to fish I. Bioassay methods for acute toxicity. *Water Res.* 3: 793-821.

ONLINE COMPUTER RETRIEVAL OF INFORMATION ON EPHEMEROPTERA: A COMPARISON OF DIFFERENT SOURCES

K. Eric Marshall

Freshwater Institute Library
Department of Fisheries and Oceans
501 University Crescent
Winnipeg, Manitoba, Canada R3T 2N6

ABSTRACT

A bibliography of 179 references on Ephemeroptera and pollution resulted from a series of online computer searches of commercially available data bases in March 1979. Eighty-three percent of the references retrieved came from two data bases, BIOSIS Previews and Selected Water Resources Abstracts, at a cost of $14.00. A further expenditure of $40.00 in searching nine other data bases retrieved the remaining references.

INTRODUCTION

The Ephemeroptera are well known as indicators of pollution. To find references on this topic a series of online* bibliographic searches were made in March 1979 on commercially available systems. The purpose of the searches was threefold:

1) references obtained would be useful to Freshwater Institute staff members preparing papers for this volume;
2) costs and efficiency of retrieval using different data bases* and the same data base on different systems would be compared; and
3) the procedure would demonstrate the service to a scientist unfamiliar with online searching.

---

* See Glossary (Appendix 1) for definition of computer science terms used.

Online bibliographic search services have been available in North America for many years. We have been using these services at the Freshwater Institute since March 1974. Currently we have access to four different systems, two Canadian and two in the U.S.A. (See Appendix II for addresses). These four systems allow access to over 100 different data bases covering most fields of knowledge. Thirty of these data bases are used regularly at the Institute. In an average month 30 searches are made in response to requests from staff.

PREPARING THE BIBLIOGRAPHY

Searches were made on all of the data bases containing references on Ephemeroptera within one week. Since the various vendors are constantly improving their services the results of the searches indicate the relative performance of the systems at that time.

As well as making a general search for references on Ephemeroptera and pollution, a specific search was made for references dealing with mercury, cadmium or selenium.

The following search strategy was used:

1. Search for: EPHEMEROPTERA$* OR MAYFLIES OR MAYFLY
2. Search for: POLLUTION
3. Combine 1 AND 2 *(General Search)*
4. Search for: MERCUR$ OR SELENIUM OR CADMIUM
5. Combine 1 AND 4 *(Specific Search)*

The bibliography obtained appears in Appendix II. Full details of the status of the data bases when the searches were made and further discussion on the principles of online searching are in Marshall (1979).

DATA BASES SEARCHED

Table 1 summarizes the data bases used and Table 2 shows the cumulative cost of production of the bibliography.

BIOSIS Previews

This data base is a machine readable version of the indexes and bibliographic references in *Biological Abstracts* and *Bioresearch Index*

---

* $ indicates truncation of a word - EPHEMEROPTERA$ will retrieve Ephemeroptera, Ephemeropteran and Ephemeropterans; similarly MERCUR$ will retrieve any word beginning mercur.

Table 1. Comments on the data bases used in compiling the bibliography.

Data Base	Comments
BIOSIS Previews	The most productive source. Full use made of Biosystematic and Concept Index terms.
Selected Water Resources Abstracts	Only available to us on QL Systems. Abstracts searchable.
Chemical Abstracts	Searches on Dialog on the full data base gave more references than CAN/OLE which lacks the subject index terms.
NTIS	A comprehensive data base on U.S. Government Research Reports.
Pollution Abstracts	A good data base to search for references on pollution. Dialog carried an updated form with searchable abstracts. Orbit lacked this form when searched and gave relatively poor retrieval.
Canada Water	Covers Canadian publications on the aquatic sciences.
Enviroline	Similar subject area coverage to Pollution Abstracts. The paper version is *Environment Abstracts*.
Excerpta Medica	Good coverage of European literature. Medical bias but one section covers environmental pollution.
Commonwealth Agricultural Bureaux	Machine readable form of all of the abstracting and indexing services published by the Bureaux. The only other source for the one reference found by this data base was *Aquatic Science and Fisheries Abstracts*.
SciSearch	Corresponds to *Science Citation Index*. Very current but subject searches not easy to carry out as no subject headings are added. Words from the title must be search and many synonyms and related terms are need for good retrieval.
Conference	Specializes in conference papers.

Table 2. Comprehensive search for references on Ephemeroptera and pollution: retrieval and costs.

Data Base	System and Years Searched*	Number of References Retrieved				Costs		
		Total for Data Base	Unique to Data Base	New to Search	Running Total of New Refs.	For Data Base $ Can.	Cumulative Total $ Can.	Per Reference ¢ Can.
1. BIOSIS Previews	Dialog 69-71 OLE 72-78; 79	120	88	120	120	9.35	9.35	7.8
2. Selected Water Resources Abstracts	QL 75-78	42	23	28	148	4.81	14.16	9.6
3. Chemical Abstracts	Dialog 72-76 and 77-79	13	6	7	155	4.30	19.46	12.6
4. NTIS	Dialog 64-79	10	5	6	161	2.83	22.29	13.8
5. Pollution Abstracts	Dialog 70-79	16	3	5	166	3.19	25.48	15.3
6. Canada Water	QL 70-78	7	4	4	170	4.33	29.81	17.5
7. Enviroline	Orbit 71-79	14	3	3	173	6.96	36.77	21.2
8. Excerpta Medica	Dialog 75-78	11	3	3	176	4.38	41.15	23.4
9. Commonwealth Agricultural Bureaux	Dialog 72-79	1	0	1	177	2.50	43.65	24.7
10. SciSearch	Dialog 78-79	9	1	1	178	7.12	50.77	28.5
11. Conference	Dialog 73-79	2	1	1	179	3.89	54.66	30.5

* For fuller details see Marshall (1979).

but excludes the abstracts. The biosystematic and concept indexes are valuable search tools for higher systematic groups (e.g. Ephemeroptera) and broad subject areas (e.g. water pollution) respectively. Searching on these index terms was efficient because BIOSIS indexing is thorough. In the general search this was done instead of using the words in the search strategy described above.

Aquatic Science and Fisheries Abstracts

This data base is not included in Table 2. Only two six month samples of this data base were online when the searches were made. Only one reference was found on each of the two different periods covered and these were not unique to this base. However, in June 1979, I repeated my serach on QL Systems which had then increased its coverage to 1978 and the first three months of 1979. Nine references on mayflies and pollution were retrieved. Clearly the value of this data base will increase as more references are entered into it.

SciSearch

Because this is the only scientific data base in which the references cited in a paper can be searched a separate search was made on this data base to demonstrate a citation search. The search was made to identify recent papers which had cited Clubb *et al.* (1975). Four references were retrieved. Two were reviews of pollution in general and two specific to mayflies and pollution. One of the latter references was not retrieved in the other searches. SciSearch is usually the most up-to-date of the online data bases.

DISCUSSION

Table 2 indicates the optimum preparation of the bibliography. None of the searches cost more than $10.00. The total cost for retrieving all of the references was less than $55.00.

Searches of the two most productive data bases yielded 83% of the references for $14.16. The remaing 17% of the references were retrieved only by expending a further $40.50 to search nine additional data bases. Therefore judgement has to be used to decide how complete a search should be. An online search is not 100% complete because most of the data bases online do not cover material published before 1968 and some government publications are not well covered. The *Zoological Record* and early issues of *Biological Abstracts*, and other indexing services relevant to the exact topic being researched have to be searched manually. Current awareness services (e.g. the lists of new publications appearing in *Eatonia* and the North American

Benthological Society annual bibliographies) should be used. Library accession lists (e.g. the Freshwater Biological Association's) also form excellent sources of references.

RESUME

En mars, 1979, on a fait une serie de recherches documentaires en direct d'ordinateur; on les a fait dans les banques de données qui sont commercialement utilisables. Une bibliographie qui contient 179 références liées aux éphéméroptères et à la pollution était le résultat. Quatre-vingts-trois pour cent des références récupérés sont venus des deux bases de données, BIOSIS Previews et Selected Water Resources Abstracts, à un coût de $14.00. Une additionnelle dépense de $40.00, en recherchant neuf autres bases de donneés, a récuperes les références qui reste.

ZUSSAMENFASSUNG

Eine Reihe von "on-line" Computer-Suchen, im März 1979, an kommerziell verfügbaren Informationsspeichern duchgeführt, erbrachte eine Bibliographie mit 179 Literatur-Quellenangaben aus dem Gebiet Eintagsfliegen und Umweltverschmutzung. Dreiundachtzig prozent dieser Quellenangaben Kamen von 2 Informationsdiensten, nämlich "BIOSIS Previews" und "Selected Water Resources Abstracts", zum Preis von 14 Dollar (Can.). Mit zusätzlich 40 Dollar (Can.) für 9 weitere Informationsspeicher wurden die restlichen Literaturangaben ermittelt.

ACKNOWLEDGMENTS

Thanks are due to Mr. J.F. Flannagan for his encouragement in this study and to Dr. D.M. Rosenberg for his comments on a draft of this paper.

REFERENCES

Clubb, R.W., A.R. Gaufin and J.L. Lords. 1975. Acute cadmium toxicity studies upon nine species of aquatic insects. *Environ. Res.* 9: 332-341.

Marshall, K.E. 1979. Online retrieval of information: a comparison of different systems used to produce a bibliography on Ephemeroptera and pollution, 1969-78. *Can. Fish. Mar. Serv. Tech. Rep.* 878: 1-18.

APPENDIX I. GLOSSARY OF COMPUTER SCIENCE AND RELATED TERMS

CAN/OLE (CANadian OnLine Enquiry). The acronym of the online search service operated by Canada Institute for Scientific and Technical Information (CISTI) of the National Research Council in Ottawa.

Data base A machine readable collection of data stored in a computer in a way in which it can be searched by a user. Most bibliographic data bases have a corresponding paper format also (e.g. *Pollution Abstracts*, *Aquatic Science and Fisheries Abstracts*).

NTIS (National Technical Information Service). A service of the U.S. Department of Commerce concerned with disseminating U.S. Government Reports.

Offline A form of communication with a computer in which no immediate response is expected. The response (e.g. a printout of references) is made later at the computer centre and mailed to the user.

Online A form of communication with a computer in which immediate responses are received on the remote terminal. In this mode a "conversation" can be carried out and additional questions asked in the light of previous responses.

QL (Quic/Law). The name given to a project involving the development of a retrieval system for legal information from which the present QL System company developed.

APPENDIX II. ONLINE RETRIEVAL SERVICE VENDORS

The following are the four systems used at the Freshwater Institute. (There are other systems available which we do not use.)

CAN/OLE: Canada Institute for Scientific and Technical Information
National Research Council
Ottawa, Ontario, Canada K1A 0S2

QL Systems: 408 Princess Anne Building
797 Princess Street
Kingston, Ontario, Canada K7L 1G1

Orbit: SDC Search Service
2500 Colorado Avenue
Santa Monica, CA, U.S.A. 90406

Dialog: Lockheed Missiles and Space Co., Inc.
Code 5020/201
3251 Hanover Street
Palo Alto, CA, U.S.A. 94304

APPENDIX III. BIBLIOGRAPHY ON EPHEMEROPTERA AND POLLUTION 1969-78.

Format

This is a format of a sample reference:

x Clubb, R.W., A.R. Gaufin and J.L. Lords. 1975. Synergism between dissolved oxygen and cadmium toxicity in five species of aquatic insects. *Environ. Res.* 9(3): 285-289.
BA 60:61371; W76-10345; PA 75-04453

The x before the author's name indicates that this reference was retrieved in the search for Ephemeroptera and mercury or cadmium or selenium. Titles of papers may not correspond exactly to the printed papers because most of the data base producers make small changes to conform with their editorial policies. I have inserted punctuation where necessary. On the line following the page numbers I have indicated where an abstract of the paper may be found. (BA = *Biological Abstracts*, BI = *Bioresearch Index* [No abstract], C = *Chemical Abstracts*, W = *Selected Water Resources Abstracts*, P = *Pollution Abstracts*, E = *Environment Abstracts*).

Further information on the data base and system from which each reference was retrieved, and for any errors noted in the data bases citing the references, see Marshall (1979).

Bibliography

Alekseev, V.A. 1973. Toxicological characteristics and symptom complex of acute phenol intoxication of aquatic insects and arachnids. *Tr. Inst. Biol. Vnutr. Vod Akad. Nauk SSSR* 24: 72-89.
C 80:91780v

Anderson, R.V. 1977. Concentration of cadmium, copper, lead and zinc in thirty-five genera of freshwater macroinvertebrates from the Fox River, Illinois and Wisconsin. *Bull. Environ. Contam. Toxicol.* 18(3): 345-349.

Anonymous. 1969. Proceedings of the conference on changes in the biota of Lake Erie and Ontario. 48 p.

Anonymous. 1972. Biological survey of Pinkham Creek and the North Platte River in the vicinity of Cowdrey, Colorado, August 24, 1972. PB 244 955; EPA Report S and A/TSB-11: 9 p.
W78-04781

Anonymous. 1974. Clark Fork River study, Montana, July-August, 1973. PB 225 580; EPA 908/2-74-001: 25 p.
W78-09532

Avery, E.L. 1970. Effects of domestic sewage on aquatic insects and salmonids of the Gallatin River, Montana. *Water Res.* 4(2): 165-177.
BA 52:7178

Balduzzi, A., E. Gaino and S. Spano. 1976. The Arroscia Centa hydrologic system, Western Liguria, in relation to its zoobenthonic populations with special attention to its Ephemeroptera and Plecoptera. [In Italian] *Boll. Mus. Ist. Biol. Univ. Genova* 44: 25-54.
BA 66:51092

Bell, H.L. and A.V. Nebeker. 1969. Preliminary studies on the tolerance of aquatic insects to low pH. *J. Kans. Entomol. Soc.* 42(2): 230-236.
BA 50:106733

Benda, R.S. and M.A. Proffitt. 1974. Effects of thermal effluents on fish and invertebrates. p. 438-447. *In:* CONF-37050. Proceedings of a symposium held at Augusta, Georgia, May 3-5, 1973. J.W. Gibbons and R.R. Sharitz (ed.). Thermal Ecology. Oak Ridge, Tenn., U.S.A.E.C.
W78-02018

Bennington, S.L. 1977. Benthic macrofauna substrate relationships in an unpolluted and a polluted stream environment. *Mauri Ora* 5: 21-38.
BA 66: 50984

Biesiadka, E. and K. Kasprzak. 1977. An investigation of the macrofauna of the River Warta within the city of Pozan. *Acta. Hydrobiol.* 19(2): 109-122.
W78-09107

Bluzat, R., F.J. Rodriguez-Ruiz and J. Seuge. 1976. Acute toxicity of some insecticides, herbicides and detergents on four freshwater ponds' invertebrate species. *C.R. Hebd. Seances Acad. Sci. Ser. D. Sci. Nat.* 283(9): 1089-1092.
BT 1978:2199

Bluzat, R. and J. Seuge. 1978. Action de produits detersifs chez les larves aquatiques de *Chaoborus* et de *Cloeon*: importance de leur etat physiologique. Action of detergent products in the aquatic larvae of *Chaoborus* sp. and *Cloeon* sp.: importance of their physiological activity. *Toxicology* 10(3): 221-228.
BA 67:18997

Braasch, D. and U. Jacob. 1976. Die Verwendung von Ephemeropteren (Insecta) der DDR als indicatoren fur die Wassergute. The use of Ephemeroptera (Insecta) of the DDR as indicators of water quality. *Entomol. Nachr.* 20(7): 101-109.

Breaud, T.P., J.E. Farlow, D.C. Steelman and P.E. Schilling. 1977. Effects of the insect growth regulator, Methoprene, on natural populations of aquatic organisms in Lousiana intermediate marsh

habitats. *Mosq. News* 37(4): 704-712.
BA 65-69590

Britt, N.W., J.T. Addis and R. Engel. 1973. Limnological studies of the Island area of western Lake Erie. *Bull. Ohio Biol. Surv.* 4(3): 1-85.
W77-11926

Brittain, J.E. 1976. The temperature of two Welsh lakes and its effect on the distribution of the freshwater insects. *Hydrobiologia* 48(1): 36-49.
W77-00878

Brooker, M.P. and R.J. Hemsworth. 1978. The effect of the release of an artificial discharge of water on invertebrate drift in the River Wye, Wales. *Hydrobiologia* 59(3): 155-163.
E 78-05540

Brown, J.S. 1971. A survey of the macrobenthic organisms in an industrial dishcarge embayment. *J. Ala Acad. Sci.* 42(3): 132-133.
BI 1973:73871

Brusven, M.A. and C. MacPhee. 1974. An evaluation of squoxin on insect drift. *Trans. Am. Fish. Soc.* 103(2): 362-365.
W76-04206

Brusven, M.A. and K.V. Prather. 1971. Effects of siltation and coarser sediments on distribution and abundance of stream-inhabiting insects. PB 207 349; W72-05024: 74 p.

Bugbee, S.L. and C.M. Walter. 1973. Responses of macroinvertebrates to gasoline pollution in a mountain stream. p. 725-731. *In:* Prevention and Control of Oil Spills, Proceedings of Joint Conference, Washington, D.C., March 1973.
W77-00268

Chaisemartin, C. 1975. Effects of chromates (potassium chromate) on respiratory metabolism and tissue water mineral balance in larvae of two species from freshwater macrobenthos. *C.R. Seances Soc. Biol. Fil.* 169(2): 384-390.
BA 62:11310

Chaisemartin, C. 1976. Sensibilite aux chromates de deux larves-nymphes du macrobenthos: *Ecdyonurus venosus venosus* (Ephemeropteres) et *Cordulia aena* (Odonatopteres). Chromate sensitivity of larval nymphs from the macrobenthos: *Ecdyonurus venosus vensus* (Ephemeroptera) and *Cordulia aena* (Odonatoptera). *Ann. Hydrobiol.* 7(1): 67-68.
BI 1977:47744

Chandler, C.M. 1978. A preliminary checklist of benthic macroinvertebrates families from basket sampling, during early summer, in the west fork of Stones River, Tennessee. *J. Tenn. Acad. Sci.* 53(1): 25-28.
BI 15:7707

Chandler, J.H. and L.L. Marking. 1975. Toxicity of the lampricide 3-trifluoromethyl-4-nitrophenol to selected aquatic invertebrates and frog larvae. *U.S. Fish. Wildl. Serv. Invest. Fish.*

Control 62: 7 p.
BI 1975:74187

Chino, Y. 1975. On the aquatic insect communities in shore zones and influents as well as effluents of lakes in Shinshu. [In Japanese] *New Entomol.* 24(1): 8-18.
W77-10949

Chutter, F.M. 1972. An empirical biotic index of the quality of water in South African streams and rivers. *Water Res.* 6(1): 19-30.
BA 54:25003

Clady, M. 1975. Comparison of the bottom fauna in 1916 and 1967 in a Bay of Oneida Lake, New York. *N.Y. Fish Game J.* 22(2): 114-121.
BA 60:42467

x Clubb, R.W., A.R. Gaufin and J.L. Lords. 1975. Acute cadmium toxicity studies upon nine species of aquatic insects. *Environ. Res.* 9(3): 332-341.
BA 60:61370; W76-10342; E 76-03160; P 75-04454

x Clubb, R.W., A.R. Gaufin and J.L. Lords. 1975. Synergism between dissolved oxygen and cadmium toxicity in five species of aquatic insects. *Environ. Res.* 9(3): 285-289.
BA 60:61371; W76-10345; P 75-04453

Craven, R.E. and B.E. Brown. 1970. Power plant discharge water and benthos in Boomer Lake, Payne County, Oklahoma. *J. Kans. Entomol. Soc.* 43(2): 122-128.
BA 51:93910

De March, B., J.F. Flannagan, M. Friesen, *et al.* 1975. Compilation of literature pertaining to the culture of aquatic invertebrates, algae and macrophytes. Vol. 2. *Can. Fish. Mar. Serv. Tech. Rep.* 567: 32 p.

Doherty, F.G. and W.D. Hummon. 1978. Preliminary study on toxic mode of action of acid mine water on aquatic invertebrates (Ephmeroptera, Plecoptera) through respirometry. (Abstract) *Am. Zool.* 18(3): 632.

Dolan, J.M. III, B.C. Gregg, J. Cairns, Jr., *et al.* 1974. The acute toxicity of three new surfactant mixtures to a mayfly larvae. *Arch. Hydrobiol.* 74(1): 123-132.
BA 59:40492

Dye, L.L. 1976. Effects of temperature on growth, fecundity and emergence of burrowing mayflies located near a power plant. Energy and Environmental Stress in Aquatic Systems, Augusta, Ga, 2-4 Nov. 1977, Selected Papers.

Echaubard, M. and A. Neveu. 1975. Qualitative and quantitative changes in the benthic fauna of a trout stream, the Couze Pavin, Puy-du-Dome, caused by agricultural and urban pollution. [In French] *Ann. Stn Biol. Beese-en-Chandesse* 9: 1-24, 1974-75.
W77-08955

Edmunds, G.F., Jr. and R.W. Koss. 1972. A review of the Acanthametropidinae with a description of a new genus (Ephemeroptera: Siphlonuridae). *Pan-Pac. Entomol.* 48(2): 136-144.
BA 55: 26510

Eidt, D.C. 1975. The effect of Fenitrothion from large-scale forest spraying on benthos in New Brunswick headwaters streams. *Can. Entomol.* 107(7): 743-760.
BA 60: 52292

Ettinger, W.S. and K.C. Kim. 1975. Benthic insect species composition in relation to water quality in Sinking Creek, Centre County, Pennyslvania. *Proc. Pa Acad. Sci.* 49(2): 150-154.
BI 1976:97203

Esterly, C.W. n.d. The effect of a thermal effluent on the benthos of Thomas Hill Reservoir, Missouri. PB 264-853
W77-05605

Fall, J. 1976. Materials concerning ephemeride Ephemeroptera fauna of the Bystrzyca River of the Lublin District Poland. [In Polish] *Ann. Univ. Mariae Curie-Sklodowska Sect. C. Biol.* 31: 211-220.
BA 66:40349

Farlow, J.E., T.P. Breaud, C.D. Steelman and P.E. Schilling. 1978. Effects of the insect growth regulator, Diflubenzuron, on non-target aquatic population in a Louisiana intermediate marsh. *Environ. Entomol.* 7(2): 199-264.
BA 66:61557

Ferguson, V.M. and R.C. Fox. 1978. A comparison of aquatic insects in natural inlets with those in the heated effluents from the Oconee Nuclear Station littoral zone. *J. Ga Entomol. Soc.* 13(3): 202-213.
BA 67:1789

Ferrarese, U. and B. Sambugar. 1976. Investigation on the interstitial hyporheal fauna in the Adige River in relation to the degree of pollution. [In Italian] *Riv. Idrobiol.* 14(1): 47-128.
BA 67: 33924

Folmer, L.C. 1978. Avoidance chamber responses of mayfly nymphs exposed to eight herbicides. *Bull. Environ. Contam. Toxicol.* 19(3): 312-318.
C 88:184242b; W78-11685; P 78-04710

Fredeen, F.J.H. 1974. Tests with single injections of Methoxychlor black fly (Diptera: Simuliidae) larvicides in large rivers. *Can. Entomol.* 106(3): 285-305.
C 81:115868d

Fredeen, F.J.H. and D.T. Spurr. 1978. Collecting semiquantitative samples of blackfly larvae (Diptera: Simuliidae) and other aquatic insects from large rivers with the aid of artificial substrates. *Quaest. Entomol.* 14(3): 411-432.
BA 64:7896

Fremling, C.R. 1970. Mayfly distribution as a water quality index. PB 198 223; EPA Water Pollut. Control Res. Ser. 16030-DQH-11/70: 44 p.

Fremling, C.R. 1975. Acute toxicity of the lampricide 3-trifluoromethyl-4-nitrophenol (TFM) to nymphs of mayflies (*Hexagenia* sp.). *U.S. Fish Wildl. Serv. Invest. Fish Control* 58: 8 p.
BI 1975:75897; C 83:54248u; P 75-04911

Friberg, F., L.M. Nilsson, C. Otto, *et al.* 1977. Diversity and environments of benthic invertebrate communities in south Swedish streams. *Arch. Hydrobiol.* 81(2): 129-154.
BA 77:7477; W79-00209

Gaufin, A.R. 1973. Water quality requirements of aquatic insects. PB 228 582, EPA-660/3-73-004: 98 p.

Gordon, C.C. and P.C. Tourangeau. 1977. Fluoride accumulation in the terrestrial and aquatic components of the Douglas Creek drainage of Western Montana. PB 288 277; *Mont. Univ. Joint Water Resour. Res. Cen. Bozeman, Compl. Rep.* 91: 45 p.
W79-00504

Grahn, O., H. Hultberg and L. Lander. 1974. Oligotrophication - a self-accelarating process in lakes subjected to excessive supply of acid substances. *Ambio* 3(2): 93-94.
W76-11403

Greenawald, R.B. and B.L. Haase. 1973. Effects of domestic and industrial pollution on the aquatic macroinvertebrate communities in Brodhead Creek. *Proc. Pa Acad. Sci.* 47: 149-152.
BI 1974:59679

Greenfield, J.P. and M.P. Ireland. 1978. A survey of the macrofauna of a coal-waste polluted Lancashire fluvial system. *Environ. Pollut.* 16(2): 105-122.
BA 66:57147; PA 78-05344

Gregg, B.C., J.M. Dolan III, J. Cairns, Jr. *et al.* 1974. Toxicity tests with surfactants and the mayfly *Isonychia* (Ephemeroptera: Isonychiidae). (Abstract) *ASB (Assoc. Southeast. Biol.) Bull.* 21(2): 58.
BI 1975:74420

Gysels, H. 1975. Electrophoretical and histochemical investigations on some noxious Lepidoptera and on the impact of pesticides upon *Ephemera danica*, a water-dwelling, innocuous mayfly. *Acta Zool. Pathol. Antverp.* 62: 129-141.
C 85:138230m

Hall, C.A.S. 1969. Mortality of the mayfly nymph *Ephemerella rotunda* at low dissolved oxygen concentrations. *J. Elisha Mitchell Sci. Soc.* 85(1): 34-39.
BA 51:29041

Hall, H.A., D.C. Eidt, P.E.K. Symons and D. Banks. 1975. Biological consequences in streams of aerial spraying with Fenitrothion against spruce budworm in New Brunswick. *Water Pollut. Res. Can.* 10: 84-88.

Harper, D.B., R.V. Smith and D.M. Gotto. 1977. BHC residues of domestic origin: a significant factor in pollution of freshwater in Northern Ireland. *Environ. Pollut.* 12(3): 223-233.
BA 64:11350; W77-12046; P 77-02736; E 77-03665

Harrison, A.D. and J.J. Rankin. 1974. The composition and density of the stream fauna of St. Vincent, West Indies, in relation to litter from rain forest. p. 80. *In* Abstracts XIX Congress International Association of Limnology, Winnipeg.
BI 1975:90336

Herricks, E.E. and J. Cairns, Jr. 1974. Rehabilitation of streams receiving acid mine drainage. *Va Polytech. Inst. State Univ. Water Resour. Res. Cen. Bull.* 66: 284 p.
BA 59:65570

Heuss, K. and W. Wichard. 1975. X-ray microanalysis of zinc and iron in the chloride cells of mayfly larvae of the River Sulz. *Oecologia (Berl.)* 19(1): 71-74.
BA 60:57645

Hopwood, A.J. 1974. Thermal effects of nuclear power plant on the Mississippi at Monticello, Minnesota. PB 235 833: 150 p.
E 75-04154

Howell, F.G. and J.B. Gentry. 1974. Effect of thermal effluents from nuclear reactors on species diversity of aquatic insects. p. 562-571. *In* J.W. Gibbons and R.R. Sharitz (ed.) Thermal Ecology. Oak Ridge, Tenn., U.S.A.E.C.
BI 1975:30152; W76-03492

Hubbard, M.D. and W.L. Peters. 1978. Environmental requirements and pollution tolerance of Ephemeroptera. PB 288 998: 469 p.

Hurlbert, S.H., M.S. Mulia and H.R. Willson. 1972. Effects of an organophosphorus insecticide on the phytoplankton, zooplankton and insect populations of freshwater ponds. *Ecol. Monogr.* 42(3): 269-299.
BA 55:42302

Ide, Y. 1971. The study of the aquatic insect fauna of the Matsukawa River and Yonako River (mineral acid polluted river). [In Japanese] *Bull. Inst. Nat. Educ. Shiga Heights* 10: 11-24.
BA 56:48265; W75-09316

Ischinger, L.S. and T.F. Nalepa. 1975. Freshwater macroinvertebrates. (Review) *J. Water Pollut. Control Fed.* 47(6): 1520-1538.
BI 1975:95448

Ito, T. and K. Kuwada. 1964. Aquatic ocmmunities in polluted streams with industrial and mining wastes. I. Effect of the papermill waste on the benthic invertebrates. *Annu. Rep. Noto Mar. Lab.* 4: 33-43.
BA 50:56847

x Jeantet, A.Y., R. Martoja and M. Truchet. 1974. Role of spherocrystals of the intestinal epithelium in the resistance of an insect to mineral pollution: experimental data obtained by use of electronic microelectrode and secondary ion emission microanalyzer. *C.R. Hebd. Seances Acad. Sci. Ser. D. Sci. Nat.* 278(10): 1441-1444.
BI 1976:52541

Jensen, C.F. 1978. Ephemeroptera in Linding A, 1964-1965. [In Danish] *Nat. Jutlandica* 20: 59-78.
BA 67:9676

Kaesler- R.L. 1974. Cluster analysis and diversity in recovery of damaged streams. PB 239 792; *Kans. Water Resour. Cen. Contrib.* 141: 77 p.
W75-04934

Kaesler, R.L., J. Cairns, Jr. and J.S. Crossman. 1974. Redundancy in data from stream surveys. *Water Res.* 8(9): 637-642.

x Kaminski, A. 1975. The use of biological tests for the examination of pesticides. *Pure Appl. Chem.* 42 (1/2): 255-284.
BI 1976:54490

Kellogg, R.L. and R.V. Bulkley. 1976. Seasonal concentrations of Dieldrin in water, channel catfish and catfish food organisms, Des Moines River, Iowa, 1971-1973. *Pestic. Monit. J.* 9(4): 186-194.
BA 62:28956

Kennedy, H.D. and D.F. Walsh. 1970. Effects of Malathion on two warm water fishes and aquatic invertebrates in ponds. *U.S. Bur. Sport Fish. Wildl. Tech. Pap.* 55: 13 p.
BA 53:12979

Kotila, P.M. 1978. Effect of Antimycin on stream insects in field and laboratory trials. PB 289 700; W79-02156: 120 p.

Kotila, P.M. and W.L. Hilsenhoff. 1978. Effects of Antimycin on stream insects. PB 288 336; W79-00525; *Wisc. Water Resour. Cent. Tech. Rep.* WIS-WRC 78-05: 55 p.
W79-00525

Kownacki, A. 1977. Biocenosis of a high mountain stream under the influence of tourism. Part 4. The bottom fauna of the stream Rybi Potok (the High Tatra Mountains). [In Polish] *Acta Hydrobiol.* 19(3): 293-312.
BA 66:26280; W79-00221; P 78-03296

Kubicek, F. and J. Helan. 1976. Chemical and biological aspects of the eutrophication of a trout brook. *Limnologica* 10(2): 399-403.
BI 1977:51462; C 86:16741z

Langford, T.E. 1971. The distribution, abundance and life histories of stoneflies (Plecoptera) and mayflies (Ephemeroptera) in a British river warmed by cooling water from a power station. *Hydrobiologia* 38(2): 339-376.
BA 52:135829; E 71-06047

Langford, T.E. 1975. The emergence of insects from a British river warmed by power station cooling-water. Part 2. The emergence patterns of some species of Ephemeroptera, Trichoptera, and Megaloptera in relation to water temperature and river flow, upstream and downstream of the cooling water outfalls. *Hydrobiologia* 47(1): 97-133.
BA 61:6945; W77-03044; E 76-03738

Langford, T.E. and R.J. Aston. 1972. The ecology of some British rivers in relation to warm water discharges from power stations. *Proc. Roy. Soc. Lond. Ser. B. Biol. Sci.* 180(1061): 407-419.
BI 1972:50391

Langford, T.E. and E.S. Bray. 1969. The distribution of Plecoptera and Ephemeroptera in a lowland region of Britain (Lincolnshire). *Hydrobiologia* 34(2): 243-271.
BA 51:122191; P 70-00249

Langford, T.E. and J.R. Daffern. 1975. The emergence of insects from a British river warmed by power station cooling-water. Part 1. The use and performance of insect emergence traps in a large spate river and the effects of various factors on total catches upstream and downstream of the cooling-water outfalls. *Hydrobiologia* 46(1): 71-114.
BA 60:18989

Larimore, R.W. 1974. Stream drift as an indication of water quality. *Trans. Am. Fish. Soc.* 103(3): 507-517.
W75-02989

Lehmkuhl, D.M. 1972. Change in thermal regime as a cause of reduction of benthic fauna downstream of a reservoir. *J. Fish. Res. Board Can.* 29(9): 1329-1332.
BA 55:30456

Lehmkuhl, D.M. 1974. Thermal regime alteration and vital environmental physiological stiganls in aquatic organisms. p. 216-222. *In* J.W. Gibbons and R.R. Sharita (ed.) Thermal Ecology. Oak Ridge, Tenn., U.S.A.E.C.
BI 1975:30122; W76-03478

Lewis, M. 1977. Aquatic inhabitants of a mine waste stream in Arizona. *U.S. For. Serv. Res. Note* RM 349: 7 p.
BI 15:18834

Lewis, P.A. 1974. Taxonomy and ecology of *Stenonema* mayflies (Heptageniidae: Ephemeroptera). PB 241 235; EPA-670/4-74-006: 89 p.
W75-07694

Lichtenberg, R. 1972. Hydrobiological investigations of a brick pool south of Vienna, Austria, Hallateich. *Sitzungsber. Oesterr. Akad. Wiss. Math-Naturwiss. Kl.* 180(8/10): 279-316.
BA 57:30445

Loeffler, H. 1977. Biological evaluation of fishing data in the Grosse Lauter, a trout brook in the Swabian Alb, West Germany. *Beitr. Naturkd Forsch. Suedwestdtschl.* 36: 81-90.
BA 67:27354

Lucas, A.M. 1971. Biological survey of Wilson Reservoir - Saline River, Kansas, August 1970. PB 259 874: 13 p.
W78-05109

Lusk, E.E., D.J. Womeldorf, R.K. Washing *et al.* 1976. Effects of nonthermal aerosol applications of Chlorpyrifos on nontarget arthropods and people in an urban area. *Proc. Pap. Annu. Conf. Calif. Mosq. Control Assoc.* 44: 78-85.
BI 1977:138

Maki, A.W., L. Geissel and H.E. Johnson. 1975. Comparative toxicity of larval lampricide 3-trifluoromethyl-4-nitrophenol to selected benthic macroinvertebrates. *J. Fish. Res. Board Can.* 32(8): 1455-1459.
BA 62:22982

Marcussi, G. and L. Nigro Faccipieri. 1976. Ecological studies on the zoocenosis of the Pescara River. [In Italian] *Riv. Idrobiol.* 13(1): 259-380.
BA 67:20874

Marshall, B.E. 1972. Some effects of organic pollution of benthic fauna. *Rhod. Sci. News* 6(5): 142-145.
BI 1972:89672

Masnik, M.T., J.R. Stauffer, C.H. Hocutt and J.H. Wilson. 1976. The effects of an oil spill on the macroinvertebrates and fish in a small southwestern Virginia creek. *J. Environ. Sci. Health Part A. Environ. Sci. Eng.* 11(4/5): 281-296.
BI 1977:3254

Mason, W.T., Jr., C.I. Weber, P.A. Lewis and E.C. Julian. 1973. Factors affecting the performance of basket and multiplate macroinvertebrate samplers. *Freshw. Biol.* 3(5): 409-436.
BA 57:65387

Masteller, E.C. 1977. An aquatic emergence trap on a shale stream of western Pennyslvania. *Melsheimer Entomol. Ser.* 23: 10-15.
BA 66:69732

Mattice, J.S. and L.L. Dye. 1977. Effect of steam electric generating station on the emergence timing of mayfly, *Hexagenia bilineata* (Say). 20th Congress International Association of Limnology, Copenhagen. CONF-77083-1: 15 p.

Mauch, E. 1976. Principal forms of saprobic organisms for the biological analysis of bodies of water. Part 5. *Cour. Forschungsinst. Senckenb.* 21(5): 564-797.
BA 63:41625

Mauck, W.L. and L.E. Olson. 1977. Polychlorinated biphenyls in adult mayflies (*Hexagenia bilineata*) from the upper Mississippi River. *Bull. Environ. Contam. Toxicol.* 17(4): 387-390.
BI 15:26334; C 87:16741z; P 77-05309; E 77-04332

McIvor, C.C. 1976. The effects of organic and nutrient enrichment on the benthic macroinvertebrate community of Moggill Creek, Queensland. *Water (J. Aust. Water Wastewater Assoc.)* 3(4): 16-21.
W77-05529

Miura, T., W.D. Murray and R.M. Takahashi. 1975. Effects of Dimilin on nontarget organisms in early spring *Culex tarsalis* larval habitats. *Proc. Pap. Annu. Conf. Calif. Mosq. Control Assoc.* 43: 79-83.
BI 1976:61928

Miura, T. and R.M. Takahashi. 1974. Insect development inhibitors - effects of candidate mosquito control agents on nontarget aquatic organisms. *Environ. Entomol.* 3(4): 631-636.
BA 59:34683

Miura, T. and R.M. Takahashi. 1976. Effects of synthetic pyrethroid SD-43775, 4-chloro-α-1-methyethylbenzeneacetic acid cyano-3-phenoxyphenylmethyl ester on nontarget organisms when utilized as a mosquito larvicide. *Mosq. News* 36(3): 322-326.
BA 64:72161

Muirhead-Thomson, R.C. 1978. Relative susceptibility of stream macroinvertebrates to Temephos and Chlorpyrifos, determined in laboratory continuous-flow systems. *Arch. Environ. Contam. Toxicol.* 7(2): 129-137.
P 78-04754; E 78-03707

Muirhead-Thomson, R.C. 1978. Lethal and behavioural impact of Chlorpyrifos methyl and Temephos on select stream macroinvertebrates: experimental studies on downstream drift. *Arch. Environ. Contam. Toxicol.* 7(2): 139-147.
P 78-04755; E 78-03708

Muirhead-Thomson, R.C. 1978. Lethal and behavioural impact of Permethrin, NRDC-143, on selected stream macroinvertebrates. *Mosq. News* 38(2): 185-190.
BA 66:61607

Mulla, M.S. and H.A. Darwazeh. 1976. Field evaluation of new mosquito larvicides and their impact on some nontarget insects. *Mosq. News* 36(3): 251-256.
BA 63:14395

Mulla, M.S., H.A. Navvab-Gojrati and H.A. Darwazeh. 1978. Biological activity and longevity of new synthetic pyrethroids against mosquitoes and some nontarget insects. *Mosq. News* 38(1): 90-96.
BA 66:51257

Napier, S. and W.D. Hummon. 1976. Survival of mayfly larvae under mine acid conditions. *Int. Rev. gesamten Hydrobiol.* 61(5): 677-682.
BA 63:62179

Nebel, C., R.D. Gottschling, J.L. Holmes and P.C. Unangst. 1978. Ozone oxidation of phenolic effluents. *Int. Ozone Inst. 2nd Symp. Ozone Technol. Montreal, May 1975:* 374-392.
E 77-02741

Nehring, R.B. 1976. Aquatic insects as biological monitors of heavy metal pollution. *Bull. Environ. Contam. Toxicol.* 15(2): 147-154.
BI 1977:25204; C 85:51316v; W78-00426; E 76-04726

Nixon, R.A. 1978. Biogeochemical analyses of the impact of the Richard B. Russell Scenic Highway on Dukes Creek, White County, Georgia. *Ga J. Sci.* 36(1): 16-24.
BI 15:54898

Olive, J.H. 1976. Chemical, physical and biological assessment of water quality in the Cuyahoga River, Ohio, 1973-1974. *Ohio J. Sci.* 76(1): 5-15.
BA 62:51596

Olive, J.H. and C.A. Dambach. 1973. Benthic macroinvertebrates as indexes of water quality in Whetstone Creek, Morrow County,

Ohio, Scioto River basin. *Ohio J. Sci.* 73(3): 129-148.
BA 59:46008

Olive, J.H. and K.R. Smith. 1975. Benthic macroinvertebrates as indexes of water quality in the Scioto River basin, Ohio. *Bull. Ohio Biol. Surv.* 5(2): 1-124.
BA 62:51574

Oseid, D.M. and L.L. Smith, Jr. 1974. Factors influencing acute toxicity estimates of hydrogen sulfide to freshwater invertebrates. *Water Res.* 8(10): 739-746.
BA 59:40477; W75-11027

Oseid, D. [M.] and L.L. Smith, Jr. 1975. Long-term effects of hydrogen sulfide on *Hexagenia limbata* (Ephemeroptera). *Environ. Entomol.* 4(1): 15-18.
C 83:202439z

Pesek, T. and G.L. Hergenrader. 1976. The influence of water quality upon the benthic macroinvertebrates of Salt Creek, Lancaster County, Nebraska. *Trans. Nebr. Acad. Sci.* 3: 45-60, 1974-76.
BA 64:7795

Pierce, S.T. 1974. The effects of thermal enrichment on the macroinvertebrate populations of the Wabash River. (M.A. Thesis) PB 237 502: 72 p.
W75-01325; E 76-00754

Prater, B.L. and M.A. Anderson. 1977. A 96-hour bioassay of Otter Creek, Ohio. *J. Water Pollut. Control Fed.* 49(10): 2099-2106.
BA 65:26294; W78-00890

Pratt, J.M. 1977. Effects of unrecorded pollution from urban stormwater runoff on benthic macroinvertebrates of the Green River, Massachusetts. Ph.D. Thesis (University Microfilms Order No. 77-30582).
W78-06321

Ramusino, M.C. and B. Rossaro. 1974. The Po River at the Island of Serafini, Italy. *Int. Lombardo Accad. Sci. Lett. Rend. Sci. Biol. Med. B.* 108: 89-128.
BA 60:48232

Ransom, J.D. and C.W. Prophet. 1974. Species diversity and relative abundance of benthic macroinvertebrates of Cedar Creek basin, Kansas. *Am. Midl. Nat.* 92(1): 217-222.
BA 58:53673

Roeser, B. 1976. The invertebrate fauna of the Broel and its tributary brooks, West Germany. *Decheniana* 129: 107-130.
BA 63:31495

Rosenberg, D.M. and A.P. Wiens. 1978. Effects of sediment addition on macrobenthic invertebrates in a northern Canadian river. *Water Res.* 12(10): 753-764.
BA 67:46772

Rueger, M.E., T.A. Olson and J.E. Scofield. 1969. Oxygen requirements of benthic insects as determined by manometric and polarographic techniques. *Water Res.* 3(2): 99-120.
BA 50:79330

Ryder, R.A. 1968. Dynamics and exploitation of mature walleyes, *Stizostedion vitreum vitreum*, in the Nipigon Bay region of Lake Superior. *J. Fish. Res. Board Can.* 25(7): 1347-1376.
BA 50:17764

Rye, R.P., Jr. and E.L. King, Jr. 1976. Acute toxic effects of two lampricides to twenty-one freshwater invertebrates. *Trans. Am. Fish. Soc.* 105(2): 322-326.
BA 62:63775

Samaan, A.A. and A.A. Aleem. 1972. The ecology of zooplankton in Lake Mariut, Egypt. *Bull. Inst. Oceanogr. Fish.* 2: 339-371.
BA 57:30442

Samman, J.T. and M.P. Thomas. 1978. Effect of organophosphorus insecticide, Abate, used in the control of *Simulium damnosum*, on nontarget benthic fauna. *Int. J. Environ. Stud.* 12(2): 141-144.
P 79-00399

Sanders, H.O. and D.F. Walsh. 1975. Toxicity and residue dynamics of the lampricide 3-trifluoromethyl-4-nitrophenol (TFM) in aquatic invertebrates. *U.S. Fish Wildl. Serv. Invest. Fish Control.* 59: 9 p.
BI 1975:75236; P 75-04912

Sarai, D.S. 1976. Total and fecal coliform bacteria in some aquatic and other insects. *Environ. Entomol.* 5(2): 365-367.
BA 62:14636

Savage, N.L. and F.W. Rabe. 1973. The effects of mine and domestic wastes on macroinvertebrate community structure in the Coeur d'Alene River, Idaho. *Northwest Sci.* 47(3): 159-168.
BA 61:1400

Schaefer, C.H., T. Miura, F.S. Mulligan III and E.P. Durpas. 1974. Insect development inhibitors - formulation research on Altosid. *Proc. Pap. Annu. Conf. Calif. Mosq. Control. Assoc.* 42: 140-145.
BI 1976:3270

Schaefer, C.H., T. Miura, F.S. Mulligan III and R.M. Takahashi. 1974 Insect development inhibitors - biological activity of RE-17565, RE-17937 and RE-18286 against mosquitoes (Diptera: Culicidae) and nontarget organisms. *Proc. Annu. Conf. Calif. Mosq. Control Assoc.* 42: 147-151.
BI 1976:3271

Sears, H.S. and W.R. Meehan. 1971. Residues in fish, wildlife and estuaries. Short-term effects of 2, 4-D on aquatic organisms in the Nakwasina River wattershed [!], southeastern Alaska. *Pestic. Monit. J.* 5(2): 213-217.
C 76:21710

Service, M.W. 1977. Mortalities of the immature stages of species B of the *Anopheles gambiae* complex in Kenya - comparison between rice fields and temporary pools: identification of predators of insecticidal spraying. *J. Med. Entomol.* 13(4/5): 535-545.
BA 64:1633

Seuge, J., R. Bluzat and F.J. Rodrigues-Ruiz. 1978. Effets d'un melange herbicide (2,4-D et 2,4,5-T): toxicite aigue sur 4

especes d'invertebres limniques; toxicite chronique chez le mollusque pulmone *Lymnea*. Effects of a mixed herbicide (2, 4-D and 2,4,5-T): acute toxicity over 4 species of freshwater invertebrates; chronic toxicity among the pulmonate mollusc *Lymnea*. *Environ. Pollut.* 16(2): 87-104.
P 78-05666

Sherberger, F.F., E.F. Benfield, K.L. Dickson and J. Cairns, Jr. 1977. Effects of thermal shocks on drifting aquatic insects - a laboratory simulation. *J. Fish. Res. Board Can.* 34(4): 529-536.
BA 64:38992

Sodergren, A. and B.J. Svensson. 1973. Uptake and accumulation of DDT and PCB (polychlorinated biphenyls) by *Ephemera danica* (Ephemeroptera) in continuous-flow system. *Bull. Environ. Contam. Toxicol.* 9(6): 345-350.
BI 1974:50080; C 79:74596b; P 74-01907; E 73-08020

Sodergren, S. 1976. Ecological effects of heavy metal discharge in a salmon river. *Inst. Freshw. Res. Drottningholm Rep.* 55: 91-131.
BA 66:44805

Solomon, J.L. and W.L. Lockhart. 1977. Rapid preparation of micro-sample and gas liquid chromatographic determination of Methoxy-chlor residues in animal tissues and water. *J. Assoc. Off. Anal. Chem.* 60(3): 690-695.

Speed, D.C. and Y.J. McGaha. 1972. A study of the macroscopic bottom fauna from selected streams in northern Mississippi. (Abstract) *J. Miss. Acad. Sci.* 17: 80 p.
BI 1974:13007.

x Spehar, R.L., R.L. Anderson and J.T. Fiandt. 1978. Toxicity and bioaccumulation of cadmium and lead in aquatic invertebrates. *Environ. Pollut.* 15(3): 195-208.
BA 66:30615; W78-08695

Srokosz, K. 1977. Phytophilous fauna in ponds fertilized with sugar refinery wastes. *Acta Hydrobiol.* 19(3): 233-242.
W79-00217

Stauffer, J.R., Jr., J.H. Wilson and K.L. Dickson. 1976. Comparison of stomach contents and condition of two catfish species living at ambient temperatures and in a heated discharge. *Prog. Fish-Cult.* 38(1): 33-35.
BA 63:7224

Surber, E.W. and W.E. Bessey. 1974. Minimum oxygen levels survived by stream invertebrates. PB 269 012; *Va Polytech. Inst. State Univ. Water Resour. Res. Cent. Bull.* 81: 52 p.
W 78-08926

Tanaka, H. 1971. Aquatic insect communities in the rivers of the Yoro Obitsu and Isumi, Chiba Prefecture. *Bull. Freshw. Fish. Res. Lab. Tokyo* 21(1): 21-45.
BA 55:40295

Taylor, M.P. 1974. Biological monitoring in Wheeler Reservoir before operation of Browns Ferry nuclear plant. p. 399-413.

In J.W. Gibbons and R.R. Sharitz (ed.) Thermal Ecology. Oak Ridge, Tenn., U.S.A.E.C.
BI 1975:30140

Thomas, N.A. n.d. Pre-impoundment biological survey of Stockton Reservoir, Sac River, Missouri, August 1968. PB 254 740.
W77-01793

x Thorp, V.J. and P.S. Lake. 1974. Toxicity bioassays of cadmium on selected freshwater invertebrates and the interaction of cadmium and zinc on the freshwater shrimp *Paratya tasmaniensis*. *Aust. J. Mar. Freshw. Res.* 25(1): 96-104.
BA 58:69634

Tomkiewicz, S.M., Jr. and W.A. Dunson. 1977. Aquatic insect diversity and biomass in a stream marginally polluted by acid strip mine drainage. *Water Res.* 11(4): 397-402.
BA 64:13615; W78-00451

Turoboyski, L. 1973. The indicator organisms and their ecological variability. *Acta Hydrobiol.* 15(3): 259-274.
W75-05055

Tusa, I. 1973. Mayfly larvae in current habitats of Bela Creek (the northwestern part of Moravia, Czechoslovakia). *Acta Hydrobiol.* 15(3): 311-320.
BA 58:60016; W75-05687

Tusa, I. 1974. Mayfly larvae (Ephemeroptera) in current habitats of three trout streams with stony bottoms, northwestern Moravia, Czechoslovakia. *Acta Hydrobiol.* 16(3/4): 417-430.
BA 60:36351

Van Dyk, L.P., C.G. Greeff and J.J. Brink. 1975. Total population density of Crustacea and aquatic insects as an indicator of Fenthion pollution of river water. *Bull. Environ. Contam. Toxicol.* 14(4): 426-431.
BI 1977:9292

Van Jaarsveld, J.H. 1970. Laboratory study on the toxicity of Dieldrin to freshwater invertebrates. *Phytophylactica* 2(4): 269-273.
C 78:67897b

Voskresenskii, K.A. and N.G. Dmitrieva. 1977. Detection of sublethal toxic pollution of bodies of water by estimating adaptive reactions of test organisms. [In Russian] *Biol. Nauki (Mosc.)* 20(10): 132-139.
BA 66:555128

Wagemann, R., N.B. Snow, D.M. Rosenberg *et al.* 1978. Arsenic in sediments, water and aquatic biota from lakes in the vicinity of Yellowknife, Northwest Territories, Canada. *Arch. Environ. Contam. Toxicol.* 7(2): 169-191.
P 78-04452

Wallace, R.R. and H.B.N. Hynes. 1975. The catastrophic drift of stream insects after treatments with Methoxychlor (1,1,1,-trichloro-2,2-bis(p-methoxyphenyl)ethane). *Environ. Pollut.* 8(4): 255-268.
W76-00327

Ward, D.L. 1972. The significance of selected physical and chemical variables in benthic macroinvertebrate distribution in a small subarctic stream. (Master's Thesis). PB 224 055: 71 p.

Warner, R.W. 1971. Distribution of biota in a stream polluted by acid mine drainage. *Ohio J. Sci.* 71(4): 202-216.
BA 53:1303

Welch, H.E., P.E.K. Symons and D.W. Narver. 1977. Some effects of potato farming and forest clear-cutting on small New Brunswick streams. *Can. Fish. Mar. Serv. Tech. Rep.* 745: 13 p.
BI 15:4931

White, D.S. and J.R. Gammon. 1976. The effect of suspended solids on macroinvertebrate drift in an Indiana creek. *Proc. Indiana Acad. Sci.* 86: 182-188.
BI 16:16707

Wichard, W. and M. Schmitz. 1975. The histochemical detection of heavy metals in the chloride cells of aquatic insects as an indicator of water pollution. *Verh. Ges. Oekol. Erlagen* 4: 155-159, 1974.
BI 1976:9564

Wielgolaski, F.E. 1975. Biological indicators of pollution. *Urban Ecol.* 1(1): 63-79.
BA 62:34265; E 76-03398

Winner, R.W., J. Scott Van Dyke, N. Caris and M.P. Farrel. 1975. Response of the macroinvertebrate fauna to a copper gradient in an experimentally polluted stream. *Int. Ver. theor. angew. Limnol. Verh.* 19: 2121-2127.
BI 197:22804

Winner, R.W., S. Van Dyke and N. Caris. Response of the macroinvertebrate fauna to a copper gradient in an experimentally polluted stream. p. 231. *in:* Abstracts XIX Congress International Association of Limnology, Winnipeg.
BI 1975:90636

Witt, A., Jr., R.S. Campbell, G.P. Busacker *et al.* n.d. The effect of water temperature on eoclogical processes in a reservoir receiving heated water. PB 264 191.
W77-05108

Wojcik, B. and L. Butler. 1977. Aquatic insects as indicators of stream environmental quality in northern West Virginia. *W. Va Agric. Exp. Stn. Bull.* 653T: 16 p.
BI 1977-45077

Wojtalik, J.A. and T.F. Waters. 1970. Some effects of heated water on the drift of two species of stream invertebrates. *Trans. Am. Fish. Soc.* 99: 782-788.

Zelenko, V. and L. Kosta. 1973. New method for the isolation of methylmercury from biological tissues and its determination at the partspermilliard level by gas chromatography. *Talanta* 20(1): 115-123.
C 78:11983q

MERMITHIDS (NEMATODA) AND MAYFLIES (EPHEMEROPTERA)

W.M. Hominick and H.E. Welch*

Imperial College Field Station
Silwood Park, Ashurst Lodge
Ascot, Berks. England

ABSTRACT

This review is intended to stimulate interest in the presently anecdotal association between mermithids and mayflies. A brief account of the biology of aquatic mermithids is followed by a review of the sparse information available on mermithid parasitism of mayflies. Finally, methods for rearing and preserving mermithids are given, so that the nematodes are in a condition useful for taxonomic purposes.

INTRODUCTION

Nematodes of the superfamily Mermithoidea are obligate parasites of invertebrates, chiefly insects. By our estimation, 650 species in 83 genera have been described, though many species are of doubtful validity. Of these, 424 species in 40 genera were described from aquatic environments where hosts, when they are known, are commonly chironomids, simuliids or culicids. Since these insects frequently coexist with ephemeropterans, it is not surprising that mayflies have also been cited as hosts. What is surprising is the scarcity of these reports. This review will attempt to stimulate interest in the presently anecdotal association between mermithids and mayflies. It is aimed at those workers familiar with ephemeropterans but unfamiliar with mermithids. Therefore, a brief account of the biology of aquatic mermithids is presented. A review follows

* Department of Zoology, University of Manitoba, Winnipeg, Manitoba, Canada, R3T 2N2

of the little that is known about mermithid parasitism of mayflies. Finally, methods for rearing and preserving mermithids are given so that when mermithids are discovered, they will be in a condition useful for taxonomic purposes. For detailed treatments of the biology and taxonomy of mermithids, the reader is referred to Nickle (1972), Poinar (1975), Rubtzov (1972, 1974, 1977, 1978) and Welch (1965).

BIOLOGY OF AQUATIC MERMITHIDS

Adult aquatic mermithids are slender nematodes, between one and three cm long, that live in or on the bottom substrate, where mating and oviposition occur. After a period of embryonic development and a moult within the egg, the preparasitic juvenile hatches. This non-feeding stage must penetrate a host within a few days if it is to survive. It penetrates directly through the cuticle and undergoes a period of massive growth and development in the haemocoele. For example, *Romanomermis culicivorax* (syn. *Reesimermis nielseni*) increases 18-fold in length during its 6-8 day parasitic existence (Gordon *et al.* 1974). Ultimately, the nematode emerges from late instar larvae or nymphs, from pupae, or from adults and takes up a free-living existence. Most species emerge as post-parasitic juveniles and mature for several days to a few weeks. The moult to the adult stage is terminated by the ecdysis of an outer thick cuticle and inner thin membranous one.

During their parasitic existence, mermithids absorb nutrients from the haemolymphs of the host. The intestine, which is modified as a food storage organ, becomes packed with food reserves which, at least in *R. culicivorax*, are predominantly lipids and glycogen (Ittycheriah *et al.* 1977). Postparasitic stages do not feed. The host, in turn, if it survives to adulthood, is sterile due to the depletion of its nutrients. Parasitized mayflies are no exception (Benech 1972a, Murphy 1922, Needham *et al.* 1935). In any case, emergence of the parasite usually signals the death of the host by a combination of a loss of haemolymph through the hole made by the parasite, mechanical injury and reduced nutritional reserves.

Another group of obligate parasites of invertebrates are the hairworms or gordian worms (Nematomorpha), that are sometimes confused with mermithids. Adults are generally found near the edges of lakes, ponds and streams, where eggs are laid. Free-living larvae that hatch are short-lived, though some species extend their survival by encysting. They develop only when ingested by a suitable invertebrate (mainly grasshoppers, crickets, beetles, roaches, mantids, centipedes and millipedes). Larvae penetrate out of the gut and develop in the body cavity. Alternatively, larvae of some species may be eaten be a unsuitable host such as a snail, mayfly, midge, mosquito, fish or amphibian (Inoue 1962, Poinar and

Doelman 1974). Then, the parasite encysts in various tissues and remains viable, but does not develop. If this transport host is eaten by an appropriate host, the parasite resumes development. As the parasitic stages, like those of mermithids, have a non-functional digestive tract, nutrition is probably attained by absorbing materials across the body wall. The parasite emerges from the host as a juvenile, and only when the arthropod is associated with water. It matures in water or damp soil, while the host is sterilized and dies when the parasite emerges.

In the present context, differentiating between mermithids and nematomorphs is not difficult. Mermithids can develop in mayflies, while nematomorphs apparently utilize mayflies only as transport hosts, so they remain as small larvae, encysted within host tissues. Poinar and Doelman (1974) presented excellent photographs and drawings of nematomorph larvae with the characteristic 'head' or presoma, evaginable proboscis, stylets and spines. By contrast, early stages of mermithid parasites are elongated, without hardened structures and usually lie free in the haemocoele. Gordon *et al.* (1974) gave photographs and drawings of a developing mermithid.

MERMITHID PARASITISM OF MAYFLIES

Table 1 lists authors who cited mayflies as hosts for mermithids, together with some notes on the ecology of the host genus. With the exception of Poinar *et al.* (1976), the authors assigned the nematodes to the family Mermithidae or to the genus *Mermis*. *Mermis* is frequently used incorrectly by inexperienced workers as a collective genus rather than as a defined taxon for certain terrestrial mermithids. Therefore, to our knowledge, no mermithid that develops in mayflies has been described though some of the many whose hosts are unknown may be parasites of mayflies. *Pheromermis pachysoma*, which Poinar *et al.* (1976) described, is a parasite of wasps that utilizes aquatic insects as transport hosts.

Although data in Table 1 are limited, they show that mermithids are not restricted to mayflies with particular habitats or habits. Mermithid literature shows that while a few terrestrial species are able to parasitize unrelated insects that occur in the same habitat, most species are more restricted and parasitize members of a single family, genus or even species. Many environmental factors affect host specificity. For example, Galloway and Brust (1976) added 50,000 *R. culicivorax* preparasites/m^2 to a snow-melt pool in Manitoba, Canada, and obtained parasitism of 8-20% of *Aedes vexans* larvae, while six other *Aedes* species and nymphs of a mayfly (probably *Paraleptophlebia* sp.) were not parasitized. It is doubtful that the unparasitized culicids escaped parasitism because of host specificity of this parasite, for it has a wide host range amongst culicids (Petersen 1976). As the authors suggested, mosquitoes

Table 1. Ephemeroptera cited as hosts for mermithid nematodes, with notes on the ecology of the host genus.

Host	Cited by	Ecology of Host Genus[1]: Habitat	Habit	Trophic Relationships
BAETIDAE				
Baetis rhodani	Degrange 1960 Benech 1972 a, b Sukop 1973	Lotic (erosional & deposional) and Lentic (vascular hydrophytes)	Swimmers Climbers Clingers	Sediment feeders (detritus, diatoms) or scrapers
Baetis vagans	Murphy 1922			
Callibaetis pretiosus *Callibaetis pictus*	Pescador[2] Poinar *et al.* 1976 (lab. infection)	Lentic (vascular hydrophytes)	Swimmers Climbers	Not Known
CAENIDAE				
Caenis sp.	Pescador[2]	Lotic (depositional) Lentic (littoral: sediments	Sprawlers	Sediment feeders (scrapers)
EPHEMERIDAE				
Ephemera vulgata	von Linstow 1878[2]	Lotic & Lentic (depositional: sand-gravel)	Burrowers	Sediment feeders (engulfers)
Hexagenia limbata *Hexagenia* sp.	Neeve 1932 Needham 1920	Lentic & Lotic (depositional: sand-silt)	Burrowers	Sediment feeders
HEPTAGENIIDAE				
Heptagenia sp.	Muttkowski 1929 Muttkowski & Smith 1929	Lotic (running water riffles)	Clingers	Scrapers Sediment feeders (engulfers)
HOST UNDETERMINED	Needham *et al.* 1935 Zakhidov 1973 Poinar *et al.* 1976			

1 from Edmunds 1978

2 cited by Arvy and Peters 1976

probably escaped parasitism because of factors such as low temperatures that prevented infection by preparasites and death of the preparasites before hatching of several of the *Aedes* species. Indeed, the life cycles of mermithids should be expected to be synchronized with those of their hosts (e.g. Hominick and Welch 1971). Chances of successful penetration and development generally decrease as the host instar increases (e.g. Bailey and Gordon 1977, Petersen and Willis 1970). Consequently, parasitism of any species except the 'natural' or 'usual' one in a particular locality will be a relatively rare occurrence. The limited data of Muttkowski and Smith (1929) support such an hypothesis. They recorded 2 of 12 *Heptagenia* sp. nymphs parasitized, while 5 *Ameletus* sp., 2 *Emphemerela* sp. and 11 *Drunella* sp. nymphs taken from the same site on the same day were not parasitized. This, and the data in Table 1, argues for a mermithid fauna (i.e. more than one species) that is unique to ephemeropterans.

The trophic relationships of 38 genera of Ephemeroptera are listed by Edmunds (1978). Assuming that the method of feeding listed first is the most common, then 22 (58%) of the genera consist primarily of sediment-feeding species. By comparison, four of the five genera (80%) in Table 1 whose trophic relationships are given by Edmunds (1978), have mainly sediment-feeders. Thus, mermithids may be more likely to occur in sediment-feeders than in predators, filterers or scrapers, with the parasite eggs or preparasites ingested by the host. Such a notion is supported by the data in Table 2, which show that the genera parasitized by nematomorphs contain mainly sediment-feeders. These parasites infect their hosts via the mouth (Dorier 1930, Inoue 1962, Poinar and Doelman 1974). Up to now, studies on aquatic mermithids show that infection occurs by penetration of the host cuticle. Thus, even if mermithids do infect sediment-feeding ephemeropterans more often than other types, this may simply reflect the fact that sediment-feeders spend a larger proportion of their time closely associated with substrates where parasites occur.

Some solid facts exist in addition to the preceding speculations. Thus, Muttkowski and Smith (1929) recorded 2 of 12 (16%) *Heptagenia* sp. nymphs parasitized, while Neeve (1932) recorded 7 of 131 (5%) *Hexagenia limbata* nymphs parasitized in Gull Bay, Lake Winnipeg. Furthermore, other collections either at different times at the same site (Muttkowski and Smith 1929), or at different times and sites (Neeve 1932), did not yield parasitized individuals. These data illustrate a basic characteristic of mermithids: they are focal in distribution and usually parasitize a small proportion of the host population, though they may cause high mortality in localized areas (Rubtzov 1977). They show sharp fluctuations in numbers which sometimes are more apparent than real, for mermithids usually delay development of their hosts. Thus, unparasitized insects emerge first, leaving the parasitized ones in seemingly large numbers

Table 2. Ephemeroptera cited as hosts for Nematomorpha, with notes on the ecology of the host genus.

Host	Parasite	Cited by	Ecology of host genus[1]		
			Habitat	Habit	Trophic Relationships
BAETIDAE					
Baetis sp.	*Paragordius varius*	White 1966	Lotic (erosional & deposional) and Lentic (vascular hydrophytes)	Swimmers Climbers Clingers	Sediment feeders (detritus, diatoms) or scrapers
	Paragordius sp.	White 1969			
B. muticus	Gordiidae	Arvy & Sowa 1976			
Cloeon dipterum	*Parachordodes tolosanus*	von Linstow 1892[2] von Linstow 1889[3]	Lotic (erosional)	Swimmers Clingers	Sediment feeders
	Chordodes japonensis	Inoue 1960, 1962 (lab. infection)			
EPHEMERELLIDAE					
Ephemerella spp.	*Paragordius* sp.	White 1969	Lotic (erosional & depositional) and Lentic (vascular hydrophytes)	Clingers Sprawlers Swimmers	Sediment feeders (detritus, diatoms) Scrapers, some shredders
EPHEMERIDAE					
Ephemera vulgata	*Parachordoes tolosanus*	Meissner 1856[2] von Linstow 1891[3]	Lotic & Lentic (depositional: sand-gravel)	Burrowers	Sediment feeders (engulfers)
LEPTOPHLEBIIDAE					
Leptophlebia sp.	*Paragordius* sp.	White 1969	Lotic (erosional: sediments & detritus)	Swimmers Clingers Sprawlers	Sediment feeders
HOST UNDETERMINED	Gordiidae	Hartmeyer 1909			
	Neochordodes occidentalis	Poinar & Doelman 1974			

[1] from Edmunds 1978

[2] cited by Arvy and Peters 1976

[3] cited by Van Zwaluwenburg 1928

(Welch 1965).

While mermithids may delay development of their hosts, they usually do not affect the number of instars. Thus, Benech (1972b) reported that two parasitized *Baetis rhodani* adult females moulted the same number of times as unparasitized ones. Both the host age at the time of infection and the number of parasites per host are probably important. Craig and Webster (1974) showed experimentally that the older *Schistocerca gregaria* was at the time of infection, the greater the number of *Mermis nigrescens* required to inhibit the next-but-one moult.

In aquatic insects, the time of infection may determine not the number of instars, but whether the parasite is carried into the adult stage of the host. Alternatively, some mermithid species, such as *Perutilimermis culicis* (Nickle 1972), may be adapted to complete their development only when the host becomes adult. In either case, mermithids were recorded from adult mayflies (Benech 1972b, Murphy 1922, Muttkowski 1929, Needham et al. 1935) and this would be an important factor in disseminating the parasites. Presumably, they would emerge when the host alights on water as occurs for some mermithids from chironomids (Hominick and Welch 1971) and simuliids (Mokry and Finney 1977).

The effect of mermithids on the behavior of infected adult mayflies remains to be determined. Murphy (1922) reported that four infected female *Baetis vagans* appeared natural in flight while Muttkowski (1929) captured an adult that was "fluttering weakly" and the "the weight of the parasite held it from rising". These apparently contradictory observations may be equally applicable. Wülker (1961) observed that some infected adult chironomids swarm apart from uninfected ones, some fly near the water surface or settle in the water in sheltered places (Wülker 1963), while others are strong enough to participate in swarms with uninfected individual (Wülker 1963).

One other effect of mermithid parasitism is noteworthy; namely, the formation of intersexes in parasitized adult chironomids, simuliids and heleidids (=Ceratopogonidae) and of intercastes in ants (Welch 1965, Wülker 1964). It is interesting that gynandromorphs, which have erroneously been called 'intersexes', have been reported in the mayfly families Baetidae, Ephemeridae, Heptageniidae, Leptophlebiidae and Potamanthidae (Agnew 1977). Gynandromorphs are individuals with a mixture of male and female tissues composed of cells that differ genetically, while intersexes have all tissues composed of genetically identical cells. As gynandromorphs and intersexes may appear similar, for example, the colour pattern of ephemeropteran gynanders is sometimes described as being atypical, neither female nor male (Agnew 1977), special attention should be given to the secondary sexual characters of mayflies found parasitized by mermithids.

REARING AND PRESERVING MERMITHIDS

To identify mermithids, it is necessary to obtain adult specimens. The Russian worker, Prof. I.A. Rubtzov, classified mermithids using juvenile characters, but his system has not been accepted by other workers. Therefore, it is necessary to detect parasitized hosts, rear them until their parasites emerge, and then rear the parasites until they moult to the adult stage.

Examination of insects for mermithids is best done with a dissecting microscope and transmitted light. Observation is facilitated by placing the nymphs in the inverted cover of a 5 cm diameter glass petri dish with a small amount of water. The bottom is then carefully placed on the specimen, which is slightly flattened between the two layers of glass. The depth of the layer of water is critical - too little and the specimen may be damaged; too much and the specimen is not immobilized.

Detection of early stages of mermithids without dissecting the host is difficult because their small size and lack of distinguishing features cause them to look like host structures such as tracheae, ducts or tubules. Later stages are more easily detected because they occupy most of the haemocoele, are coiled on themselves several times, and having depleted most of the host's nutrient reserves, cause the body wall to be more transparent than normal. These stages will be found in the large (late) instars.

The method for rearing parasitized nymphs will depend on the species and its habitat, which would be simulated as closely as possible. J.F. Flannagan (pers. comm.) routinely rears burrowing nymphs in a layer of mud that has been sterilized and covered with dechlorinated water to a depth of about 10 cm. The water is gently aerated and pulverized Tetramin fish food provides nutrients. With such a system, the water and substrate would have to be gently washed through a series of sieves (smallest aperture approximately 0.1 mm) weekly to monitor the progress of the hosts and to obtain any parasites that had emerged.

Adult hosts are more easily maintained, as they can be kept in a small container with some water in the bottom. When the mayfly rests on the water, the parasites should emerge.

Mermithids that have emerged should always be kept and examined in a layer of water. They can be picked up with pasteur pipettes, but should be handled as little as possible as they are easily damaged. Although the morphology of mermithids from mayflies is unknown, aquatic mermithids have postparasitic juveniles with a posterior tail appendage. As the adult has no such appendage, its absence serves as a simple indication that the final ecdysis has occurred. Also, adult females are characterized by a ventral vulva and muscular vagina in the midregion of the body, while adult males have one or two cuticularized, refractile spicules at the terminus.

If the final ecdysis has not occurred, the parasites should be maintained in a layer of fine sand covered with dechlorinated tap water. During this time, a common problem is infection and rapid destruction of the nematodes by aquatic fungi, Hence, it is prudent to keep the specimens in more than one container. Best survival is usually obtained by using large culture vessels (approximately 500 mL) and xenic rather than axenic methods (see Bailey *et al.* 1977). In our experience with mermithids from blackflies, a layer of autoclaved sand about 0.5 cm deep is covered to a depth of about 5 cm with dechlorinated tap water that is changed every two days. Nematodes are not surface sterilized before addition to the vessel, and strict aseptic techniques are not followed. Flaming instruments is a good precaution against transfer of microorganisms between vessels. A temperature of about 15° is usually satisfactory.

The method by which the specimens are killed and fixed is critical as some methods cause distortions of taxonomic characters. Detailed experiments by a colleague at Imperial College Field Station, J. Curran, show that the nematodes should be killed by plunging them into 65° water for three seconds. Hotter or longer results in distortions The nematodes are immediately fixed in TAF (7 mL formalin, 2 mL triethanolamine, 91 mL distilled water) for at least three days. They can then be processed to glycerol by the Seinhorst method (see Poinar (1975), page 17) and mounted permanently. If they belong to a described genus, two recent keys to genera of mermithids (Poinar 1977, Rubtzov 1978) will be useful. However, mermithid taxonomy is in a state of flux, so the specimens are best sent to an authority for identification. In all likelihood, they will belong to at least one species that is new to science.

RESUME

Cette étude a pour but de stimuler l'intérêt a l'égard de l'association entre les mermithidés et les éphéméroptères présentement au stade anecdotique. Un apercu biologique des mermithidés aquatiques précède l'étude des renseignements fort peu nombreux qui existent sur le parasitisme des éphéméroptères au dépens des mermithidés. Enfin, on suggère des méthods de culture et de préservation des mermithidés propres à rendre les nématodes utiles à des fins taxonomiques.

ZUSSAMENFASSUNG

Dieser Bericht beabsichtigt das Interesse an der gegenwärtig anekdotenhaften Verbindung zwischen Mermithiden und Eintagsfliegen anzuregen. Einem kurzen Rückblick auf die Biologie der aquatischen Mermithiden folgt ein Überblick über die spärliche Information, die wir in Bezug auf das Mermithiden Parasitentum der Eintagsfliegen besitzen. Zum Abschluß werden Methoden für die Zucht und Konservierung von Mermithiden angegeben, damit die Nematoden sich in einem für taxonomische Zwecke brauchbaren Zustand befinden.

REFERENCES

Agnew, J.D. 1977. Gynanders and sex determination in Ephemeroptera. *Eatonia* 24: 1-2.

Arvy, L. and W.L. Peters. 1976. Liste des Ephéméroptères-hôtes de parasites, de commensaux et autres associés. *Ann. Parasitol Hum. Comp.* 51: 121-141.

Arvy, L. and R. Sowa. 1976. Recherche sur les parasites internes des Éphéméroptères de la région de Cracovie. *Ann. Parsitol Hum. Comp.* 51: 111-120.

Bailey, C.H. and R. Gordon. 1977. Observations on the occurrence and collection of mermithid nematodes from blackflies (Diptera: Simuliidae). *Can. J. Zool.* 55: 148-154.

Bailey, C.H., R. Gordon and C. Mills. 1977. Laboratory culture of the free-living stages of *Neomesomermis flumenalis*, a mermithid nematode parasite of Newfoundland blackflies (Diptera: Simuliidae). *Can. J. Zool.* 55: 391-397.

Benech, V. 1972a. La fecondité de *Baetis rhodani* Pictet. *Freshw. Biol.* 2: 337-354.

Benech, V. 1972b. Le polyvoltinisme chez *Baetis rhodani* Pictet (Insecta: Ephemeroptera) dans un ruisseau a truite des Pyrénées-Atlantiques, Le Lissuraga. *Ann. Hydrobiol.* 3: 141-171.

Craig, S.M. and J.M. Webster. 1974. Inhibition of moulting of the desert locust *Schistocerca gregaria* by the nematode parasite *Mermis nigrescens*. *Can. J. Zool.* 52: 1535-1539.

Degrange, C. 1960. Recherches sur la reproduction des Éphéméroptères. *Trav. Lab. Hydrobiol. Piscicult. Univ. Grenoble* 50-51: 7-193.

Dorier, A. 1930. Recherches biologiques et systématiques sur les Gordiacés. *Trav. Lab. Hydrobiol. Piscicult. Univ. Grenoble* 22: 1-184.

Edmunds, G.F. 1978. Ephemeroptera. p. 57-80. *in*: R.W. Merritt and K.W. Cummins, eds. An introduction to the aquatic insects of North America. Kendall Hunt, Iowa.

Galloway, T.D. and R.A. Brust. 1976. Field application of the mermithid nematode, *Romanomermis culicivorax* Ross and Smith, for the control of mosquitoes, *Aedes* spp., in spring in Manitoba. *Manit. Entomol.* 10: 18-25.

Gordon, R., C.H. Bailey and J.M. Barber. 1974. Parasitic development of the mermithid nematode *Reesimermis nielseni* in the larval mosquito *Aedes aegypti*. *Can. J. Zool.* 52: 1293-1302.

Hartmeyer, R. 1909. Gordiidae. p. 86-88. *in*: A. Brauer ed. Die Süsswasserfauna Deutschlands. Heft 15: Nematodes, Mermithidae und Gordiidae. G. Fischer, Jena.

Hominick, W.M. and H.E. Welch. 1971. Sychronization of life cycles of three mermithids (Nematoda) with their chironomid (Diptera) hosts and some observations on the pathology of the infections. *Can. J. Zool.* 49: 975-982.

Inoue, I. 1960. Studies on the life history of *Chordodes japonensis*, a species of Gordiacea. II. On the manner of entry into the aquatic insect-larvae of *Chordodes* larvae. *Annot. Zool. Jpn.* 33: 132-141.

Inoue, I. 1962. Studies on the life history of *Chordodes japonensis*, a species of Gordiacea. III. The mode of infection. *Annot. Zool. Jpn.* 35: 12-19.

Ittycheriah, P.I., R. Gordon and W.J. Condon. 1977. Storage material of the nematode *Romanomermis culicivorax*, a mermithid parasite of larval mosquitoes. *Nematologica* 23: 165-171.

Mokry, J.E. and J.R. Finney. 1977. Notes on mermithid parasitism of Newfoundland blackflies, with the first record of *Neomesomermis flumenalis* from adult hosts. *Can. J. Zool.* 55: 1370-1372.

Murphy, H.E. 1922. Notes on the biology of some of our North American species of mayflies. *Bull. Lloyd Lib.* 22, *Entomol. Ser.* 2: 1-46.

Muttkowski, R.A. 1929. The ecology of trout streams in Yellowstone National Park. *Roosevelt Wild Life Ann.* 2: 155-240.

Muttkowski, R.A. and G.M. Smith. 1929. The food of trout stream insects in Yellowstone National Park. *Roosevelt Wild Life Ann.* 2: 241-263.

Needham, J.G. 1920. Burrowing mayflies of our larger lakes and streams. *Bull. U.S. Bur. Fish.* 36: 269-292.

Needham, J.G., J.R. Traver and Y.C. Hsu. 1935. The biology of mayflies. Comstock Publishing Company, Ithaca, New York. 759 pp.

Neeve, F. 1932. A study of the mayflies (*Hexagenia*) of Lake Winnipeg. *Contrib. Can. Biol. Fish.* 7: 177-201.

Nickle, W.R. 1972. A contribution to our knowledge of the Mermithidae (Nematoda). *J. Nematol.* 4: 113-146.

Petersen, J.J. 1976. Status and future of mermithid nematodes as control agents of mosquitoes. p. 236-240. *in*: Proceedings of the First International Colloquium on Invertebrate Pathology and IXth Annual meeting of the Society for Invertebrate Pathology. Queen's University, Kingston, Canada.

Petersen, J.J. and O.R. Willis. 1970. Some factors affecting parasitism by mermithid nematodes in southern house mosquito larvae. *J. Econ. Entomol.* 63: 175-178.

Poinar, G.O. Jr. 1975. Entomogenous nematodes. E.J. Brill, Leiden. 317 p.

Poinar, G.O. Jr. 1977. C.I.H. key to the groups and genera of nematode parasites of invertebrates. Sheila Willmott ed. Commonwealth Agricultural Bureaux, Farnham Royal, Bucks, England. 43 p.

Poinar, G.O. Jr. and J.J. Doelman. 1974. A rexamination of *Neochordodes occidentalis* (Montg.) comb.n. (Chordodidae: Gordioidea): larval penetration and defence reaction in *Culex pipiens* L. *J. Parasitol.* 60: 327-335.

Poinar, G.O. Jr., R.S. Lane and G.M. Thomas. 1976. Biology and redescription of *Pheromermis pachysoma* (v. Linstow) n. gen. n. comb. (Nematoda: Mermithidae), a parasite of yellowjackets (Hymenoptera: Vespidae). *Nematologica* 22: 360-370.

Rubtzov, I.A. 1972. Aquatic Mermithidae of the fauna of the U.S.S.R. Vol. I. "Nauka", Leningrad. 254 p. (Translated for the Dept. Agric., Agricultural Research Service. TT74-52046, National Technical Information Service, Springfield, Virgina.)

Rubtzov, I.A. 1974. Aquatic Mermithidae of the fauna of the U.S.S.R. Vol. II. "Nauka", Leningrad. 222 p. [In Russian].

Rubtzov, I.A. 1977. Mermithids: Origin, biology and distribution. "Nauka", Leningrad. 191 p. [In Russian].

Rubtzov, I.A. 1978. Mermithids: Classification, Importance and Utilization. "Nauka", Leningrad. 207 p. [In Russian].

Sukop, I. 1973. Annual cycle of mayflies (Ephemeroptera) in a karstic stream. *Acta. Entomol. Bohemoslov.* 70: 81-85.

Welch, H.E. 1965. Entomophilic nematodes. *Annu. Rev. Entomol.* 10: 275-302.

White, D.A. 1966. A new host record for *Paragordius varius* (Nematomorpha). *Trans. Am. Microsc. Soc.* 85: 579.

White, D.A. 1969. The infection of immature aquatic insects by larval *Paragordius* (Nematomorpha). *Great Basin Nat.* 29: 44.

Wülker, W. 1961. Untersuchungen über die Intersexualität der Chironomiden (Dipt.) nach *Paramermis*-Infektion. *Arch. Hydrobiol. Suppl.* 25: 127-181.

Wülker, W. 1963. Prospects for biological control of pest Chironomiden in the Sudan. WHO/EBL/11: 1-23.

Wülker, W. 1964. Parasite-induced changes of internal and external sex characters in insects. *Exp. Parasitol.* 15: 561-597.

Zakhidov, M.T. 1973. Mermithids-parasites of some organisms that serve as fodder for fish. *Hydrobiol. J.* 9: 57-60.

Zwaluwenburg, R.H. Van. 1928. The interrelationships of insects and roundworms. *Bull. Exp. Sta. Hawaiian Sugar Planters' Assoc. Entomol. Ser.* 20: 1-68.

NUMERICAL ABUNDANCE VALUES OF MAYFLY NYMPHS FROM THE HOLARCTIC REGION

Hugh F. Clifford

Department of Zoology
University of Alberta
Edmonton, Alberta
Canada T6G 2E9

ABSTRACT

Members of seven Holarctic families, Heptageniidae, Baetidae, Ephemerellidae, Leptophlebiidae, Ephemeridae, Caenidae, and Siphlonuridae, accounted for 96% of the reported abundance values. *Baetis* and *Ephemerella* were the most reported taxa for the entire Holarctic region. In North America, the order was *Ephemerella*, *Baetis*, and *Hexagenia*. *Baetis rhodani*, a Palaearctic species, was the most reported species. For the total mayfly fauna, the mean of all average yearly abundance values was 375/m^2. Of these average yearly abundance values, 21% of the values were less than 100/m^2 and 93% were less than 700/m^2.

INTRODUCTION

Since 1966 I have been studying limnological features of a brown-water stream in west-central Alberta, Canada. Numerical estimates of abundance, especially for immature mayflies, have been an integral part of certain studies of this brown-water stream. The absolute values have been important and valid facets of temporal and spatial distribution studies for a single area within this stream. But what is the importance of abundance estimates when culled from the literature for comparative purposes? There now exists a voluminous literature dealing in part with the abundance of aquatic invertebrates per unit of substrate area. Indeed, estimating number of invertebrates per unit area seems to be a routine procedure in stream work. However, numerous confounding factors would seem to make these estimates almost impossible to evaluate when used for comparative purposes.

Chief among these are the large variability in numbers per unit area over short stretches of the substrate and the very large seasonal changes in numbers for certain taxa. Hynes (1970) and Pennak (1977) discuss many of the factors affecting the validity of quantitative sampling of stream benthos. Both agree that some measure of abundance is essential for understanding the biology of stream organisms. The prodigious number of reported density values for certain taxa, in itself, commands respect; and it would be instructive to attempt some sort of comparative analysis of these reported abundance values.

METHODS

I surveyed 152 articles that reported abundance values of mayfly nymphs from the Holarctic region. The factors that make these estimates difficult to evaluate when used for comparative purposes were disregarded. The values were separated into two groups: (1) number per unit area at any one time, and (2) where samples were taken throughout the year, average yearly abundance values. The values at "any one time" were either from a single unit area sample (e.g. 0.09 m^2) or from the average of several samples from the same site. Occasionally, the values of each of several unit area samples for a single site would be reported; in these cases, I simply took the maximum value, on the premise that abundance estimates generally tend to be conservative. The data were treated mainly in a relative sense, with as much emphasis on the frequency that a taxon was reported as on the absolute abundance value. For some of the relative compilations, the Nearctic (i.e. U.S.A. and Canada, there being no reports from Mexico) reports were treated separately. Most of the Palaearctic reports came from Poland (26%), Great Britain (24%), Czechoslovakia (17%), Japan (12%), and Sweden (11%).

A similar analysis was also attempted for biomass (standing crop), but there were not enough values derived by the same weighing procedure to make comparisons meaningful. For nymphal wet weights of individual taxa per unit area, there were too many weighing procedures (resulting in very wet weight to almost dry weight). And of the few dry weight values derived by comparable procedures, about 75% dealt with a single taxon, *Hexagenia*. Hence, the ensuing analysis is restricted to numberical abundance, i.e. numbers per square meter.

RESULTS

Taxa Frequency

Taxa within each family. Abundance values were available for 15 of the 18 Holarctic families. For the entire Holarctic region, taxa within two families, Heptageniidae and Baetidae, accounted for about half of the reported values; and taxa within only seven families accounted for 96% of all reported abundance values (Table 1). In North America, taxa within Heptageniidae and Ephemerellidae were most important, and taxa of only 5 of the 16 Nearctic mayfly families accounted for 83% of the Nearctic values.

Table 1. Percent number of times that numerical abundance values were reported for taxa within each family.

Holarctic	%	Nearctic only	%
Heptageniidae	25	Heptageniidae	25
Baetidae	23	Ephemerellidae	21
Ephemerellidae	18	Baetidae	15
Leptophlebiidae	11	Ephemeridae	12
Ephemeridae	8	Leptophlebiidae	10
Caenidae	6	Siphlonuridae	6
Siphlonuridae	5	Caenidae	5
Tricorythidae	2	Tricorythidae	2
Polymitarcyidae	<1	Polymitarcyidae	<1
Potamanthidae	<1	Potamanthidae	<1
Oligoneuriidae	<1	Oligoneuriidae	<1
Baetiscidae	<1	Baetiscidae	<1
Metretopodidae	<1	Metretopodidae	<1
Ametropodidae	<1	Ametropodidae	<1
Prosopistomatidae	<1		

Genera and species. *Baetis* and *Ephemerella* were, by a considerable margin, the most reported genera for the entire Holarctic region (Table 2). In the Nearctic region, *Ephemerella* accounted for twice as many of the reported values than did any other mayfly genus. *Baetis rhodani* Pictet, a Palaearctic species, was the most reported species. Although the North American fauna was not dominated by any one species, there were slightly more abundance values for *Hexagenia limbata* (Serville), *Ephemerella doddsi* Needham and *Baetis vagans* McDunnough than for other Nearactic species.

In North America, Ephemerellidae (single genus *Ephemerella)* was the only major family exhibiting a large percentage of abundance values at the species level (Table 3). In contrast, for the Palaearctic fauna, abundance values of all families except Caenidae and Siphlonuridae were reported at the species level at least 80% of the time. Most of the Palaearctic values pertained to mayflies of Europe, where in many areas mature nymphs can be identified to species for almost all families. In North America, the only inclusive nymphal keys to the species level for a major taxon are those for *Ephemerella* by R.K. Allen and G.F. Edmunds, Jr. This would seem to be mainly responsible for the relatively high percentage of ephemerellid abundance values at the species level and certainly appears to be good ecological evidence for the usefulness of such keys.

Table 2. Number of times numerical abundance values were reported for the 10 most reported genera.

Holarctic	No.	Nearctic only	No.
Baetis	64	*Ephemerella*	39
Ephemerella	63	*Baetis*	16
Rhithrogena	24	*Hexagenia*	16
Hexagenia	16	*Stenonema-Stenacron*	14
Stenonema-Stenacron	14	*Paraleptophlebia*	10
Caenis	13	*Rhithrogena*	8
Paraleptophlebia	13	*Caenis*	7
Ecdyonurus	12	*Tricorythodes*	5
Epeorus	10	*Isonychia*	5
Isonychia	8	*Cinygmula*	5

Table 3. Percent number of abundance values within each of the major families reported at the species level.

	Holarctic %	Palaearctic %	Nearctic %
Ephemerellidae	85	95	80
Baetidae	69	94	29
Heptageniidae	58	81	37
Leptophlebiidae	47	92	16
Siphlonuridae	41	57	30
Ephemeridae	28	100	18
Caenidae	26	36	11

Lentic habitats. Only 5% of all values pertained to mayflies from lentic habitats. Taxa of Ephemeridae, mainly *Hexagenia*, accounted for 65% of these abundance values, followed by taxa of Caenidae (12%), Leptophlebiidae (12%), and other families (11%).

Absolute Values

Taxa exhibiting large densities. Of all abundance values reported as > 1000/m^2 at one sampling time, three genera, *Baetis* (41%), *Ephemerella* (19%), and *Caenis* (16%), accounted for 76% of the values for the entire Holarctic region. In North America, *Baetis* (25%), *Ephemerella* (20%), *Hexagenia* (20%), and *Caenis* (15%) accounted for 80% of the values. Rarely, mayfly nymphs, usually the total mayfly fauna, but sometimes a single taxon, were reported as occurring in very large numbers, over 10,000/m^2. The mayflies of most of these reports were either associated with aquatic macrophytes or the collecting equipment sampled deep into fairly loose substrates. Hence, numbers per unit area would probably not be an accurate expression of abundance in these cases. Numbers per unit volume of substrate would seem more appropriate.

Average yearly abundance values. These are values derived mainly from semimonthly, monthly, or bimonthly samples throughout the year and then averaged. For the total mayfly fauna, 21% of the average yearly values were < 100/m^2 and 93% were less than 700/m^2 (Table 4). The mean was 375/m^2 and the median was 320/m^2. The largest average yearly abundance value for the total mayfly fauna was 1448/m^2, but there were indications from some of the non-yearly reports that had samples been taken throughout the year the maximum average yearly value would have been considerably higher than 1448/m^2.

Table 4. Percentage numbers and cumulative percentages of all average yearly abundance values (entire Holarctic region) for each of eight categories. Values represent the total mayfly fauna.

Numbers/m^2	%	Cumulative %
< 100	21	21
100-199	16	37
200-299	9	46
300-399	14	60
400-499	7	67
500-599	12	79
600-699	14	93
> 700	7	100

Means of average yearly abundance values for genera having mean values > $10/m^2$ were: *Baetis* ($325/m^2$), *Ephemerella* ($130/m^2$), *Hexagenia* ($83/m^2$), *Caenis* ($32/m^2$), and *Stenonema-Stenacron* ($14/m^2$). Although one must treat these absolute values cautiously, they do indicate the relatively great abundance of *Baetis* and, to a lesser extent, *Ephemerella* nymphs in Holarctic regions.

DISCUSSION

Although there are 18 families of known Holarctic mayflies, only a small number of genera within a few families contribute much to the reported numerical abundance values. In many aquatic systems there are only a few relatively abundant mayfly species when abundance is considered in terms of the entire year. This seems especially truc in northern waters. Ulfstrand (1975) studied the mayfly fauna of several streams in Swedish Lapland and found that one species, *Baetis rhodani*, accounted for over 50% of all the mayfly nymphs collected. In a brown-water stream of Alberta, Canada, which I have been studying since 1966, there are 18 mayfly species; but two species, *Leptophlebia cupida* (Say) and *Baetis tricaudatus* Dodds, account for 80% of the nymphs, and this is when averaged over several years (Clifford 1978). Perhaps, in a sense, this is fortunate for ecologists studying stream communities, since we can put our efforts into studying the few abundant mayfly species and be quite certain we are obtaining data on the important mayfly contributors to community structure.

Why are nymphs of the genus *Baetis* relatively so abundant? In respect to my survey, *Baetis* leads all taxa in number of reported values, number of values > $100/m^2$, and average yearly abundance values. Perhaps species richness within the genus is important. Also, Macan (1958) suggests that catchability might have something to do with the large numbers of *Baetis* nymphs that are often reported in abundance studies. Certainly it is well-documented that *Baetis* nymphs exhibit a great propensity to drift, at least in temperate regions. The *Baetis* phenomenon does pose interesting ecological problems and suggests, especially in North America, important lines of taxonomic research. For a mayfly ecologist in North America, a species key to *Baetis* nymphs would be a most valuable tool.

RESUME

Les insectes appartenant à sept familles holarctiques, les heptagéniidés, les baetidés, les leptophlébiidés, les éphemeridés, les caenidés et les siphlonuridés comptent pour 96% des quantités d'insectes signalés. Les *Baetis* et les *Ephemerella* sont les taxons le plus souvent relevés dans la région holarctique tour entière. En Amérique du nord, *l'Ephémérella* vient en premier lieu, suivi de la *Baetis* puis de *l'Hexagénie*. La *Baetis rhodani*, espèce paléarctique, est l'espèce le plus souvent signalée. La moyenne proportionnelle de l'agrégat des quantités annuelles moyennes de toute la faune d'éphéméroptères s'établit à $375/m^2$. De ces quantités annuelles moyennes, 21% étaient inférieures à $100/m^2$ et 93%, a $700/m^2$.

ZUSSAMENFASSUNG

Mitglieder von sieben holarktischen Familien: *Heptageniiden*, *Baetiden*, *Ephemerelliden*, *Leptophlebiiden*, *Ephemeriden*, *Caeniden* und *Siphlonuriden* lieferten 96% der aufgeführten Abundanzwerte. *Baetis* und *Ephemerella* waren die häufigst gemeldeten Taxa für die gesamte holarktische Region. In Nordamerika lautete die Reihenfolge dagegen: *Ephemerella*, *Baetis* und *Hexagenia*. *Baetis rhodani*, eine palaearktische Art, wurde am meisten genannt. Für die gesamte Eintagsfliegen Fauna war das Medium aller durchschnittlichen, jährlichen Abundanzwerte $375/m^2$. Davon beliefen sich 21% auf weniger als $100/m^2$ und 93% auf weniger als $700/m^2$.

REFERENCES

Clifford, H.F. 1978. Descriptive phenology and seasonality of a Canadian brown-water stream. *Hydrobiologia* 58: 213-231.

Hynes, H.B.N. 1970. The ecology of running waters. Univ. Toronto Press, Toronto.

Macan, T.T. 1958. Methods of sampling the bottom fauna in stony streams. *Mitt. Int. Ver. theor. angew Limnol.* 8: 1-21.

Pennak, R.W. 1977. Trophic variables in Rocky Mountain trout streams. *Arch. Hydrobiol.* 80: 253-285.

Ulfstrand, S. 1975. Diversity and some other parameters of Ephemeroptera and Plecoptera communities in subarctic running waters. *Arch. Hydrobiol.* 76: 499-520.

THE MAYFLY FAMILY BAETISCIDAE (EPHEMEROPTERA). PART I

Lewis Berner and Manuel L. Pescador*
Department of Zoology, University of Florida
Gainesville, Florida 32611

ABSTRACT

The history of the taxonomy and phylogeny of *Baetisca*, the single genus in the family Baetiscidae, is reviewed beginning with the description of *Prosopistoma* as a crustacean. The relationship of the Prosopistomatidae to the Baetiscidae is treated as well as the concurrence of the authors with Edmunds' conclusion that the two families are derived from a common ancestor but parallel evolution must explain some of their similarities. There is a summary of presently recognized species and characteristics by which nymphs are differentiated. A brief treatment of the ecology and life history of various species is included.

INTRODUCTION

If Latreille had had nymphal specimens of *Baetisca* in 1833, would he have been as puzzled by them as he was by *Prosopistoma* when he described these curious nymphs as a new genus of crustacean? Certainly, *Baetisca* nymphs are sufficiently different from other Ephemeroptera to cause one to conclude that he might have done so. The superficial similarities of the mesonotal shields and the gills were enough to lead early entomologists to the misconception that the two are very closely related. Only in recent years, through

---

* Laboratory of Aquatic Entomology, Florida A&M University, Tallahassee, Florida 32307

modern phylogenetic studies of all stages, has it been recognized that there are significant differences between them.

Say (1839) described a new species of *Baetisca*, which he called *obesa*, because of its short, robust body. Next, Walker (1853) characterized *Baetis obesa* both in Latin and English. Hagen (1861) also briefly redescribed the subimago following Say's earlier descriptions, and the species was recognized by Walsh (1862) as being a member of his new genus *Baetisca* after he secured male imagos. In 1863, he remarked (p. 178) "*B. obesa* Say, Walsh. Both genus and species new to me; very curious." Say's description was based on the subimago. Walsh (1864), after four years, successfully reared adults from the nymph and was then able to relate the two stages. He had sent a specimen of the nymph to Hagen, who wrote to Walsh "The larva No. 66 is the most extraordinary animal I have seen, so that I asked myself whether it really belonged to Insecta." After careful analysis he concluded that it was an insect assigned to the "Ephemerina." Even more surprising, he was able to relate the nymph to an adult *Baetisca obesa* which Walsh had sent to him earlier. Walsh (1864), with Hagen's confirmation in hand, gave an excellent description of the nymph and discussed his observations of its habitat and emergence.

Eaton (1871), in his *A Monograph on the Ephemeridae*, redescribed the genus and the single, known species, *B. obesa* in Latin. He illustrated the forewing (pl. II, Fig. 6) and the male genitalia (pl. V, Fig. 9), both rather poorly.

Provancher (1876) reported a species from Quebec as *Clöe unicolor* Hagen, but in 1877 decided that his specimens represented a new species, which he named *Clöe rubescens*, nov. sp. Subsequently, McDunnough (1921), after studying Provancher's male specimen, assigned it to *Baetisca* making *rubescens* the second species to be included in the genus.

Joly (1880) published a French translation of Walsh's (1864) description of *Baetisca* and added comments in footnotes. Here he compared features of *Baetisca* with those of *Prosopistoma punctifrons* (= *P. foliacium* Fourcroy) and *P. variegatum* Latrielle. Vayssière, (1882) in his very comprehensive paper on Ephemeroptera, reported that he saw analogies between *Prosopistoma* and *Baetisca* nymphs in their respiratory apparatus and he discussed their morphology at length as well as providing excellent illustrations. He concluded that *Baetisca* is to North America what *Prosopistoma* is to Europe and Madagascar.

Eaton (1885) described and illustrated both adult and nymphal stages of *Baetisca*. Subsequently, Needham (1905) reported that *B. obesa* had been taken in New York State and in Indiana and he included photographs of the nymph and subimaginal wings. Ulmer

(1920) merely listed the family Baetiscidae with only a single genus, *Baetisca*, included in it. The next reports of the genus were those of Clemens (1913, 1915) in his study of mayflies of the Georgian Bay area of Lake Huron.

Traver (1931) added two new species of *Baetisca*, *carolina* and the enigmatic *callosa* from North Carolina and West Virgina, respectively. Soon afterwards, McDunnough (1932) reported two additional species, *B. laurentina* and *B. lacustris*, from Canada, and also provided a key to the known nymphs. Traver again reported in 1932 on *B. carolina* in North Carolina and in 1937 she described *B. thomsenae* from that State. The latter species was synonymized with *B. carolina* by Berner (1955).

The classic work by Needham, Traver, and Hsu (1935), *The Biology of Mayflies*, gave a comprehensive summary of the taxonomic knowledge of *Baetisca*. Neave (1934) described *B. bajkovi* from Manitoba and Spieth (1933, p. 359) concluded that "*Baetisca* seems a distinct entity in the phylogenetic arrangement of mayflies." He erected the superfamily Baetiscoidea to accommodate the family.

Baetisca rogersi was described from northwestern Florida by Berner (1940) and further details were added by him in 1950. He summarized the knowledge of the southeastern species of the genus in 1955 and described two new ones at that time, *B. escambiensis* and *B. gibbera*. Burks treated the Illinois species in 1953, and in 1962 the late Justin Leonard and his wife Fannie did the same for Michigan. Daggy (1945) reported that the genus occurs in Minnesota.

Following World War II, with the great upsurge of interest in aquatic insects, *Baetisca* was frequently encountered in stream studies and reports of its occurrence grew. In 1959, Berner summarized our knowledge of the genus to that date in tabular form. Until Edmunds' (1960) report no species was known positively to occur in western North America and in that paper he described *Baetisca columbiana*, a species not subsequently reported. Schneider and Berner (1963) described another Florida species, *B. becki*, from the northwestern part of the State. Later, Pescador and Peters (1971) described the imago of *becki* reared from the nymphal stage. The most recently described species, *B. berneri*, is that of Tarter and Kirchner (1978) from West Virginia. The latest treatment of *Baetisca* was given by Edmunds, Jensen, and Berner (1976) in their book on North and Central American mayflies.

PHYLOGENETIC RELATIONSHIPS OF THE BAETISCIDAE

Few papers have been published dealing with the relationships of Baetiscidae and Prosopistomatidae. In 1917, Lameere included two subfamilies in his Prosopistomidae, Hexagenitinae, which is composed of the fossil genus *Hexagenites*, and Prosopistominae, divided into two tribes, the Baetiscini and Prosopistomini.

In a discussion of the systematic position of *Prosopistoma*, Gillies (1954) concluded that both *Baetisca* and *Prosopistoma* are derived from a common stock of great antiquity, but since then evolution and specialization have proceeded independently. His conclusion was based on the nymphs having many structural features in common, notably a mesonotal carapace, branchial chamber, and cephalization of the central nervous system, and the adults having the similarly unique arrangement of the cubital veins of the wings.

Fontaine (1958), studying the affinities of *Prosopistoma*, noted the nymphal similarities with *Baetisca*. She proposed two causes that might result in the similarities: (1) *Prosopistoma* and *Baetisca* possess true systematic affinities and come from the same ancestor and (2) *Prosopistoma* and *Baetisca* are two evolved genera--*Prosopistoma* having the higher degree of evolution (the mesonotal shield forming a true respiratory chamber, the labium completely encircled in its basilar portion, the nervous system with the thoracic and abdominal ganglia fused into a single mass). They are parallel evolutionary lines and are distinct from the remainder of the Ephemeroptera.

Fifty-two years after the appearance of his study of the morphology of immature Ephemeroptera, Vayssière published in 1934 an anatomical study of *Baetisca obesa* and *B. carolina*. He concluded that the internal anatomy of *Baetisca* is much like that of *Prosopistoma* and that his ideas put forward in 1882 about the relations of the two had now been authenticated.

Landa's (1969) investigation of the comparative anatomy of mayfly nymphs indicated that the Neoephemeridae showed close relationships to the Caenidae, Prosopistomatidae, and particularly to the Baetiscidae. In the Caenidae, he concluded that *Brachycercus* is more closely related to *Baetisca* than is *Caenis*. His phylogeny, based on the comparative anatomy of the tracheal system, Malpighian tubes, and the nerve band of nymphs and illustrated in a 1973 paper, represented the Neoephemeridae and the Baestiscidae as being related and the Caenidae related to the Prosopistomatidae. The four families are derived from a common stem.

Demoulin (1956 and 1969) presents cogent arguments in support of his contention that the Baetiscidae are derived from the Oniscigasterinae and the Prosopistomatidae from the Ameletopsinae and that both families should be included in the superfamily Siphlonuroidea. He has given an interesting analysis of morphological evidence leading him to his conclusions; however, we remain in agreement with Edmunds' interpretation of the phylogeny of these

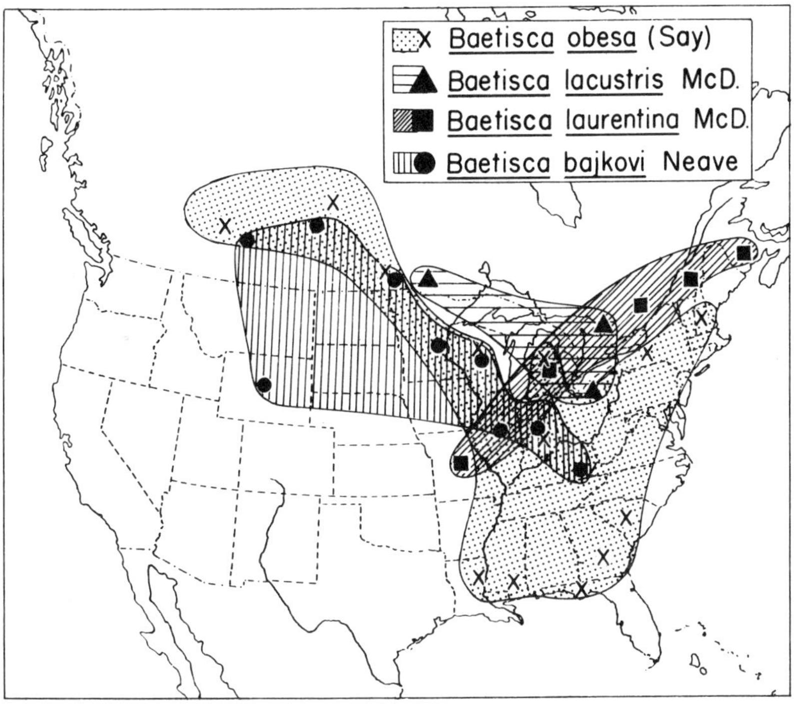

Figure 1. Presently known distribution of the species of *Baetisca* in North America.

two families.

In Edmunds' comprehensive study (1972) of the evolution of Ephemeroptera, a diagram of a probable phylogeny for the families and subfamilies was given. He discussed in detail his reasoning for the derivation as pictured and postulated that parallel evolution must explain some of the similarities existing between the Baetiscidae and Prosopistomatidae. Similar phylogenetic relationships were given by Koss (1973) and Riek (1973) using egg and nymphal characters respectively. Riek is somewhat more reserved, however, about the relationships of these two groups. He stated that although the Prosopistomatidae and Baetiscidae are referred to the same superfamily at present, the differences both in nymphs and adults are such that a more-detailed study may indicate that they are not as closely related as at present assumed. He pointed out that the Prosopistomatidae, in which the venation is very reduced, are the only Recent mayflies in which there is no evidence of a triadic branching of MA. The Baetiscidae, on the other hand, have the most generalized venation of all Recent species;

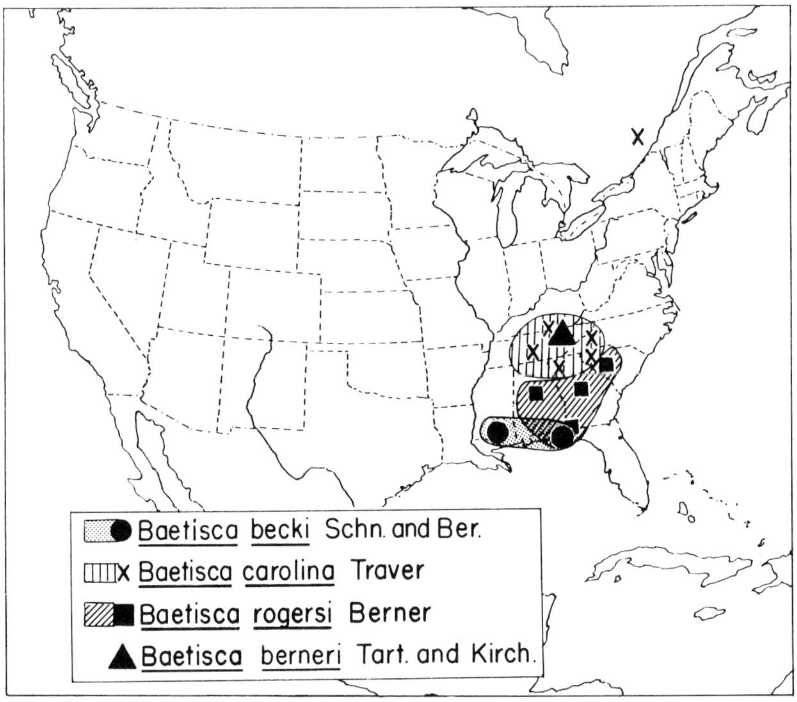

Figure 2. Presently known distribution of the species of *Baetisca* in North America.

the venation differs from that of the Lower Permian Misthodotidae mainly in the basal alignment of the veins. There is also a slight reduction in the anal field.

Edmunds *et al.* (1976) again diagrammatically represented the Baetiscidae and the Prosopistomatidae as being derived from a common ancestor, which are, in turn, related to the Neoephemeridae and the Caenidae. Our conception of phylogenetic relationships of the Baetiscidae are in accord with these. The presumed common origin of the Baetiscidae and the Prosopistomatidae are reflected in the assignment of the two families to the same superfamily, Prosopistomatoidea. McCafferty and Edmunds (1979), in proposing the suborders Schistonota and Pannota, have retained the superfamily Prosopistomatoidea for the Baetiscidae and Prosopistomatidae. We find no convincing data to contradict Edmunds' argument as to the kinship of these two families.

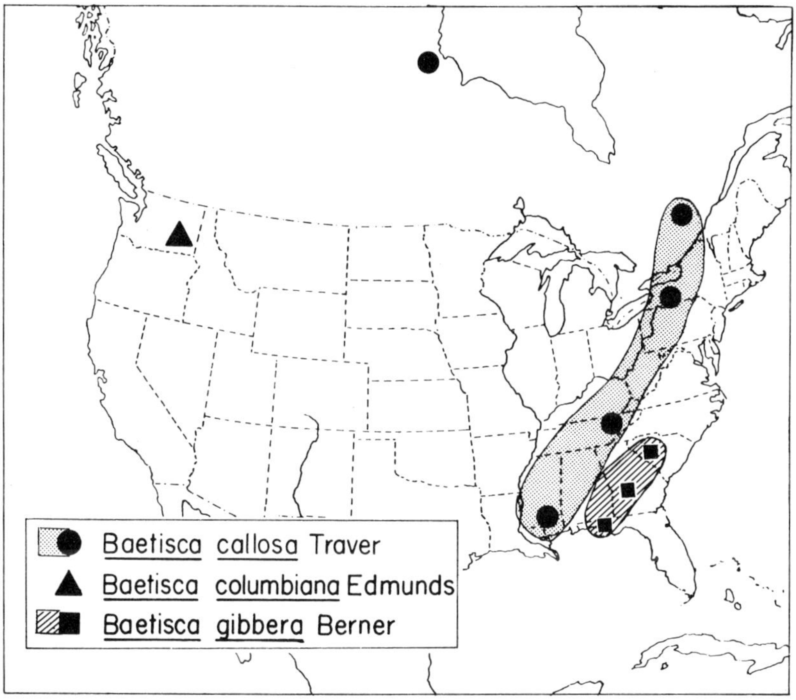

Figure 3. Presently known distribution of the species of *Baetisca* in North America.

TAXONOMY AND DISTRIBUTION

Presently, there are thirteen described species in the genus *Baetisca*, each of which will be treated in detail in Part II of our report on the family. The adults are most readily separated on the basis of wing pigmentation, with the male being the more distinctive sex. There is some likelihood that we shall be able to separate species of subimagos by differences in color pattern in the wings, but further study is necessary before we can determine the validity of this characteristic as a distinguishing one.

The various species of *Baetisca* are most easily characterized by nymphal differences and similarities. The structures we have employed for differentiation include the following:

Head: Shape and size of the frontal projections; shape and size of the genal projections; surface sculpturing.

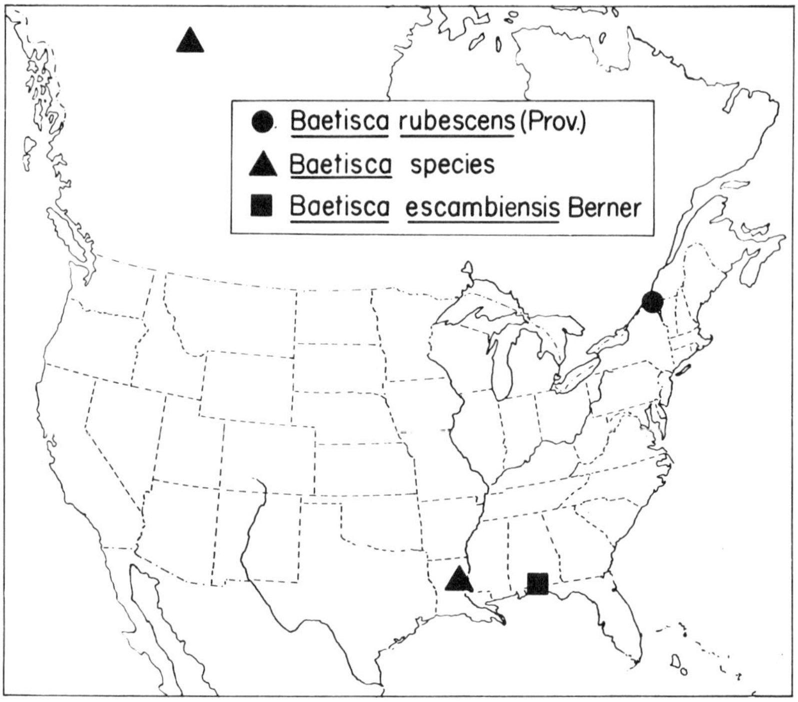

Figure 4. Presently known distribution of the species of *Baetisca* in North America.

Thorax: Lateral and dorsal projections of the mesothoracic shield; sculpturing and mottling pattern dorsally and ventrally.

Abdomen: Postero-median elevations and postero-lateral spines; mottling pattern

Based on our preliminary data, other nymphal characters that show considerable taxonomic value for species identification include: Head - marginal setation; eyes with or without stripes. Thorax - ratio between maximum length and width (including lateral projections) of mesonotal shield; legs, form and shape of femoral setae, setation of tibiae, and color pattern. Abdomen - form and shape of tergal setae and color pattern of caudal filaments.

The species, using nymphal morphology alone, fall roughly into four major groups as shown below. As the nymphal stage of *B. rubescens* is unknown, we have not listed it.

GROUP			
A	B	C	D
B. bajkovi	*B. becki*	*B. callosa*	*B. escambiensis*
B. lacustris	*B. berneri*	*B. columbiana*	
B. laurentina	*B. carolina*	*B. gibbera*	
B. obesa	*B. rogersi*		

The various species are distributed widely over most of the eastern part of North America and into the Great Plains and northwestward to the Northwest Territories of Canada (Wiens *et al.* 1975). No species has yet been reported from the southwestern United States but one, *B. columbiana*, was described from the State of Washington. A single, dubious report of a California collection of *Baetisca obesa* was given by Eaton (1885) but there has been no further substantiation of the occurrence of the genus in this area (Day 1956). We have, therefore, concluded that Eaton's record was in error. Specific distributions of the species will be given in Part II but in general the known species, other than *B. obesa*, have rather restricted ranges. Some of this restriction reflects inadequate collecting rather than localization of the species (Figs. 1 - 4).

ECOLOGY AND LIFE HISTORY

There are only few detailed studies of the ecology and life histories of any species of *Baetisca*. Lehmkuhl (1972) reported on his study of the biology of *B. bajkovi* in Saskatchewan. Here, nymphs hatched in late summer, spent the winter under ice, and adults emerged in June and July of the following year. Pescador and Peters (1974) provided a detailed report of the life history and ecology of the southeastern *B. rogersi* in Florida. They concluded that eggs were dormant during the hot summer months, nymphs appeared in September and continued to be present through June of the following year. Emergence of adults began in April and continued to early July with only a single generation each year. The most recent life history study is that of Chaffee and Tarter (1979); working with *B. bajkovi* in West Virginia, they found it to be univoltine with greatest growth occurring from July to August. Subimagos emerged during May.

Most species of *Baetisca* are strictly rheophilic, except for *B. lacustris* and *B. obesa*. The nymphs of *B. lacustris* were described as lake dwellers while *B. obesa* can live in both running waters and ponds. Traver (1931) described the nymphal habitats of *B. carolina* and *B. callosa* and stated that the *carolina* nymphs were mostly collected from streams with their beds composed of loose gravel and sand near the quiet edge. Quoting Needham's account of his collecting of *callosa* nymphs in Johnny Cake Run near Mt. Storm,

West Virginia, Traver stated that they were found in a stream only a few inches deep with water flowing among the small stones and gravel. Traver also described the habits of nymphs, subimagos and imagos of *B. carolina*.

A paper by Berner (1955) on the Southeastern species of *Baetisca* gave accounts of the nymphal habitats of *B. escambiensis*, *B. gibbera* and *B. obesa* and included brief notes on the nymphal habits of *escambiensis* and *obesa*. Berner observed that *escambiensis* nymphs were collected most abundantly from river sandbars where an admixture of clay in the sand prevailed, the water current was relatively slow, and algal growth over the bottom was minimal. The nymphs of *gibbera* were taken, from rather swiftly flowing streams, clinging to the underside of logs anchored at the bank or in areas where there was a thin overlaying layer of small pebbles. Compared to Walsh's observation of *obesa* nymphs in which he noted that they frequent rapidly-running rivers and attached themselves in repose to the undersurfaces of submerged stones, Berner observed a totally different sort of habitat of the species. He found the nymphs to occur almost invariably in very slow to almost stagnant water where they may be very common in moss which is attached to submerged trunks of trees, or other vegetation. Recently Pescador collected several nymphs of *obesa* from residual pools in flooded areas adjacent to the Chipola River in northwestern Florida.

The nymphs of *B. rogersi* usually inhabit gravel beds in sand bottom creeks (Berner 1940, 1950). Pescador and Peters (1974) observed that mature *rogersi* nymphs were typical members of a lithophilous association, living in the exposed, stony substratum section of the stream. Younger nymphs in the fourth through seventh instar live only in areas with a thick growth of filamentous algae and filamentous moss. Nymphs approaching emergence moved to quiet shallow sections of the stream, presumably to search for objects or places to leave the stream and emerge. Pescador and Peters also comprehensively discussed the habits of nymphs, subimagos and imagos of the species.

Nymphs of *B. becki* were observed by Schneider and Berner (1963) to occur in a swiftly flowing, clear, shallow, sand-bottomed stream under six to ten inches of water. Pescador and Peters (1971) collected *becki* nymphs from swifter, deeper water with more shifting, cleaner sand than nymphs of *rogersi* living in the same river. In a study of mayflies of Illinois, Burks (1953) observed that the nymphs of *B. bajkovi* developed in fairly rapid creeks and moderate-sized rivers. The early instar nymphs occur in the swift current, while the late instar nymphs migrate to the comparatively still eddies along the banks. Lehmkuhl (1972) stated that while nymphs of *B. bajkovi* are frequently collected in silty areas, their normal microhabitat appears to be solid substrate.

Except for McDunnough's (1932) brief comments on collecting numerous nymphal skins of *B. lacustris* found along the shores of Lake Erie and a mature nymph found in a pool, no other published accounts of the nymphal habitat of this species are known.

Leonard and Leonard (1962) briefly described the habitat of *B. laurentina* nymphs. The nymphs seem to prefer medium-sized to large rivers, where they partially submerge themselves in the flocculent surface layer of silt beds in quiet areas near the bank. In Wisconsin, Hilsenhoff (1975) found the nymphs of *Baetisca* in sandy streams where they inhabit sand and silt near the banks.

Nymphs of the recently described *B. berneri* seemed to prefer slow to medium water speeds with depths ranging from 15 to 46 cm, and a substrate mixture of sand, gravel, and small stones (Tarter and Kirchner 1978).

RESUME

Les auteurs examinent l'histoire de la taxonomie et de la phylogénie de la *Baetisca*, l'unique genre de la famille des baetiscidés, en décrivant d'abord le *Prosopistoma* en tant que crustacé. Ils traitent de la parenté qui existe entre la famille des prosopistomatidés et celle des baetiscidés et se déclarent d'accord avec la conclusion d'Edmund a l'effet que les deux familles descendent d'un ancêtre commun et que les similitudes sont sans doute dues au fait que leur évolution s'est effectuée parallèlement. L'ouvrage contient un précis des espèces présentement connues et des traits qui différencient les larves. Il contient également un aperçu de l'écologie et du cycle de vie des diverses espèces.

ZUSSAMENFASSUNG

Die Taxonomie und Phylogenie von *Baetisca*, der einzigen Gattung der Familie Baetiscidae, wird im geschichtlichen Überblick dargestellt, beginnend mit der Beschreibung der Nymphe von *Prosopistoma* als Krebstierchen. Es wird sowohl die verwandtschaftliche Beziehung der Prosopistomatiden zu den Baetisciden behandelt, als auch die Übereinstimmung der Autoren mit Edmunds' Schlußfolgerung, daß die beiden Familien auf eine gemeinsame Ahnenform zurückzuführen sind. Einige ihrer Ähnlichkeiten müssen jedoch durch parallele Evolution erklärt werden. Daran schließt sich eine Übersicht der zur Zeit anerkannten Arten und Charakteristiken, nach denen die Nymphen unterschieden werden. Auch eine kurze Behandlung der Ökologie und Lebensgeschichte verschiedener Arten ist beigefügt.

REFERENCES

Berner, L. 1940. *Baetisca rogersi*, a new mayfly from northern Florida. *Can. Entomol.* 72: 156-160.

Berner, L. 1950. The mayflies of Florida. Univ. Fla. Press, Gainesville. (*Biol. Sci. Ser.* IV, No. 4).

Berner, L. 1955. The southeastern species of *Baetisca* (Ephemeroptera: Baetiscidae). *Q. J. Fla Acad. Sci.* 18(1): 1-19.

Berner, L. 1959. A tabular summary of the biology of North American mayfly nymphs (Ephemeroptera). *Bull. Fla State Mus.* 4(1): 1-58.

Burks, B.D. 1953. The mayflies or Ephemeroptera, of Illinois. *Bull. Ill. Nat. Hist. Surv.* 26(1): 1-216.

Chaffee, D.L., and D.C. Tarter. 1979. Life history and ecology of *Baetisca bajkovi* Neave in Beech Fork of Twelvepole Creek, Wayne County, West Virginia (Ephemeroptera). (Abstract). *ASB (Assoc. Southeast. Biol.) Bull.* 26(2): 87.

Clemens, W.A. 1913. New species and new life histories of Ephemeridae or mayflies. *Can. Entomol.* 45: 246-262; 329-341.

Clemens, W.A. 1915. Rearing experiments and ecology of Georgian Bay Ephemeridae. *Contr. Can. Biol.* 4: 113-128.

Daggy, R.H. 1945. New species and previously undescribed naiads of some Minnesota mayflies (Ephemeroptera). *Ann. Entomol. Soc. Am.* 38: 373-396.

Day, W.C. 1956. Ephemeroptera pp. 79-105. *in* R.L. Usinger, ed. Aquatic insects of California. Univ. Calif. Press, Berkeley.

Demoulin, G. 1956. Les Baetiscidae dans la classification des Ephéméroptères. *Bull. Inst. R. Sci. Nat. Belg.* 32(35): 1-4.

Demoulin, G. 1969. Remarques critiques sur la position systématique des Baetiscidae et des Prosopistomatidae (Ephemeroptera). *Bull. Inst. R. Sci. Nat. Belg.* 45(17): 1-17.

Eaton, A.E. 1871. A monograph of the Ephemeridae. *Trans. Entomol. Soc. Lond.* pp. 1-164.

Eaton, A.E. 1885. A revisional monograph of recent Ephemeridae or mayflies. *Trans. Linn. Soc. Lond. Ser.* 2, 3: 153-230.

Edmunds, G.F., Jr. 1960. The mayfly genus *Baetisca* in western North America (Ephemeroptera: Baetiscidae). *Pan-Pac. Entomol.* 36: 102-104.

Edmunds, G.F., Jr. 1972. Biogeography and evolution of Ephemeroptera *Annu. Rev. Entomol.* 17: 21-42.

Edmunds, G.F., Jr., S.L. Jensen, and L. Berner. 1976. The mayflies of North and Central America. Univ. Minnesota Press, Minneapolis.

Fontaine, J. 1958. Les affinités systématiques du genre *Prosopistoma* Latrielle (Ephéméroptère). Deuxieme note: Le genre *Baetisca* Walsh: les formes Ailees. *Bull. Mens. Soc. Linn. Lyon* 27: 133-140.

Gillies, M.T. 1954. The adult stages of *Prosopistoma* Latreille (Ephemeroptera) with descriptions of two new species from Africa. *Trans. R. Entomol. Soc. Lond.* 105: 355-372.

Hagen, H. 1861. Synopsis of the Neuroptera of North America with a list of South American species. *Smithson. Misc. Collect.* 4: 38-55.
Hilsenhoff, W.L. 1975. Aquatic insects of Wisconsin with generic keys and notes on biology, ecology, and distribution. *Tech. Bull. Dep. Nat. Resour., Wisc.* 89: 1-53.
Joly, E. 1880. Sur la nymphe du genre d'Ephémérines *Baetisca* Walsh. *Bull. Soc. Etudes Sci. Angers.* 8: 157-173.
Koss, R.W. 1973. The significance of the egg stage to taxonomic and phylogenetic studies of the Ephemeroptera. pp. 73-78. *in:* Proc. First. Intern. Confer. on Ephemeroptera. E.J. Brill, Leiden.
Lameere, A. 1917. Etude sur l'évolution des Ephémères. *Bull. Soc. Zool. Fr.* 42: 41-48.
Landa, V. 1969. Comparative anatomy of mayfly larvae (Ephemeroptera). *Acta Entomol. Bohemoslov.* 66: 289-316.
Landa, V. 1973. A contribution to the evolution of the order Ephemeroptera based on comparative anatomy. pp. 155-178. *in:* Proc. First Intern. Confer. on Ephemeroptera. E.J. Brill, Leiden.
Latreille, M. 1833. Description d'un nouveau genre de Crustacés. *Nouv. Ann. Mus. Hist. Natur.* 2: 23-24.
Lehmkuhl, D.M. 1972. *Baetisca* (Ephemeroptera: Baetiscidae) from the western interior of Canada with notes on the life cycle. *Can. J. Zool.* 50: 1015-1017.
Leonard, J.W., and F.A. Leonard. 1962. Mayflies of Michigan trout streams. Cranbrook Inst. Sci. Bloomfield Hills, Mich.
McCafferty, W.P., and G.F. Edmunds, Jr. 1979. The higher classification of Ephemeroptera and its evolutionary basis. *Ann. Entomol. Soc. Am.* 72: 5-12.
McDunnough, J. 1921. Two new Canadian mayflies. *Can. Entomol.* 53: 117-120.
McDunnough, J. 1932. New species of North American Ephemeroptera II. *Can. Entomol.* 64: 209-215.
Neave, F. 1934. A contribution to the aquatic insect fauna of Lake Winnipeg. *Int. Rev. gesamten Hydrobiol.* 31: 157-170.
Needham, J.G. 1905. Mayflies and midges of New York. *N.Y. State Mus. Bull.* 86: 17-62.
Needham, J.G., J.R. Traver, and Yin-Chi Hsu. 1935. The biology of mayflies. Comstock Publ. Co., Ithaca.
Pescador, M.L. and W.L. Peters. 1971. The imago of *Baetisca becki* Schneider and Berner (Baetiscidae: Ephemeroptera). *Fla. Entomol.* 54: 329-334.
Pescador, M.L. 1974. The life history and ecology of *Baetisca rogersi* Berner (Ephemeroptera: Baetiscidae). *Bull. Fla. State Mus., Biol. Sci.* 17: 151-209.
Provancher, L. 1876. Petite faune entomologique du Canada. Néuroptères. Fam. III. Ephémérides. Ephemeridae. *Nat. Can. (Que.)* 8: 264-268.
Provancher, L. 1877. Petite faune entomologique du Canada et

particulament de la Province de Quebec. *Nat. Can.* (Que.) 9: 80-84.
Riek, E.F. 1973. The classification of the Ephemeroptera pp. 160-178. *in:* Proc. First Intern. Conf. on Ephemeroptera. E.J. Brill, Leiden.
Say, T. 1939. Descriptions of new North American neuropterous insects, and observations on some already described. *J. Acad. Nat. Sci. Phila.* 8: 9-46.
Schneider, R.F., and L. Berner. 1963. A new southeastern species of *Baetisca* (Ephemeroptera: Baetiscidae). *Fla. Entomol.* 46: 183-187.
Spieth, H.T. 1933. The phylogeny of some mayfly genera. *J. N.Y. Entomol. Soc.* 41: 55-86: 327-391.
Tarter, D.C., and R.F. Kirchner. 1978. A new species of *Baetisca* from West Virginia (Ephemeroptera: Baetiscidae). *Entomol. News* 89: 209-213.
Traver, J.R. 1931. The ephemerid genus *Baetisca*. *J. N.Y. Entomol. Soc.* 39: 45-66.
Traver, J.R. 1932. Mayflies of North Carolina. *J. Elisha Mitchell Sci. Soc.* 47: 85-161.
Traver, J.R. 1937. Notes on mayflies of the southeastern States (Ephemeroptera). *J. Elisha Mitchell Sci. Soc.* 53: 27-86.
Ulmer, G. 1920. Übersicht über die Gattungen der Ephemeropteren, nebst Bemerkungen über einzelne Arten. *Stett. Entomol. Zeit.* 81: 97-144.
Vayssière, A. 1882. Recherches sur l'organisation des larves des Éphémérines. *Ann. Sci. Nat.* (6): *Zool.* 13(1): 1-137.
Vayssière, A. 1934. Etude anatomique des larves nymphales des *Baetisca obesa* et *carolina*. *Ann. Sci. Nat. Zool. Ser.* 10, 17: 381-406.
Walker, F. 1853. List of the specimens of neuropterous insects in the collection of the British Museum. Part III. (Termitidae-Ephemeridae). British Museum (Natural History)., London. (pp. 533-585).
Walsh, B.D. 1862. List of the Pseudoneuroptera of Illinois contained in the cabinet of the writer, with the description of over 40 new species, and notes on their structural affinities. *Proc. Acad. Sci. Phila. Ser.* 2: 361-402.
Walsh, B.D. 1863. Observations on certain N.A. Neuroptera, by H. Hagen, M.D., of Koenigsberg, Prussia; translated from the original French Ms., and published by permission of the author, with notes and descriptions of about twenty new N.A. species of Pseudoneuroptera. *Proc. Entomol. Soc. Phila.* 2: 167-179.
Walsh, B.D. 1864. On the pupa of the ephemerinous genus *Baetisca* Walsh. *Proc. Entomol. Soc. Phila.* 3: 200-206 .
Wiens, A.P., D.M. Rosenberg, and N.B. Snow. 1975. Species list of aquatic plants and animals collected from Mackenzie and Porcupine river watershed from 1971-1973. *Can. Fish. Mar. Serv. Tech. Rep.* 557: 1-39.

PREMIERE DESCRIPTION D'UN EPHEMEROPTERE EN ANGLETERRE : L'EPHEMERON DE HARRIS (1730 - 1788)

Lucie Arvy

Laboratoire d'Histo-enzymologie de la Faculté de
Médecine de Paris, 45 rue des Saints-Pères
75006 Paris, France

RESUME

Il ya a deux siècles, M. Harris remarqua un éphéméroptère posé contre sa fenêtre et le signala sous le nom d'éphéméron; cinquante quatre ans plus tard, J. Curtis le décrivit sous le nom de *Brachycercus harrisella.*

"The writings of Moses Harris are, on the whole, much scarcer and much less known than is generally imagined".

Hagen, 1861

Admirateur de son oncle, le jeune Moses s'habitua, dès l'âge de 12 ans, à collectionner et décrire les insectes londoniens; ses travaux sont restés longtemps si peu répandus que les bibliothèques des Sociétés, linnéenne et royale, ne les possédaient pas, ils ont pourtant été imprimes à répétition; la première édition date de 1766: *"The natural history of English insects namely Moths and Butterflies together with the plants on which they feed".* Dès 1776, la monographie de Harris parut avec un titre francais, le texte étant disposé sur deux colonnes: une colonne gauche en anglais et une colonne droite en français: *Une exposition des Insectes anglais, avec des observations et des remarques curieuses dans lesquelles chaque Insecte est particulièrement décrit; ses parties et ses propriétés sont considérés, leurs sexes distingués, et leur histoire naturelle fidèlement récitée. Le tout enrichi des tailles douces, dessinées, gravées et colorées par l'auteur:* (Robson et Dilly édit.); dès la page 24, Harris écrivait: "Le dessus deploye ses ailes environ un demi-pouce. Cet insecte est grossi à la Fig. 1 comme vu au microscope. Il n'a point d'antennes, langue, machoires ou trompe que je pouvais découvrir. La tête est plate et fort mince. Les yeux sur le dessus de la tête paraissent en forme de croissants. La tête est immobile. Le corcelet est bossu et gros. L'abdomen est composé de neuf anneaux à l'extrémité desquels il se trouve trois queues comme des poils. Les pieds au nombre de six sont longs et déliés. Les ailes sont deux, grandes et étendues, d'une couleur blanche obscure, fort minces et transparentes. Je ne scai pas à quel genre remettre cet insecte comme les caractères generiques ne répondent à aucun dans le système de Linné mais en mon opinion, il est un ephemeron. Je lay pris volant sur ma fenêtre". (Fig. 1 et 2).

De fréquentes éditions suivirent: en 1778, en 1782, en 1789 et en 1794 (cette dernière sous la direction de la "Société entomologique" de Londres). Beaucoup plus tard, Westwood (1840) veilla encore à une édition de l'histoire naturelle de Harris. L'examen de ces éditions successives persuade que Harris fut le premier entomologiste anglais à faire une étude appliquée "of the museum beetle and its depradations" (Wilkinson 1968).

Très généralement ignoré, Harris fut remarqué par Curtis (1834), lors de sa descrition des éphéméroptères anglais, en *Ephemera* Linn., *Baetis* Lea., *Cloëon* Lea., et *Brachycercus* Curt.; Curtis dédia à Harris "l'éphéméron" qu'il avait vu sur le bord de sa fenêtre. Il donna de *Brachycercus harrisella* la description suivante

"Head short; eyes small and remote, at least in one sex, thorax large and ovate; abdomen not longer than the thorax, terminated by 3 very short setae, thick at the base : wing 2, rather short but ample the form of a Musca, with many longitudinal nervures and 3

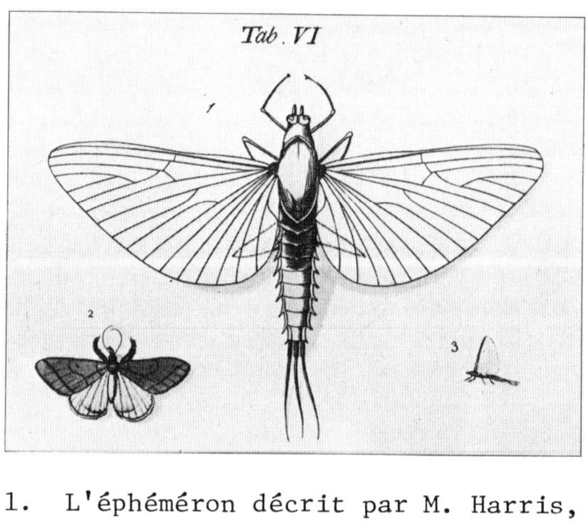

Figure 1. L'éphéméron décrit par M. Harris, en 1766.

EXPOSITION
OF
English Insects
(Including the several Classes of Neuroptera, Hymenoptera, & Diptera, or Bees, Flies, & Libellulae)
EXHIBITING on 51 COPPER PLATES near 500 FIGURES,
accurately drawn, & highly finished in Colours, from Nature.
The Whole
minutely Described, Arranged, & Named according to the
LINNEAN SYSTEM with REMARKS.
The Figures of a great number of MOTHS, not in the Aurelian Collection, formerly published by the same Author, and a Plate with an explanation of Colours, are likewise given in the Work.
BY
Moses Harris.
London:
Sold by Mr White, Bookseller, in Fleet street, & Mr Robson, in New Bond street.
MDCCLXXXII

Figure 2. Harris (1782), Page de titre.

transverse ones between the costa and disc; legs short, at least in one sex. Obs. These characters, although imperfect for want of specimens, are sufficient to distinguish this group for the present."

En 1839, Burmeister fit de l'éphéméron de Harris un nouveau genre *Oxycypha* et Bengtsson (1917) l'imita, en créant le genre *Eurycaenis;* cependant Eaton (1892) ignore l'éphéméron de Harris (Fig. 1).

Lorsque Curtis eut créé son genre *Brachycercus* pour une forme à cerques courts, Stephens (1835) ne soupçonna pas qu'une forme à

Figure 3. Moses Harris

cerques longs pouvait être le ♂ de la précédente et il en fit un *Coenis*, de sorte que le *Brachycercus* de Curtis fut abandonné. Par la suite, Campion (1923) fit la critique de cette systématique et Lestage (1924) proposa d'accepter une famille de Brachyceridae.

ABSTRACT

Two centuries ago M. Harris noticed a mayfly resting on his window and classified it as an "ephemeron"; fifty-four years later, J. Curtis described it under the name *Brachycercus harrisella*.

ZUSSAMENFASSUNG

Vor zweihundert Jahren bemerkte M. Harris eine Eintagsfliege an seinem Fenster und bezeichnete sie mit dem Namen "ephemeron". Vierundfünfzig Jahre später beschrieb J. Curtis diese Fliege unter dem Namen *Brachycercus harrisella*.

BIBLIOGRAPHIE

Bengtsson, S. 1917. Weitere Beitrage zur Kenntnis der nordischen Eintagsfliegen. *Entomol. Tidskr.* 38, 186.

Burmeister, H. 1839. Handbuch der Entomologie. Enslin édit., Berlin, 2: 796-797.

Campion, H. 1923. On the use of the generic name *Brachycercus*. *Ann. Mag. nat. History,* 11(64), 515.

Curtis, J. 1834. Descriptions of some nondescript British species of Mayflies of Anglers. *Philos. Mag. J. Sci.,* 4, 1-11.

Eaton, A.E. 1892. Insecta-Neuroptera-Ephemeridae. *Biologia Centrali-Americana,* 1-16.

Goffe, E.R. 1946. The Syrphidae (Diptera) of Moses. *Entomol. Mon. Mag.,* 82, 67-86.

Hagen, H., and M. Harris. 1861. *Entomol. Weekly Intel.,* 10: 181-3.

Hagen, H., and M. Harris. 1862. *Bibl. entomol.,* Leipzig p. 341-2.

Harris, M. 1766. The Aurelian : or Natural History of English Insects: namely, Moths and Butterflies together with the plants on which they feed...London.

Harris, M. 1776. An exposition of english Insects with curious observations... Robson édit.,

Harris, M. 1782. *An exposition of English Insects.* White (éd).

Harris, M. 1840. The Aurelian: or natural history of English Insects... New Edition J.O. Westwood (ed.) London.

Lestage, J.A. 1924. A propus du genre *Coenis* Steph. *Brachycercus* Curt. (Ephemeroptera). *Ann. Soc. entomol. Belg.* 64, 61-62.

Lisney, A.A. 1960. A bibliography of British Lepidoptera 1608-1799, p. 156-75, 6 fig. Chiswick édit., London.

Needham, J.G., J.R. Traver and Y.C. Hsu. 1969. *Brachycercus* (Baetidae), p. 639-42. *in:* The biology of mayflies with a systematic account of North American species. Classey, Hampton.

Stephens, J.F. 1835. Illustrations of British Entomology, or a synopsis of indigenous insects: containing their generic and specific dictinctions. Baldwin Cradock, London.

Wilkinson, R.S. 1968. English entomological methods in the seventeenth and eighteenth centuries. Part III: Moses Harris' the Aurelian. *Entomol. Rec.,* 80, 193-200.

PICTET ET EATON: LES PREMIERS SPECIALISTES DES EPHEMEROPTERES.

PICTET AND EATON: THE FIRST MAYFLY SPECIALISTS

Janice Peters, Lucie Arvy*, and William L. Peters

Laboratory of Aquatic Entomology
Florida A&M University
Tallahassee, Florida 32307 USA

ABSTRACT

Brief biographies of François Jules Pictet and Alfred Edwin Eaton are presented. These two distinguished 19th century workers were first to specialize in the study of Ephemeroptera and their monographs are the basis for all modern work.

INTRODUCTION

Dans les revues qui rapellent l'histoire de l'entomologie, (telle celle de Lindroth 1973), il est admis que M.P. Rambur devait être reconnu comme le premier spécialiste des Névroptères (en accord avec Linné). Néanmoins, Rambur ne fut pas réelement intéressé par les éphémérides, car il les trouvait trop inconsistants et trop fragiles pour les manipuler, de sorte qu'il ne leur consacra que 8 pages dans sa monographie. Il suggéra qu'ils devaient être étudiés vivants et devaient représenter plusieurs genres car ils étaient très différents les uns des autres. Introduisant la systématique des Ephémérides dans son "Histoire naturelle des Insectes Névroptères" (1842) il précisa: "...je me bornerai à faire connaître un très-petit nombre d'espèces, dans la crainte de donner des descriptions imparfaites et inexactes."

* Laboratoire d'Histoenzymologie, Faculté de Médecine, 45, rue des Saints-Pères, 75006 Paris, France.

En fait, ce fut F.J. Pictet qui trouva que ce groupe négligé était non seulement curieux mais encore important, de sorte qu'il publia la première monographie au monde sur les éphéméroptères (texte en 1843 et planches et 1845). Pictet s'intéressa à l'anatomie et à la structure de l'adulte et des nymphes; il fut le premier à distinguer les caractéristiques des nymphes d'après leurs modes de vie (larves: fouisseuses, plates, nageuses, rampantes). Cependant la section systématique de la monographie de Pictet se limita aux espèces qu'il avait lui-même étudiées.

Un peu plus tard Eaton commença l'étude des éphéméroptères anglais; peu après il étudia la nomenclature dans le monde entier, en correspondant avec tous les chercheurs connus et en visitant la plupart des Musées, les plus importants; ainsi de nombreuses fois, il fit une description très claire de toutes les espèces connues au monde, en une seule monographie (1883-1888).

Ni Pictet, ni Eaton ne partirent de la promotion linnéenne des Névroptères et crurent que cette famille pouvait rentrer dans cet ordre. Cependant, Pictet (en accentuant leur originalité morphologique) et Eaton (en mettant en ordre leurs sections et leurs séries) ont dressé un ordre moderne et une classification en familles acceptable. Il semble donc tout-à-fait souhaitable que les réalisations de Pictet et de Eaton à la IIIème conférence internationale sur les Ephéméroptères soient bienvenues.

FRANÇOIS JULES PICTET

François Jules Pictet appartenait à une famille d'intellectuels; il naquit à Genève le 27 septembre 1809 et y mourut le 15 mars 1872, assez brusquement, des suites d'une chute sur le verglas, à Berne. L'usage était alors répandu à Genève de distinguer les personnes d'une même famille, quand elle était nombreuse, comme celle des Pictet, par le nom de la femme; ainsi le père de François Jules, qui avait épousé une Baraban, s'appelait Jean Pierre Pictet-Baraban, alors que notre Pictet, qui avait épousé une De La Rive, s'est appelé Pictet de la Rive, depuis son mariage. Son père, qui était magistrat, était ami du naturaliste De Candolle. En 1823, F.J. Pictet se fit inscrire à l'Académie de Genève et en sortit bachelier ès-lettres et en langues anciennes; il suivit alors les cours de la Faculté des Sciences, où de Candolle l'orienta vers la Zoologie et le fit nommer répétiteur du Prince Charles Antoine de Hohenzollern-Sigmarigen. En 1829, il fut docteur ès-sciences et étudia le droit, pendant un an. Sa première publication fut faite en collaboration avec son père; ce dernier, Jean Pierre, avait publié (en 1808) un itinéraire des vallées autour du Mont-Blanc, qui avait eu du succès et, en 1829, aidé de son fils, François Jules, il le remania; très précisément, la carte géographique de cet itinéraire fut dessinée par François Jules. En 1830, il partit

Figure 1. François Jules Pictet. (d'après Soret.)

pour Paris, où pendant 6 mois et apparemment d'accord avec Audoin, il étudia les moeurs des Névroptères, leur anatomie, leur classification. Peu après, il succèda à De Candolle et, en 1859, Claparède lui succèda.

Son premier travail marquant (1841-1842) porta sur les métamorphoses des perles et des phryganes; il y décrivit 120 espèces dont 34 étaient nouvelles pour la Science; ce travail lui valut d'une part le prix Davy et, d'autre part, sa nomination de Professeur à l'Académie de Genève; il publia ensuite la suite de ses recherches sur les Névroptères: "Les Ephémérines" (1843-1845). A la mort d'Audouin, il abandonna cette sorte de recherches et s'orienta vers la Paléontologie. Il s'intéressera d'abord aux Insectes de l'ambre, puis à tous les fossiles: de 1844 à 1846, il publia, en 4 volumes, une énumération de tous les genres et espèces fossiles connus, depuis les infusoires jusq'aux Mammifères; ils eurent tant de succès qu'ils furent réédités de 1853 à 1857. Les études sur les gisements fossilifères furent faites avec de nombreux collaborateurs: Renevier, Gaudin, De la Harpe, Humbert, De Loriol, Jaccard. F.J. Pictet adopta la théorie des créations successives, qui suppose que chaque époque géologique s'est terminée par un anéantissement complet des espèces qui la caractérisaient.

Outre ses recherches scientifiques, F.J. Pictet eu de nombreuses autres activités economiques et politiques; il était membre des Commissions de l'instruction publique, membre de l'Assemblée constituante, recteur de l'Académie de Genève (de 1847 à 1850 et encore de 1866 à 1868). Pour tous les éphéméroptérologistes, sa monographie sur les éphémérines reste, après 140 ans, d'une importance fondamentale; F.J. Pictet a bien mérité de l'éphéméroptérologie, il en a souligne l'importance et les difficultés; les grandes différences, qui existent parfois entre les stades subimago et imago, ou entre les mâles et les femelles, font qu'il est presque fatal d'établir, à tort, des espèces nouvelles si on ne connait pas toutes les especes, sous toutes leurs formes; c'est pourquoi Pictet a donné une illustration non seulement des insectes mais des pièces (buccales, oculaires, antennaires, génitales, etc...) utilies aux systématiciens.

ALFRED EDWIN EATON

Alfred Edwin Eaton (1 December 1844 - 23 March 1929), eulogized as the "monographer of the may-flies," was a clergyman by profession. Born and raised in Little Bridy, Dorset, he showed an early interest in natural history and joined the Entomological Society of London while a student at Cambridge University. He obtained his B.A. degree in 1868 (M.A. in 1871), was ordained deacon in 1869 and priest in 1870, served in 2-4 year posts as curate or vicar of 6 parishes from 1869 to 1892, and then obtained a holiday chaplaincy to Algeria from 1892 to 1897.

He was, at the same time, an entomologist and naturalist, accompanying the arctic explorer B. Leigh Smith on an expedition to Spitzbergen in the summer of 1873 and describing its birds and animal life. His work on the Spitzbergen expedition apparently led to his selection as naturalist for the 1874-1875 British Transit of Venus Expedition to Kerguelen Island and the Cape of Good Hope; results of the expedition were published in the *Philosophical Transactions* (vol. 168, 1879) "in which Eaton's copious and valuable field-notes are presented in detail." Eaton did not work the collections from this expedition, but gave the material to specialists; for Eaton, however, the discovery of wingless Diptera (eventually described by G.H. Verrall) led to a life-long interest in that order.

Eaton was generally interested in all Neuroptera and published his first note on a mayfly in 1865 *(Entomol. Mon. Mag. 2:14)*. Exactly how he came to specialize on mayflies is unknown, but Hagen's paper on British species *(Entomol. Ann.* 1863: 1-36) and Pictet's monograph (1843-1845) probably had an influence, as did his friend in the Entomological Society of London, Robert McLachlan, who gave Eaton his mayfly collections for study. (To McLachlan, mayflies were "unsatisfactory insects from a collector's point of view."). In 1867, McLachlan proposed that the Society compile a catalog of British insects. Only 3 parts of the catalog were completed before the project ended in disputes over taxonomy of Coleoptera. Eaton's "Monograph on the Ephemeridae" (1871), part of this series, was later expanded into the "Revisional monograph..." (1883-1888), the standard for all later mayfly systematics.

Eaton continued to publish occasionally on mayflies, but concentrated most of his research on the moth-flies (Psychodidae), especially during the last 30 years of his life. He had published a revision of the British species in 1894-1895 and intended (as he had done with the mayflies) to produce a monograph on the family. According to Edwards, Eaton last visited the British Museum in 1919 and expressed regret that ill health had prevented him from completing this project. After his death, his wife gave the collection to the British Museum, an acquisition termed "the largest and most important in existence." Eaton had earlier deposited his mayflies in the same museum. His library was given to the Entomological Society of London.

Eaton was well known for his careful attention to detail, and the biological literature of his day is filled with acknowledgments to Eaton as collector (lichens, algae, Thysanura, Hymenoptera, Diptera, etc.). Active in the Entomological Society of London, he served as a member of its council from 1877 to 1879. He seemed always to take notes and was a frequent contributor to the *Entomologist's Monthly Magazine* from 1865 to 1913, a fact that makes it easy to follow his extensive travels and the wide field of his entomological interests. Two notes deserve further mention for the personal character of their information. One is the 1906 paper with Saunders which describes the activity of insects, entomologists, and local residents of Old Castile during a solar eclipse, and the other is an 1896 note on Eaton's arrest in Algeria for collecting without a permit. From the 1896 paper, here is an autobiographical description of the previer systematicist of mayflies: "Dressed like a dirty tramp, without a collar, wearing a hat that had seen better days" ...with "six feet of 'sugar cane' (bamboo) with brass fittings, a sun-umbrella, square straw basket, muslin nets not strong enough for fish... clearly a suspicious character, or wandering lunatic."

RESUME

Cette ouvrage présente des biographies courtes du François Jules Pictet et Alfred Edwin Eaton. Ces deux entomologistes éminents du XIX siècle étaient les premiers spécialistes des ephéméropterès et leur monographies forment le base pour toutes ouvrages modernes.

ZUSSAMENFASSUNG

Die Biographien von François Jules Pictet und Alfred Edwin Eaton werden in kurzer Form dargestellt. Diese hervorragenden Forscher des 19.Jahrhunderts waren die ersten Ephemeropteren-Spezialisten, und ihre Monographien sind die Grundlage aller modernen Untersuchungen.

BIBLIOGRAPHY

Anonymous. 1872. [obituary, F.J. Pictet] *Entomol. Mon. Mag.* 8: 294-295.

Anonymous. 1929. [obituary, A.E. Eaton] *Entomol. Mon. Mag.* 65: 112.

Anonymous. 1929. Recent acquisitions of the British Museum. *Science (Wash. D.C.)* 70: 573.

Anonymous. 1930. [obituary, A.E. Eaton] *Entomol. News* 41: 63.

Eaton, A.E. 1871. A monograph on the Ephemeridae. *Trans. Entomol. Soc. Lond.* 1871: 1-164.

Eaton, A.E. 1883-1888. A revisional monograph of recent Ephemeridae or mayflies. *Trans. Linn. Soc. Lond., Sec. Ser. Zool.* 3: 1-352, 65 plates.

Eaton, A.E. 1896. An English entomologist arrested in Algeria. *Entomol. Mon. Mag.* 32: 185-186.

Eaton, A.E. and E. Saunders. 1906. Aculeate Hymenoptera from Burgos, Old Castile, collected during the total eclipse of the sun, August 30th, 1905, with a list of the species obtained. *Entomol. Mon. Mag.* 42: 96-98.

Edwards, F.W. 1929. [obituary, A.E. Eaton] *Entomologist (Lond.)* 62: 167-168.

Lindroth, C.H. 1973. Systematics specializes between Fabricius and Darwin: 1800-1859, p. 119-154, In: Smith, R.F., T.E. Mittler and C.N. Smith (ed.), History of Entomology. Annual Reviews, Inc. Palo Alto.

Neave, S.A. 1933. The History of the Entomological Society of London, 1883-1933. Entomological Society London, 224 p.

Pictet, F.J. 1843-1845. Histoire naturelle générale et particulière des Insectes Névroptères. Famille des Ephémérines. Baillière édit., Paris (Kessmann et Cherbuliez édit., aussi à Genève). 300 p. (1843), 49 planches (1845).

Pictet, F.J. 1846. Traité élémentaire de Paléontologie, ou Histoire naturelle des animaux fossiles considérés dans leurs rapports zoologiques et géologiques. Baillière edit.

Pictet, F.J. et H. Hagen. 1856. Die im Bernstein befindlichen Neuropteren der Vorwelt, p. 41-126, tab. III-VI, In: Berendt,

G.C. (ed), Die im Bernstein befindlichen Organischen reste der Vorwelt. Bd. 2, Abt. 2. Comm. Nicolaischen Buchhandlung, Berlin.

Rambur, M.P. 1842. Histoire naturelle des Insectes. Névroptères. Libr. Encyclopéd. de Roret, Paris.

Soret, J.L. 1872. François Jules Pictet. Notice biographique. Imp. Ramboz et Schuchardt, Genève. p. 77-86 (avec un portrait et la liste des ouvrages). [Mus. Hist. nat. Paris: β2079, 172463, 172177]

Venn. J.A. 1944. Alumni Cantabrigienses. Part II. from 1752 to 1900. Cambridge Univ. Press. Vol. II, p. 377.

Registrants at Conference

1. D.G. Cobb 2. J. Fontaine 3. K.E. Marshall 4. P.M. Grant
5. I. Campbell 6. A. Suter 7. F.P. Ide 8. R.E. Peters
9. H.F. Clifford 10. D.D. Williams 11. P.T.P. Tsui
12. J.F. Flannagan 13. J. Peters 14. A. Kownacki 15. H.J. Schoonbee
16. C. Edmunds 17. L. Berner 18. I. Müller-Liebenau 19. P.J. Suter
20. L.D. Corkum 21. K.E. Gibbs 22. A.P. Wiens 23. M.D. Hubbard
24. K.J. Aanes 25. M.K. Friesen 26. B. Nagell 27. S.L. Leonhard
28. J.V. Ward 29. P.M. Laufersweiler 30. P.S. Maitland
31. R.S. Harvey 32. J.S. Mattice 33. R.K. Allen 34. G.F. Edmunds, Jr. 35. S. Cianciara 36. R. Sowa 37. M.C. Zimmerman
38. K.F. Whelan 39. J.A. Mathias 40. J.D. Agnew 41. M.L. Pescador
42. D.M. Rosenberg 43. W.L. Peters 44. R.W. Flowers 45. D.T. Mulvey
46. R.D. Wickstrom 47. B.W. Sweeney 48. J.J.H. Ciborowski
49. R. Hartland-Rowe 50. T.J. Fink 51. J.E. Brittain

NOT IN PHOTOGRAPH: D. Allan, L. Arvy, E.C. Gyselman, W. Hominick, Kathleen Maitland, B.E. Townsend

Registered but not present: M. Keffermüller, D.R. Towns

AUTHOR INDEX

This index covers only contributors to this volume. Authors cited are not indexed.

Aanes, K.J. 419-442
Agnew, J.D. 353-366
Allen, R.K. 71-91
Arvy, L. 525-529, 531-537

Berner, L. 169-177, 511-524
Bishop, J.E. 381-403
Brittain, J.E. 179-186

Campbell, I.C. 297-308
Cianciara, S. 265-276
Ciborowski, J.J.H. 321-330
Clifford, H.F. 331-341, 503-509
Cobb, D.G. 155-166
Corkum, L.D. 321-330, 331-341

Edmunds, C.H. 277-285
Edmunds, G.F., Jr. 277-285

Fink, T.J. 367-380
Flannagan, J.F. 155-166, 287-296, 443-455, 457-465
Flowers, R.W. 93-102
Fontaine, J. 201-210
Friesen, M.K. 287-296, 457-465

Gibbs, K.E. 167-168
Gyselman, E.C. 309-319

Harvey, R.S. 211-230
Hominick, W.M. 491-502
Hubbard, M.D. 19-31

Keffermüller, M. 115-121
Kownacki, A. 405-418
Kukalová-Peck, J. 19-31

Laufersweiler, P.M. 287-296
Lawrence, S.G. 457-465
Leonhard, S.L. 457-465

Maitland, P.S. 123-140
Marshall, K.E. 467-489
Müller-Liebenau, I. 103-114

Nagell, B. 259-264

Pescador, M.L. 43-56, 511-524
Peters, J. 531-537
Peters, W.L. 33-41, 43-56, 57-69, 531-537

Rosenberg, D.M. 443-455

Schoonbee, H.J. 343-352
Sowa, R. 141-154
Suter, P.J. 381-403
Swanepoel, J.H. 343-352
Sweeney, B.W. 211-230

Towns, D.R. 57-69

Vannote, R.L. 211-230

Ward, J.V. 169-177
Welch, H.E. 491-502
Whelan, K.F. 187-199
Wiens, A.P. 443-455
Williams, D.D. 1-17
Wissing, T.E. 231-257

Zimmerman, M.C. 231-257

TAXONOMIC INDEX

This index covers only taxa of Ephemeroptera. For other taxa see Subject Index. Pages bearing illustrations are printed in italics.

Acentrella 117,119
  sinaica 117
Acerella 71,82,86
  longicauda 82
  percultα 73
  uenoi 73
Adenophlebiodes
  burgeoni 282
Afronurus 94,98
Ameletopsinae 514
Ameletopsis 304
Ameletus 162,172,173,282,409, 443,446,447,449,451,495
  inopinatus 128-131,145,146, 179,405,408-410,413,416, 436-438
  sparsatus 162
  velox 172
Ametropodinae 505
Ametropus
  fragilis 142,151
Analetris 22
Aprionyx 45,47
Arthroplea
  bipunctatum 162
  congener 127,128
Arthropleinae 162
Atalomicria 45,47,66
Atalonella 43-47,*48*,*49*,50,52
  inconspicua 384,385,387
Atalophlebia 45,47,66
  australasica 387
Atalophlebiinae NEW SUBFAMILY 38,39
Atalophlebioidea 60
Atalophlebioides 47,50,53,58-61, *63*,*64*,65
  auklandensis 58
  cromwelli 58
  sepia 47,50,58-61,*63*,*64*,66
Attenella 71,84
  attenuata 84
Attenuatella 84
Baetidae 4,5,7,34,115-121,128, 162,171,294,446,494,496, 497,503-506
Baetis 2,4,5,8-11,103,113,115, 116,119,120,162,204,207,282, 294,305,327,331-339,408,413-415,435,439,443,446,447,449, 451,496,503,505,506,508
  alpinus 145,146,148,405,407-416
  atrebatinus 128,130
  beskidensis 148,149
  bicaudatus 162,172,173
  buceratus 128-131,149,151
  bundyae 155
  calcaratus *116*, 149-151
  digitatus 127,128
  flavistriga 160,162
  foemina 155,158,162
  frondalis 7
  fuscatus 128,148,149,151,201
  gracilis 147,149-151
  hageni 310-314,316,317
  harrisoni 343,344,348,349
  horti 294
  insignificans 172
  intercalaris 287,289,293,294, 310-313,315-317
  intermedius 172
  lapponicus 119,155,160-162, 436,437
  levitans 7
  lutheri 146,148,149,410
  macani 179-184
    bundyae 161,162
  melanonyx 146,148,410,412,416
  *muticus*128-131,146,148,149,405, 407-410,413-416,436-438,496
  niger 128-131,147,149,151,436, 437
  parvus 8,10,11,160,162,172
  posticatus 369
  propinquus 162
  pygmaeus 7,160,162,289-294

rhodani 127-131,136,145-149, 151,180,247,305,326,405,407-410,413-416,434-437,439,494, 497,503,505,508
scambus 128-130,146,148,149, 407,408,410,435-437
sinaicus *117,118,119,*120,148, 149,407,408,410
soror 381,384-387,394-401
spinosus 293,294
subalpinus 435-437
tracheatus 147
tricaudatus 172,293,331,508
tricolor 151
vagans 7,162,322-327,339,494, 497,505
vardarensis 147-151
vernus 128,130,145-149,369, 405,408-410,414-416,435-437
Baetisca 511-514,517-519
bajkovi 513,515,519,520
becki 513,516,519,520
berneri 513,516,519,521
callosa 5,513,517,519-520
carolina 513,514,516,519,520
columbiana 513,517,519
escambiensis 513,518,519,520
gibbera 513,517,519,520
lacustris 513,515,519,521
laurentina 513,515,519,521
obesa 163,512,514,515,519,520
rogersi 367,369,370,372,373, 513,516,519,520
rubescens 512,518
thomsenae 513
Baetiscidae 505,511-524
Baetiscoidea 513
Baetodes 71
Baetopus 116
tenellus 151
wartensis 147
Behningia
ulmeri 142,147
Behningiidae 280
Brachycercus 163,282,526-528
harrisella 128,130,151,526
minutus 150,151
pallidus 150,151
Caenidae 128,163,280,283,294, 494,503,505,506,514,516
Caenis 5,163,282,343,344,348,349 408,410,494,506-508,514,528
cuniana 369,373-374
forcipata 294
horaria 127-132,136,149,151
luctuosa 151
macrura 128,130-132
moesta 128-132
rivulorum 128-131,146,148,149
robusta 128-130
tardata 294
Calliarcys 38
Callibaetis 162,172
floridanus 369
pictus 494
pretiosus 494
Campsurus 71,282
Campylocia 282
Caudatella 71,77,78
heterocaudata 77
Centroptilum 71,162
luteolum 128,130,131,146,148, 149,151,410,436-438
nana 150,151
pennulatum 128,130,149
pulchrum 149-151
rufostrigatum 294
shadini 142,147
Cheirogenesia
decaryi 282
Chitonophora 74
sp. *nympha unicolorata* 74
Chloeon
dipterum 247
Choroterpes 38
inornata 172
mexicanus 372,373
picteti 147,149-151
Choroterpides 38
Chromarcyinae 280
Cincticostella 71,82-84,86
nigra 82
insolta 83
thani 83
Cinygma 8,10
Cinygmina 94

Cinygmula 162,172,173,506
mimus 173
par 173
ramaleyi 173
tarda 162
Clöe
rubescens 512
unicolor 512
Cloeon 4,162,526
alamance 7
cognatum 151,261-262,369
dimidiatum 369
dipterum 128,130,131,142,151, 244,247,259-264,265-276,368, 436,496
inscriptum 259-264
rubropictum 294
simile 128-136,151,368,369,374 375
Compsoneuriella 94
Crinitella 71,81
coheri 81

Dactylobaetis 72,280
Dannella 71,85-88,*87*
bartoni 85-86,88
lita 86
provonshai 86
simplex 85,86
Deleatidium 58-61,66
Dentatella NEW SUBGEN. 71,86, 88
Dipterophlebiodes 38
Dolania 282
americana 21,211-230
Drunella 71,79-81,495
coloradensis 80
conestee 81
doddsi 80
flavilinea 73
fuscata 80
gilliesi 73
grandis 79
kohnoae 80
kohnoi 80
pelosa 80,86
trispina 80
tuberculata 80
walkeri 80-81
wayah 81
Eatonella 71,80-81
Eatonigenia 282
Ecdyonurus 5,94,98,142,143,305,327, 408,413,415,506
affinis 149,151
aurantiacus 147,149-151
carpathicus 146,148
dispar 128-131,147-149,151,410
fascioculatus 149,150
insignis 128,130,147-149,151
lateralis 146,148,149,409,410
macani 147,149,151
quadrilineatus 146
starmachi 148,149
subalpinus 145,146
submontanus 146,148,149,410
torrentis 128,130,131,146,148, 149,305,407,408,410
venosus 21,128,130,131,146,148, 305,326,369,370,375,407,410 415
Elassoneuria 353,355,357,361,*362*, 363,*364*
Epeorus 142,143,162,407,446,450 506
albertae 172
deceptivus 162,172
longimanus 172
sylvicola 146,148,149,408,410
Ephemera 526
danica 128,130,146,148,149, 187-199,410
lineata 128,130,148,149,151
simulans 7,163,370
vulgata 128,130,494,496
Ephemerella 2,4,5,8,71,73-76, 163,169,173-174,282,327,407, 495,496,503,505-508
attenuata 7
aurivillii 74,163,436,438,443, 446,448,449,451,452
berneri 74
bicolor 7,163,443,446,449,451
coloradensis 172,174,505
doddsi 172,174,505
excrucians 7,74,163
g. grandis 172,174
ignita 127-131,146,148,149, 151,305,408,410,437
infrequens 172,174

invaria 322-326,328
krieghoffi 74,146,148,149,408,410
major 146,148,149,151,410
margarita 172,173
mesoleuca 151
micheneri 172,174
mucronata 74,437
needhami 7,74
notata 128-130,149
septentrionalis 74
simplex 7,163
subvaria 7,322-328
temporalis 7
tibialis 10,172,174
unicolorata 74
Ephemerellidae 7,71-91,129,163, 171,496,503,505,506
Ephemerellinae 71-91
Ephemerellini 71-88
Ephemeridae 5,7,128,294,494,496, 497,503,505,506
Ephemeroidea 37
Ephemeron 525-529,*526*
Epheron 280,282
album 294,310,312,317
virgo 142,149,151
Eurycaenis 527
Eurylophella 71,84-85,86,*87*,163
bartoni 86
brunnescens 85
karelica 84,85
lithuanica 73
Euthraulus
bugandensis 343,344,347-349
Euthyplocia 72
Euthylociidae 280

Gilliesia 38

Habroleptoides 38,407
modesta 146-149,151,408-410, 415,416
Habrophlebia 38
fusca 128,130
lauta 146,148,149,408,410
Habrophlebiodes 38
Hapsiphlebia 43-47,*48*,*49*,50,*51*, 52,66
Heptagenia 71-72,93-102,161,162, 294,494,495
aphrodite *101*
coerulans 149-151
criddlei 97,*101*,172
diabasia 7
elegantula 172
flava 149,151
flavescens 94,*95*,*99*,*101*,162, 294,443,446,448-452
fuscogrisea 128,161
hebe *95*,*99*,162,294
jewetti 294
julia 162
lateralis 128,130,131,305
longicauda 128,149,151
lucidipennis 7,*101*
maculipennis 162,310,312,317
pulla 7
simplicioides 97,*99*,101,172
sulphurea 5,128-131,149,151, 201,436-438
Heptageniidae 5,7,93-102,128,162, 171,294,446,494,497,503-506
lucidipennis group *101*
NEW GENUS A 96-98
hebe *95*,*99*
juno *101*
thetis *99*
umbratica 97
NEW GENUS B 96-98,*99*
criddlei 97, *101*
horrida 99
inconspicua *101*
kennedyi 98
salvini 98
simplicioides 97,*99*,101
NEW SUBGENUS 1 96,97-98
NEW SUBGENUS 2 96,97-98
Heptageniinae 162
Heterocloeon 4
curiosum 7
Hexagenia 4,6,196,249,282,494, 495,503,504,507-508
limbata 7,194,195,231-257, *248*,*249*,*251*,*252*,294,368,494, 495,505
munda 247
recurvata 247
rigida 457-465
Hexagenites 514
Hexagenitinae 514

Hyrtanella 71,86,89
christinea 89
Hyrtanellini NEW TRIBE 71,88-89

Iron 8,10,71,446
Ironopsis 71
Isca 280,283
Isonychia 72,506
bicolor 294
harperi 294
ignota 150,151
rufa 294

Jarmila 28,29
elongata 28,29
Jarmilidae 29,30
Jappa 45,47,52
Jubabaetis NEW GENUS 103-114
pescadori NEW SPECIES 103-104,*105*,*106*,*107*,*108*

Kirrara 38
Kukalova 29
americana 21,*22*,25,26,*27* 28,29
moravica 28,29
Kukalovidae 28,29

Lachlania 72,282
Leptohyphes 71,86
dicinctus 86
melanobranchus 86
mirus 86
undulatus 86
Leptophlebia 38,158,161,163,496
cupida 158,160,331-333,335-339,446,450,508
marginata 128-131,151,179-184
nebulosa 158,160,163,450
vespertina 128-131,179-184,369
Leptophlebiidae 34-41,43-56,57-69,128,147,163,171,280,294,446,496,497,503,505,506
NEW GENUS A 43-47,*48*,*49*,50 *51*,52-53
NEW GENUS B 47,50,53
NEW GENUS C 47,52
NEW GENUS D 47,52
NEW GENUS E 47,50
NEW GENUS F 50
Leptophlebiinae 38,39
Leptophleboidea 37
Litobrancha
recurvata 247

Massartella 45,47
Massartellopsis 47,*48*,*49*,50,53,66
Meridialaris 43-47,*48*,*49*,50,*51*,53,66
Metretopidae 162,505
Metretopus
borealis 158,162
Mirawara
aapta 299
megaloprepia 299
purpurea 297-308
Misthodotidae 516
Myllonella NEW SUBGENUS 71,80

Neoephemera 4
youngi 369,370,373
Neoephemeridae 514,516
Neozephlebia 45,47,52,58-61,*63*,*64*,65,66
Neurocaenis
discolor 343,344,347-349

Oboriphlebia NEW GENUS 29,30
moravica 29
quarta NEW SPECIES 29,30
quinta NEW SPECIES 29,30
tertia NEW SPECIES 29,30
Oboriphlebiidae NEW FAMILY 29,30
Oligoneuriella 282,355
keffermuellerae 142,147
mikulskii 142,150,151,355
pallida 142,150,151,355
rhenana 147-151,355,410
Oligoneuriidae 353-366,505
Oligoneuriinae 280
Oligoneuriopsis 353,355,357,361,363
elisabethae 357,*358*,*360*
lawrencei *354*,*356*,357
Oligoneurisca
borysthenica 142,147
Oniscigasterinae 514
Oxycypha 527

Palingenia 282
longicauda 142,147
Palingeniinae 280
Pannota 516
Paraleptophlebia 10,38,163,282, 327,436,446,450,493,506
cincta 128,130,149,151
guttata 163
heteronea 172
mollis 294,322,323,325,326,328
praepidita 310,312,317
submarginata 128,130,131,146, 148,149,151
tumida 128,130
Parameletus 162
Penaphlebia 43-47,*48*,*49*,50,51,52
Petersophlebia 47,53
Platybaetis NEW GENUS 103-114
edmundsi NEW SPECIES 104,*109*, *110*,*111*,*112*,*113*
uenoi 112
Plethogenesia 282
Polymitarcyidae 280,294,505
Polyplocia 280,283
Potamanthidae 128,283,294,497,505
Potamanthodes 283
Potamanthus
luteus 128,149,151
myops 294
Povilla 282
Proboscidoplocia 280
Procloeon
bifidum 128,130,148,149,151
ornatum 149,151
Prosopistoma 201-210,*207*,280,283, 511-512,514
foliaceum 201-204,206,512
punctifrons 512
variegatum 512
Prosopistomatidae 505,513-516
Prosopistomatoidea 516
Prosopistominae 514
Protereisma 28,29
latum *24*,25-26
Protereismatidae 28,29
Protereismatoidea 19,29,30
Pseudocloeon 103,113,115,116,119, 120,162,172,443,446,447,449, 451
dubium 7,289,293,294
hyalopterum 117,118
inexpectatum 116,*117*,*118*,*119*, 120,149-151
myrsum 310-317
punctiventris 289,293,294

Raptobaetopus 116,201-210
tenellus 201-203,205
Rhionella NEW SUBGENUS 71
Rhithrogena 5,10,72,98,142,143, 162,282,408,409,411-413,446, 450,506
diaphana 149,407,408,410,413
ferruginea 146,148,149,410
germanica 128,149,410
gorganica 145,146,148
hageni 172
hercynia 148,410
hybrida 146,407-411,413,415, 416
iridina 145-146,148,407-411, 413,415,416
loyolaea 145,146,405,407-414, 416
pellucida 293,294
robusta 172
semicolorata 5,128,130,131,148, 149,305,326,407,408,410,413
undulata 162
Rhoenanthus 283

Schistonota 516
Serratella 71,73,75-76,86
carolina 75
deficiens 75
levis 75
micheneri 73,76
serrata 75
spiculosa 75
subsolana 76
teresa 75
thailandensis 73,75-76
tibialis 75
Siphlonisca
aerodromia 167-168
Siphlonuridae 22-23,128,162,167- 168,171,294,446,503,505,506
Siphlonuroidea 514
Siphlonurus 147,162,167,446,450
aestivalis 151,179

alternatus 147,162
armatus 128
lacustris 128,130,131,148,149, 179-184,410,436
linnaeanus 128
occidentalis 171
inflatus 162
spectabilis 162
Siphloplecton 162
interlineatum 162
Stenacron 506,508
canadense 367,370
interpunctatum 294,370
canadense 310,312,317
pallidum 446,450
Stenonema 4,72,162,506,508
bicaudatum 294
exiguum 7
frontale 5
fuscum 7,446,450
integrum 294
interpunctatum 294,339
ithaca 446,450
nepotellum 294,310,312,317
pulchellum 446,450
rubrum 7
terminatum 7,294
tripunctatum
tripunctatum 162
vicarium 162,443,446,448,450-452

Tasmanocoenis
tillyardi 381,384-394,400-401
Teloganella 280,283
Teloganopsis 71,77,78
media 77
Thalerosphyrus 94
Thraulodes 71
Thraulus 38,65,66
Timpanoga 71,86,88,89
hecuba 88
Torleya 71,76-77,86
major 73,76
nepalica 73
padunica 73
Tortopus 72
Traverella 72,280,282
Tribrochella NEW SUBGENUS 71,80
Tricorythidae 171,280,294,505
Tricorythodes 4,5,72,86,163, 280,506
allectus 287,289,294
atratus 289
bullus 86
cristatus 86
dimorphus 86
minutus 172,173
Tricorythus 280
Triplosoba 23

Ulmerophlebia 47,52
Unirhachella NEW SUBGENUS 71,80

Vietnamella NEW COMBINATION 71,82-84

Zephlebia 45,47,50,52,58-66,*63*, *64*
cruentata 45,47,58-61
NEW GENUS I 62,66
NEW GENUS II 62,66

SUBJECT INDEX

This index includes taxa of groups other than Ephemeroptera as well as other subjects.

Aberfeldy River (Victoria, Australia) 297-308
Abundance 503-509
Acroneuria
  abnormis 339
  californica 337
Activity 297-308
Angling 1-4
Applied biology 1-17
Archodonata 19,26-28
Artificial substrates 443-455
Australia, Aberfeldy River 297-308

Bats 6,278
Bibliography on pollution 474-489
Big Qualicum River (Vancouver Island, Canada) 8
Bioassay 6,457-465
Biotic index 6
Birds 6,277,278
Boothia Peninsula (NWT, Canada) 156-159,162-163
Brandenberg Pond (Ohio, USA) 232,233

Cadmium, toxicity 457-465
Canada,
  Big Qualicum River (BC) 8
  northern 155-166
  Rat River (Manitoba) 309-310
  Roseau River (Manitoba) 287-296
Carpathia (Poland) rivers 141-154
Chemical composition, eggs 227
Colonization 8,443-455
Computer retrieval 467-474
Conservation in Poland 141-154

Distribution *see* Zoogeography
Distributional zones 72-73
Dragonflies 6,277,278
Drift 309,313,321-330,331-341

Eaton, Alfred Edwin 531,532, 534-535
Eire 187-199
Emergence 222-224,277-285,287-296
Energetics 211-230
Erasipteridae 28

Fecundity 183,224-226
Fish 2,4,6-13,339
Flyfishing 1-4
Folla River (Norway) 419-442
Food 201-210,231-257,343-352
Fossil nymphs 19-31
France, Rhône 201-203

Great Britain, synoptic survey 123-140
Gynandromorphs 497

Habitat analysis in Great Britain 123-140
Harris, Moses 525-529

Information retrieval 467-474
Instar determination 367-380
Ireland 187-199
Iron pyrites mining 419-442
Isogenoides elongatus 331-338

Keewatin District (NWT, Canada) 156-159,162-163

Life histories 179-186,187-199, 211-230,259-264,265-276, 277-285,287-296,309-319, 381-403
Lithoneura 27-28
  lameerei 27
Lithophlebia 27-28

Mackenzie River (NWT, Canada) 156-157,160,162-168
Maine (USA) 167-168
Malaysia 278-279,288

Megasecoptera 28
Mermithoidea 491-502

Nematoda 491-502
Nematomorpha 496
New Caledonia 34
New Zealand 57-69
Norway
  mountain stream 419-442
  subalpine lake 179-186

Odonata 6,277,278
Oil contamination 443-455
Online information retrieval 467-474
Øvre Heimdalsvatn (Norway) 179-186

Palaeodictyoptera 27,28
Paragnetina media 339
Permian nymphs 19-31
Pictet, François Jules 531-534
Phylogeny 33-41,43-51,57-65
Plectrocnemia conspersa 339
Poland
  Tatra mountains 405-418
  zoogeography and ecology 141-154
Pollution
  bibliography 474-489
  indicators 4-7,141-154
Population stability 309-319
Porcupine River (Yukon, Canada) 156,157,160,162-163
Pteronarcys californica 247
Pyrite mining 419-442

Radio-isotope techniques 343-352
Rat River (Manitoba, Canada) 309-319
Rhône (France) 201-203
Rocky Mountain stream (USA) 169-177
Roseau River (Manitoba, Canada) 287-296

Salmon Enhancement Program 6, 8-13
Scanning electron microscopy 353-366
Scotland, synoptic surveys 124-128,134-136
Sediments, remains in 14
Sheelin, Lough (Ireland) 188, 189,195-197
Shetland Islands, synoptic survey 124-127,134-136
South America, southern 43-56
St. Vrain Creek (Colorado, USA) 169-177
Swarm markers 282-283
Sweden, overwintering *C. dipterum* 259-264

Tatra Mountains (Poland) 405-418
Tayside (Scotland) synoptic survey 125-128
Trail River (NWT, Canada) 443-455

Uganda 279,282
Upper Three Runs Creek (S.C., USA) 211-230
U.S.A.
  Brandenberg Pond (Ohio) 232, 233
  Maine 167-168
  St. Vrain Creek (Colorado) 169-177
  Upper Three Runs Creek (SC) 211-230

Venation 23-26,*24*,*27*,*48*
Vistula River (Poland) 141-154

Xenophlebia 27-28

Zoogeography
  Ephemerellidae 71-91
  Leptophlebiidae 52-53,65-67
  Northern Canada 155-166
  Polish species 141-154
  zones 72-73